IEE MATERIALS AND DEVICES SERIES 9

Series Editors: Professor D. V. Morgan
Dr N. Parkman
Professor K. Overshott

ELECTRICAL DEGRADATION AND BREAKDOWN IN POLYMERS

Other volumes in this series:

ELECTRICAL DEGRADATION AND BREAKDOWN IN POLYMERS

L. A. Dissado and
J. C. Fothergill

Peter Peregrinus Ltd. on behalf of the Institution of Electrical Engineers

Published by: Peter Peregrinus Ltd., London, United Kingdom

© 1992: Peter Peregrinus Ltd.

Peter Peregrinus Ltd.,
Michael Faraday House,
Six Hills Way, Stevenage,
Herts. SG1 2AY, United Kingdom

British Library Cataloguing in Publication Data

A CIP catalogue record for this book
is available from the British Library

ISBN 0 86341 196 7

Printed in England by The Redwood Press, Wiltshire

To our families:

Enid and Morwenna
and
Barbara, Peter and Rosie

Preface

The study of electrical breakdown is not a trivial pursuit. Whilst the twentieth century has seen great advances in the science of breakdown and pre-breakdown processes there is still much to be understood. F. W. Peek's book, 'Dielectric Phenomena in High-Voltage Engineering' (McGraw Hill Inc.) published in 1920 described engineering observations of breakdown in commercially-used materials and systems but very little was understood of the breakdown mechanisms. S. Whitehead's two books on dielectric breakdown published in 1932 and 1951* show the considerable advances made during this period. He noted that it was not until engineers and theoretical scientists cooperated closely that the various, fundamentally different, types of breakdown could be classified. J. J. O'Dwyer's two books† published in 1964 and 1973 took the analytical models of both conduction and breakdown in solid dielectrics considerably further.

Since the publication of these books, the scientific engineering understanding of breakdown has moved on considerably. The 'classical' models of breakdown were purely deterministic relying on a chain of causes and effects to result in breakdown through the system being unable to sustain an energy balance above a critical stress. These models usually assumed homogeneous insulation and resulted in a single-valued breakdown strength at which the whole of the insulation would breakdown. Generally however measured breakdown strengths are found to be different every time they are measured (at least in non-crystalline solids) and may be described by a statistical distribution. Furthermore breakdown usually results in a channel of destruction through the insulation rather than its complete destruction across a broad front. Also since the publication of O'Dwyer's books extensive studies have been made on the effects of long-term aging such as electrical-tree and water-tree degradation. These have been found to be particularly detrimental in synthetic polymeric insulation which is becoming increasingly used in low and high voltage stress applications. Such degradation results in a distribution of times-to-breakdown and the 'tree-type' formations are elegantly described by the new mathematical formalism of fractals. These advances in understanding the stochastic nature of breakdown and the science and mathematics that describe it have important implications for engineers responsible for designing and maintaining insulating systems. In view of these advances we decided there was a requirement for a new, comprehensive review of the field.

---

* 'Dielectric phenomena, Vol. 3: breakdown of solid dielectrics', ed. WEDMORE, E. B. (Ernest Bern Ltd., 1932) and 'Dielectric breakdown of solids' (Clarendon Press, Oxford, 1951)
† 'The theory of dielectric breakdown of solids' and 'The theory of electrical conduction and breakdown in solid dielectrics' (Clarendon Press, Oxford, 1964 and 1973 respectively)

Our book is divided into five parts. *Part 1* serves as an introduction to cater for a broad range of backgrounds of readership. It introduces the chemical and physical structure of polymers, solid-state physics pertinent to charge movement through polymers, and categories of aging and break-down mechanisms. *Part 2* is a comprehensive review of electrical degradation in polymers and, in particular, electrical-tree and water-tree degradation. *Part 3* is an introduction and review of conduction and deterministic break-down in solids as applied to polymers. This part is not intended to be as in-depth as O'Dwyer's 1973 book but assumes less initial knowledge of the subject. It includes advances made since his book such as filamentary (chan-nel) breakdown, a section on partial discharges, and also examples of the breakdown mechanisms as they relate to polymers. *Part 4* discusses the stochastic nature of breakdown both from the empirical and from the physical-modelling viewpoints. *Part 5* indicates the implications for engineers of the increased scientific understanding presented in earlier parts.

In the book we have chosen to concentrate on polymers as they are commonly used, however much of the discussion applies to non-crystalline materials generally. We have attempted to make the book both an introduc-tion and a comprehensive review and we appreciate that this has, in places, made some of our introductory arguments rather concise but hopefully complete. We envisage that readers using this book as an introductory text will find the complementary material cited in the references useful and will be able to also use this book as a guide for further study of the subject. During the writing of the book we have noticed the large amount of research which has been carried out in Japan in the last twenty years and we hope that this book will serve as a review of this as well as other international research effort.

The style we have adopted reflects the nature of breakdown and degrada-tion in polymers. Not only are such phenomena distributed widely on a basis which is not predictable for an individual sample, but materials given the same name often differ from one another in terms of composition and morphology. Since breakdown is always an extreme condition for a par-ticular sample such differences can have a significant effect on the results reported in the literature particularly when undisclosed experimental test factors such as laboratory temperature, relative humidity, and the forming method and prior history of the sample are taken into account. On top of all this it is a sad fact that very often different features of the same process, such as tree inception time and number density, are often discussed as though they were measures of the same thing. It is not surprising therefore that significantly different values are reported for reputedly the same features, and sometimes the results are even more contradictory. We have therefore adopted an approach in which an 'identikit' picture of the pro-cesses under consideration is built up by means of mutually-complementary pieces of information. Wherever possible only results and data corroborated by a number of studies, preferably from different laboratories, have been used. The picture deduced therefore does not rely on single 'definitive' experiments which it might be justifiably argued do not exist in breakdown studies. Of necessity such an approach requires a large number of references, and we hope that the reader will bear with us in this, and if necessary forgive us if the reference list, though substantial, is not exhaustive.

We would like to thank STC Technology Ltd. for acting as the original stimulus for our interest in this field and, together with our University departments, for actively encouraging that interest over several years. Thanks are also due to our many colleagues including Sue Wolfe, Simon Rowland, Robert Hill, John Houlgreave, and Annabel Eccles who have helped through their discussions. We would particularly like to thank Gary Stevens for his careful and constructive criticism of our manuscript throughout its preparation and rescind our occasional threats of murder as unfounded reactions to his useful comments. We would like to thank Noreen Berridge, Joyce Meredith and Doug Pratt in the Leicester University Engineering Department Drawing Office for their patient preparation of the diagrams. We would like to thank our families, to whom this book is dedicated, for their understanding and patience with our long periods of inattention to family life during the preparation of the manuscript. We also acknowledge the many teachers who sowed the seeds of scientific enquiry in us, among them, Mr. Brace, Mr. Coates, Professor Craig, Mr. Day, Mr. Smith, Mr. Dickinson, Professor Pethig, and Professor Lewis. Finally one of us (LAD) would like to thank the typist for a valiant effort in the face of much trial and tribulation.

Canterbury, April 1991   L. A. Dissado
                                                         J. C. Fothergill

Editors preface

This book began life as a single chapter in a collection of reviews dedicated to examining advances in understanding the relationship between the chemical, morphological and defect structure of polymers and their physical properties and performance. As with many publishing ventures, that book was never completed. However, that early chapter demonstrated that many significant advances had occurred in our understanding and description of electrical aging and breakdown in polymers which could barely be explored or appreciated within the constraints of a single review article, and the notion of a book solely dedicated to this purpose was born.

The breadth of this undertaking appeared at first to be staggering in view of the many disciplines involved and the distance travelled since the appearance of the authoritative books of J. J. O'Dwyer in 1964 and 1973. Since that time a more complete view of the structure of semicrystalline and amorphous polymers had been gained and progress made in understanding the nature of charge injection, electronic excitations and transport in polymers. Advances in band theory and the appearance of molecular electronics also contributed to and continues to feed this development. Perhaps the most radical influence has been the move away from a purely deterministic view of electrical breakdown to one centred on the stochastic nature of failure processes. This leads directly to the concepts of spatially distributed, temporally progressive and cumulative aging processes which manifest themselves in specific statistical descriptions of failure. Fortunate coincidences have also contributed; including the development of fractal descriptions of damage formation processes, the development of reliable methods to assess the statistical characteristics of breakdown observations and the appearance of powerful numerical modelling methods and the computing power to implement them.

For the first time it now appears that it may be possible to connect the microscopic physics and chemistry of electrical aging with failure occurring in the bulk and to extend that to real electrical insulation systems where the statistical nature of failure has been recognised for many years but not understood. Hence the benefits are not confined to the satisfaction felt by the physicist when finding a model that truly seems to fit the facts but also help the engineer who should be able to define appropriate insulation assessment methods and acceptance criteria which suit his application and have a sound physical basis.

This journey is far from complete and much remains to be done before electrical and water treeing and partial discharge degradation and failure in polymers are fully understood, mathematically described and translated to better insulating materials and practical systems. What is encouraging is that the journey has begun. It is in this context that this book plays a key role in setting a background against which further advances can be compared and in directing the novice and the practitioner alike to new avenues of thought and research.

I am grateful to the authors for their endurance and also their patience in meeting this demanding task. I am confident that the rewards of their efforts will be seen for many years as the international community benefit from the wealth of information in this book and the lead that they have set.

G. C. Stevens
March 1992

Contents

Part 3
DETERMINISTIC MECHANISMS OF BREAKDOWN

Part 4
THE STOCHASTIC NATURE OF BREAKDOWN

Part 5
ENGINEERING CONSIDERATIONS FOR BREAKDOWN TESTING AND DEGRADATION ASSESSMENT

PART 1

INTRODUCTION TO POLYMERS AND ELECTRICAL BREAKDOWN

Introduction

> 'Electrical breakdown is a subject to which it is difficult to apply our usual scientific rigour. A well-designed insulator (in the laboratory) breaks down in service if the wind changes direction or if a fog descends.'
>
> Solymar and Walsh (1984)[1]
> Copyright © 1984, Oxford University Press

Lightning, the electrical breakdown of air, was probably the first electrical phenomenon to be observed. However, despite the tremendous effort which has gone into understanding electrical breakdown since the end of the eighteenth century[2], the subject is still one of the least understood. Polymers are often the best and most economic electrically insulating construction materials. They generally exhibit very high breakdown strengths (typically up to $\sim 10^9 \ V \cdot m^{-1}$), they have low dielectric losses (typically $\tan \delta < 10^{-3}$) and high DC resistivities (typically $> 10^{16} \ \Omega \cdot m$). These electrical characteristics coupled with the wide range of mechanical strengths and stiffnesses available from plastics, their high corrosion resistance, ease of forming and manufacture, and their reasonably low cost[3] has led to their increasing use as electrical insulators following the development of synthetic polymer systems in the 1930s.

However this widespread use is not reflected by a good physical understanding of their mechanisms of electrical breakdown or the way in which an applied voltage may cause eventual degradation after a period of perhaps years. Nor has such breakdown and degradation been well characterised as a function of parameters such as voltage, time of voltage application, thickness and temperature to name but a few. Recently there have been some major advances in both the physical understanding of these processes by the scientist and their characterisation as a function of parameters associated with their operating conditions by the design engineer. Such advances in understanding should continue so that the behaviour of polymers can be predicted in untried environments and critical applications. Better understanding of failure mechanisms will also lead to better component designs to prevent failure.

Some of the problems encountered, whilst being frustrating to the engineer, may be fascinating to the scientist! For example one of the reasons for choosing polyethylene as an insulator for power cables in the early 1960s was that it was hydrophobic and it was therefore assumed that there would be no problems associated with water ingress into the insulation material. However, after a few years in service many of these early cables started to breakdown due to insulation failure as a result of a phenomenon known as 'water treeing'. This degradation mechanism, which is still a major form of insulation degradation in power cables, was not predicted by scientists at the time. Whilst this phenomenon is still not completely understood, its investigation has led scientists to a much greater awareness of the complex physical, electrical, and chemical interactions taking place under such conditions. The response of engineers was to introduce triple extrusion techniques to prevent moisture ingress and improve the dielectric–electrode interface, and to use crosslinked polyethylene for the manufacture of such cables[4]. This has considerably reduced the water treeing problem although it is difficult to accurately predict the long-term performance of these cables.

Throughout this book we hope to consistently consider electrical degradation and breakdown from both the scientific and engineering point of view. Inevitably many of our examples are often drawn from the world of power cables since they fall directly within our experience. This is not too restrictive because power cables are one of the largest users of polymers as insulators and they display a wide variety of electrical breakdown and degradation phenomena which can be carried over to linear and network polymers and other component applications. Since the readership of this book is likely to be drawn from diverse fields this first part is designed to give introductory overviews of: the nature of polymers; solid-state physics pertinent to polymers; and electrical degradation and breakdown.

Chapter 1
Polymer structure and morphology

1.1 Structure of polymers

The purpose of this chapter is to give a summary of the terminology used in describing the chemical and physical structure of polymers and to identify the structure of polymers which are commonly used as electrical insulators. For a comprehensive introductory description of polymer structure the reader is referred to the excellent texts by Mills[5], Hall[6], and McCrum *et al*[7]. Nomenclature is set in *italics* as it is introduced.

1.1.1 Chemical structure

Polymers consist of long-chain macromolecules with repeating *monomer* (or *mer*) units, an idea first proposed by the Nobel laureate Hermann Staudinger in 1920 and which laid the foundation for systematic polymer synthesis. A polymer is usually named by putting the prefix 'poly-' in front of the name of the monomer from which it is derived. For example the monomer *ethylene* is the repeated monomer in *polyethylene* (eqn. 1.1)

$$
\begin{array}{c}
\underset{\underset{\displaystyle H}{|}}{\overset{\overset{\displaystyle H}{|}}{C}} = \underset{\underset{\displaystyle H}{|}}{\overset{\overset{\displaystyle H}{|}}{C} } \\
\text{ethylene}
\end{array}
\qquad
\begin{array}{c}
\left[\begin{array}{c} \overset{\overset{\displaystyle H}{|}}{} \quad \overset{\overset{\displaystyle H}{|}}{} \\ -C-C- \\ \underset{\underset{\displaystyle H}{|}}{} \quad \underset{\underset{\displaystyle H}{|}}{} \end{array}\right]_n \\
\text{polyethylene}
\end{array}
\qquad (1.1)
$$

('Polythene' is the ICI trade name for polyethylene.) A selection of polymers which are commonly used in electrical insulation, together with their monomers is given in Table 1.1. Different molecular units are attached to the ends of the chains (CH_3 in polyethylene), however, since n, the *degree of polymerisation*, is very large (typically in the range 10^3—10^5) the end units do not usually influence the physical properties of the polymer. It is more common to quote the *relative molecular mass* ($\equiv$ molecular weight) of the polymer than the degree of polymerisation.

Table 1.1 includes many important polymers which are based on a $\left(-\overset{|}{\underset{|}{C}}-\overset{|}{\underset{|}{C}}-\right)$ linkage along the length of the polymer 'backbone'. These are known as '*homopolymers*'. '*Heterochain*' polymers, in which carbon atoms in the backbone have been replaced by other elements, are placed into categories depending on their characteristic chemical linkages. For example *polyesters* are formed from *glycols* and *dicarboxylic acids* (eqn. 1.2*a*) to form

Table 1.1   Molecular structure of selected polymers

Generic structure		Name (abbreviation)
$\begin{array}{c} X \quad X \\ | \quad | \\ -C-C- \\ | \quad | \\ X \quad X \end{array}$	$X = H$ $X = F$	polyethylene (PE) polytetrafluoroethylene (PTFE)
$\begin{array}{c} H \quad X \\ | \quad | \\ -C-C- \\ | \quad | \\ H \quad H \end{array}$	$X = CH_3$ $X = Cl$ $X = C_6H_5$ $X = OCOCH_3$	polypropylene (PP) poly(vinyl chloride) (PVC) polystyrene (PS) poly(vinyl acetate) (PVA)
$\begin{array}{c} H \quad X \\ | \quad | \\ -C-C- \\ | \quad | \\ H \quad X \end{array}$	$X = Cl$ $X = F$ $X = CH_3$	poly(vinylidine chloride) (PVDC) poly(vinylidine fluoride) (PVDF) polyisobutylene (butyl rubber)
$\begin{array}{c} H \quad X \\ | \quad | \\ -C-C- \\ | \quad | \\ H \quad Y \end{array}$	$X = CH_3$ $Y = COOCH_3$	poly(methyl methacrylate) (PMMA)
$\begin{array}{c} H \qquad\quad X \\ \backslash \qquad / \\ C=C \\ / \qquad\; \backslash \\ -CH_2 \qquad CH_2- \end{array}$	$X = H$ $X = CH_3$	polybutadiene (BR) polyisoprene (natural rubber)
$-(CH_2)_n-O-\overset{\overset{\displaystyle O}{\|}}{C}-\bigcirc-\overset{\overset{\displaystyle O}{\|}}{C}-C-$		$n = 2$ poly(ethylene terephthalate) (PET)
$-(CH_2)_n-\overset{\overset{\displaystyle O}{\|}}{C}-\underset{\underset{\displaystyle H}{\|}}{N}-$		$n = 5$ polyamide 6 (PA6, nylon 6) $n = 10$ polyamide 10 (PA10, nylon 10)
$-(CH_2)_n-\underset{\underset{\displaystyle H}{\|}}{N}-\overset{\overset{\displaystyle O}{\|}}{C}-(CH_2)_m-\overset{\overset{\displaystyle O}{\|}}{C}-\underset{\underset{\displaystyle H}{\|}}{N}-$		$m = 4$, $n = 6$, polyamide 6.6 (PA6.6, nylon 6,6)
$-\bigcirc-\overset{\overset{\displaystyle CH_3}{\|}}{\underset{\underset{\displaystyle CH_3}{\|}}{C}}-\bigcirc-O-\overset{\overset{\displaystyle O}{\|}}{C}-O-$		polycarbonate (PC)

Table 1.1 continued

Generic structure	Name (abbreviation)
	poly(ether ether ketone) PEEK

Table 1.2 Families of heterochain polymers

Characteristic linking group	Polymer family
$-\overset{\text{H}}{\underset{\text{O}}{\overset{\|}{\text{C}}}}-\text{N}-$	poly*amide*
$-\overset{\text{O}}{\underset{\|}{\text{C}}}-\text{O}-$	poly*ester*
$-\text{O}-$	poly*ether*
	poly*imide*
$-\overset{\text{X}}{\underset{\text{X}}{\text{Si}}}-\text{O}-$	poly*silicone*
$-\overset{\text{O}}{\underset{\text{O}}{\overset{\|}{\text{S}}}}-$	poly*sulphone*
$-\text{O}-\overset{\text{H}}{\underset{\text{O}}{\overset{\|}{\text{C}}}}-\text{N}-$	poly*urethane*

chains of the form shown in eqn. 1.2*b* in which the chemical groups R_a and R_g are linked by *esters*. Common families of polymers and their characteristic linking groups are shown in Table 1.2.

$$
\begin{array}{ccc}
& \quad \overset{\displaystyle O}{\underset{\displaystyle \|}{}} \quad \overset{\displaystyle O}{\underset{\displaystyle \|}{}} & \\
HO-R_g-OH & HO-C-R_a-C-OH & \quad (1.2a)
\end{array}
$$

glycol   dicarboxylic acid

$$
\begin{array}{cc}
-\underset{\displaystyle \underset{\|}{O}}{C}-R_a-\underset{\displaystyle \underset{\|}{O}}{C}-O-R_g-O- & \qquad -\underset{\displaystyle \underset{\|}{O}}{C}-O- \qquad (1.2b)
\end{array}
$$

polyester   ester group

The asymmetric homopolymers such as the vinyls are further classified according to their *stereoregularity* (sometimes referred to as *tacticity*) which refers to the regularity of side group ('X') positions if the molecule could be 'laid out flat'. *Isotactic* polymers have all the 'X' groups on the same side (see eqn. 1.3*a*) so that all the monomers have the same orientation; in *syndiotactic* polymers the orientation alternates consecutively (eqn. 1.3*b*); and in *atactic* polymers the orientation is random (eqn. 1.3*c*).

$$
\begin{array}{ll}
\begin{matrix}
H & X & H & X & H & X & H & X \\
| & | & | & | & | & | & | & | \\
-C & -C & -C & -C & -C & -C & -C & -C- \\
| & | & | & | & | & | & | & | \\
H & H & H & H & H & H & H & H
\end{matrix}
& \text{isotactic} \qquad (1.3a)
\end{array}
$$

$$
\begin{array}{ll}
\begin{matrix}
H & X & H & H & H & X & H & H \\
| & | & | & | & | & | & | & | \\
-C & -C & -C & -C & -C & -C & -C & -C- \\
| & | & | & | & | & | & | & | \\
H & H & H & X & H & H & H & X
\end{matrix}
& \text{syndiotactic} \qquad (1.3b)
\end{array}
$$

$$
\begin{array}{ll}
\begin{matrix}
H & X & H & X & H & H & H & X \\
| & | & | & | & | & | & | & | \\
-C & -C & -C & -C & -C & -C & -C & -C- \\
| & | & | & | & | & | & | & | \\
H & H & H & H & H & X & H & H
\end{matrix}
& \text{atactic} \qquad (1.3c)
\end{array}
$$

Simple polymer chains may form branches off the main chain; this is commonly found in polyethylene (e.g. eqn. 1.4).

$$
\begin{matrix}
H & H & H & H & H & H & H & H & H \\
| & | & | & | & | & | & | & | & | \\
-C & -C & -C & -C & -C & -C & -C & -C & -C- \\
| & | & | & | & | & | & | & | & | \\
H & H & H & H & & H & H & H & H
\end{matrix}
$$

$$
\begin{matrix}
H-C-H \\
| \\
H-C-H \\
| \\
H-C-H \\
|
\end{matrix} \qquad (1.4)
$$

Such branches can occur every 30—100 monomer units along the backbone and result in side-branches which can be short (e.g. up to several monomer units long) and also long (e.g. as long as the main chain). Branching can be produced or inhibited to a large extent by altering the polymerisation conditions. Branching reduces the potential for regular molecular packing and so lowers the density producing, in the case of polyethylene (PE), what is commonly called *low-density polyethylene (LDPE)*. LDPE is mechanically inferior to *high-density polyethylene (HDPE)*, its non-branched counterpart, but nonetheless it has excellent dielectric properties and is commonly used in high-voltage power cables. The terms 'high density' and 'low density' are rather ambiguous since the density is directly related to the degree of 'crystallinity' in semicrystalline polymers. This degree of crystallinity can be expressed as the volume or mass fraction of polymer present in the crystalline phase.

As well as polymers having branches, they may also have *'cross-links'* in which the polymer chains are joined by short, long or, even polymeric molecules which effectively form connecting branches. A crosslinked polymer is therefore in principle essentially one gigantic molecule and so above its melting point (should it be semicrystalline) or above the so-called glass transition temperature (if it is amorphous) it becomes rubber-like rather than liquid. Crosslinking is usually achieved in one of three ways:

(i)   A catalyst (or 'initiator') may be incorporated into the polymer so that after the polymer has been moulded or cast into the required geometry, it is heated to initiate the crosslinking reaction. Such polymers are known as *'thermosets'*, since their shape is 'set' and irreversible, and the crosslinking process is generally known as *'curing'*. In some cases the initiators are sufficiently reactive that crosslinking can occur efficiently at room temperature. *Rubbers* are the subset of crosslinked polymers which are above their melting or glass-transition temperatures at room temperature. Natural rubber is only very lightly crosslinked but the polymer chains are extremely long and therefore intimately entangled resulting in similar physical properties to cross-linked polymers. (This is what makes it 'tacky'; crosslinking or *'vulcanisation'* using sulphur bridges overcomes this problem.) Polymers which are not crosslinked, and can therefore be remoulded to other shapes, are known as *'thermoplastics'*; polyethylene, polypropylene and poly(etheretherketone) are examples. Polyethylene cables are usually crosslinked by incorporating 1—2% of a peroxide (e.g. dicumyl peroxide) which does not react when the polymer is extruded but reacts subsequently when the cable is treated with super-heated steam or high-pressure nitrogen gas at elevated temperatures in a long tube. In an alternative technique, silane crosslinking, the crosslinking reaction consumes water and until this is supplied it cannot occur. After the cables have been formed they are put in a 'sauna' to crosslink. Polyethylene crosslinks by means of polymer free radicals produced via the initiation process. Since these are usually generated in low concentration, the product is normally only lightly crosslinked.

(ii)   Radiation may be used to promote crosslinking. The process can only be realistically used on thin sections and may cause other degradation such as reduction in chain length. As the apparatus required is usually expensive, the technique only finds limited application.

(iii) A chemical hardener may be added to cause crosslinking. *Epoxy resins* are a family of thermoset polymers in which two components are mixed to eventually form a glassy product at room temperature which has good electrical insulating properties and is highly impermeable to water. Whilst they have, for a long time, found applications in high voltage switchgear and electrical machine insulation and the 'potting' of electrical components, they are now finding applications in low and high voltages transformers and other systems. Epoxies are polymers in which the end groups contain the three-membered epoxide ring:

$$
\begin{array}{c}
\overset{\displaystyle O}{\overset{\displaystyle /\ \backslash}{CH_2-CH-}}
\end{array}
\tag{1.5a}
$$

A common configuration is that of the diepoxide structure:

$$
CH_2-CH-R-CH-CH_2
\tag{1.5b}
$$

where R is commonly bisphenol-A:

$$
\begin{array}{c}
\left[CH_2-O- \bigcirc -\overset{\displaystyle CH_3}{\underset{\displaystyle CH_3}{C}}- \bigcirc -O-CH_2 \right]_n
\end{array}
\tag{1.5c}
$$

where n is typically about 10. The action of the curing agent or 'hardener' is to open and join onto the epoxide rings. Diamine compounds $(H_2N-R'-NH_2)$ are commonly used and provide four sites for attachment:

$$
\begin{array}{cc}
\overset{\displaystyle OH}{-R-CH-CH_2} & \overset{\displaystyle OH}{H_2C-HC-R-} \\
 & \\
 \searrow \quad N-R'-N \quad \swarrow & \\
 & \\
-R-CH-CH_2 & H_2C-HC-R- \\
\underset{\displaystyle OH}{} & \underset{\displaystyle OH}{}
\end{array}
\tag{1.5d}
$$

Because of the high chemical reactivity and potentially large number of epoxide rings, crosslinking in cured epoxies can be very high and an extensive network of connections with high mechanical rigidity is produced. These materials are usually highly loaded with inorganic particulate fillers to improve their physical properties and reduce the cost of components.

Polymers may also be formed from more than one type of monomer unit: *copolymers* are manufactured from two different types of monomer and *terpolymers* from three different types. Copolymers are classified into four groups according to the way in which the two monomers are ordered along the length of the polymer chain: (i) *alternating copolymers* in which the two monomer types alternate with one another; (ii) *random copolymers* in which the monomers are randomly ordered along the 'backbone'; (iii) *block copolymers* in which the 'backbone' consists of a sequence of many consecutive monomers of one type and then of the other type; and (iv) *graft copolymers* in which chains of one type of monomer are grafted or linked onto chains of the other type. *Ethylene-propylene rubber* (EPR or EPM), an *elastomer* or synthetic rubber, is a typical example of a block copolymer with sequences of ten to twenty ethylene units followed by similar length sequences of propylene units. This arrangement is a consequence of the significant difference between the reaction rates for the two monomers to join the chain end. The *reactivity ratio* for ethylene[5], that is the ratio of the reaction rates of ethyl monomers joining ethyl chain ends to those joining propylene chain ends is 17·8. The reactivity ratio for propylene is 0·065. The use of EPR in high-voltage power cables is increasing as it may be more resistant to water-tree degradation than polyethylene. Typical commercial EPRs contain weight fractions of between 40% to 60% ethylene resulting in an amorphous material with a glass transition temperature of −58°C to −50°C. However the product used contains high concentrations (~50%) of additives such as kaolin. Ethylene-propylene-diene terpolymers (EPDM) are also used, with dienes such as dicyclopentadiene which provide sites for vulcanisation with sulphur unlike thermoplastic elastomers such as EPR which require the use of peroxides. It should be noted that *linear low-density polyethylene* (*LLDPE*) is the rather misleading name given to the (branched) copolymer of ethylene with a comonomer such as butene hexane or octane. It differs from LDPE primarily in containing no long-chain branches, only short ones.

1.1.2 Physical structure

Polymeric systems can be found having varying degrees of structural complexity. The lowest level in this structural hierarchy is that of *molecular conformation*. In polyethylene the most stable conformation of the polymer chain is a simple planar zig-zag since the separation of adjacent hydrogen atoms (0·254 nm) is slightly greater than their van der Waals' diameter (0·239 nm) and this so-called *trans* conformation gives rise to the lowest Gibbs free energy. For other simple homopolymer chains with larger side group molecules such as PTFE or PVC this is not possible since the planar zig-zag pattern would leave insufficient spacing for the side-group atoms to fit. Such molecules therefore adopt a lowest energy configuration resulting in a helical chain structure, which is also found in some biological molecules. More complicated chemical structures, including most heterochain molecules tend to have less apparent regularity to their molecular configuration.

For the so-called *amorphous* polymers (such as poly(methyl methacrylate)) this chain conformation and its long range configuration may be the highest level of structure that is generally realized. Some polymers are generally considered as amorphous but may in fact display limited crystallinity if their chains are partially aligned by strains introduced during manufacture (e.g. poly(vinyl chloride)). The long molecules in amorphous polymers exist in very coiled-up states unless they have been partially aligned in this way. By using a freely-jointed chain model (with the joints being the covalent bonds) it can be shown for a random or 'Gaussian' chain that the mean distance between the molecule chain ends is only $l \cdot n^{1/2}$ where l is the length of each rigid segment between joints and n is the number of randomly orientated segments. For example a chain with 10^4 segments each 1·5 nm long would have a chain length of 15 μm but the mean distance between the chain ends would be only 150 nm.

As an amorphous uncrosslinked polymer is cooled through its glass-transition temperature (Section 3.1.1) the large segmental motions of the backbone stop and the polymer becomes a *super-cooled liquid*, i.e. a *glass*. In this state the polymer can be thought of as cross-bonded by van der Waals' attractions between molecules whereas above the glass transition temperature thermal fluctuations are too great for these bonds to be effective. For amorphous uncrosslinked polymers above their glass transition temperatures the polymer is a viscous liquid. For crosslinked amorphous materials, and those with very long chains in which a great deal of entanglement takes place chains become constrained by their neighbours and true liquid behaviour cannot be established above the glass transition temperature, T_g, where they behave as rubbers.

The stiffness of the material, measured in terms of its Young's modulus is approximately proportional to the covalent crosslinking concentration[8] where this is defined as the proportion of covalent bonds which participate in crosslinking. For uncrosslinked materials the change in modulus occurring across the glass transition can be several orders of magnitude but as the crosslinking density (and the glass transition temperature) increases the magnitude of this change decreases. For very heavily crosslinked materials (*network polymers*), with more than about 1 in 50 of the covalent bonds required for crosslinking, the Young's modulus does not change significantly as the material is taken through its glass transition temperature since segmental motions are highly restricted by the crosslinking network. Solids in the glassy state usually have a Young's modulus $>2 \times 10^9$ Pa so that nylons and epoxies with covalent crosslinking concentrations of approximately 1 in 30 and Young's moduli of $10—30 \times 10^9$ Pa above and below their glass transition temperatures may always be thought of as solids by this definition. Raising the temperature further will eventually cause such polymers to degrade. (Diamond is the limiting case with a covalent crosslinking concentration of 100% and a Young's modulus of 1000×10^9 Pa.)

1.2 Factors affecting crystallinity

Degradation, electrical or any other kind, necessarily involves the rearrangement of the chemical and/or physical structure of a material such that the properties of interest are deleteriously affected. In Section 1.1 we summarised the chemical structure of polymers and we also described amorphous polymers, which do not exhibit significant crystallinity. In this section we wish to describe the crystalline state that polymeric molecules may adopt so that we can effectively discuss the influence of polymer morphology and properties on electrical degradation processes. It should be stressed that few polymers have a completely regular crystal structure. Neither can they be thought of as poly-crystalline as in the case of metals since in most crystalline polymers the long polymer molecules may traverse a number of discrete crystallites. Most 'crystalline' polymers are therefore always semi-crystalline with the result that their specific volume* is always greater than would be calculated on the basis of the physical dimensions and relative molecular mass of their crystallographic unit cell.

Fig. 1.1 shows the decrease of specific volume with temperature for amorphous and semi-crystalline polymers. As a semi-crystalline polymer is cooled from its liquid state through the melting temperature, its specific volume drops abruptly reflecting the closer and regular packing of molecules into crystalline structures. (Fast cooling rates will not allow extensive crystallisation to take place and will result in more residual amorphous regions and a consequently higher specific volume.) If the polymer is amorphous then no abrupt drop in specific volume will be observed at the melting temperature, it will simply continue to decrease linearly with decreasing temperature until the glass transition temperature is reached. Both amorphous and semi-crystalline polymers are mechanically rubbery at temperatures between the melting and glass-transition temperatures. In the rigid glassy state below the glass transition temperature the specific volume of both amorphous and semi-crystalline polymers continues to decrease, but at a slower rate.

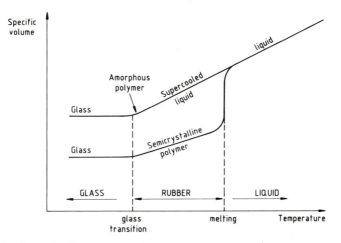

Fig. 1.1 A schematic diagram showing the variation of specific volume with temperature for an amorphous and a semicrystalline polymer

* Specific volume is the reciprocal of density.

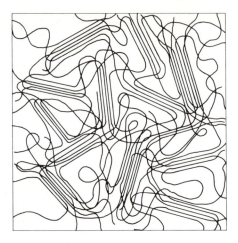

Fig. 1.2   In the fringed micelle model several long polymer chains run parallel to each other through part of the volume before diverging again. This gives rise to crystalline and amorphous regions

The first model to explain crystallinity in polymers was the '*fringed micelle*' model, Fig. 1.2. In this model several long polymer chains run parallel to each other through part of the volume before diverging again into an amorphous region. Individual molecules run through several crystalline volumes thereby linking them together. This is, however, thermodynamically unlikely to occur on a large scale and so only occurs in polymers of extremely low crystallinity such as poly(vinyl chloride). Generally data from scanning electron micrographs do not agree with the fringed micelle model. Here we briefly examine what causes and prevents crystallisation and the types of morphologies resulting from crystallisation.

1.2.1 Chemical structure
Solids are held together by stabilising forces between atoms and molecules. The inter-molecular forces in polymers are of the secondary, van der Waals' type (which explains their generally low melting points) for which there is an optimum inter-molecular spacing at which their potential energy is a minimum. Thus the total internal free energy of a system of molecules is lower when the molecules adopt a regular structure rather than a random arrangement; it is for this reason that most solids crystallise spontaneously (for many this is not obvious from their macroscopic properties as they form a polycrystalline structure rather than a single crystal).

In polymers one might expect the long chain molecules to align in structures containing parallel molecules. Since the definition of a crystal is in terms of its well-ordered periodicity it is not surprising however that polymers without a regular structure (such as random copolymers and atactic polymers) do not crystallise. We have already seen in Section 1.1.2 that molecules with regular large side groups may form helical structures and we might expect these to crystallise. Complicated molecules however, or those with large side groups, may not easily be able to arrange themselves in a close-packed fashion during the crystallisation process (especially as this most commonly takes place during very quick solidification from the melt during forming or moulding operations) and may display little or no crystallinity. In some cases it may be possible to promote crystallisation in these materials by altering the crystallisation conditions.

1.2.2 Crystallisation conditions

As well as requiring regular and reasonably simple molecular structures, crystallisation requires that molecules (or parts of polymer molecules) are able to align themselves so as to form part of the growing crystal. Thus the temperature must be below that for spontaneous melting but not so low as to hinder molecular freedom of movement. Furthermore the initiation or *nucleation* of crystal structures is required; this may occur intrinsically or it may occur extrinsically by the deliberate addition of nucleating agents or incidentally through the presence of contaminants.

The formation of polymer crystal structures requires molecular chain folding (since they are long this will be the case even in chain-extended crystal structures). The increase in internal free energy required to produce the chain folding and the associated crystal surface is compensated for by the decrease in energy released in crystal formation (a reduction in entropy and specific volume). The equilibrium melting temperature, T_m^0, is defined as the temperature at which the rate of energy released by one process is equal to that required by the other, i.e. it is the temperature at which the polymer neither increases nor decreases its degree of crystallinity. For polymer crystals to grow there must be a reduction in the Gibbs free energy during their formation, and this will be the case when the melt is supercooled to below the equilibrium melting temperature. Because the surface free energy is virtually independent of temperature, as the isothermal crystallisation temperature increases and approaches the equilibrium melting temperature increasingly large volumes and hence increasingly large lamella thicknesses are required to cause a net reduction in the Gibbs free energy. This results in an increase in crystal size with crystallisation temperature. For example for the 'lamella' crystal structure described in Section 1.3, consideration of the kinetics and thermodynamics of polymer crystallisation leads to the temperature dependence of their thickness, l_l, on the isothermal crystallisation temperature, T, with the approximate form[9]

$$l_l \propto \frac{\sigma_e}{\Delta H_f} \frac{T_m^0}{T_m^0 - T} \tag{1.6}$$

where σ_e is the fold surface free energy per mole of crystal; ΔH_f is the enthalpy of fusion per mole of crystal; and T_m^0 is the ideal infinite-crystal melting temperature. Thus the smaller the supercooling, $\Delta T \ (= T_m^0 - T)$ the thicker and more stable the lamella. The greater the supercooling, the higher the rate of crystallisation. Thus if the degree of supercooling is high, rapid crystallisation will occur at many nucleation sites simultaneously resulting in many small lamellae. It is found empirically that the fastest rate of crystallisation occurs at a supercooling of $\Delta T \simeq 0 \cdot 2 \times T_m^0$.

Nucleation may be homogeneous or heterogeneous. Homogeneous nucleation occurs due to the chance event of correct alignment of molecules or possibly due to sites which have retained part of their crystallinity from a previous solidification. Homogeneous nucleation results in poor control over the crystallisation conditions and so nucleating agents may be added to cause heterogeneous nucleation and thereby control the nucleation density and rate. In many cases heterogeneous nucleation dominates as crystals nucleate on physical or chemical imperfections such as dust particles, catalytic residues, pigments and surfaces.

1.2.3 Mechanical flow

Many forming processes, such as the extrusion of cables or thin films, involve large strain deformations or flow which induces preferential orientation of the long-chain molecules. Extension in the melt followed by rapid cooling of the orientated melt tends to align molecules so as to enhance crystallinity in that direction (it may even permit crystallisation in otherwise non-crystalline polymers). The resulting anisotropy may even bestow some crystal-like properties on amorphous polymers such that, in the direction of alignment, the physical vector properties tend to approach the properties of the covalent bonds whereas in the orthogonal directions they tend towards those of the van der Waals' bonds. Crystallisation followed by plastic extension (drawing) may produce anisotropic crystal structures with greatly improved unidirectional mechanical properties. For example[7] the Young's modulus and tensile strength of isotropic poly(ethylene terephthalate) can be increased by factors of more than five and ten respectively with this technique. Either of these anisotropic improvements will be destroyed by melting.

1.3 Morphology and factors affecting it

All forms of polymer crystallinity involve alignment of long polymer chains. This may take the form of a lamella 'sheet' or long fibres and 'super structures' consisting of more complicated formations of the basic structures. The crystal structures or morphology which occurs depends upon the factors affecting the crystallinity described in Section 1.2 and the method of manufacture. In this section we describe the structures according to the methods of solidification and subsequent mechanical deformation.

1.3.1 Crystallisation from solution

In dilute solutions polymer molecules become untangled and isolated from one another. They have a similar coiled shape however to that which they would have in the solid. Crystallisation from solution is normally only used for the preparation of laboratory samples and the lamella is the most commonly observed morphological unit. From solution these take the form of thin plates which can only be observed with a scanning electron microscope. In lamella the alignment of the molecules is through the plates (i.e. at a right angle to the surface) and folded back and forth at the surface[10], Fig. 1.3.

Lamellae are typically 10—20 nm thick, about 1 μm long and about 0·1—1 μm wide and therefore have the geometry of a wide ribbon or a sheet. The time required for a randomly coiled molecule to form a single

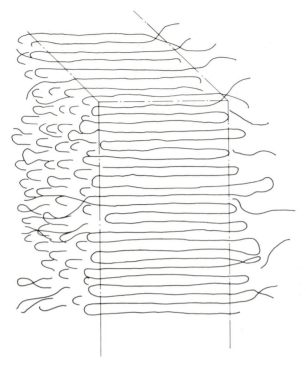

Fig. 1.3 Lamella formation. The diagram shows polymer chains folded back and forth on themselves to form a semicrystalline, ribbon-shaped region known as a lamella. Polymer chains leaving the lamella may be replaced by others joining it. Such chains pass through the amorphous regions and connect different lamellae

lamella is much longer than that usually available during crystallisation so that the same molecule may fold several times in a number of neighbouring lamellae during crystallisation. Once crystallisation is complete a single polymer chain may run from one lamella through the intervening space to the next lamella and so on thereby forming *intercrystalline links*. Since longer molecules tend to form more of these links the degree of polymerisation is important in determining the strength of a polymer. As molecules leave a lamella they may fold back on themselves to re-enter the lamella at the next adjacent site, they may re-enter at sites further away (this is sometimes referred to as switchboard since it resembles the old-fashioned plug-and-socket telephone switchboard), they may join the intercrystalline space, or they may form intercrystalline links. It is uncertain exactly what factors determine this fine lamella surface structure and the occurrence of inter-crystalline links.

1.3.2 Crystallisation from the melt
In crystallisation from molten polymers many lamellae may grow from a single nucleus and lamellae may branch thereby forming a spherulite[11,12,13],

Fig. 1.4. Between the lamellae are amorphous regions and a spherulite behaves more as a collection of lamella crystallites than a single crystal. A typical polymer melt comprises molecules which can vary greatly in relative molecular mass, branching, and, perhaps, stereoregularity (tacticity). The high relative molecular mass components crystallise preferentially reject-ing[14,11] low relative molecular mass, atactic and highly branched molecules as impurities. During the formation of the lamellae the impurity concentra-tion builds up at the crystal melt interface to a level higher than that in the bulk of the melt thereby causing an impurity rich layer to form around it. Such a layer may completely surround the lamellae crystals and its thickness will depend upon impurity diffusion rates and the time taken for the crystal to form. The growth rate will generally slow down as it has to extract pure material from the surrounding impurity layer whose impurity concentration is progressively increasing. The exact mechanism of spherulite formation is unclear but it is clear that they continue to grow until they impinge upon each other, i.e. they are volume filling. Thus for an initially reasonably homogeneous mixture of molecular size, tacticity, branching, purity etc. in the melt, regions of crystallites appear which are reasonably uniform in these respects, separated by amorphous regions containing more non-uniform and impure molecules, Fig. 1.5.

It may be conjectured that as a crystalline region grows the layer of less-crystallisable material which builds up around it is thinner or more-crystallisable in some patches than others. For this reason one would expect projections to form on the surface of the growing crystal through such patches which would then tend to protrude into and through the impurity layer into material which is purer. As they found this purer material they would be able to grow faster and project themselves further into the bulk

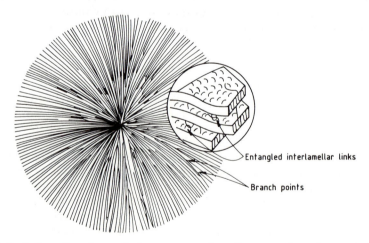

Entangled interlamellar links

Branch points

Fig. 1.4 Spherulite formed from many lamellae growing from a nucleus. Spherulites continue to grow by crystallisation of the amorphous material surround-ing them until they either impinge on another spherulite or the surrounding material is too irregular or impure to crystallise

(a)

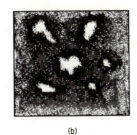
(b)

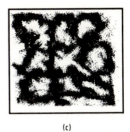
(c)

Fig. 1.5

(*a*)   During the initial stages of crystallisation the higher relative molecular mass components crystallise rejecting lower molecular mass components and inhomogeneities. The light areas represent areas in which crystallisation has started and they are surrounded by the material they have rejected — the 'less pure' material is shown as darker.

(*b*)   Projections of crystalline material are initiated randomly and grow preferentially as they protrude into less impure, more regular material which is more easily crystallised.

(*c*)   Finally reasonably crystalline parts are separated from each other by impure and amorphous regions

of the material. The growth front of the crystal would therefore either degenerate into a series of fibrils (one-dimensional projections) or into a growing layer (a two-dimensional projection), i.e. a lamella, which is approximately equal to the thickness of the impurity layer. In many cases the most suitable material for continuing the growth of the crystallite would be the remainder of the long molecules of which it is already partially made up and these may then fold back on themselves. Because many lamellae may grow from a single site, spherulites may form and continue to grow, in a volume-filling way, until they impinge on other spherulites or interfaces. Between the lamellae in a spherulite, and between spherulites, are not only amorphous material and impurities, but more lamellae formed out of lower relative molecular mass material and which form at a slightly lower temperature (a higher degree of supercooling). It should be emphasized that most of the ideas in this paragraph are very much conjectural, but it is hoped that they will lead to some insights into the crystallisation process.

1.3.3 Crystallisation during flow
During crystallisation in polymer flow for which there is a high degree of shear, e.g. in thin film extrusion, the flow-induced molecular orientation may affect the morphology. In particular the chains may become extended rather than adopting their more typical, coiled up structure. In this case the growth of lamellae is less likely as single chains will have to be disentangled from a considerable distance in front of the growing crystal. Chain-extended structures (rather than the chain-folded lamellae structures) are therefore more likely to result in crystallisation. This may result in *sheave*-like structures, or *fibrils*, or, if chain-folded crystals grow off the side of the chain-extended structures, *shish-kebab*-like structures may be formed. Fibrils

may consist of a series of crystalline units, roughly cube shaped of side ~10 nm with molecules aligned along the fibril axis, intimately connected by many inter-crystalline links. The fibrils themselves may only be weakly bonded and may themselves be aligned more accurately by plastic deformation after solidification (i.e. by drawing).

1.3.4 Crosslinking

Crosslinking a linear polymer prior to crystallisation may cause a normally semi-crystalline polymer to lose its crystallinity. The most common method of crosslinking is to trigger a thermally-activated catalyst by raising the temperature. Since this temperature must be well above the melting temperature if crosslinking is to take place in a reasonably short time, this form of crosslinking always takes place when the polymer is in its liquid state, i.e. in the *amorphous* melt phase. If the temperature is raised very quickly then it may be possible for the crosslinking to start before the lamellae structure has completely dispersed and this level of structure may be retained, at least in part. Because the degree of molecular rearrangement is severely restricted by the crosslinks, it seems likely that the lamellae will not be able to attain a full spherulitic structure in the solid state. Under these conditions in polyethylene, ordered lamellae micro-domains[15], spherulitic precursor lamellae sheaves[16,17], or extended lamellae structures[18] (slow cooled HDPE) may be the only form of crystalline arrangement possible. However spherulites are sometimes observed in compression-moulded thin-film specimens[19,15] which have been crosslinked in this way. In such specimens the temperature throughout the bulk of the material may be raised very quickly from the melting point to that required for the initiation of crosslinking and macro-structural rearrangement may also be hindered by the thin section of the film (typically only a few spherulite diameters thick) and by cohesive forces between the polymer and the mould surface.

Not all crosslinking techniques require the use of temperatures in excess of the melting point. For example in the silane-crosslinking process developed for polyethylene power cable insulation, the polymer is cured at approximately 100°C. This is just below the melting point of the more dominant and thicker lamellae (~105°C) and any spherulite formed during moulding would not be destroyed at this temperature. However spherulites are not normally observed in this material and it seems likely that the silane, which is grafted at temperatures above the melting point, inhibits the production of long dominant lamellae required for spherulitic growth. Radiation may also be used in crosslinking the solid state below the melting point. At lower doses this form of crosslinking does not destroy the crystallinity and any original spherulitic structure will be retained in these materials provided they are not heated above their melting point.

1.4 Bulk defects and free volume

The usual crystal defects, such as dislocations, may be present in polymer crystals but are difficult to observe since the crystallinity is only partial and they are not therefore likely to be dominant. Many crystal defects are unique to polymers. For example chain ends may be present within the crystal, it is possible that there may be buried loops (i.e. folding within the crystal), or the regular chain may be interrupted with an incorrect mer (for example a polypropylene unit may occur in a polyethylene chain giving rise to a methyl group in a polyethylene crystallite. Such defects act to increase unit cell dimensions and specific volume and lower the melting point.

The specific volume is also increased by the apparently inevitable presence of microvoids in polymers. These may be formed during manufacture by the evaporation of volatile decomposition products from various chemicals reactions such as those used for crosslinking and those associated with antioxidants. They may also be formed from impurities and additives which may decompose, migrate, and outgas from crosslinking inhomogeneities in network polymers[20,21], and atmospheric gases which have not diffused out. Since diffusion is the *net* movement of particles down a concentration gradient due to the *random* movement of individual particles, there will always be some (individual) voids which do not leave the polymer during manufacture. For example Stevens *et al.*[22] have found, using light scattering techniques, that an epoxy resin they manufactured and de-gassed under extremely stringent conditions still contained 10^{12}—10^{13} voids m^{-3} with diameters of 2 to 5×10^{-7} m (i.e. about 10^{-7} of the total volume). At the other extreme, voids in the 'halos' of steam-cured crosslinked polyethylene cables may be as large as $4 \, \mu$m and occupy as much as 1% of the total volume[23]. Typically microvoids are in the range 10^{-8}—3×10^{-7} m. They are likely to grow in the presence of mechanical and electrically-caused (e.g. by Maxwell forces) mechanical stresses[22,24]. The upper size limit of microvoids (i.e. when microvoids become voids) is simply a matter of definition. In practice the large end of the microvoid distribution can be observed using optical microscopic techniques and the prefix *micro* suggests that the border-line might lie at diameters of a *micro*meter. The lower size limit is defined in terms of microvoids which are permanent. Below a few nm ($< \sim 3 \times 10^{-9}$ m) there is likely to be space which continuously appears and disappears due to molecular motion, resulting in *thermal density fluctuations* and which can be measured using laser light scattering[21] and observed as electrical noise[25] in dielectric measurements if the molecules are polar[26,27].

1.5 Techniques for characterising crystallinity and morphology

1.5.1 Calculating crystallinity from density measurements
For a semi-crystalline polymer it is possible to estimate the fraction of the solid that is crystalline, i.e. its *crystallinity*, by determining the curve of specific volume with temperature if the dimensions of the unit cell are known (this may be determined by X-ray diffraction). For example at 20°C polyethylene has an orthorhombic unit cell with dimensions 736 pm × 492 pm × 254 pm = 9.20×10^{-29} m^3 containing two monomers units and therefore having a relative molecular mass of 56.1. This results in a specific volume for the pure crystalline parts of the polyethylene of $v_x = 0 \cdot 987 \times 10^{-3}$ m^3 kg^{-1}. The

specific volume of the amorphous fraction may be estimated using the bulk specific volume versus temperature curve and extrapolating the straight line from the liquid phase down to 20°C; this yields a value of $v_a = 1.160 \times 10^{-3} \, \text{m}^3 \, \text{kg}^{-1}$ at 20°C. If the measured specific volume is v_m then the crystallinity, x, is defined as:

$$x = \frac{v_a - v_m}{v_a - v_x} \times 100\% \qquad (1.7)$$

So for a typical HDPE with a density of $962 \cdot 5 \, \text{kg} \cdot \text{m}^{-3}$ at 20°C, and therefore a specific volume of $1 \cdot 039 \times 10^{-3} \, \text{m}^3 \, \text{kg}^{-1}$, the crystallinity would be:

$$x = \frac{1 \cdot 160 - 1 \cdot 039}{1 \cdot 160 - 0 \cdot 987} \times 100\% = 70\%$$

As will be seen in Section 1.5.3 different techniques can be used to establish the crystallinity and may yield slightly different results since they sample the structure in different ways.

1.5.2 Permanganic etching

In the early 1980s, permanganic etching techniques became a popular method for revealing the spherulitic structure of polymers[21,28-32] particularly of polyethylene and crosslinked polyethylene. However there is considerable evidence from small-angle light scattering (SALS), scanning electron microscopy (SEM), optical microscopy, and chlorosulphonic etching that the 'spherulites' observed using early permanganic etching techniques were, in fact, artifacts which Bamji *et al.*[19] termed 'nodules'. It is suggested that the production of these nodules may be explained in terms of shielding by crystals which precipitate out of the etching solution onto the polyethylene surfaces[33], the collapse of swollen partially-dissolved polyethylene when etchant is removed[19], or the formation of radial crack boundaries[34]. More recently improved parmanganic etching techniques have been developed which overcome this artifact problem[20].

Although nodules were found to display many of the characteristic trends expected from spherulites (e.g. size dependence on preparation procedure, surface and interfacial effects such as transcrystallisation[35,36], water tree[30,31] and electrical breakdown[37] delineation along nodule boundaries) there are considerable discrepancies in absolute size between them and spherulites observed using other techniques. For example Bamji *et al.*[19] used compression-moulded 20 μm thick films of peroxide-crosslinked polyethylene in which they found spherulites of 4—5 μm diameter using small-angle light scattering and optical microscopy whereas permanganic etching by the same workers revealed structures with an average diameter of about 11 μm. Similar sized spherulites were also found in compression-moulded polyethylene specimens by Stevens and Swingler[15]. In the case of polypropylene the situation was reversed with the true spherulites being much larger (100 μm) than nodules found by etching (~30 μm). In References 38, 39, and 20, it has been shown that water tree boundaries are delineated by the *true* spherulite boundaries (~2 μm diameter) whereas the extent to which the nodule boundaries are followed depends on the extent of staining.

1.5.3 Other techniques

Different techniques give rise to different measures of the crystallinity. Four techniques are commonly used: infra-red absorption and Raman spectroscopy (also used to determine lamellae thickness via LAM modes), differential scanning calorimetry (DSC), X-ray diffraction, and the standard density measurement technique described earlier. Infra-red spectroscopy can distinguish the *all-trans* chain conformations present in the crystal phase from the *trans–gauche–trans* conformations present at the fold surface and in the amorphous phase. So if all-trans vibration peak absorptions are used to assess crystallinity these will be weighted by the core of the crystal and avoid the surface or fold contribution to the crystallinity. Since the *all-trans* conformation only occurs in the truly parallel polymer molecules inside and away from the surfaces of the lamella crystals this technique tends to give the most conservative value for crystallinity. Crystallinity may be determined using the DSC technique by comparing the measured enthalpy of fusion with known values for a pure crystal fraction. This technique probably gives

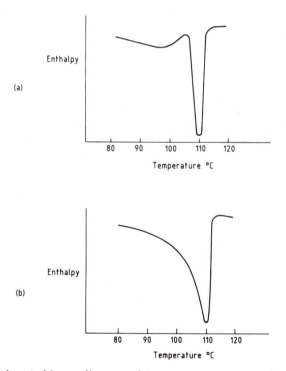

Fig. 1.6 Enthalpy (arbitrary linear scale) versus temperature from differential scanning calorimetry (DSC) experiments: (*a*) isothermally-crystallised low-density polyethylene; (*b*) non-isothermally-crystallised low-density polyethylene

(a)   (b)

Fig. 1.7 Polarised light micrographs showing morphology of polyamide 12: (*a*) slow cooled; (*b*) fast cooled

the most realistic value for crystallinity as it takes into account the contribution to crystallinity of the folds at the surfaces of the lamellae which require thermal energy to form. The density and X-ray diffraction methods are usually combined when assessing crystallinity. X-ray diffraction is used to calculate the unit cell dimensions to obtain a 'correct' crystal density (this is necessary because short chain branching can cause the unit cell to be larger than the branch-free case). This crystal density is combined with the macroscopic density measurement to assess the crystal fraction. This value is generally close to the DSC value but is naturally biased to a higher estimate of crystallinity (provided voids are absent).

The diversity of structure, including the range of relative molecular mass, the range of size of spherulites, but in particular the range of size of lamellae, give rise to a melting range rather than a discrete melting temperature which is also dependent upon the crystallisation conditions (in particular the rate of cooling). Fig. 1.6 shows DSC traces[40] of enthalpy versus temperature for two specimens of the same material, low-density polyethylene, which had been prepared under different crystallisation conditions (the width of the curve is indicative of the range of melting temperatures). In curve (*a*) the polymer was allowed to remain at its equilibrium melting temperature for a long time so that big lamellae and spherulites were able to form whereas in curve (*b*) the polymer was rapidly quenched which resulted in a marked increase in the prevalence of smaller microstructures. Fig. 1.7 shows that spherulitic structure is found in nylon 6,6[40]. The spherulites can be seen to be much larger when the nylon was slowly cooled, Fig. 1.7*a*, than was the case when it was cooled much faster, Fig. 1.7*b*, as expected from the crystallisation conditions.

1.6 Additives

Polymers are rarely used in their pure state, even for the purposes of high quality electrical insulation, so that the structures described above may be somewhat different to those encountered in practice.

Saetchling[3] divides additives into three categories:

(i) *Auxiliaries*, which are used up during the manufacturing process and thereafter only their residues remain. Catalysts used in crosslinking and as emulsifying agents are usually included in this category (although they are not used up and therefore remain in the insulation after the manufacturing process).

(ii) *Additives*, added in small concentrations (<10%) which do not substantially alter the structure of the polymer but which affect its behaviour during use[41]. These include: *lubricants* and *parting agents* which improve the flow characteristics and facilitate mould opening and release; *stabilizers* to afford protection from heat and ultra-violet radiation; *antioxidants* which act to inhibit oxidation; *flame retardants* which may act to either inhibit the supply of oxygen when heated or which quench the fire by dissociation into water (e.g. $2Al(HO)_3 \rightarrow Al_2O_3 + 3H_2O$); *pigments* (not generally used in high-quality insulating applications); *flexibilizers* for increased toughness; and others such as *antimicrobials* and *antistatic* agents.

(iii) *Compounding ingredients* which are used in large concentrations (10— 70%) to change the properties of the polymer during manufacture and use. These fall into two sub-categories which have complementary functions: *fillers* and *plasticisers*. Fillers may be used to reduce cost by filling the polymer out and they usually improve the mechanical characteristics such as the Young's modulus, impact strength, dimensional stability and heat resistance. They may take the form of particulates, fibres or even micro-spheres. They may also improve the electrical insulation qualities, for example they have been used to improve the treeing resistance (see Chapter 8). Plasticisers are generally used to reduce brittleness and aid flow in manufacture, their classic use is in the extrusion of poly(vinyl chloride). They shift mechanical loss peaks to lower temperatures and decrease the hardness. They may influence the structure since they dissolve and work by reducing inter-chain binding forces.

Chapter 2
Polymers as wide band-gap insulators

The word *insulator* comes from the latin *insula* meaning *island* suggesting that it is isolated from its surroundings and that it should prevent electric current flow through it. It may therefore come as a surprise that, in this chapter, we should want to consider the electronic states in insulators and the charge transport processes through them. However these states and processes do exist and some appreciation of them is necessary for an understanding of breakdown phenomena. In this chapter we introduce the basic electrical terminology and account for the wide variation of observed electrical conductivities and charge transport processes. Our treatment is introductory and necessarily simplified; the reader is referred to more advanced texts during the discussion.

2.1 Conductivity

When a voltage, V, is applied between two points on a material an *electric field* (sometimes known as a *voltage stress*) $E = -\nabla V$, is set up. (The word *dielectric* comes from the Greek *dia* meaning *through* + *electric*, i.e. a material through which an electric field can penetrate.) Depending upon the electrode geometry and other factors this electric field may or may not be uniform. For example in Fig. 2.1a an electrically homogeneous material of thickness, l, is sandwiched between parallel electrodes and it can be seen from the equipotential lines that the field is uniform in the centre away from the edges and therefore has a magnitude, $|E| = V/l$. In the needle-shaped electrode arrangement shown in Fig. 2.1b however, the field is highly divergent at the needle tip resulting in a greatly enhanced field in that region.

Electrically-charged particles in the material will experience an electrostatic force in the direction of the electric field equal to the product of their charge, e (in Coulombs, C) and the electric field, which, if they are free to move, causes them to *drift* so that charge is transported through the material. Charge transport is also possible by *diffusion* whereby the random movement of individual charged particles results in a net movement of particles from a region of higher concentration to a region of lower concentration. The rate at which charge is transported $(C \cdot s^{-1})$ is known as the electric *current*, I (in amperes, A). This is a system property whose material equivalent is the *current density*, J $(A \cdot m^{-2})$.

Ohm's law, the result of empirical observations, states that the current though a conductor *due to drift* is proportional to the voltage across it:

$$I = \left(\frac{1}{R}\right) V$$

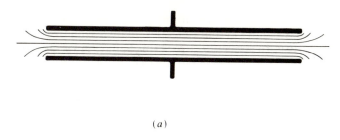

(a)

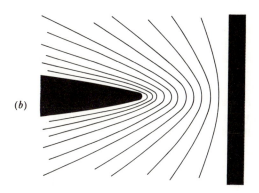

(b)

Fig. 2.1 Equipotential lines in homogeneous dielectrics subject to a static electric field for two electrode configurations: (a) parallel-plate; and (b) point (hyperboloid)-plane

where R (in ohms, Ω) is the *resistance* of the conductor to current flow. Ohm's law can also be expressed in terms of the material properties: field, E, and current density, J, as:

$$J = \left(\frac{1}{\rho}\right) E = \sigma E \tag{2.1}$$

where ρ is the *resistivity* of the material ($[\Omega \cdot m]$) and $\sigma = (1/\rho)$ is the *conductivity* ($[\Omega \cdot m]^{-1}$). This implies that the mean velocity of the charged particles in the direction of the field, the so-called *drift velocity*, is proportional to the applied field.

Materials display an extremely wide range of resistivities (and hence of conductivities also), possibly wider than any other measurable physical parameter. Superconductors have resistivities $<10^{-25} \, \Omega m$, typical resistivities of metallic conductors are $\sim 10^{-8} \, \Omega m$, semiconductors display resistivities in the approximate range $10^{-4} - 10^{10} \, \Omega m$ and insulators in the approximate range $10^{10} - 10^{20} \, \Omega m$. This range of 10^{45} orders of magnitude must be one of the largest of any physical parameter.

Conductors exhibit a linear ('Ohmic') current-density dependence upon electric-field up to high current densities, many exhibiting it up to the point at which they fuse. In semiconductors this behaviour is also observed at low fields. However above a critical field the drift velocity, and hence the current, reaches a limiting value in semiconductors, resulting in an apparent increase of resistivity with field. In most insulators a small continuous current (DC) is found to flow after the application of a steady-state electric field and at low fields this is often found to be Ohmic. The low-field conductivities of various polymers are given in Table 2.1. This current may be extremely difficult to measure since: (i) it may be extremely small; (ii) there is usually a large polarisation current, which may take a very long time to decay, due to the average movement of *localized* charges (i.e. those not completely free to move throughout the material) which are disturbed by the field; and (iii) special precautions must be taken to avoid the effects of surface currents. Since the conductivity of insulators is also highly dependent upon temperature, sample preparation and history, and other parameters, there may be large discrepancies between measured conductivities from different workers.

In this section we intend to outline those factors which influence the conductivity of materials, and insulators in particular, and hence explain the enormous variation of observed conductivities.

The conductivity of a material due to a charge carrier species, *i*, is given by:

$$\sigma_i = n_i e_i \mu_i \tag{2.2}$$

where n_i is the concentration (m^{-3}) of the charge carrier type *i*, e_i is the electric charge (C) on each carrier, and μ_i is known as its *mobility* $(m^2\,V^{-1}\,s^{-1})$ which is defined as the drift velocity per unit field:

$$\mu = \frac{v_d}{E} \tag{2.3}$$

In general more than one charge-carrier species (e.g. electrons, holes, protons, ions) may contribute to the conductivity. The overall conductivity for *N* contributing charge-carrier species is then given by:

$$\sigma = \sum_i^N \sigma_i = \sum_i^N |n_i e_i \mu_i| \tag{2.4}$$

Table 2.1   Low-field conductivities of various polymers

Material	Temperature	Conductivity	Reference
PVC	67°C	$10^{-14}\,\Omega^{-1}\,m^{-1}$	42
PET	80°C	$<10^{-19}\,\Omega^{-1}\,m^{-1}$	43
PE	40°C	$<2\times10^{-18}\,\Omega^{-1}\,m^{-1}$	43
PA 6·6	24°C	$<10^{-15}\,\Omega^{-1}\,m^{-1}$	44
PMMA	50°C	$3\times10^{-18}\,\Omega^{-1}\,m^{-1}$	45
PS	50°C	$10^{-16}\,\Omega^{-1}\,m^{-1}$	45
PEEK	60°C	$<5\times10^{-16}\,\Omega^{-1}\,m^{-1}$	46

Normally however, under given conditions, one charge carrier is likely to dominate. The range of conductivities must therefore be explained in terms of large variations in n, e, or μ.

In solids and liquids, large highly ionised molecules, if they exist, are unlikely to be free to move or contribute to the conductivity. The values of e of *delocalised* charge carriers are therefore likely to be *small* integer multiples of q, the charge on an electron ($q = 1 \cdot 602 \times 10^{-19}$ C). The small variations in e cannot therefore account for the large observed variation in conductivity.

The concentration of charge carriers can vary from almost zero to more than one per atom ($< \sim 10^{29}$ m^{-3}) and this range is largely responsible for the large range of conductivities. *Energy band theory*, introduced in Section 2.2.2, has generally been used to explain the large range of values of n and hence σ. The theory was originally developed to explain the observed electron energies in crystalline materials, it was subsequently used extensively in semiconductor theory, and has evolved, not entirely satisfactorily, to encompass semi-crystalline and amorphous materials.

The apparent value of mobility, μ, may also vary considerably between materials. In fact the range of true mobilities is probably only a few orders of magnitude, but under certain conditions the product $n \times \mu$ may be observed to vary greatly and it is often convenient to assume a constant carrier concentration, n and give explanations in terms of changing mobility. An example of this is trap-limited mobility described briefly in Section 2.5 and in more detail in Section 9.1.1(b).

In this section we have restricted discussion to Ohmic behaviour. However, because of the high resistivity of insulators, it is possible to apply very high fields without destructive Joule heating and above comparatively low fields (typically $\sim 10^6$ V m^{-1}), the conductivity of insulators is observed to rise with field; this is sometimes referred to as *super-Ohmic* behaviour. In this case the product $n \times \mu$ can increase for a variety of reasons and these are described in Chapter 9.

2.2 Energy band theory: Basic concepts

In this section we introduce the basic terms associated with energy band theory and intend to give some insight into how the theory accounts for the large differences in carrier concentrations observed between materials. However we state the results of energy band theory rather than derive it in any way. There are many texts which give good introductions to band theory, (e.g. References 47, 48 and 49). For more rigorous quantum mechanical derivations see References 50, 51, 52, 53.

First we describe the basic concepts of band theory as applied to covalent or ionic crystals (i.e. those bonded by extremely-strong covalent bonds or reasonably-strong electrostatic interactions). Silicon is an example of a covalently-bonded crystal; band theory is extensively used to explain the operation of silicon-based electronic semiconductor devices. Polymers however are a type of molecular crystal, albeit with extremely large molecules, and therefore have very strong, covalent, intra-molecular bonding but only very weak, van der Waals, inter-molecular binding forces. The relatively-simple band theory, described in this section, used for covalent and ionic crystals (i.e. conventional strong-bonding 'crystalline' materials) must therefore be modified when it comes to polymers and the resulting differences are discussed in Section 2.3.

2.2.1 Bonding

In an isolated atom, electrons are only allowed certain discrete energies which correspond to orbitals in the Bohr model. These *energy levels* can be probed using a variety of spectroscopic techniques in which photons are either absorbed by the material and excite electrons to higher energy levels or are emitted by the material as electrons 'drop' to lower energy levels.

The electron energies of atoms in condensed matter may well be different from those of isolated atoms since the electrons may take part in interatomic bonding. Atomic bonds can be divided into primary and secondary classes. Primary bonds are the strong bonds ($\gg \sim \frac{1}{2} \text{eV} \simeq 50 \text{ kJ} \cdot \text{mol}^{-1}$) in which valence electrons from the s- or p-orbitals are shared or exchanged. Secondary bonds are much weaker ($\ll \frac{1}{2} \text{eV}$). Typically intramolecular cohesion is due to primary bonding whereas intermolecular cohesion is due to secondary bonding.

Primary bonding may be heteropolar (ionic) or homopolar (covalent). Many bonds are not clearly hetero- or homo-polar but a combination of both. Metallic bonding is usually considered as a third category of primary bonding because, although it is an intermediate between the other two, metals form a major class of materials.

In ionic bonding, anions and cations are attracted to form stable units. Atoms in groups I or II of the periodic table may lose electrons and become cations with completed outer shells and similarly those in classes VI or VII may gain electrons to form anions with completed outer shells. The ionic bond results from the coulombic attraction between such ions. Common examples of ionic bonding are Na^+Cl^- with a bond energy of $7 \cdot 8 \text{ eV}$ ($752 \text{ kJ} \cdot \text{mol}^{-1}$) and LiF ($10 \cdot 4 \text{ eV} = 1003 \text{ kJ} \cdot \text{mol}^{-1}$). Ionic bonds are not stereo-specific since an ion will attract all oppositely charged ions in its vicinity. The number of contacting neighbours is determined by the ratio of the ionic radii. For example the large cesium ion (Cs^+) can attract eight chlorine (Cl^-) neighbours.

Covalent bonding occurs when atoms lying in the centre of the periodic table share pairs of valence electrons of opposite spin in a single joint orbital. The electrons tend to spend more time between the atoms than elsewhere and so one can imagine, in a simplified way, that a covalent bond exists as the attraction of positive ions to the intervening shared electrons. Since the covalent bond arises by the sharing of electrons between two specific atoms the bond has a specific direction and strong covalent attraction only occurs between first neighbours. The number of neighbours is therefore limited by the number of electrons available for sharing rather than the ionic radii (e.g. carbon will form four bonds and oxygen two). Intramolecular bonding in virtually all polymers and most organic materials is covalent. Examples of covalent bonds and their approximate bond energy in electron volts (multiply by 96·4 to convert to $kJ \cdot mol^{-1}$) include: C—C (3·8 eV), C=C (7·0 eV), C—H (4·5 eV), C—F (4·7 eV), C—Cl (3·5 eV), C—O (3·7 eV), C=0 (5·6 eV), O—O (2·3 eV), N—H (4·5 eV), N—O (2·6 eV), and H—H (4·5 eV).

Metallic atoms possess very loosely held valence electrons and so within a matrix of metal atoms the outer electrons are free to move throughout the structure as an 'electron gas'. The metallic bond can be thought of as covalent or ionic or both depending upon the material. In a sense each metal atom 'takes turns' forming covalent bonds with each of its neighbours whereas, in another sense, one can imagine the negative electron gas ionically bonding together the positive metallic ions. The metallic bond is non-directional and generally the number of nearest neighbours is high (typically 8—12).

Secondary (van der Waals) bonding is due to electric dipole-dipole interaction between nearby molecules or atoms and occurs in all systems. The dipoles may be permanent (as in asymmetric molecules where the centres of positive and negative charge are not coincident such as water and hydrogen fluoride) or temporary, occurring only when the electron distribution in a symmetric molecule is displaced due to quantum fluctuations in electron position and energy. The van der Waals bonding energies of polar atoms may be as high as $\sim0·5$ eV ($\sim50 kJ \cdot mol^{-1}$) but for non-polar atoms they may be as little as 0·01 eV ($1 kJ \cdot mol^{-1}$).

A common relatively-strong secondary bond is the *hydrogen bond* (sometimes referred to as the hydrogen bridge). Hydrogen is frequently bound such that it is on the end of a molecule (e.g. H—F) or the 'outside' of large organic or polymeric molecules. In this case it can be thought of as a (positive) proton which is covalently bound. Since the positive charge of the proton is not shielded by other electrons, it may strongly attract the electrons of another atom. This results in secondary bonds with energies typically in the range 0·05–0·5 eV.

In covalently and metallically bonded molecules and crystals, the electrons involved in the bonding are not associated with individual atoms but are delocalised over those atoms which are bound together (i.e. those which have overlapping electron orbitals). From the quantum-mechanical viewpoint, the electron can be considered as a particle or as a standing wave which interacts with the crystal lattice nuclei and electrons in its vicinity. The electron state is defined by its *wave function* (of time and position) whose square is proportional to the probability of finding the electron at that point. The lower-energy electrons in deeper orbitals, closer to the nucleus, do not interact significantly with neighbouring atoms. These retain most of the characteristics seen in the isolated neutral atom, their energies only being weakly affected by their surroundings.

2.2.2 Electron energy bands

The *Pauli exclusion principle* states that no two electrons occupying the same space can have exactly the same energy, although two electrons can occupy the same energy level if they have opposite 'spins'. For the case of the isolated atom this manifests itself as consecutive pairs of electrons occupying higher and higher energy levels (orbitals). By considering isolated atoms being brought closer together to form crystalline solids or molecules, it can be seen that electrons from similar orbitals, i.e. with the same initial energies, will start to interact when their wave functions begin to overlap significantly. (These energy states are said to be *degenerate*.) This interaction causes them to acquire slightly different energies from their original orbital energy (thereby 'lifting their degeneracy'). In general when there are a large number of atoms with strongly overlapping orbitals such as in metals or ionic crystals, a *band* of allowed electron energies will result. The band will be centred on the original orbital energies and will contain a discrete energy level for each of the N interacting atoms. When N is small, as in molecules, the separation between the levels will be substantial ($\sim$eV), but when N is large the levels are very 'finely spaced'. This is shown in Fig. 2.2. Because the energy levels within these bands are so closely spaced, an electron with an energy within a band, can easily move to an adjacent level within the band provided it is not already occupied. For all intents and purposes, therefore, the bands can be treated as a continuum of states with electrons being allowed to have any energy within the band as long as the average energy remains constant. This can be observed using spectroscopic studies since the observed absorption or emission 'lines', corresponding to electron energy transitions between discrete energy levels in isolated atoms, are seen to broaden into bands as the atoms are brought closer together.

In a covalently-bonded crystal at a temperature of absolute zero all the electrons will occupy the lowest possible energy levels, with two electrons of opposite spins in each level. If each atom has an even number of valence electrons, then the highest occupied band will be exactly full, if it has an

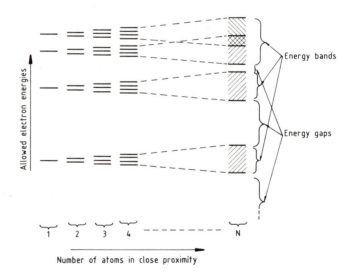

Fig. 2.2   As large numbers of atoms are brought together they interact causing the discrete electron energy states of the individual atoms to form quasi-continuous bands separated by well-defined 'band gaps'

odd number of valence electrons it will be exactly half-full. In the latter case, as the temperature is increased, electrons can easily acquire kinetic energy from electrostatic interaction with vibrating nuclei and move into higher energy levels within the same half-filled band. However in the former case the band is already completely full and for electrons to acquire more energy they must be able to jump to an unoccupied level in a higher energy band. Whether this is thermodynamically feasible depends on the temperature and the energy gap between the bands. The probability, P, that an electron 'occupies' an energy level with an energy, $\mathbb{E}$, at a temperature, T, is given by the *Fermi–Dirac distribution function*:

$$P(\mathbb{E},\, T) = \left[1 + \exp\left\{ \frac{\mathbb{E} - \mathbb{E}_f}{k_B T} \right\} \right]^{-1} \tag{2.5}$$

where k_B is the Boltzmann constant and $\mathbb{E}_f$ is known as the *Fermi energy*; if an energy level exists at the Fermi level it has a 50% probability of being occupied, and $\mathbb{E}_f$ would be the highest level occupied at absolute zero. The Fermi level also corresponds to the thermodynamic concept of chemical potential so that if two materials with different Fermi levels are joined then electrons will flow from one to another until the Fermi levels coincide.

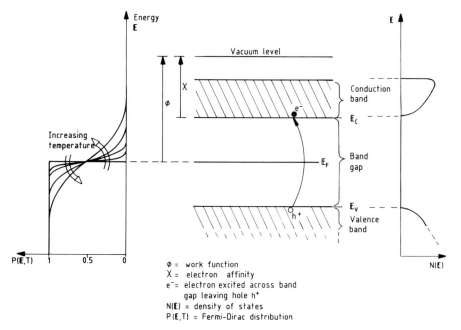

Fig. 2.3 An energy-band diagram for a covalently-bonded crystal. On the left of the diagram is the Fermi–Dirac distribution indicating the probability of occupancy of an energy level, $\mathbb{E}$, as a function of temperature

A so-called *energy-band diagram* is shown in Fig. 2.3 for a covalently-bonded crystal with an even number of valence electrons; we will see that all insulators fall into this class. The Fermi–Dirac distribution is also shown for different temperatures. The highest filled band at absolute zero is known as the *valence band* (for an ionic-bonded crystal the corresponding bands are named after the ions), and the first unoccupied band as the *conduction band*. (For molecular materials the equivalent terms are the highest filled molecular orbital, the HFMO, and the lowest unfilled (or empty) molecular orbital, the LUMO or LEMO.) Between the bottom of the conduction band, $\mathbb{E}_c$, and the top of the valence band, $\mathbb{E}_v$, is the *band gap*, sometimes known as the *forbidden band gap* since electrons cannot acquire energies in this region. The horizontal axis in this simple picture has no general meaning but is often used to represent distance or, more usually, '*k*-space' — where *k* is the electron *wave vector*, the reciprocal of the electron wavelength, λ, in a given state of the band, i.e. $k = 2\pi/\lambda$ in a specific direction. The reason for using *k*-space is that the electron wave vector determines the phase of the electron-wave at any site, *x*, through the product $k \cdot x$ and is proportional to the momentum of the electron–particle since $k = 2\pi \cdot p/h$ where *h* is the Planck constant ($6{\cdot}6261 \times 10^{-34}$ Js) and *p* is the momentum. Band diagrams with *k*-space horizontal axes are often known as $\mathbb{E}$-*k* diagrams. The band energy is strongly *k*-dependent and the bands may show strong curvature (cf. Fig. 2.5).

All the energies on the energy-band diagram are relative; in order to compare energies between materials, for example at interfaces, an absolute reference energy level, the *vacuum level* is defined. This is the energy of an electron at rest which has been completely removed from the material. The energy required to remove an electron from the highest occupied level of a neutral atom or molecule (i.e. the minimum energy state) is known as its *ionisation energy* and corresponds to the energy difference between the electron's energy level and the vacuum level. Sometimes the *binding energy* (or *valence energy*) *of an electron* is referred to; this is the energy required to raise the electron's energy to the Fermi level — i.e. the energy required to dissociate electron and nucleus within the material. The *electron* binding energy should not be confused with the binding energy required to dissociate atoms or molecules bonded with covalent, ionic, or van der Waals forces.

The *work function* and *electron affinity* characterize the material rather than electrons within it. The *work function*, ϕ, is defined as the energy required to remove an electron from the Fermi level to the vacuum level; it represents the minimum energy required to take an electron out of the material at absolute zero and thus determines the minimum quantum energy ($h\nu_{min}$) needed to produce photoemission. The *electron affinity*, χ, is the energy released when an electron is placed at the bottom of the conduction band, $\mathbb{E}_c$, from the vacuum level; it determines the escape condition of an electron from negatively-charged bulk material. The electron affinity also represents the energy released when an electron is added to a neutral atom. The greater the electron affinity, the more tightly bound is the added electron. In general, electron affinities decrease going down any group of the periodic table and increase going from left to right across any period.

2.2.3 Band transport

Electrons in the valence band of covalently-bound crystal lattices are not available for conduction since the extent of their delocalisation is restricted to the stereospecific covalent bonds. Electrons in the conduction band however have sufficient energy to break away from specific atoms or bonds and, since their energies are in excess of the Fermi energy, they can leave the crystal lattice through appropriate contacts thereby producing an electric current. Electrons which are excited across the band-gap to the conduction band leave behind half-occupied energy levels. It is quite easy to show the rather surprising result that electrons occupying these partially-filled levels near the top of the valence band apparently have a *negative effective mass*, m^*. (The electric field experienced by an electron is the applied electric field modified by the internal periodic field of the lattice. The kinetic energy of the electron is therefore also modified so that it *appears* to have a different mass, known as the effective mass, m^*, such that its energy, $\mathbb{E} = \frac{1}{2} m^* v^2$ or $m^* = 2\mathbb{E}/v^2$. This can be related to the $\mathbb{E}$-k diagram since momentum $p = m \cdot v = h \cdot k/2\pi$ so that

$$m^* = \frac{(h/2\pi)^2}{d^2\mathbb{E}(k)/dk^2}$$

implying that for bands with steep curvature m^* is small whilst for band with shallow curvature m^* is large.) A negative effective mass implies that the negatively-charged electrons in half-filled levels of the valence band are driven towards the negative terminal. They therefore act as if they were *positive* charged particles and are known as *holes* since they correspond to missing electrons. These are also shown in Fig. 2.3.

In general the conductivity is the sum of the components due to both holes and electrons which can be calculated using eqn. 2.4. The mobility, μ, in this equation is limited by *lattice scattering* which can be thought of as either the collision of electron/hole particles with the thermally vibrating nuclei or the diffraction of electron/hole waves when the thermally-oscillating inter-nucleus spacing coincides with an integral multiple of their wavelength. This, and other scattering processes such as electron-electron scattering, limits the mobility to:

$$\mu = \frac{e\tau}{m^*} \tag{2.6}$$

where τ is mean time between scattering events and m^* is the effective mass of the carrier (electron or hole). The mean distance between scattering events is usually referred to as the *effective mean free path*.

Whether electrons and holes are available for conduction depends on whether electrons can be thermally excited across the band gap. By using the Fermi–Dirac distribution function to calculate the probability of electrons occupying the bottom of the conduction band it is shown below that a small band gap of $< \sim 0.2$ eV (1 eV $= 1.602 \times 10^{-19}$ J $= 96.49$ kJ $\cdot$ mol^{-1} = 23.05 kcal $\cdot$ mol^{-1}) or a half-filled valence band will result in the material being a conductor, a band gap of ~ 0.2 eV to ~ 2.0 eV will result in a semi-conductor and larger band-gaps result in insulators. These three

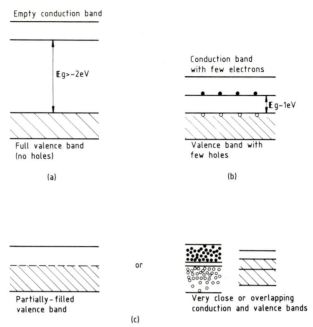

Fig. 2.4 Energy-band diagrams for (*a*) an insulator; (*b*) a semiconductor; and (*c*) for a conductor

situations are shown in Fig. 2.4. The assumption here is that the mobility of the carriers in each case is comparable. The distinction between semiconductors and insulators is somewhat arbitrary and defined more by the material's normal application than its properties. However semiconductors are usually thought of as materials whose conductivity can be controlled by substitutional doping (as in electronic semiconductor devices) whereas such doping in insulators, particularly non-crystalline insulators, is not likely to result in effective control of the conductivity since trapping is likely to dominate (see Section 2.3.2).

In order to calculate the actual concentration of electrons within a given band of energies (e.g. in the conduction band) one needs to know the concentration of states. This is known as the *density of states*, $N(\mathbb{E})$, and gives the number of states per unit volume with energies in the range $\mathbb{E}$ to $\mathbb{E} + d\mathbb{E}$. This is also shown in Fig. 2.3 for a crystalline structure. The concentration of carriers is then obtained by integrating the product of the density of states and the probability of occupancy, given by the Fermi–Dirac distribution, over the range of energies of interest. For example the number of free electrons, n_c, i.e. those between the bottom of the conduction band with energy $\mathbb{E} = \mathbb{E}_c$ and the vacuum level with energy $\mathbb{E} = 0$ would be calculated by substituting appropriate forms of the functions, $P(\mathbb{E}, T)$ and $N(\mathbb{E})$ into:

$$n_c = \int_{\mathbb{E}_c}^{0} P(\mathbb{E}, T) \cdot N(\mathbb{E}) \, d\mathbb{E} \qquad (2.7)$$

Table 2.2   Approximate calculations of carrier concentrations and conductivities in crystalline conductors, semiconductors and insulators

Band gap (eV)	type	$P(\mathbb{E}_c)$	$n_c = N_{\text{eff}}P(\mathbb{E}_c)$ (m^{-3})	$\sigma = n_c q \mu$ (Ω^{-1} m^{-1})
0	conductor	1	10^{29}	$1 \cdot 6 \times 10^7$
0·2	semiconductor	$1 \cdot 8 \times 10^{-2}$	$1 \cdot 8 \times 10^{27}$	$2 \cdot 9 \times 10^5$
2·0	insulator	$4 \cdot 3 \times 10^{-18}$	$4 \cdot 3 \times 10^{11}$	$6 \cdot 9 \times 10^{-11}$
∞		0	0	0

It should be noted that some texts confuse the density of states with the *effective* density of states, N_{eff}. The latter is a number which, if multiplied by the probability of occupancy of a state at the bottom of the conduction band, gives the total concentration of free electrons.

To get some idea of the implications of band theory in terms of the effect of band gap on conductivity we can make some approximate calculations. Making the crude assumption that the effective density of states is similar in order of magnitude to the number of atoms per unit volume in a crystalline material, $N_{\text{eff}} \sim 10^{29}$ m^{-3}, then the ranges of concentration of free electrons at room temperature ($k_B T \simeq 0 \cdot 025$ eV) can be evaluated for a conductor, semiconductor, and insulator using the ranges of band gaps given above. Furthermore if a typical mobility of 10^{-3} m^2 V^{-1} s^{-1} is assumed then the ranges of typical conductivities can also be calculated. These calculations are shown in Table 2.2 and it can be seen that the ranges of conductivities are not dissimilar to those given earlier.

2.3 Band theory applied to polymers

2.3.1 Theoretical considerations
In this section we now develop the 'crystalline' or periodic potential band theory for molecular crystals and polymers in particular.

The discussion so far has considered the energy bands found in perfect periodic crystalline structures. These are the simplest structures to consider since they have infinite periodicity in all directions and the nuclei interact with the (3-dimensional) electron-wave in the same way in any unit cell. Polymers are characterised by very long chain molecules formed of typically 10^3—10^5 strongly covalently-bonded monomer molecules each with their own electronic states. As the polymer chain is built up, degenerate monomer molecular orbitals form a series of extended electronic states, i.e. energy bands within the molecule. Note however that these bands are now essentially 1-dimensional, if free carriers exist in these bands they can only move along the length of the backbone. We can illustrate this by postulating a one-dimensional metal consisting of an infinitely-long chain of equally-spaced carbon atoms each containing one π electron:

$$-\dot{C}-\dot{C}-\dot{C}-\dot{C}-\dot{C}-\dot{C}- \qquad (2.8a)$$
$$\ \ | \ \ \ | \ \ \ | \ \ \ | \ \ \ | \ \ \ |$$

This structure would be conducting and this is consistent with the chemist's concept of 'resonant' structures and also the diamagnetic properties of benzene-like chemical structures that result from circulating π-electron currents[48]. The band structure for this is shown in Fig. 2.5a where the vertical axis is electron energy (as in the energy-band diagram) and the horizontal axis indicates the electron wave vector, k, in the direction of the polymer 'backbone'. In this case we can define $k = n \cdot \pi/(N \cdot a)$ for $n = 0, 1, 2, 3, \ldots, (N-1)$ where a is the length of a unit cell consisting of two adjacent carbon atoms, and N is the number of such monomer units in the chain. For the infinite structure considered, electrons can take any corresponding values of $\mathbb{E}$ and k given by the lines on the diagram. At absolute zero the electrons would only occupy the states given by the lower line whilst at higher temperatures some may be excited to states given by the upper line. For structures with a finite number of atoms, only certain points on these lines would represent allowed $\mathbb{E}$-k states of the electrons, so that there will always be a finite energy gap.

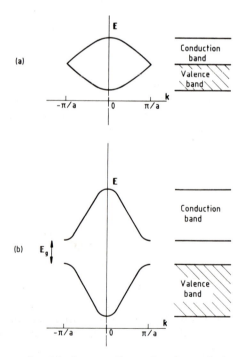

Fig. 2.5 $\mathbb{E}$-k diagrams for (*a*) the one-dimensional metal given by eqn. 2.8*a* and (*b*) the saturated σ-bonded carbon chain given in eqn. 2.8*c*

The structure shown in eqn. 2.8*a* is, in fact, unstable, and Peierl's distortion produces a bond-alternating polyene structure of the form:

$$\cdots -\overset{\displaystyle |}{C}=\overset{\displaystyle |}{C}-\overset{\displaystyle |}{C}=\overset{\displaystyle |}{C}-\overset{\displaystyle |}{C}=\overset{\displaystyle |}{C}-\cdots \tag{2.8b}$$

Because the bond lengths are now unequal there is π-electron localisation around the double bond and a band gap of around 2 eV results, Fig. 2.5*b*, thereby making it semiconducting rather than conducting.

In contrast André[54] has calculated the band structure of a saturated σ-electron bonded carbon chain without bond alternation of the form:

$$\cdots -\overset{\displaystyle |}{\underset{\displaystyle |}{C}}-\overset{\displaystyle |}{\underset{\displaystyle |}{C}}-\overset{\displaystyle |}{\underset{\displaystyle |}{C}}-\overset{\displaystyle |}{\underset{\displaystyle |}{C}}-\overset{\displaystyle |}{\underset{\displaystyle |}{C}}-\overset{\displaystyle |}{\underset{\displaystyle |}{C}}-\cdots \tag{2.8c}$$

such as might be found in polyethylene. The band structure for this has the form shown in Fig. 2.5*b*. However, in this case, a substantial band gap results producing an insulating structure.

For infinite structures these $\mathbb{E}$-$\boldsymbol{k}$ states correspond to bands of energies equivalent to the valence and conduction bands. For the one-dimensional metal case there is no band gap, for the conjugated structure a relatively small band gap results, and for the saturated case the band gap is large. The substantial band gap of the latter case results from the strong binding between electrons and nuclei within the polymer molecule; the electrons reside in σ-orbitals and are localised to their bonds. The tight bonding characteristic of the σ-bond system, implies that very high energies are required to dissociate electrons from the nuclei and form extended orbitals along the chains. For inter-chain delocalisation even higher energies are required since the conduction electron must overcome the polymer molecule's ionisation energy before it can transfer to another chain. Furthermore the wave functions of lower-energy electrons (i.e. those with orbitals closer to the nucleus) interact with each other much less and consequently the width of energy bands associated with them is very much narrower. For band theory to be applicable, carriers must remain within the energy constraints of the band for a time greater than the mean time between scattering, τ. For this to be the case, the *Heisenberg uncertainty principle* implies that the width of the band, W, should be such that:

$$W > \frac{h}{2\pi\tau} \tag{2.9}$$

Typically τ is $\sim 10^{-14}$ s in polymers[55] so that $W > \sim 0.066$ eV for band conduction to be realistic. This is generally not the case for lower-energy electrons.

Other assumptions that we have made for the formation of energy bands along the polymer molecules are that the monomer molecular orbitals are degenerate and that the monomers are regularly spaced along a conformationally regular and infinite chain containing no defects. None of these are likely to be perfectly true in practice and McCubbin[56] has pointed out some of the consequences of this for the formation of energy bands.

Structural disorder can take various forms. The real polymer chain is never rigid and above the glass-transition temperature segmental motions may be strong. In the majority of cases, polymer chains, whether in amorphous or crystalline phases contain non-periodic sequences (e.g. Gaussian chains in amorphous phases and chain folds in crystal phases). This lack of periodicity in the lattice spacing interferes with the formation of ideal or classical bands for which a periodic lattice structure is essential. Polymers with large pendant groups are likely to have an even more distorted backbone. In these systems charge carriers do not exist in extended energy band states but are localised by polarisation and relaxation effects that create a distribution of energy states which exceeds the band width due to the overlap of the originally degenerate molecular orbitals. Chain ends are likely to be a relatively insignificant problem for the formation of band structures if sufficiently long chains exist but these will be a major factor in the consideration of bulk conduction as carriers must be transported from one chain end to another. Other wave phenomena such as plasmons (correlated valence electron oscillations) and phonons (thermally-generated inter-molecular vibrations including acoustic waves) also interfere to cause irregularities in the band structure. As well as structural disorders there are likely to be chemical inhomogeneities from additives such as antioxidants and also from impurities such as crosslinking agent residues and oxidation products (Section 1.6). It may be surprising then that there is a recognizable band structure in polymers at all! However, as will be seen later, there is reasonable agreement in many respects between theoretical and experimental data.

In an ideal crystal the lattice spacing is perfectly periodic, at least at a temperature of absolute zero. At higher temperatures there will be a distribution of interatomic spacings because of lattice vibrations and this will cause temporal changes in the energy band gap. In technologically-utilised amorphous semiconductors such as amorphous silicon there will be significant changes in interatomic spacing due to non-crystallinity and this will cause both spatial and temporal changes in the band gap. Furthermore incompletely-bound atoms at crystal defects give rise to *dangling bonds* which can be satisfied by either the removal or donation of an electron (or both) and thus behave as states within the band gap. However these energy states are not extended throughout the crystal, they only exist in the vicinity of the chemical or structural defect. Electrons or holes entering these *localised states* are therefore not available for conduction and may have to acquire considerable energy from phonons or photons before they can leave. These states are therefore called *traps*. (In amorphous silicon the density of states in the conduction and valence bands is $\sim 10^{28} \mathrm{~m}^{-3}$ and typically $\sim 10^{26} \mathrm{~m}^{-3}$ in the centre of the band gap. By 'satisfying' the dangling bonds with hydrogen this may be reduced to $\sim 10^{23} \mathrm{~m}^{-3}$.)

The traps may take the form of donors (hole traps) with energy levels immediately above the top of the valence band, acceptors (electron traps) with energy levels immediately below the bottom of the conduction band, or sites where the field from an electron or hole can reorientate the local structure thereby creating a potential well from which escape may be difficult (self-traps). The time which carriers spend in traps depends on the 'depth' of the trap, i.e. the energy required to remove the carrier, the temperature and possibly other factors such as field. Self-trapped charge is a form of *space charge* in which a region of an insulator (or semiconductor) contains localised charge of one polarity which is not compensated by an equal concentration of opposite polarity charge. This causes local field enhancements due to Poisson's equation:

$$\nabla E = \frac{\rho_c}{\varepsilon} \qquad (2.10)$$

where ρ_c is the charge density (C m^{-3}) and ε is the permittivity.

In practice the density of trap states in the energy 'gap' can be high and the edges of the conduction and valence bands very ill-defined. A schematic diagram showing localised states as squares is given in Fig. 2.6. Electrons can escape from localised states given sufficient energy and drift to other localised states under the influence of the local electric field. However conduction in the band gap by this method is very slow since carriers spend most of their time in the traps and very little travelling between them. Further from the centre of the band gap the localised states are in much closer proximity and it may be possible for electrons to tunnel or hop between sites which are close to each other (see Sections 2.5 and 9.1.1(*b*) for more detail). The inter-state distance is very critical for the occurence of this form of conduction and the probability of hopping or tunnelling

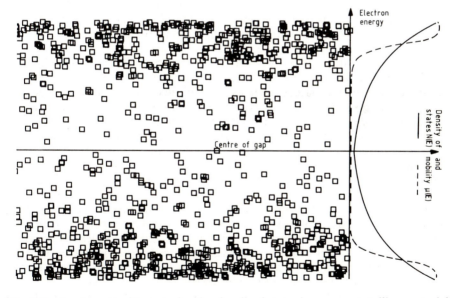

Fig. 2.6 A schematic diagram showing localised states in a non-crystalline material (indicated by squares) as a function of electron energy together with the density of states, $N(E)$, and the mobility, $\mu(E)$. Although the band edges are ill-defined the effective mobility decreases sharply near the centre of the gap where there is a low concentration of states and it is unlikely that hopping or tunnelling can take place between them

between states increases rapidly as the distance decreases below a few nanometers. In the diagram overlapping squares schematically show states between which electron or hole transport is likely. Fig. 2.6 also shows a possible space-averaged density of states as a function of electron energy together with the effective mobility. It can be seen that although the band edges and band gap are not well defined there are reasonably well defined energies at which the mobility changes. Mott[57] first pointed out that states in the band tails of non-crystalline materials would be localised and that energies of localised and extended states would be separated by sharp boundaries which Cohen *et al.*[58] termed the *mobility edges*; these delineate the so-called *mobility gap*. This concept is useful in separating out those electron energy states responsible for charge transport from those influencing other properties, e.g. optical absorption. For reviews of electrical conduction and carrier traps in polymeric insulation and non-crystalline materials see Ieda[59] and Mott[60].

Ritsko[61] has shown that the minimum mobility required by band theory is

$$\mu_{min} = \frac{2\pi e a^2}{8h} \tag{2.11}$$

where a is the lattice spacing. In polyethylene, for example, with $a = 0.254$ nm, $\mu_{min} = 1.2 \times 10^{-5}$ m^2 V^{-1} s^{-1}. Estimates of mobility from bulk conduction measurements are much less than this (typically 10^{-10}–10^{-14} m^2 V^{-1} s^{-1}) and such conduction processes are likely to be limited by the mechanisms of inter-chain rather than intra-chain transport.

2.3.2 Correlation of calculated and measured electronic energy bands in polymers

Many electronic state calculations have been carried out on ideal polymer structures. For example André[54] lists references of calculations carried out on polyethylene; derivatives of polyethylene including PTFE, PVDF, and PVDC; and other types of polymers including LiF chains, polyglycine, and phosphite, boron-nitrogen and ladder polymers. More recently Ritsko[61] has reviewed the current understanding of electronic states in polymers and Ieda[62] has also commented upon it. Such calculations can be compared with experiments which probe electronic states such as UV and X-ray photo-emission (sometimes referred to as UV and X-ray photo-electron spectroscopy, UPS and XPS), optical absorption and inelastic electron scattering (see Ritsko[61] and references therein). It was found that calculations yield estimates of the density of states in the valence band in reasonable agreement with such experiments but tend to overestimate the band gap and fail to accurately predict the density of states in the conduction band[63].

Many comparisons have been made between theoretical predictions and experimental measurements on polyethylene since this is both chemically simple and technologically important. The valence bands in polyethylene have been studied using photoemission by various workers including Fabish[64] and by photo-electron spectroscopy[65]. Their experimental results have been shown to be in reasonable agreement with various types of electronic state calculations[61,62,63,66]. Indeed André[54] has noted a correlation of 0·997 between the experimentally determined positions of the valence bands and values from *ab initio* calculations. It should be emphasised that these calculations are for single non-interacting polymer chains. Since intermolecular interactions are weak and intramolecular distances may be large in polymers such calculations can be representative of the valence band structure for the material as a whole but an accurate description could only be expected for crystalline systems.

The band gap of polymers can be considerably overestimated by over-simplistic calculations. For example Ieda[62] quotes reported values for the energy gap of polyethylene as ranging from 7·6 to 9·0 eV (experiments) and from 7·7–18·96 eV (calculations). The likely reason for this is that many of the simpler calculations do not account properly for the effects of wave function overlap and electron correlation.

The calculations do not generally agree well with measurements of the conduction band because even minor chemical and physical imperfections significantly alter the conduction band structure. In order to investigate the rôle of conduction electrons in polyethylene and other organic dielectrics, the Swiss group of Cartier, Pfluger, Zeller and others has developed a technique which they call 'internal photoemission for transport analysis' (IPTA)[67,68,69,70]. The method is based on the photo-injection of *hot electrons* into the conduction band of insulators at a metal-dielectric interface and the measurement of their energy distributions after transmission through insulating films of varying thickness. (Hot electrons have high thermal/kinetic energies and are therefore high up in the conduction band.) The technique gives various information including the position of the conduction band, the effective mean free path length, and the scattering mechanisms and has been used[68,70] on the linear alkane hexatriacontane, n-$C_{36}H_{74}$, because of its similarity to polyethylene.

Hexatriacontane has a band gap of 8·8 eV and the electrode material used in the IPTA experiments was polycrystalline platinum which has a Fermi level which exactly matches that of the hexatriacontane, i.e. exactly half way up the band gap and therefore 4·4 eV below the conduction band. The conduction band edge is slightly above the vacuum level, i.e. the material has a negative electron affinity, so that electrons are spontaneously emitted into the vacuum by a process generally referred to as exoelectron emission. Typical results[68] in the form of a graph of emitted electron intensity versus energy above the Fermi level, $\mathbb{E}_f$, are given in Fig. 2.7. The intensity essentially reflects the bulk mobility, and the mobility shoulder corresponding to the conduction band edge, can be clearly seen at 4·1 eV above the Fermi energy. The peak near this shoulder may be attributed to thermalised electrons scattered down to the bottom of the conduction band during transport through the film. Inelastic electron-phonon scattering was dominant for low-energy electrons (0·5 eV—1·5 eV) but elastic electron-phonon scattering became significant for electrons with higher energies (1·5 eV—8·8 eV). The results are particularly interesting since Cartier and Pfluger were able to show directly the effects of trapping. They found that degradation was caused by hot electrons with energies as low as 3 to 4 eV

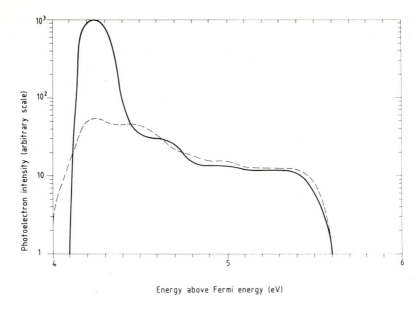

Energy above Fermi energy (eV)

Fig. 2.7 The electron energy spectrum for the emission of photo-electrons after traversing a 150 Å n-$C_{36}H_{74}$ paraffin film deposited on a platinum electrode. Solid line before, and dashed line after irradiation with 17 eV electrons. Data taken from Cartier and Pfluger[68]

and that bombardment of the film with high-energy (17 eV) electrons caused (C—H)-bond scission, polymerisation, and possible cross-linking and that it increased the trap concentration from 10^{16} m^{-3} to 10^{22} m^{-3}. This strongly suppressed the transport of the electrons at the bottom of the conduction band and it is supposed that this is due to their getting trapped in irradiation-induced defects.

It is generally found that the conductivity of polymers is thermally activated according to the Arrhenius Law such that

$$\sigma = \sigma_0 \exp\{-\mathbb{E}/k_B T\} \tag{2.12}$$

It is tempting to ascribe this to the thermal excitation of electrons across the band gap thereby providing free electrons and holes (as in thermistors and conventional semiconductor devices). However this cannot be the principal mechanism since the band gaps of polymers are simply too large. For example in polyethylene with a band gap of 8·8 eV, the probability of an electron occupying the lowest level in the conduction band due to thermal excitation at 100°C, i.e. in the vicinity of its melting temperature, is ~3·7 × 10^{-60} and is therefore negligible. The mechanisms responsible for the observed conductivity dependence upon temperature are therefore more complicated than the simple creation of carriers by thermal excitation across the band gap.

2.3.3 Summary
Conventional energy-band theory may be used to explain the wide range

of carrier concentrations and conductivities observed in materials with covalently-bound crystal lattices. This may be developed to describe the electron energies found within large molecules such as those found in polymers. However, in this case the valence and conduction bands are less clearly defined and temporal and spatial inhomogeneities cause states to occur in the band gap. Electron transport may take place between closely-spaced localised states away from the centre of the band but is much less likely between the more widely-spaced localised states in the gap. This gives rise to a 'mobility gap'.

Conventional electron and hole transport may be possible in a limited way along the length of many polymer molecules. Whilst a form of conduction band may exist throughout bulk-polymers for the transmission of hot electrons, it is unlikely that well-defined valence and conduction bands exist throughout the bulk of most polymers. However, hot electrons are not likely to be the 'normal' method of charge transport between polymer molecules as they are likely to cause damage to the material possibly leading to breakdown.

2.4 Ionic conduction

In Section 2.3 it was shown that band theory may be able to explain charge transport, both as holes and electrons, along the length of polymer molecules or parts thereof. Between polymer molecules however, there is insufficient electron wave-function overlap to form a continuous valence band. (Because electrons can only move easily along the polymer chain they are sometimes said to exhibit *anisotropic mobility*.) Whilst high-energy electron states may connect different molecules, these states would be occupied by hot electrons (i.e. electrons with kinetic energy significantly greater than $k_B T$) and these would be quickly scattered and are therefore more likely to be responsible for breakdown than conduction. In this book, it is not our intention to describe current thinking on polymer conduction mechanisms in any detail. This field is active and many review articles exist (e.g. References 59, 71, 72). Similarly there has been a considerable interest recently in the manufacture of electrically-active conducting polymers such as the polyacetylenes[73], but our discussion here of charge transport in polymers will be restricted to those used as insulators. In this section and Section 2.5 we wish to give an overview of the types of carriers and typical mechanisms of conduction in solids. Electrode effects, the effects of high fields, and the way in which traps and space charges affect the conductivity will be left until Chapter 9.

Charge transport may be classified according to the type of charge carrier. It is convenient to distinguish between ionic (including protonic) transport and electronic (including hole) transport. The type of charge carrier primarily responsible for conduction depends not only on the material's chemical and physical composition but also on the frequency of the applied field and the temperature. Other factors such as electric field, electrode material, electrode configuration, and environmental factors such as hydrostatic pressure, humidity and atmospheric gases may also be influential. It is not suprising then that the understanding of the mechanisms of charge transport through polymers is still incomplete.

In any system through which current flows and which is contacted by metal electrodes, DC ionic conduction cannot continue *ad infinitum*. Ionic transport is mass transport and metallic electrodes can only provide electrons or recombination sites for holes. Although some ions may already exist in the material, ionic transport must eventually rely on the creation of ions, their movement, and, if space charge is not to build up, their destruction. These ions may be created by electrolytic action at the electrode/insulator interface (e.g. by hydrolysis of ambient moisture) or by decomposition of the insulator or even of the electrodes. Ionic conduction is likely to be the normal surface conduction mechanism, although this (together with tracking, i.e. surface breakdown) is specifically out of the scope of this book.

We may distinguish between two types of ionic conduction: (i) intrinsic ionic conduction which proceeds by the dissociation of main-chain or side groups followed by proton or/and electron transfer through hydrogen-bonded networks (this is possible both in polymers which contain ions such as ionomers and polyelectrolytes and also in polymers which contain groups capable of ionisation); and (ii) extrinsic ionic conduction in which ions which are no part of the chemical structure of the polymer being present as an addition or impurity (e.g. from material added to the insulator and from electrolytic reactions) percolate through the structure.

An example of intrinsic ionic conduction is in polyamide 6.6 (nylon 6,6) which exhibits both protonic and electronic conduction[44] above 120°C. A possible mechanism for this is the transfer of protons through weakly-bonded hydroxyl groups as two adjacent chains come into close proximity through random segmental motion. A series of such transfers is shown in eqn. 2.13 in which a positive charge can be seen to move from left to right and a negative charge from right to left. Whilst it is possible for electrons to leave through the anode, it is not possible for protons to leave through a metallic cathode and so space charge will eventually build up there which may prevent further conduction.

For this type of conduction to occur, the groups must take up energetically favourable positions for the charge transfer to take place. For example in the crystalline poly(methylene oxide), ionic conduction cannot take place but if more methylene groups are added the chains become flexible and ionic conduction is observed[44]. Intrinsic ionic conductivity should therefore increase with temperature (this has been observed in polyamide 6.6[74]) and be virtually non-existent below the glass-transition temperature (as observed in PMMA and unsaturated polyester[45]). Ionic conduction has also been observed in PVF[59] and poly(olefin oxides)[75].

Extrinsic ionic conduction has been observed in some polymers (e.g. sulphonated polystyrene[76,77,78], polystyrene-phenylmethane-tetra-*n*-butyl-ammonium thiocyanate[79]) and some biopolymers[48]. Whilst it may seem reasonable to expect this type of conduction in technologically important insulators because of the presence of dissociable additives (e.g. antioxidants and crosslinking agents) and of free radicals such as alkyl and carbonyl radicals (perhaps from radiation crosslinking or partial discharges) or even

ANODE $\xrightarrow{\text{electric field}}$ CATHODE

$$
\begin{array}{c}
\vdots \\
N-H \\
| \\
N-H\cdots O=C \\
| \qquad\qquad | \\
O=C \qquad\quad \vdots \\
| \\
\vdots
\end{array}
$$

$\downarrow$

$$
\begin{array}{c}
\vdots \\
N^{+}-H \\
\| \\
N^{-}\ H-O-C \\
| \qquad\qquad\quad | \\
O=C \qquad\qquad \vdots \\
| \\
\vdots
\end{array}
$$

$\downarrow$

$$
\begin{array}{c}
\vdots \\
N-H \\
| \\
N^{+}-H_2\cdots O=C \\
| \qquad\qquad | \\
O=C \qquad\quad \vdots \\
| \\
\vdots
\end{array}
$$

$$
\begin{array}{c}
\vdots \qquad\qquad N \\
\qquad\qquad\quad \| \\
N-H\cdots O^{-}-C \\
| \qquad\qquad\quad | \\
O=C \qquad\qquad \vdots \\
| \\
\vdots
\end{array}
$$

$\downarrow$ $\qquad\qquad$ $\downarrow$

$$
\begin{array}{c}
\vdots \\
N^{+}-H \\
\| \\
N-H \qquad H-O-C \\
| \qquad\qquad\qquad\quad | \\
O=C \qquad\qquad\qquad \vdots \\
| \\
\vdots
\end{array}
$$

$$
\begin{array}{c}
\qquad\qquad\quad N \\
\qquad\qquad\quad \| \\
N^{-}\ H-O-C \\
| \qquad\qquad\quad | \\
O=C \qquad\qquad \vdots \\
| \\
\vdots
\end{array}
$$

$\downarrow$ $\qquad\qquad$ $\downarrow$

$$
\begin{array}{c}
\vdots \\
N^{+}-H_2 \\
\| \\
N-H \qquad O=C \\
| \qquad\qquad\quad | \\
O=C \qquad\qquad \vdots \\
| \\
\vdots
\end{array}
$$

$$
\begin{array}{c}
\qquad\quad N \qquad\qquad\quad \vdots \\
\qquad\quad \| \qquad\qquad N-H \quad O=C \\
O^{-}-C \qquad | \qquad\qquad | \\
| \qquad\qquad O=C \qquad \vdots \\
\vdots \qquad\qquad\quad |
\end{array}
$$

... etc.

Eqn. 2.13 *Ionic Conduction in Polyamide 6.6*
Self-dissociation of hydrogen–bonded amide groups and subsequent bond rearrangement leads to the possibility of electron
and proton transport.

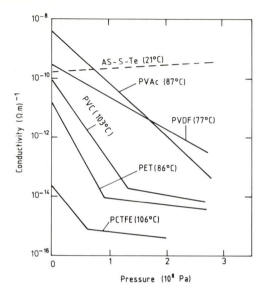

Fig. 2.8 The dependence of conductivity on hydrostatic pressure for various polymers. Taken from Saito *et al.*[81]
Copyright © 1968 John Wiley and Sons

of water ions, direct confirmation is not commonly observed. Water is known to increase the surface conductivity of many polyelectrolytes[80] and the bulk conductivity of biopolymers[48] but its effects may be quite diverse since it can act as a source of ions, as a high permittivity impurity, as a plasticiser and it may modify local structure. A change in conductivity with hydration does not therefore necessarily imply ionic conduction.

It is commonly thought that an increase in pressure should result in a decrease in conductivity if ionic conduction is the operative mechanism since the decrease in free volume inhibits ion percolation. Electronic conduction should be enhanced by pressure due to the increase in electron wavefunction overlap. This is, however, rather oversimplified since the effect of increased pressure on intrinsic ionic conduction may also enhance charge transfer between cooperating inter-chain groups. Also considerable increases in pressure would be required to increase orbital overlap significantly. Saito *et al.*[81] have shown that the conductivity of various polymers used for insulation does decrease with increasing pressure, Fig. 2.8, suggesting that extrinsic ionic conduction predominated under the conditions of the test but more work is needed in this area to be definitive.

2.5 Electronic conduction

Electron and hole transport through a band structure has been described earlier in this chapter. This type of conduction may exist along the length of polymer chains where a regular chain conformation is maintained although the band gap is too large for normal thermal excitation to be responsible for the creation of free carriers. Between molecules and at chain conformational defects, electrons and holes are likely to be trapped in localised states. Electrons may be trapped by acceptors or ionised donors and holes by donors or ionised acceptors. The presence of a trapped charge can lead to local polarisation and distortion of the lattice structure resulting in local energy-band deformation making it more difficult for the trapped charge to escape. Transport of the charge must be accompanied by transport of its associated polarisation; this composite package of charge and its lattice distortion is known as a *polaron*[82]. In order to leave traps the carriers have to overcome large potential barriers with the aid of thermal/phonon excitation. As we discuss in Chapter 9 this may be aided by high electric fields and other factors. As the liberated carrier approaches the next trap it will be attracted towards and then into its 'potential well'.

A mechanism for electron/hole transport then is *thermally activated hopping* from one trap to the next whereby the electron/hole gains sufficient energy from the lattice by means of thermal fluctuations to overcome the potential barrier. An alternative mechanism by which an electron/hole can move from one trap to a close adjacent trap is by quantum mechanical *tunnelling*. This occurs because, according to quantum mechanics, an electron is not spatially well defined like a particle; one can only define the probability of its presence at a particular point in terms of the square of its wave function. If there is only a very narrow barrier, typically less than 1 nm, then the probability that it exists on the other side of the barrier is finite. In this way the electron-wave can *tunnel* through narrow barriers simply 'appearing' on the other side. Obviously there is no classical analogue to quantum mechanical tunneling in the normally observable world. The dynamics of hopping and tunnelling are well documented in standard texts (e.g. References 52, 53) and also discussed further in Chapter 9.

Whilst the inter-trap mobility of electrons and holes within intramolecular energy bands is likely to be quite high ($>10^{-4}$ m^2 V^{-1} s^{-1}), their mobility in inter-molecular regions is probably much lower (typically 10^{-7}—10^{-14} m^2 V^{-1} s^{-1}) because they are likely to remain in traps for extended periods. However it should be noted that carriers travelling along chains are also likely to encounter traps at chain folds and polar groups. Indeed it may be that the concepts of energy bands are not applicable to polymers with monomers containing dipoles since these could lead to local variations in energy level of as much as 1 eV[83] providing 'periodic trapping'.

Mort *et al.*[84] have shown that hole transport via phonon-assisted hopping between traps in the amorphous region is responsible for conduction in poly(*N*-vinylcarbozole) and its charge-transfer complex with trinitrofluorenone. In this material the backbone of the molecule plays no part in the conduction. The transit time of holes across a sample decreases with increasing temperature indicating an increase in conductivity. Since transport in this and similar materials is essentially stochastic it is not very meaningful to specify a mobility, the mean velocity of carriers per unit field, since this is a function of time (the Scher–Montroll theory[85]).

The effect of crystallinity in polymers is to lower the conductivity irrespective of the conduction mechanism. If conduction is ionic, ion mobility through crystalline regions will be low; if electronic it may be faster but the crystalline-amorphous interface experiences Maxwell–Wagner interfacial polarisation and acts as a strong trapping region. (A heterogeneous medium exhibits dielectric dispersions which its individual components do not. These are associated with the non-uniform distribution of free charges across the interface between the dissimilar dielectric components. These polarisations are known as Maxwell–Wagner[86,87,88,89] or interfacial polarisations. For a review of interfacial dielectric phenomena see Pethig[48].) For example conduction in polyethylene is thought to be by electrons and is constrained by their trap-limited transport through amorphous regions[90,91]. However some features are more in keeping with ionic transport, for example the conductivity of both LDPE and HDPE increases as the temperature is raised through the melting point[92], and its activation energy is proportional to the cohesive energy density[93]. In either case the conduction is not by free carriers but rather through a frequency-dispersive hopping transport[85,90] mechanism.

Chapter 3
Overview of electrical degradation and breakdown

3.1 Low level degradation in polymers

In addition to those forms of degradation for which an electric field is necessary we shall include in this chapter a brief description of physical[94] and chemical[95] aging. These two latter processes are important not only because they can influence the probability of breakdown itself, but also because they may be involved in electrical degradation when driven by an electrical field during service. Consequently some of the microscopic mechanisms of aging feature prominently in postulated mechanisms for electrical degradation phenomena.

3.1.1 Physical aging

First let us examine physical aging. In essence this arises because the facility of polymer chain segmental motions in amorphous regions (be they linear or crosslinked) decreases catastrophically (i.e. with an increasing rate of decrease) as the temperature drops towards the glass transition temperature. This behaviour causes a slowing down in the return to equilibrium of induced mechanical strains and dielectric polarisation following removal of the appropriate stress. This can be followed through the mechanical and dielectric responses of the polymer chains (the so-called α response associated with main-chain motion) at a number of different constant temperatures. As the temperature of the measurement is reduced the observed relaxation time increases and eventually becomes sufficiently long for a non-equilibrium state to exist for an observable period. This can be followed conveniently in dielectric response measurements by examination of the dielectric loss as a function of frequency and noting the change in loss peak position as the temperature is varied.

Dielectric response measurement results for PVC are shown in Fig. 3.1a. It can be seen that there is a high-temperature absorption with a high activation energy (the α-response) accompanied by an absorption at higher frequencies which is only clearly distinguishable at lower temperatures when the polymer is in a glass-like condition. This latter absorption is the so-called β-response and is characteristically due to local intra-segmental and side-chain relaxations[96]. When the system is cooled at a constant rate a change in the magnitudes of several physical properties is observed at the temperature for which the non-equilibrium state becomes metastable. This temperature is termed the glass transition temperature, T_g. Its value however depends on many factors including the cooling rate, and thus T_g

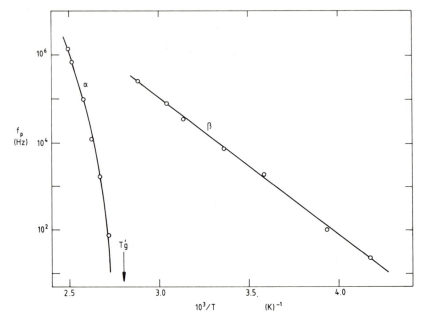

Fig. 3.1

(a) An Arrhenius plot showing the variation of the α and β dielectric loss peak frequencies in PVC as a function of temperature. The temperature T'_g at which the α-relaxation frequency becomes zero and the system formally becomes rigid is indicated. Data taken from Ishida[100]

is rather ill-defined and somewhat arbitrary. As the temperature is lowered in this region the macroscopic viscosity and the relaxation rate asymptotically approach infinity at a temperature[97,98,99] T'_g, see Figs. 3.1b and 3.1c. The rigid state of the system represented by T'_g is a well-defined physical concept, and, although it is unrealizable in real materials, their state at any chosen temperature can be defined through the closeness of their approach to rigidity. This approach follows a power law trajectory in reduced temperature (i.e. viscosity $\eta \propto (T - T'_g)^{-s}$; relaxation time $\propto (T - T'_g)^{-r}$).

Between T_g and T'_g the non-equilibrium state is metastable, i.e. it will exist long enough to be observable but will allow a slow structural relaxation via segmental motion in the amorphous region which thereby increases in density. This time-dependent change as the system moves towards thermal equilibrium is what is known as physical aging. The steep increase of relaxation time with temperature indicates that a few degrees below T_g the metastable state will exist for many months or years depending on the material in question. Below T'_g the material is essentially rigid except for intra-segmental (β) motions and aging can be expected to be practically non-existent. It has been suggested[94] that semicrystalline polymers will also exhibit this form of aging at temperatures above their bulk T_g because the amorphous regions adjacent to the crystalline lamella will have higher glass

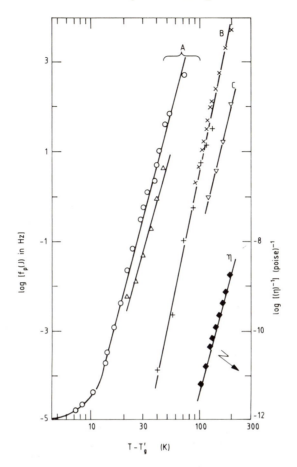

Fig. 3.1

(b) Power-law dependence upon reduced temperature $(T - T'_g)$ of the compliance relaxation frequency $f_p(J)$, (relaxation time $= (2\pi \cdot f_p(J))^{-1}$) and viscosity, η, of the *same* polyisobutylene material. The different data were obtained in different laboratories, with A, B, and C referring to different compliance relaxations (left-hand scale) and reported in Reference 99. Note that the temperature at which the relaxation times and viscosity become infinite, T'_g, in $\eta = (T - T'_g)^{-s}$ and $f_p(J) = (T - T'_g)^r$ is the same value, 183 K in all cases, in contrast to a definition of T_g based on a chosen value of the relaxation time
Copyright © 1984 John Wiley and Sons

transition temperatures than that of the rest of the amorphous body. Filled materials such as rubber may also behave similarly with the filler particles taking the place of crystal lamella. Although evidence for this effect is limited the plethora of weak relaxations observed in such materials, which involve segments projecting from the crystal surfaces, lends support to this contention.

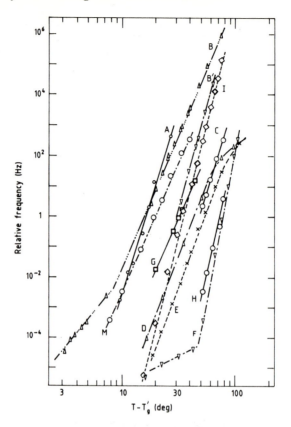

Fig. 3.1

(c) Power-law dependence of the relaxation frequency ($f_p = \omega_p/2\pi$) for a number of glass forming systems, showing $f_p \propto (T - T'_g)^r$. Data taken from Reference 98 and the systems referred to are:

Material		T'_g (K)	r
A	43·4% chlorobenzene in pyridene	122	13
B	Polypropylene oxide	198	4·6 and 8·8
B′	Polypropylene oxide (10^8 Pa)	200	17·4
C	Polypropylene	223	12·3
D	Polypropylene oxide solution in methyl-cyclohexane	200	9·8
E	Eugenol	185	9·9
F	Polyvinyl Acetal	290	1·9 and 19·6
G	3 methyl-3 heptanol	155	8·9
H	Poly-γ-benzyl-L-glutamate	199	17
I	Polymethyl acrylate	263	14·4
M	Polystyrene	370	8·3

A physical description of the aging (or structural relaxation as it is sometimes called) is usually given in terms of a reduction in free volume, although analytical free-volume models have not yet been proved correct in detail[94]. Free volume is the difference between the specific volume (inverse of specific gravity) of the actual material and that of the close-packed equilibrium form at the same temperature (cf. Fig. 3.3). It is a measure of the inefficiency of the molecular packing and in the case of amorphous polymers arises because chain structure and steric hindrances prevent all potential lattice sites from being occupied, Fig. 3.2. The structural 'holes' that comprise the free volume range from molecular size upwards and may well have a number density and shape that is self-similar to scale changes[97]. Free volume should be clearly distinguished from microscopic voids which have radii of 0·1 μm or larger and smooth surfaces. Such voids are generated by chemical action during processing either through the production of trapped bubbles of gas or chain scission and crosslinking due to excess curing agents.

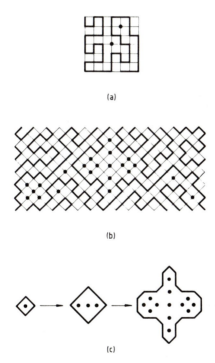

(a)

(b)

(c)

Fig. 3.2 Schematic representation of free volume. The thick lines represent the bonds of a polymer chain constrained to join the sites of a regular lattice, and the spots are the unoccupied sites that comprise the free volume.

(*a*) One chain with single unoccupied sites.
(*b*) Many chains giving multiply unoccupied sites, and a range of free volume sizes.
(*c*) Schematic representation of self-similar free volumes of different sizes

The disorder in the molecular packing of this type of material can be regarded as the result of the release of the internal energy (ΔU) stored in the appropriate ordered form. Since the empty lattice sites can be chosen in a number of different ways the free volume can be represented by the configuration entropy, ΔS, of the disordered form which is related to ΔU via $T \cdot \Delta S = \Delta U$. In Fig. 3.3 the entropy of the glassy state is compared to that of the thermal equilibrium state, with the change on aging indicated by an arrow. Because segmental motion which is required for the chain re-arrangements during aging, slows down as the free volume is reduced, the approach to equilibrium is self-retarded and in general these relaxations are non-exponential. A similar effect occurs in cross-linking and swelling; thus these are processes which can be expected to continue, albeit at an ever-decreasing rate, well beyond the time at which they were initiated[101].

The effect of the reduction of free volume during aging is to increase the relaxation times of both linear and non-linear creep, however the effect is not so pronounced at high mechanical stresses. This has led to the conjecture[94], which is consistent with the available evidence, that mechanical stresses which drive the segmental motions can generate free volume irrespective of the presence of a tensile or compressive component. Since segmental motions can usually be driven dielectrically as well as mechanically, such a process allows for an electric coupling to free volume degradation on a sub-microscopic scale[102].

Although little work has been carried out on physical aging under the influence of purely mechanical forces[103], some of its features have been elucidated[104,105,106]. In both thermosetting polymers[105] (epoxy resins) and semicrystalline polymers[104] (polyethylene) the effect of mechanical stresses of a sufficiently high level was to increase the microvoid density and size. Whereas the mechanical forces were externally applied in the case of the epoxy resin[105], the microvoid production in polyethylene was observed in

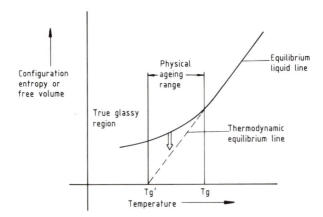

Fig. 3.3   Schematic representation of the temperature dependence of the free volume or equivalently configuration entropy in a glass forming system. After Struik[94]

cable samples that had been unused and exposed only to a service environment[104]. Under these circumstances the mechanical stresses occur internally as a result of a slow crystallisation in the amorphous regions; i.e. secondary crystallisation (see Chapter 1 for a fuller discussion of polymer morphology). This process can continue for several years at ambient temperatures and, because the polymer is solid and its volume essentially constant, local density increase due to the secondary crystallisation (usually isothermal crystal thickening) is compensated by a generation of microvoids and discontinuities. An increase in the number of interfaces between amorphous and chain-fold regions may also occur giving a greater density of traps and defects, and consequently an increase in local space charge concentrations[106,107]. The effect of such aging is to increase the propensity of the polymeric insulation to a range of electrical degradation processes such as partial discharges, electrical treeing, and water treeing, see Section 3.1.3 (and subsequent chapters for detailed discussion).

During service polymeric insulation can experience temperatures substantially in excess of the ambient value (e.g. up to 90°C in power cables). Under these conditions small and imperfect crystalline lamella in semicrystalline polymers can melt. Subsequent cooling allows a variety of processes to occur, including recrystallisation and lamella thickening as a result of the extension of folded chains[108]. As long as the temperature is not too high (<90°C in polyethylene) such heat treatment will accelerate the production of microvoids via increased mechanical stresses[109]. If the polymer is softened however the crystal lamella arrangement can be reorganized[110,111]. When the initial arrangement is either by design[111] or serendipity[112] such as to maintain a high breakdown strength (i.e. lamella axis orientated perpendicular to the field[113]) the thermal disarrangement will be deleterious.

Among other aging processes which can produce damaging structural defects (i.e. microvoids and microcracks) via local mechanical stresses mention must be made of those arising from differential thermal expansion in heterogeneous systems. These may be particularly dangerous in the filled systems such as the mica-loaded epoxy resins used in high voltage machine insulation[108,114].

During service a combination of all the above aging effects may occur, generally leading in polyethylene insulation to an increase of free volume and local strains[102,104] as well as lamella rearrangement[110]. However it should be noted that such long-term aging does not appear to be reproduced in combined thermal and field accelerated laboratory tests[102].

3.1.2 Chemical aging

Chemical aging[95] usually proceeds via the formation of polymer free radicals $R^{\cdot}$ (or radical ions) following an initiation step X, i.e.:

$$X + R_{a+b} \rightarrow R_a^{\cdot} + R_b^{\cdot} \tag{3.1}$$

Free radicals are very reactive chemically and lead to propagating chain scission or cross-linking network formation via chain reactions. Two types of chain scission sequence may occur. Either the bond breaking is random in space with free radical transfer between chains or it unzips a chain by the ejection of volatile monomers or sidegroup products. The former case produces degradation products containing large molecular weight fragments and is favoured by polyethylene, whereas the latter case is typical of poly(α-methylstyrene) where it results in a large monomer fraction and poly(vinyl chloride) which dehydrochlorinates producing hydrogen chloride.

The initiating step may be thermal, oxidative, caused by UV absorption or ionising radiation, or mechanical. Thermal aging clearly controls the temperature range of the chemical stability of polymeric insulation. Since local electric currents can be expected to raise the temperature through Joule heating, the chemical degradation can also be driven electrically when a combination of aging and high fields allow filamentary currents to occur.

Oxidative initiation may occur chemically via polymer processing by-products and is catalysed by metal ions. When present in high concentration, however, such ionic species may instead be drastically inhibiting resulting in a strong reduction in the rate of oxidation[95]. Photo-oxidation may also take place and this requires optically and UV absorbing molecular moieties for its initiation which are usually processing by-products containing double bonds or carbonyl groups. The energy available is $> \sim 3 \cdot 5$ eV per photon which is sufficient to cleave a carbon-carbon bond, and both antioxidants and UV quenchers are routinely added to polymeric insulation as preventative measures. These processes will be particularly sensitive to electric fields if charged species such as radical ions are formed as intermediates, or if they bring about changes in the dielectric constant by altering the concentration of polar species. In these cases potential barriers and reaction free-energy differences will alter in the presence of an electric field thereby changing the reaction rate.

Ionising radiation, X-ray, γ-ray and high energy electrons have energies up to 10^7 eV per particle and usually produce rapid degradation to low molecular weight fragments in polymers which easily unzip such as polyvinyls and polystyrenes. However those polymers such as polyethylene with high heats of polymerisation will cross-link[115] and the process may be used to advantage. Electric fields may be involved in this process if they can produce sufficiently high energy electrons on injection or through acceleration across a void.

Mechanical initiation may occur in the processing of the polymer and can be caused by all forms of mechanical forces though the sites where bond scission takes place depends upon the type of force. This mechanism shows an interesting way in which electric-field-driven free-volume generation and chemical-bond scission can be interlinked to produce polymer degradation[116].

Hydrolysis is another form of chemical aging. Here polymer ions are generated by reaction with hydrogen or hydroxyl ions produced by the dissociation of water, leading to considerable chemical degradation in the case of formaldehyde resins for example. Such reactions will be favoured in highly alkaline or acid conditions and can be expected to couple strongly to an electric field through the enhancement of the ionic dissociation of neutral species.

As noted in section 3.1.1 crosslinking belongs to the class of processes which is progressively decelerated and hence can continue at a slow rate for a long period of time[94,101]. This is particularly the case with radiation crosslinking (electron beam) which has been observed to continue in HDPE for up to 2·5 years following initiation[101]. During this period the crystallinity continues to increase, possibly via crystallisation of broken chains[101], and at ambient temperatures microstresses are produced which are likely to generate defects causing the material to become brittle. We can expect structural changes of this type to increase the susceptibility of the insulation to electrical degradation, as discussed in Section 3.1.3.

One other type of chemical aging deserves mention here as it is rather unexpected in character. In order to prevent oxidation of the polymer antioxidants are routinely added to semicrystalline polymeric insulation for cables (see Section 1.6). However it has been noted that under some circumstances the antioxidant may either migrate out of the polyethylene[104], or aggregate[117] into clusters of about 25 μm in size. Such clusters can both lead to stress cracking and promote oxidation instead of suppressing it[118], a combination which is particularly dangerous, and there is evidence to relate their incidence to that of the treeing failure of cables in service[117].

3.1.3 Electrical aging

From the above it can be seen that although many physical and chemical aging processes occur in the absence of an electric field[41,103], they can be both accelerated by such fields, and driven in combination with them. Electric fields (particularly DC) can also be responsible for dissociation and transport of ionised and ionisable by-products, causing a deterioration of insulation performance through increased losses and local stress enhancements, without visible degradation. The more severe forms of degradation are however visible and specifically electric in origin although sometimes requiring other essential components such as water in water trees[119]. Surface processes, such as tracking and erosion, have been excluded from consideration in this work which concentrates on bulk degradation and failure mechanisms. In addition to water trees the main aging processes are partial discharges[120] and electrical trees[121]. The degradation associated with all these aging processes has been termed low level because the voltages required to cause them can be orders of magnitude below the breakdown strength.

Partial discharges occur in the gaseous contents of microscopic voids when the field exceeds a threshold value[120,122] depending upon the void size and gas pressure (about 3×10^6 V $\cdot$ m^{-1} in the largest air-filled voids at atmospheric pressure). Erosion of the internal surfaces of the void by high energy ions and molecules produced in the discharge[123,124,125] may be followed by the formation of a filamentary pattern of channels penetrating the polymer[120,126,127]. This defect then acts as an electrical tree with repetitive discharges causing it to grow, possibly amalgamating with other such discharge centres and eventually initiating a runaway breakdown process.

In good quality cable insulation it is water trees and electrical trees which lead to the most damaging deterioration during service[128]. As their names indicate, their visual appearance under the microscope is tree-like, with water trees exhibiting a diffuse bushy growth and no distinct structure, Fig. 3.4, whereas electrical trees show a branched spiky structure, Fig. 3.5.

Electrical trees may be generated by voltages as low as ~3 kV, although field enhancement to $> \sim 3 \times 10^8 \text{ V} \cdot \text{m}^{-1}$ at metallically conducting centres is an essential requirement[129,130] in void-free materials. When electrical trees are examined in detail it is found that they consist of connected channels (i.e. hollow tubules) a few microns in diameter, with branches tens of microns long. The walls of the channels are not always carbonised and only weakly conducting; thus although a short circuit may occur if the tree spans the insulation[131] this is not always immediately the case. However it is also possible for a breakdown to be initiated before electrical trees completely cross the insulation. Thus electrical trees are not the visual aspect of an intrinsic runaway breakdown mechanism. Instead they are better regarded as contributing to the cumulative damage of the insulation under electrical stress thereby leading to an enhancement of the failure probability[131,132,133].

Water trees occur at much lower fields than those required for electrical trees[115], and have even been reported to grow[134,135] at fields as low as $1.9 \times 10^6 \text{ V} \cdot \text{m}^{-1}$ although a considerable time is required for their observa-

10 μm

Fig. 3.4 A typical water tree found in XLPE cable insulation. Taken from Dissado *et al.*[31]

Fig. 3.5 Electrical tree generated in an XLPE cable during testing under DC conditions

tion. The insulation must however be in contact with an aqueous electrolyte. There is no evidence for the presence of connected channels within water trees[16,38], which are non-conducting[136]. Water trees can cross the insulation without causing a short circuit. However they can initiate an electrical tree, either in the bulk[119] or from the central conductor[137], Fig. 3.6. Water trees therefore are a numerous but very low level of degradation for insulation in moist conditions, nevertheless sufficient deterioration may accumulate to start the system on the path to eventual breakdown.

3.1.4 Combined mechanical and electrical degradation

In general any or all of the previously mentioned aging mechanisms can be combined in a pre-breakdown degradation process, and the reader will see that many of the topics covered in this book fall into this category. However two processes deserve to be mentioned at this point so as to prepare the ground for a fuller discussion later in Part 3 of the book.

The first of these may be termed mechanically-assisted electrical breakdown, although ever since the work of Stark and Garton[138] inter-relating the two processes and identifying which is primarily responsible for breakdown has proved difficult. In this case it is found that for the polymeric materials polyethylene terephthalate (PET: semicrystalline) and FRP (an amorphous epoxy/glass composite), the breakdown strength went through a maximum with increasing compressive stress[139,140] and decreased monotonically although sometimes weakly (as in PET) with increasing tensile stress up to the yield point where it dropped sharply, Fig. 3.7. The composite epoxy which appears to be rather rigid and undergoes brittle fracture exhibits a sharply dropping breakdown strength for small percentage (~3%) elongations which corresponds to high (~150 MPa) stresses. Essentially the

(a)

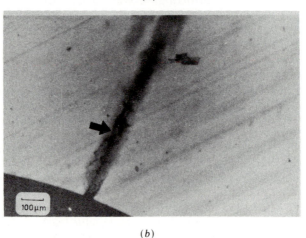

(b)

Fig. 3.6   (a) An electrical tree (indicated by arrow) found growing from the inner conductor of a failed XLPE cable following initiation by a water tree, taken from Fothergill *et al.*[137]. (b) Same tree magnified for clarity
Copyright © 1984 The Institution of Electrical Engineers

maximum found for compressive stresses can be understood on the basis of an initial increase of density which closes up voids and increases molecular packing. The compressive stresses however will also cause the breakage of tie-bonds and generate defects. At low concentrations these will trap acceler-

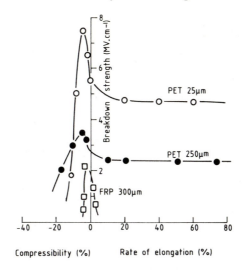

Compressibility (%)   Rate of elongation (%)

Fig. 3.7   Effect of uniaxial mechanical compression and tensile (elongation) stresses on the breakdown strength of polyethylene terephthalate (PET) and FRP (epoxy-glass composite) films (thickness shown). Figure taken from Reference 139 Copyright © 1982 The IEEE

ated charges and inhibit breakdown, but as the stress rises their concentration will increase and the breakdown strength will be progressively lowered. In contrast tensile stress can lead to crack initiation and growth, either as brittle fracture (as in FRP), or at yield (as in PET), where molecular chains can rotate, translate, unfold, and disentangle. Defects and traps can also be created at the amorphous-lamella interfaces in semicrystalline polymers (e.g. polyethylene and polypropylene) which are exposed during stretching. These may trap space charge and thereby increase the local fields[106,107]. In this region of mechanical stress the electric field can aid crack propagation up to the critical length[140] and thereafter a mechanical failure occurs[141]. On the other hand polyethylene cable was noted to exhibit a breakdown voltage which decreased with increasing compressive stress[139,142] and changed in a complicated manner with tensile stress[139,143]. One possible explanation for this is the reorganisation by the applied mechanical forces of the lamella arrangement initially present in the cable insulation[112].

The second type of process is that of electrically-assisted mechanical breakdown. A prime example is the mechanical yielding of a thermoplastic polymer under the physical pressure produced by electrostatic forces[138]. Electrofracture can also be placed under this heading[141]. Here space charge or charge deposited in a crack by discharges[106] can generate electrostatic forces which lead to failure via shear yield or by accelerated crack propagation[144]. Alternatively the presence of cracks in a surface onto which a discharge impinges will facilitate its eroding effect by exposing weak bonds[145] to the ballistic impact of heavy ions. This will be particularly the case if the surface is in tension.

In all of the processes outlined in this section it has been difficult to separate out the initiating factor from that which actually causes failure to occur, and a more complete discussion is left until Parts 3 and 4 of the book.

3.2. Categories of electrical breakdown

Models of breakdown mechanisms may be divided into three categories: (i) low-level degradation models, in which the insulating system's characteristics are deleteriously affected by the electric field possibly in conjunction with other agents; (ii) deterministic models, in which the ultimate breakdown event is the direct effect of some earlier causal event or condition produced by the exceeding of a critical electric field; and (iii) stochastic models in which either local physical conditions are considered to be constantly changing or there are local electric field variations caused by inhomogeneities such that there is a finite probability at any time that breakdown may occur. The last Section (3.1) introduced low-level degradation, this Section (3.2) introduces deterministic mechanisms, and the effects of the stochastic nature of breakdown will be discussed in the next (3.3).

This book places emphasis on low-level degradation (Part 2) and on stochastic models (Part 4) although deterministic mechanisms are also described in some detail (Part 3). We have decided to devote particular attention to low-level degradation and stochastic models for various reasons. Recent major advances have been made in these two categories, whereas this does not appear to be the case for deterministic models. Many other reviews restrict themselves to deterministic models (see Part 3). Furthermore, in Part 5, we are able to consider in some detail the problems faced by the engineer in predicting the lifetime of an insulation system, a problem which would not arise if the operating voltage could always be kept below a hypothetical deterministic breakdown voltage. Whilst deterministic models predict a specific breakdown voltage or time to breakdown[82,146], they are, nevertheless, important to consider since stochastic versions of the models essentially only modify the results.

Breakdown in polymeric insulators is always 'catastrophic' in the sense that it is irreversible and destructive resulting in a narrow breakdown channel between the electrodes. The term 'breakdown' is used elsewhere to describe processes in which a considerable current increase results from a small voltage change. Such a process may or may not be catastrophic depending upon the extent to which the input power is limited. These processes are well known to occur in, for example, reverse-biased semiconductor junctions. Whilst a similar situation could occur in polymers, perhaps under trap-filled space-charge-limited current conditions[147] or due to the mechanisms of Schottky[148] or Poole–Frenkel[149] (see Chapter 9), it would not normally be described as breakdown unless it led directly to a positive-feedback process resulting in destruction of at least a local area.

All catastrophic breakdown in solids is electrically power driven and ultimately thermal in the sense that the discharge track involves at least the melting and probably the carbonisation or vapourisation of the dielectric[150].

Deterministic models of breakdown are therefore categorised according to the process(es) leading up to this final stage. We have chosen to subdivide these processes into: electric, thermal, electromechanical, and partial discharge breakdown. These are introduced under these headings below. Both these deterministic and low-level degradation mechanisms are shown schematically in Fig. 3.8 as a function of field and time, although, of course, other factors also influence which mechanism prevails. It is interesting to note that the difference between breakdown and degradation is somewhat tenuous; perhaps breakdown is an experimental phenomenon and degradation an in-service phenomenon with a time scale of about a day to a month representing the border between the two. Partial discharge is a case of localised breakdown which causes degradation and which will eventually lead to system breakdown. However the other forms of breakdown can also lead to local damage which may accumulate[133] and also eventually lead to macroscopic breakdown (Section 15.4).

3.2.1 Electric breakdown

Because of their well-characterised and extensive use in semiconductor junctions, the two most famous examples of electric breakdown are Zener breakdown[151] and avalanche breakdown. The former was proposed as a bulk breakdown mechanism for which, in fact, it does not apply[1]. It only occurs in semiconductor junctions in which electrons may tunnel from the valence to conduction bands when the field is strong enough and the junction is sufficiently narrow. For band gaps typically observed in polymers ($> \sim 7$ eV) fields of $> \sim 10^{10}$ V$\cdot$m^{-1} would be required to cause Zener breakdown; however polymers would invariably have broken down by other mechanisms before such a field could be reached.

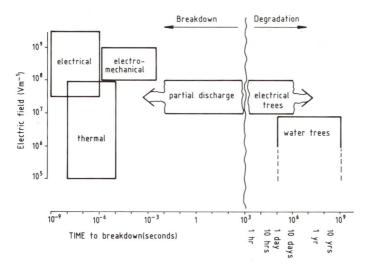

Fig. 3.8 Times and electric fields at which various electrical breakdown mechanisms are operative. Note the lack of a clear distinction between 'breakdown' and 'degradation'

Avalanche multiplication (or carrier impact ionisation as it is known when the carriers are not necessarily electrons) occurs at high field strengths when carriers acquire sufficient kinetic energy between collisions (scattering events) with the matrix to give a high probability of ionisation, with the generation of more carriers, without suffering recombination themselves. Thus a small current can be greatly multiplied by an increase in the number of carriers until it is sufficiently large to cause irreversible damage[152,82]. This type of mechanism has been used to explain breakdown in oxide films[153,154]. In polymeric materials it is more controversial whether such a mechanism usually prevails although suitable conditions may exist. This mechanism requires the existence of hot electrons whose kinetic energy, $\mathbb{E}_{kin} \gg k_B T$. Evidence for hot electrons exists (Chapter 2) such as electro-luminescence[155,165] and free-radical generation[157]. The high-mobility ($> 10^{-4}$ m^2 V^{-1} s^{-1}) states required for avalanche formation in polyethylene have been shown to exist both theoretically[67] and experimentally[61]. In thick polymeric insulation the applied (average) field cannot usually be high enough for this breakdown mechanism to occur. However there may be very high local fields, for example near the tips of electric trees (Chapter 5) or due to the build up of hetero space-charge near the electrodes (Chapter 17). In this case the breakdown may be localised.

In very thin polymer specimens at short times, Jonscher and Lacoste[133] have noted that the breakdown field is limited to magnitudes of the order of 10^9 V · m^{-1}. They have shown that this can be explained in terms of the tunnel-injection of electrons from the electrodes through or across an image-force-lowered barrier, i.e. the Fowler–Nordheim process (Section 9.2.2). By following the simplified argument presented by Solymar and Walsh[1] they have shown that a massive lowering of the barrier is to be expected for fields of the order of 10^9 V · m^{-1} which is in accordance with experimental observations[158,159,160]. It seems likely that the ultimate, or intrinsic breakdown strength, which occurs at short times, is governed by this electric breakdown mechanism which is primarily associated with the electrode-polymer interface rather than properties intrinsic to the insulator.

3.2.2 Thermal breakdown

Thermal breakdown occurs when the heat input cannot be balanced by the heat losses from the insulation either macroscopically or, more usually, in a small area. As power is dissipated by the insulation, heating occurs which usually causes an exponential increase in the electrical conductivity as more carriers become available for conduction. Alternatively the increased segmental motion may increase the mobility for intrinsic ionic conduction (Section 2.4). If the electrical stress is maintained, the current density increases in the area of elevated temperature. This serves to further increase the local temperature through Joule heating and hence the conductivity, and so 'thermal runaway' may occur. In general this leads to a highly localised filamentary breakdown path, however under appropriate conditions, breakdown may occur on a broad front[161]. The mechanism is described in more detail in Chapter 10 and is the usual mechanism in polyimides and possibly poly(vinylidene-fluoride)[162].

3.2.3 Electromechanical breakdown

Electromechanical breakdown occurs due to the electrostatic attraction of the electrodes which decreases the width of the insulation by an amount depending on the Young's modulus. If the applied voltage is maintained, the field increases due to the decrease in thickness thereby increasing the attraction ($\propto E^2$) further. The effect is exacerbated by the local heating and consequent softening which is likely to occur in this region. Garton and Stark[138] developed this model by equating the work done on the dielectric by the attractive forces with the strain energy stored; it is discussed in more detail in Chapter 11.

Whilst electromechanical breakdown of this type may seem likely, it does not seem to be commonly considered to occur in polymers and it appears that the appropriate breakdown model for softened polymer may also involve some thermal and mechanical processes localised to the amorphous regions[163,164]. Furthermore this breakdown mechanism is likely to be of only limited technological significance since polymers used in insulating systems are either not usually used above their softening point, or, if they are (such as may be the case in a polyethylene power cable under very high load conditions), then they are usually crosslinked and sufficiently thick for this effect to be negligible.

3.2.4 Partial discharge breakdown

Small voids inevitably occur in polymeric insulation even in the most-carefully prepared materials[105]. Since these are filled with gas they have a lower permittivity than the surrounding polymer and so field intensification occurs at the end walls in the direction of the local field. Depending on the gas pressure and other factors the enhanced field may cause the gas to become ionised; i.e. breakdown or discharge within the void. Because this does not necessarily cause the whole polymer insulation to breakdown (and discharge) immediately this is known as *partial discharging*. The carriers produced by the ionisation are accelerated across the void and, if they acquire sufficient energy from the field, may cause erosion as they impact on the opposite wall of the void. In thin insulating systems this may quickly lead to failure. However in thicker insulators, in which larger voids are more likely to occur but less likely to be detected, these partial discharges may produce electrical trees as described in Chapter 5. In this case the partial discharging acts more as a degradation mechanism with breakdown not occurring immediately; indeed in some systems the eventual breakdown due to partial discharges is so delayed that it is beyond the economic life. For example bakelised paper bushings survive for thirty years or more provided that the discharges are below the audible limit of about 100 pC (Day in Alston[165]). This compares with a typical maximum allowable discharge in international standard tests of 5 pC[166] and most manufacturers (and customers!) would be unhappy with more than 1 pC in polymeric cables.

In polymers there are likely to be a number of voids, in addition to the free volume, which occur naturally in the amorphous regions as a result of strains induced during crystallisation or moulding[167,168,169,104]. Although such voids will be very small ($\sim$nm) some cavities may be large enough to support partial discharges at close to operating stresses[120] (i.e. 10^7 V $\cdot$ m^{-1} for 100 μm). The precise condition under which partial discharges can be initiated will depend upon a variety of factors, such as the gaseous contents, gas pressure, and void shape as well as void size. Because of the free volume of the polymeric material itself voids of a similar size may exist in a large number of different local environments due to different local molecular packing. Thus partial discharge inception is likely to be a distributed process. Large cavities may also be formed by gas bubbles in the melt, which in polyethylene could appear as a result of a crosslinking procedure. Additionally other voids may be unintentionally introduced around inclusions such as impurities or induced and enlarged by additive diffusion, by electrostrictive forces during the application of an applied field, and by electrochemical effects such as water treeing. Such large voids can be directly observed in polymers[170,171] and will assist the free volume in facilitating the permeation of liquids and gases through the material. Partial discharge breakdown is discussed further in Chapter 13.

Under accelerated laboratory conditions or those perhaps in a poorly manufactured insulating system in which high localised divergent fields exist, it is clear that partial discharges quickly lead to the formation of electrical trees which span the insulation and lead to breakdown as described in Chapter 5. However under service conditions and uniform fields electric trees do not form immediately. It seems likely that under such conditions partial discharges give rise to defect centres which may combine to form clusters elongated in the direction of the electric field. Breakdown may then occur through the weakest link offered by the alignment of such extended defects. This is the basis of the cumulative model of breakdown proposed by Jonscher and Lacoste[133] and described in Section 15.4. Electrical trees may also form by incorporating such defects and these will assist the breakdown process. Water trees which may grow over long periods under service conditions will lead to electric field inhomogeneities and provide an extra source of voids, thereby also influencing the partial discharge breakdown process.

3.3 Engineering aspects of electrical degradation and breakdown

Theoretical approaches to dielectric breakdown have concentrated[82] upon the development of models in which a runaway conduction process occurs only when the applied field exceeds a critical value. However spatially static inhomogeneities[172] and dynamic structural fluctuations[173,174] which are inherent in polymeric materials force us to adopt a stochastic approach when assessing the dielectric lifetime of polymeric insulation. It is, therefore, not possible to determine a breakdown field and time to breakdown which will be the same for all samples of a given material. Instead a number of samples of a given material must be tested to breakdown and the results statistically analysed. This procedure will serve to identify the form of the failure distribution through shape parameters and characteristic values of the breakdown field and time to breakdown.

Development tests undertaken prior to the introduction to service of insulation systems are usually carried out under accelerating conditions which serve to reduce the time-scale of the test programme to manageable proportions. Extrapolations based on the statistics obtained are then used to estimate a working lifetime for the system. However a number of long-time changes may occur during service which alter those factors that determine both the breakdown strength and statistics[175]. Such aging normally takes place too slowly to influence accelerated tests, and hence the in-service quality of the insulation can only be fully assessed by means of additional investigations of the aging processes.

Physical[94] and chemical[95] aging will occur independently of the application of a field and may even be advantageous as is, for example, the cross-linking process in polyethylene which may continue very slowly through a service lifetime. However chemical degradation involving oxidation generally acts locally either to enhance the electric field or to reduce the breakdown strength. As a result the breakdown statistics may be considerably altered to the disadvantage of the insulation quality. Prevention of these damaging processes is usually attempted by a suitable choice of material morphology and the addition of chemicals such as antioxidants.

Electrical degradation however occurs as a result of the service conditions themselves and is not easily prevented. This class of aging includes water trees[128], partial discharges[120], and electrical trees[121] all of which degrade the breakdown resistance of the insulation. It is therefore an essential part of any test programme to determine the generation of such electrical degradation over a period of time, and its influence upon the breakdown characteristics of the insulation. This latter feature is determined by carrying out a set of breakdown tests[175,176,177] on material with a known extent of aging. In this way the change in the failure distribution and the residual life can be determined as a function of the amount of electrical degradation. A combination of these results and the time development of the degradation will then serve to allow reliable predictions to be made of the service lifetime of a chosen system. More fundamental investigations[119,178] of the mechanisms of electrical degradation are aimed at improving the insulation quality through a reduction of degradation and are a major part of the research programme of the insulation industry which is actively pursued by materials' manufacturers, systems' manufacturers, and utilities.

With the ever increasing diversity of uses for polymers as insulators, from micro-electronic device encapsulation to power cables, a thorough understanding of their electrical degradation and breakdown characteristics is vital in the formulation of new designs especially where they are to be used in untried critical environments such as space[179,180,96] or inside the human body. Furthermore it is only through a better understanding that a systematic formulation of better materials can be made. For example, unless the effect of morphology and free volume on the breakdown and electrical degradation characteristics can be understood or at least characterised, it is difficult to know how to alter these relatively easily changed features for the better.

There are also many other cases where an improvement in the insulation characteristics would be extremely beneficial. For example many capacitors are made using polymer films (polypropylene, polycarbonate, polyesters, polystyrene) sandwiched between metal electrodes. In order to decrease the size of the capacitor whilst maintaining its voltage and capacitance rating it is necessary to either increase the relative permittivity which results in a smaller electrode area, or increase its breakdown strength which results in a thinner dielectric requirement. It seems unlikely that it will be possible to increase the relative permittivity significantly without increasing the dielectric loss to an unsatisfactory extent. An increase in the breakdown strength by a factor of as much as 10^3 may however be obtainable perhaps by using thinner but multi-layered dielectric structures.

Another area of importance is that of cables where the use of polymers as the solid dielectric has revolutionised their performance. Since the 1950s the adoption of extruded poly(vinyl chloride) (PVC) for low voltage cables has enabled thinner coatings to be used and subsequent developments have improved their resistance to mechanical fatigue[103], chemical[41] and radiation attack[181], fire[182] etc. Power cables with voltage ratings greater than about 10 kV now have a very large market. In 1986 it was estimated to be $7020 million worldwide, including $1720 million in the European Economic Community, $1080 million in Japan and $1620 million in North America[183]. Since about 1965 there has been a widespread transition from oil-filled paper insulation to the use of cross-linked polyethylene for voltages up to 300 kV in AC cables[184]. An understanding leading to improvements in the insulation characteristics of such cables is critical to their economic design for reliable operation over long periods of time. A reduction of only 1 mm in the insulation thickness can save $500/km in material costs alone.

PART 2

TREEING DEGRADATION
IN POLYMERS

Introduction

Electrical treeing degradation has a long history[121,128,185], which goes back to the early development of paper/oil insulation systems. Since then it has been observed in inorganic as well as polymeric materials and must therefore be a process whose mechanism is independent of the chemical nature of the insulation. Water trees however were not recognised until reports in the early 1970s[185] identified their presence in the polyethylene cable insulation introduced in the 1960s, and associated them with premature failures. Since then water trees have been shown to occur in a wide variety of polymers[119] as well as polyethylene and its lightly crosslinked derivative (XLPE). These include rubbery copolymers (e.g. EPR), side-group (chain) polymers (e.g. PVC and polycarbonate), aromatic derivatives (polystyrenes), and network polymers (epoxy resins) both above and below their glass transition. As yet water trees have not been reported in inorganic insulation although there is evidence that ZnSe crystals can suffer local damage when attacked by vented water trees grown in coating films of polyethylene[186]. It is possible therefore that chemical factors may be involved in their formation. Over the years considerable evidence has been accumulated relating breakdown in cable insulation to the occurrence of trees[119,187]. Although the largest proportion of cable failures can be traced to mechanical or connector faults (90%), treeing is now acknowledged to be the major cause of potentially avoidable electrical failures[121].

Examination of cables removed from service showed that nearly all electrical trees were found together with water trees. Where this was not the case the origin of the electrical tree could often be traced to asperities on the conductor or minute metallic particles embedded in the polymer during extrusion. The use of semiconducting tapes to protect the insulation layer did not substantially improve the situation[188] and in fact introduced new initiation sites at embedded contaminant particles[189] such as amber[190] silica gel and ferric sulphate[191]. However after the adoption of a triple extrusion technique[192] in which semiconducting layers are simultaneously laid down over the inner conductor and outer insulation, Fig. II.1, the density of electrical trees initiated in this way was substantially reduced, leaving only water trees as a major problem[187].

In early publications water trees were commonly characterised by their shape, e.g. broccoli, bow-tie, streamer, micro, dendritic etc.[193], and termed either water or electrochemical trees depending upon their chemical composition. Water trees which were deemed to consist mainly of water, were

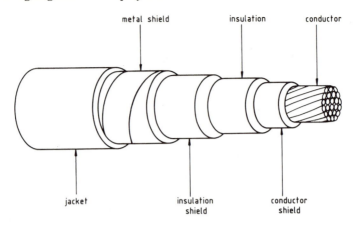

Fig. II.1 Cut-away section of a power cable with the components as labelled

opaque and disappeared on drying out. In contrast electrochemical trees were often coloured and still visible after drying, as for example the dendritic sulphide trees[194]. However it has now been recognised that these are differences in degree only depending upon the ion content and type rather than the treeing mechanism. Thus the term 'water tree' has been adopted to apply to all trees with an opaque electrolyte content.

It is now usual to distinguish just two sub-categories, vented trees and bow-tie trees. Vented trees have a stem joining them to the surface of the insulation, and are therefore in direct contact with a reservoir of aqueous electrolyte. In contrast bow-tie trees, Fig. II.2, originate from contaminants[195], boundary surfaces[196], or water-filled voids[197], within the insulation and thus have only limited access to an aqueous reservoir. Although this categorisation was based purely on visual grounds, there is evidence to show that there are significant differences particularly in their influence upon breakdown[175]. A similar nomenclature has been applied to electrical trees, with bow-tie trees originating within the bulk insulation at metallic impurities and vented trees from surface asperities and imperfections. In this case however there is no evidence for a difference in their effect upon dielectric breakdown.

Because the initiation and growth of both electrical and water trees in insulation systems is a long time process under service conditions, their laboratory investigation requires some form of acceleration. This is most conveniently supplied by the use of needle electrodes to enhance applied fields of kV/mm to nearly MV/mm. Thus metal electrodes in a double needle[198] or point-plane[129] geometry are used to initiate electrical trees, whilst needle-shaped depressions filled with aqueous electrolyte[199] perform the same role for water trees. These geometries fulfil the dual role of high reproducibility combined with field intensification but have the drawback that the polymeric material investigated does not necessarily possess either the physical properties or the morphology that exists in engineered insulation systems[15]. For this reason water tree research in cables in particular

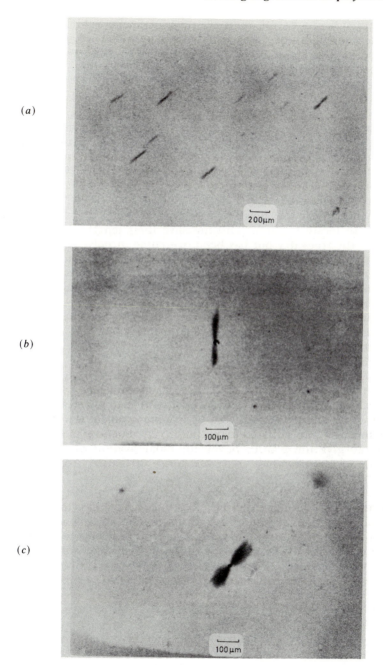

Fig. II.2 Bow tie water trees found in an XLPE cable removed from service
(Rowland, unpublished data). (*a*) In a group with no obvious initiating defect, (*b*)
initiated by a void, (*c*) initiated by a contaminant

has adopted a number of alternative test geometries, Fig. II.3. The most prominent among these uses miniature cables which are bent into a U shape and immersed in a bath of electrolyte. In this case the inner conductor may either be removed to allow conditions at the inner surface to be controlled[115] or retained to simulate service conditions more closely[200]. This latter aim is now closer to realisation through the recent development of procedures[201,202] which allow sections of production line power cable to be tested in the laboratory, Fig. II.3. One other geometry of interest makes use of Rogowski profiled[203] plaques which are circular parallel-plate specimens with a specially-profiled increasing thickness around the edge to eliminate the influence of divergence in the applied field. Here the polymer is compression moulded to the required shape with the electrolyte forming the profiled electrode[204].

All of these water tree investigations involve the need to slice the material in order to locate the trees and measure their lengths. The problem of fading as the trees dry out has been overcome in polyethylene and cross-linked polyethylene by boiling with methylene blue dye which renders them permanently visible. Until recently the opacity of EPR has made it impossible to see water trees clearly in this material, however a new staining technique combined with an oblique illumination of the microscope has now removed this limitation. Both methods are given in detail by Shaw and Shaw[119], and by their aid it is possible to determine the number density and length distribution of water trees produced during the time of application of the voltage[205].

Such features and especially the characteristic tree length, when obtained in a fixed time under standard conditions, are often used for a comparative assessment of the treeing resistance of different materials. In addition growth curves can be constructed by examining materials that have been under test for different periods of time[115,206]. Such plots however do not necessarily give the growth behaviour of an individual tree or even an ensemble of trees, because of the existence of sample-to-sample variation which is present even in the standard needle geometry[199]. The best that can be hoped for is that standard production techniques will reduce the variation to a minimum, thereby allowing the constructed plot to be a reasonable approximation to the time development of an ensemble of trees. A recent advance in technique now allows these results to be checked by comparison with the direct observation of the growth of individual trees[207]. Here a television camera is used to view growing trees in a needle point geometry, with accurate measurement being made on the image transmitted to a television screen. A simpler version used first by Fournie *et al.*[208] is to remove the system from test transiently while the tree is photographed. These latest developments place the study of water trees on a level with that of electrical trees whose growth has been directly observed both in thin polyethylene films[209] and transformer oil (see Forster[210] and references therein) for some time.

This wide range of procedures has been developed because the aim of the experiments is not always the same. For example non-destructive direct observations are designed to answer fundamental questions concerning the

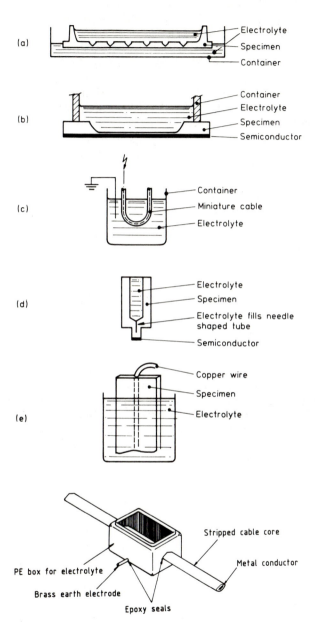

Fig. II.3 Standard geometries used for water treeing measurements in the laboratory. (*a*) Ashcraft geometry; (*b*) Rogowski plaque; (*c*) Miniature cables; (*d*) Needle-plane; (*e*) Disc; (*f*) Lengths cut from production line cables·

mechanism of the treeing phenomena, whereas tree formation in uniform field plaques gives an assessment of the material quality. The testing of complete insulation systems requires the use of cable sections and additionally a statistical analysis. A complete research programme would thus ideally embrace each type of technique, using direct observation to investigate the growth pattern of individual trees, together with destructive measurements to determine the influence that the material morphology and the insulation system has on initiation and growth rates.

Chapter 4
Water treeing degradation

4.1 What are water trees?

Water trees in crosslinked and uncrosslinked polyethylene are easily visible under the microscope, and often appear to show some structure[119]. Thus regions of greater water density have sometimes been termed channels[211]. However this terminology cannot be taken to indicate the presence of hollow tubes since no cracks or crazes have been revealed by scanning electron microscopy (SEM)[119,212]. Even though some isolated cracks have been observed[38] by transmission electron microscopy (TEM) the typical physical feature marking the path of a water tree is the presence of a large density (approximately $10^6 \, mm^{-3}$) of spherical microvoids whose radius (a) is $1 \, \mu m$—$5 \, \mu m$ (Shaw and Shaw[119] and references therein).

An illustration of the voids is given in Fig. 4.1 where they are shown imposed on a polymeric cable structure thought at the time to be spherulitic. It now appears that the apparent spherulitic structure is an artefact of the etching process used to prepare the sample for SEM[19]. Similar secondary fracture patterns are formed when microvoids act as fault centres[34]. Spherulites (size ~5 μm) have been shown to exist in moulded crosslinked (XLPE) and low density (LDPE) polyethylene[15,18], but in extruded cable insulation it is probable that only LDPE is spherulitic[17] and it is better to regard the morphology as a mosaic of crystal lamella[17,15]. The possibility

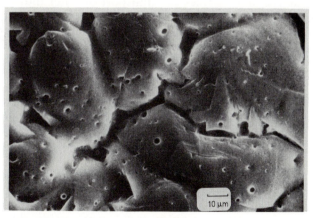

Fig. 4.1 Scanning electron microscope (SEM) picture of a water tree in field-aged cable insulation, showing the large microvoid density characteristic of water trees, taken from Dissado *et al.*[206]
Copyright © 1983 IEEE

that the voids are also an artefact has however been eliminated by their observance following freeze fracture[212,19], and their location has been identified[18] by means of TEM as the amorphous regions between the lamella.

No void pattern is discernible within the tree except where it approaches the inner conductor. Here a radial 'string of pearls' formation has been reported[212]. On drying, the voids close up and cannot be seen, however re-wetting the tree by immersion in water re-opens them as the tree is reformed. Boiling in water not only stabilises existing voids permanently but produces extra small voids $(< \sim 0.5 \ \mu m)^{[212]}$ and dielectric measurements[137] indicate that in this case the water is restricted to the voids unless a further treeing stress is applied.

It is possible that void formation is not the only modification to the polymer morphology resulting from treeing, since the voids apparently deform adjacent lamella[18] in semi-crystalline material, and small angle X-ray scattering (SAXS) has shown that in the treed regions of crosslinked polyethylene the range of lamella thicknesses is increased and their surface structure disordered[213].

No conclusive evidence has yet been published for the existence of an interconnecting channel network between the voids. Investigations used include the SEM of freeze fractured materials, with a resolution of better than $0.05 \ \mu m^{[212]}$, a helium permeability test[214], ion diffusion during dyeing[213] and the monitoring of the tree conductivity during growth[136]. However TEM has shown that void coalescence can lead to disconnected channels in water trees[38,16,39], particularly around spherulite boundaries[39], or other weak paths[16].

Although water trees may be grown using liquids other than water (see Shaw and Shaw[119]), their principal constituent in cables is water, and it has been found that the water concentration in the treed region increases linearly with the applied voltage up to around 10%[211]. The evidence that electrolyte material accompanies the transport of water into the insulation during tree formation is overwhelming. In Fig. 4.2 we show the chlorine concentration found in vented trees formed in crosslinked polyethylene cables during service. Other elements identified in trees are Fe, Al, S, Na, Cl, K, and C in bow-tie trees[204] and Cu, Si, S, Cl, K, Ca, and Fe in vented trees[134], and the copper has been shown by ESR to be present in ionised form[215]. Where water trees have been grown from $KMnO_4$, neutron activation analysis[216] has shown that not only is the permanganate concentration in the tree of the same order as in the needle electrode[217] but that it extended beyond the visible tree tip[216]. It has also proved possible[218] to show that inorganic elements, principally sodium and calcium are concentrated within the microvoids of trees grown from salt water.

The evidence for any accompanying chemical change in the polymer during treeing is not so clear-cut. The ability of cationic dyes to permanently stain the tree indicates that some oxidation must have taken place[213,219]. It has also been suggested that the action of different organic additives is based on oxidation inhibition and promotion[118,220] however the evidence here is contradictory[119]. So far IR spectroscopy has provided the best evidence for oxidation of the polymer in water treed material[221,222] where it seems to be

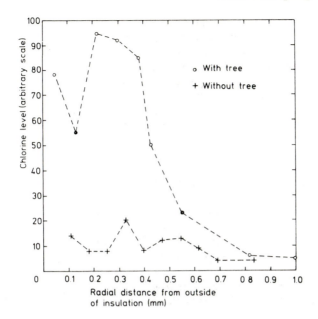

Fig. 4.2 Chlorine level as a function of radial distance from outer insulation shield in treed and untreed cable[137]. Measurements were made using EDX (energy dispersive X-ray analysis) on an SEM, and are quoted in arbitrary units of chlorine count scaled to the background level

Copyright © 1984 The Institution of Electrical Engineers

confined to loose chain ends and lamella fold surfaces. FTIR spectroscopy has also been used to identify oxidation products in the electrolyte bath from which the water trees are grown[223]. These interpretations of IR difference spectra are not however unique, and it has been suggested[218,224] that some features of the insulation spectra at least originate with contaminants such as silicates, sulphates and carboxylates[295]. In this case their observed concentration at the tree tip would suggest that they have an ionic form. Nonetheless shorter oxidation induction times for the treed regions suggest that either the contaminants accelerate oxidation or that some polymeric oxidation has taken place during tree growth[295]. In this latter case the oxidation would have to be confined to the interior void surfaces where its observation would lie at the limit of the FTIR technique.

Chemical modification of the polymer during water treeing is further corroborated by the observation of fluorescence following excitation by a 480 nm laser[225]. The emission which occurs at wavelengths greater than 520 nm proceeds from the whole tree as revealed later by staining, but is essentially featureless and hence no identification of chemical species has been made.

The picture of a water tree that has emerged experimentally is thus that of a large number of electrolyte filled voids possibly with oxidised or chemically modified inner surfaces, and in semi-crystalline polymers some lamella disordering. For the known void concentration and a radius of $2 \cdot 5 \ \mu$m the voids must contain around half the maximum quantity of water that water trees have been observed to possess, and diffuse scattering from the water in the voids will produce the tree shaped outline. Connection between the voids must be maintained by the excess aqueous electrolyte dissolved in the amorphous regions. This is likely to function as a sub-percolation water network partially enveloping high density regions such as crystalline lamella or filler particles in amorphous polymers. Contaminants and desolvated electrolyte will be found at or in advance of the tree tip.

4.2 The time dependence of water tree growth

In the most general case two or possibly three stages are involved in water tree growth[206,226]. These are: an initial inception stage; and a rapid but decelerating growth stage which is possibly followed by a stage in which growth rate is even more rapidly reduced.

This pattern of behaviour is summarised for vented trees in Fig. 4.3. Here the inception stage is characterised by an inception time, t_d, the rapid growth stage by a growth rate proportional to $(t + t_1 - t_d)^{-n}$, and the long time behaviour by a more rapidly reducing growth rate proportional to $(t + t_1 - t_d)^{-(1+m)}$ with $0 < n < 1$ and $m > 0$. Here t_1 is a parameter introduced to allow for the fact that the growth rate is finite at the time $(t = t_d)$ when the tree is initiated. The phenomenological description is completed by a cross-over time (k^{-1}) between the two growth stages, and a length parameter, L_∞. All three stages are not always observable however, and so it is often possible to reduce the number of parameters by combining some and setting others to zero.

Bow-tie growth differs from this empirical description in appearing to rapidly saturate. It has been suggested[227] that in this case the parameter n is greater than unity, but not enough work has been done yet to verify this. In some circumstances vented trees also behave in a similar manner, for example directly observed trees grown in an environment of mechanical strain[228]. However it is its observance in EPR[205] that is of most importance where it implies that vented tree growth is extremely limited although tree inception appears to be unbounded. Some examples contrasting the different growth behaviours are given in Fig. 4.4.

4.2.1 Inception

(a) Vented trees
The existence of an inception stage is typical of the formation of vented trees[118,230,231] in all experimental geometries, even including that of the needle electrode[232,233]. In the case of miniature cables the field (E) and frequency (f) dependence of t_d has been determined[206] to be

$$t_d \propto f^{-1} E^{-3 \cdot 5} \tag{4.1a}$$

by performing experiments at a range of frequencies and voltages. For point-plane geometries the time t_d is usually very short, of the order of

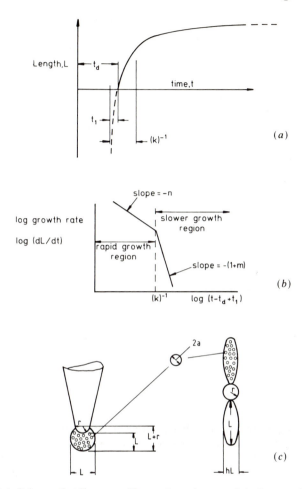

Fig. 4.3 (*a*) Schematic diagram illustrating the empirical growth behaviour of vented water trees in an ensemble, showing an inception time, a fast-growth phase, and a slow-growth phase. (*b*) The equivalent growth rate as a function of displaced time, $t + t_1 - t_d$. (*c*) Schematic representation of water trees grown from an aqueous needle electrode or void defining the tree length, L; width L or Lh; the radius of curvature, r, of the initiating site; and the radius, a, of the microvoids in the tree

minutes to hours. Here inception is often investigated by stepping up the applied AC voltage of a given frequency until a tree is observed. The field applying when a tree is initiated has been termed[232,233] 'the *minimum* field for tree inception' by analogy with electrical tree investigations. This terminology has been taken by some workers[233] to imply that a frequency dependent critical field has been determined below which water trees will not initiate. Experiments in uniform fields show however that water trees initiate at fields an order of magnitude smaller. An alternative view of this experiment is to assume that what is measured is in fact the field that is

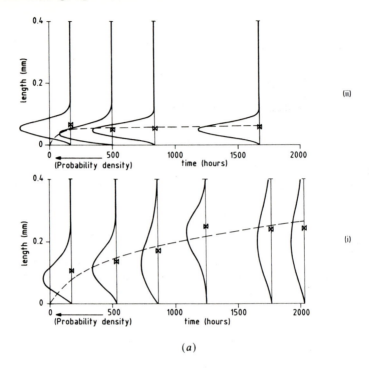

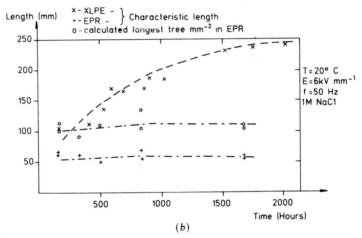

Fig. 4.4

(a) Water tree length versus time relationships for (i) XLPE and (ii) EPR. Individual
     points represent characteristic lengths from the Weibull distribution; superim-
     posed curves represent the full Weibull distribution of lengths[205]
     Copyright © 1986 The Institution of Electrical Engineers

(b) Water tree growth in Rogowski plaques of EPR, with data from XLPE shown
     for comparison[229]
     Copyright © 1986 The IEEE

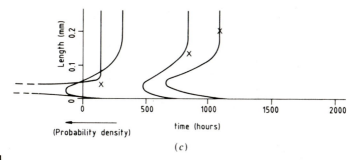

(Probability density)   time (hours)

(c)

Fig. 4.4
(c)   Measured distribution of tree lengths at various times for bow-tie trees in
unstripped cable[201]. Here × denotes the longest tree observed.
Copyright © 1988 The Institution of Electrical Engineers

required to reduce the initiation time to below the time t^* over which the
voltage is held constant in the stepped increase. By substituting t^* for t_d in
eqn. 4.1a this interpretation predicts a simple relationship between inception
field and frequency that can be used to check its validity, i.e.

$$f \propto (t^*)^{-1} E^{-3.5} \tag{4.1b}$$

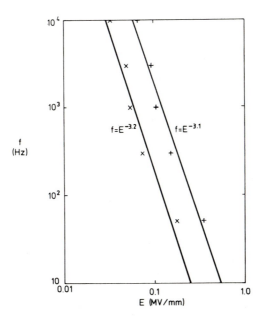

Fig. 4.5   The relationship between the field at water tree inception in polyethylene
reached by stepping at a constant rate, and its frequency. The experimental data
have been taken from Noto[232] (+); and Hossam-Eldin and El Shazli[233](×). In neither
case were details given concerning the nature of the electrolyte, however a voltage
step rate of 500 V · min^{-1} was quoted by Hossam-Eldin and El-Shazli[233]. The results
suggest that Noto used a faster step rate

which can be compared with the experimental relationship, shown in Fig. 4.5 for two cases,

$$f \propto E^{-3 \cdot 2} \tag{4.1c}$$

Since t^* is a constant for a given set of experiments the observed relationship, eqn. 4.1c, effectively confirms the interpretation based on t_d. Errors due to the accumulated stressing and the differing times within the step at which initiation occurred are not large because of the strong field dependence and are likely to account for the small difference of field exponent. The inception mechanism must therefore be the same in the different geometries, even though the needle electrode increases the local field by at least an order of magnitude.

The inception process can also be followed by counting the number density of trees observed after different times under stress, $(N(t))$. This technique is usually applied to destructive measurements and care must be taken to reduce sample-to-sample variations which would strongly influence the results. If the inception of vented trees were a random statistical process[119],

$$N(t) = N(\infty) \cdot (1 - \exp\{-t \cdot k_i\}) \tag{4.2a}$$

and

$$d[N(t)]/dt = k_i[N(\infty) - N(t)] \tag{4.2b}$$

where k_i is the inception rate constant. In this case t_d can be defined as the time required to initiate 63% of the trees and is equal to k_i^{-1}. However $N(t)$ for vented trees typically shows a sigmoidal curve, Fig. 4.6, with an observable

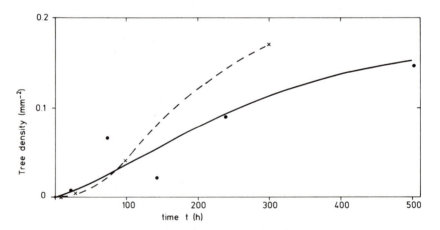

Fig. 4.6 Number density of vented trees observed in Rogowski profiled plaques of a peroxide crosslinked polyethylene at room temperature, for $E = 6 \times 10^6 \text{ V} \cdot \text{m}^{-1}$, $f = 50 \text{ Hz}$, with electrolytes; $\times$ 1 M NaCl[234], ● 1 M CaCl$_2$ (Matwiejew, unpublished data). Trees which originated at a detectable surface defect have been deleted from the count

inception time[218,231,232] corresponding to a maximum in the rate of initiation. It can therefore be concluded that for vented trees inception is a deterministic process, with the inception time t_d, defining the most probable time required for an inception site to reach the necessary and sufficient conditions required for rapid growth.

During the inception stage the incipient tree sites are likely to exhibit a visual aspect and will appear as extremely small trees (microtrees) along the interface between the insulation and the electrolyte, Fig. 4.7. In a typical miniature cable or Rogowski plaque experiment many vented trees will be found to initiate at surface scratches or contaminants. A comparison between artificially scratched and undamaged specimens shows that the inception time is considerably less for trees initiated at defects than for trees for which there are no observable defects at the inception site[115,206]. In practice the inception site environments will be distributed about the most probable site typical in form of the particular material examined. The inception times will therefore have a peaked distribution the shape of which can be expected to depend upon material morphology, surface perfection, and preparation history, while the most probable inception time must be identified with the observed value of t_d. These considerations alone will lead to the peaked distribution of tree lengths characteristic of vented trees[175,115,231,235,229,234] irrespective of any additional distribution in growth rates[236]. However the influence of the inception time upon the length of vented trees will diminish at long times as integration of the empirical growth rate up to $t \gg t_d - t_1$ will show. Thus the length distribution would be expected to narrow with time[236], and the observed persistence of its shape[234], an example of which is shown in Fig. 4.8, implies that a peaked distribution of growth rates also occurs[236].

2 μm

Fig. 4.7 Vented 'microtrees' observed at the electrolyte–XLPE interface in Rogowski plaques. Unpublished data presented at the Annual Conference of Electrical Insulation and Dielectric Phenomena (1984) by Dissado and Wolfe

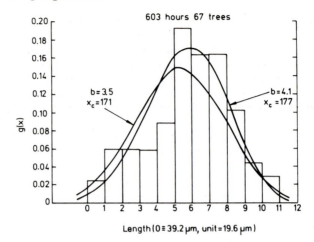

Length (0 ≡ 39.2 µm, unit = 19.6 µm)

Fig. 4.8   A typical length distribution of vented trees observed in Rogowski-profiled plaques of XLPE[234] shown as a histogram, together with the fitted Weibull distribution (continuous line) $g(x) = b \cdot x^{b-1} x_c^{-b} \cdot \exp\{-(x/x_c)^b\}$. Two alternative Weibull distributions are shown so as to illustrate the extent to which the observed data is described by the Weibull form

(b) Bow-tie trees

The reported inception behaviour of bow-tie trees is significantly different to that of vented trees. Typically their number density follows eqn. $4.2a$[119,227] although some workers have reported a sigmoidal curve[237]. These results have been taken to imply that bow-tie inception is a random statistical process. It has also been suggested that $N(\infty)$ increases with temperature cycling[238] and is essentially a linear function of the voltage[227]. The initiation rate constant has not been investigated systematically to any great extent, though it appears to be relatively insensitive to the voltage[227] and possibly increases with frequency[119,197]. An alternative interpretation of the number density versus time curves has been given by Shaw and Shaw[119] who suggested that all bow-tie trees start to grow at zero time but with a distribution of growth rates. A lower limit to the size resolution then causes the number observed to increase with time. This interpretation is consistent with reports of instantaneous initiation[239]. The observation of an exponential distribution[119,237,238] of tree lengths has also been used as circumstantial evidence for Shaw and Shaw's contention[119]. Here the number of trees per unit volume with a length greater than L is given by

$$N(L) = N_0 \exp(-L/L_0) \qquad (4.3)$$

where L_0 is the characteristic length at a given time, and the number density N_0 (of all trees) was found to be almost independent of time over the range covered by the experiment. However since the observed number density reaches saturation faster than the bow-tie tree length[119,240,227] the behaviour of N_0 can be explained, without recourse to an instantaneous initiation, if the times of measurement are longer than $(k_i)^{-1}$. In view of the large amount of evidence in favour[119] of eqn. 4.2*a* it seems likely that bow-tie initiation is statistical.

The exponential distribution of tree lengths is usually taken to reflect the distribution of growth rates, however it is also possible that propagation itself is probabilistic, i.e. growth proceeds by sequential additions to trees at random. In this case if the number of inception sites is finite, a peak can be expected in the length distribution at times sufficiently long for all sites to have undergone at least one of the stochastic events constituting growth. Such peaked length distributions have been observed in the long time ($> \sim 10^3$ hours) experiments of Naybour[200]. A peak in the bow-tie length distribution has also been observed when the optical resolution is sufficiently good for small trees ($< 10 \, \mu$m) to be observed[239] and has recently been confirmed in cable sections[201,241] (see Fig. 4.4*c*). These results indicate that the distribution of growth rates may be peaked in the bow-tie case as well as in vented trees. Such results can be reconciled with eqn. 4.3 if the distribution of trees longer than those at the peak follows an exponential form.

4.2.2 Propagation

(a) The growth law

Following inception, water trees enter into a rapid growth stage. The time power law growth rate commonly ascribed[199] to this stage is not the only form that has been adopted (Shaw and Shaw[119] and references contained therein). However the alternative expressions that have been suggested often rely upon small data sets from destructive measurements[115], and it has been shown that groups of these sets can be combined to give a better outlined plot whose growth law does in fact follow a power law behaviour[206,226]. This growth law has now been confirmed by the direct observation of growing vented trees[207], however the form quoted by some authors[199], i.e.:

$$\frac{dL}{dt} = L_\infty k^{1-n} (t - t_d)^{-n} \qquad t \geqq t_d \tag{4.4}$$

implies an infinite initial growth rate which is physically impossible. For this reason the growth law has been modified[206,208] to

$$\frac{dL}{dt} = L_\infty k^{1-n} (t + t_1 - t_d)^{-n} \qquad t \geqq t_d \tag{4.5}$$

in which the initial rate of growth is finite and given by:

$$\frac{dL}{dt} = L_\infty k^{1-n} t_1^{-n} \qquad t = t_d \tag{4.6}$$

From eqn. 4.5, t_1 can be regarded as the time over which the propagation rate has essentially its initial value.

This empirical description of tree growth can be given a theoretical foundation if it is assumed that the growth rate is proportional to a power, p, of the local electrical field $E(L)$ at the tree tip[242]. When the tree growth,

L, is spherically symmetric, and much less than the insulation thickness, D, the exact rate equation

$$\frac{dL}{dt} = Q_a E^p(L) \qquad (4.7a)$$

where Q_a is the rate constant in Ashcraft's[199] terminology, becomes

$$\frac{dL}{dt} \simeq Q\left[\frac{V}{L+r}\right]^p \qquad L \ll D \qquad (4.7b)$$

Here the tree has been assumed to grow from the surface of an initiating site which provides a divergent field. A general expression is used for $E(L)$, which defines the field in terms of the applied voltage, V, the tree length, L, and a parameter, r, which ensures that the local field is finite prior to growth (i.e. at $L = 0$). Q is an effective rate constant which includes a structure factor for the initiating site which may depend upon D and r, but is independent of L for $L \ll D$. Thus for inception in the divergent field of an aqueous needle electrode[228] (Fig. 4.3c) r is the radius of curvature and Q is given by $Q_a[\ln(1+4D/r)]^{-p}$. Similarly for bow-tie trees initiated at electrolyte filled voids r may be taken to be proportional to their radius. In general however r will be determined by the inception site stress enhancement factor which cannot always be given a structural interpretation, and Q_a will be multiplied by a geometrical factor to give Q. Integration of eqn. 4.7b gives

$$(L+r)^{p+1} - r^{p+1} = (p+1) \cdot Q \cdot V^p \cdot \tau \qquad (4.8a)$$

where τ is the period over which growth occurs, i.e. $t - t_d$. The growth rate, obtained by substitution of $(L+r)$ from eqn. 4.8a into eqn. 4.7b is found to have the form of the empirical expression, eqn. 4.5, i.e.:

$$\frac{dL}{dt} = \left[\frac{QV^p}{(p+1)^p}\right]^{1/(p+1)} \left[\tau + \frac{r^{(p+1)}}{\{(p+1)QV^p\}}\right]^{-p/(p+1)}$$

In this way the power law index can be identified as

$$n = p/(p+1) \qquad (4.9a)$$

the time parameter, t_1, as

$$t_1 = r^{p+1}/[(p+1) \cdot Q \cdot V^p] \qquad (4.9b)$$

and

$$L_\infty \cdot k^{1-n} = [Q \cdot V^p/(p+1)^p]^{1/(p+1)} \qquad (4.9c)$$

Thus by substituting eqn. 4.9b in eqn. 4.8a the tree length can be scaled by r, and the time τ by t_1 to give the dimensionless form

$$\left[\frac{L}{r}+1\right]^{p+1} - 1 = \tau/t_1 \qquad (4.8b)$$

A similar scaling formalism has been shown to apply to bow-tie trees[119] where the void diameter is equivalent to r, and the time for water to diffuse out of the void to t_1.

(b) Field dependence

The field dependent growth law, eqn. 4.7a, has been verified by fitting the numerically integrated growth rate to the observed tree lengths. An expression for the field, $E(L)$, which assumed the tree to be a conductor[199] (eqn. 4.10a) was used, and the power found to be 2. In the light of subsequent work showing the tree to be essentially a capacitor of low conductivity a re-assessment was made using directly observed tree growth in a divergent field[207]. A typical set of results given in Fig. 4.9 show that a square law dependence was still applicable[236]. The sole disposable parameter, Q_a, which was fitted at 500 hours was able to describe subsequent growth of the *same* tree up to 3000 hours. Strictly these results[199,228] do not identify the field power but rather the inverse *length* dependence of the growth rate since the voltage was not varied. Unfortunately it is not possible to deduce a field-squared growth rate in this way because water tree dependent modifications to the insulation impedance may give an extra length dependent contribution to the growth rate.

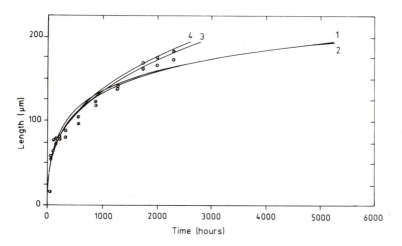

Divergent field model A=0.028 B=1.6 Q=900
Conducting tree model A=0.105 B=2.2 Q=80

Fig. 4.9 Directly observed water tree growth in a needle-plane geometry. Curves 3 and 4 were obtained from eqn. 4.7a with $p = 2$, while curves 1 and 2 were determined from the exponential growth rate $B \cdot \exp\{-A/E(L)\}$ for comparison with A and B as constants[236]. Two expressions were used for the local field, one (curves 1 and 3) is the divergent field due to the needle electrode, i.e. $E(L) = 2VD/[(rD - L^2 + 2LD) \log_e (1 + 4D/r)]$ with $A = 0.028$, $B = 1.6$, and $Q_a = 900$; and the other curves (2 and 4) treat the tree as conducting, $E(L) = V(1 + R/d)^{1/2}/[R \tanh^{-1}\{1/(1 + R/d)^{1/2}\}]$, with $R = r + L$ and $d = D - L$; and here $A = 0.105$, $B = 2.2$, $Q_a = 80$

In Klinger's model[243] construction the field at the tree tip is reduced from that of a metallically conducting tree with hyperboloid shape,

$$E_m(L) = \frac{V[1+(L+r)/d]^{1/2}}{(L+r)\tanh^{-1}[1/(1+\{L+r\}/d)^{1/2}]} \qquad (4.10a)$$

with $d = D - L$, to

$$E(L) \simeq E_m(L)/F_\omega^{1/2} \qquad (4.10b)$$

where F_ω is a frequency ($\omega = 2\pi f$) dependent factor with asymptotic values

$$F_\omega|_{\omega \to 0} = \left(1 + \frac{\sigma_p L}{\sigma_T r}\right)^2 \qquad (4.11a)$$

and

$$F_\omega|_{\omega \to \infty} = \left(1 + \frac{\varepsilon_p L}{\varepsilon_T r}\right)^2 \qquad (4.11b)$$

Here σ_p and σ_T are the conductivities, and ε_p and ε_T are the relative permittivities of the polymer and treed area respectively.

Eqn. 4.11 predicts that all the frequency dependence of the initial growth rate ($L=0$) arises from the rate constant, Q. The growth rate of trees with $L > r$ will however have an increased inverse length dependence and a frequency dependence which will be reduced or increased from that of Q according to whether $\varepsilon_p/\varepsilon_T$ is greater or less than σ_p/σ_T. Since the tree is mainly composed of water droplets weakly connected by water paths the maximum value for $\varepsilon_T/\varepsilon_p$ is ~40 whereas σ_T can be orders of magnitude greater than σ_p, and thus it can be expected that $F(\omega)$ will reduce the frequency acceleration due to Q. The frequency range over which F_ω varies however will depend upon the value chosen for σ_p. Using the realistic estimations of tree conductivities and permittivities given by Densley *et al.*[244] ($\varepsilon_T \approx 3$, $\sigma_T \simeq 1\cdot35\sigma_p$; for 10% volume fraction of water droplets in the tree), see for example the AC dielectric measurements of Fothergill *et al.*[137], it can be shown that for $\sigma_p = 10^{-14}$ $(\Omega m)^{-1}$ all variation in F_ω occurs below the service frequency (50 Hz). Only when σ_p is as large as 10^{-10} $(\Omega m)^{-1}$ does the change in F_ω occur in the test frequency range, and even then it will not be large if the ratio σ_p/σ_T is close to that of $\varepsilon_p/\varepsilon_T$ (~0·75) as derived by Densley *et al.*[244]. The most important feature of this model is therefore the altered length dependence of $E(L)$, and substitution in eqn. 4.7a shows that for $L > r$ the experimentally observed inverse square law (~L^{-2}) dependence would now originate with a *linear* field dependence. A choice between these two possibilities can only be made on the basis of experiments in which the voltage dependence has been obtained.

In experiments on scratched miniature cables[115], sample-to-sample variations led to inconclusive results with the initial growth rate being determined[206] to have a voltage power dependence between 1 and 2. Trees grown from wedge-shaped defects of controlled depth introduced into production line power cable segments[201], exhibited growth rates linearly dependent upon the voltage, Fig. 4.10. This applied to all cases of tree length dependence which varied from L^{-2} to L^{-1} and even weaker. These results should not be confused with the linear voltage dependence quoted for the tree length grown in needle-plane[245] or other geometries[128] during a given time. Re-arrangement of eqn. 4.8a shows that this measure is not very sensitive to the field power, p, since for $L \gg r$:

$$L \simeq [(p+1) \cdot Q \cdot V^p]^{1/(p+1)} \cdot t^{1/(p+1)} \qquad (4.12a)$$

and the voltage dependence rapidly approaches linearity for any power, $p \geqq 2$. The equivalent expression in Klinger's[243] model is

$$L \propto [(2p+1) \cdot Q \cdot V^p]^{1/(2p+1)} \cdot t^{1/(2p+1)} \qquad (4.12b)$$

with an asymptotic $V^{1/2}$ behaviour.

This measure of tree growth will not therefore yield the dependence of the rate constant, Q, in eqn. 4.7b upon the controlled variables, but rather a fractional power of them. Thus results quoted in this manner should not

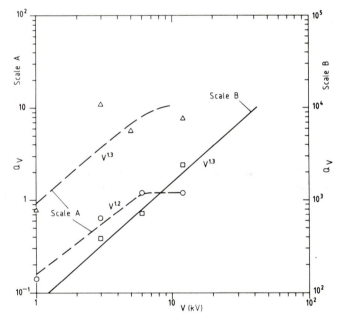

Fig. 4.10 Water tree growth rate constant, Q_V, defined by $dL/dt = Q_V \cdot f(L)$, as a function of voltage. The dimensions of Q_V are dependent upon the functional dependence of $f(L)$ upon L, and three sets of data are shown. These are taken from the tree growth curves obtained for slits located at different sites and of different depths in XLPE cables[201].

□: slit depth = 1·5 mm, (RH scale), Q_V in units of $(\mu m)^3 \cdot$ hour^{-1} (i.e. $f(L) \propto L^{-2}$)
△: slit depth = 1·5 mm, (LH scale), Q_V in units of $(\mu m)^2 \cdot$ hour^{-1} (i.e. $f(L) \propto L^{-1}$).
○: slit depth = 2 mm (LH scale), Q_V in units of $(\mu m)^{3/2} \cdot$ hour^{-1} (i.e. $f(L) \propto L^{-1/2}$).
In eqn. 4.7b Q_V is given as $Q \cdot V^p$, however here in all cases Q_V is approximately linearly dependent upon V regardless of the form of $f(L)$

be compared with theoretical predictions. Instead, either the behaviour of the initial growth rate (i.e. at $L = 0$) or the time to grow a tree of chosen length, should be used. When plotted in the latter way, Fig. 4.11, Yoshimura *et al.*'s[245] data give a voltage power, p, of ~2·3, which taken together with the voltage linearity reported for L at 5 hours, suggests a square law dependence. Since a linear relationship of L^3 with τ was also identified, this field squared behaviour suggests that Klinger's modification was inapplicable over the time scale of the experiment.

However direct measurement of individual trees in the initial growth region[246] ($L < \sim r$) as a function of voltage and radius of curvature lead to different conclusions. Here p was identified as unity, Fig. 4.12a, by plotting the LHS of eqn. 4.8a as a function of time. The growth rate at 30 minutes was taken as the initial growth rate and plots, Figs. 4.12b and 4.12c, showed that

$$\left. \frac{dL}{dt} \right|_{t=1/2\,\mathrm{hr}} \propto \frac{V}{(r+L)\ln(1+4D/r)} \approx E(L) \qquad L \lesssim r \qquad (4.13)$$

which agreed with the identification of a linear field dependence ($p = 1$) at these short times. When taken together with the observation of an inverse square law (L^{-2}) for $L > r$ in the same type of experiment[228], these results are consistent with a linear dependence of the growth rate upon a local field whose form is that given by Klinger, eqn. 4.10b.

The data available on the field dependence of the water tree growth rate is thus still somewhat inconclusive. Yoshimura *et al.*'s[245] results are obtained at short times under accelerated conditions, as are those of Filippini and

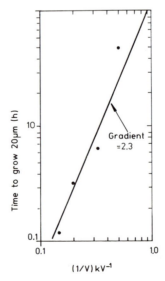

Fig. 4.11   The time required to grow a tree of a chosen length plotted as a function of voltage, taken from the data of Yoshimura *et al.*[245], which were obtained in a point-plane geometry

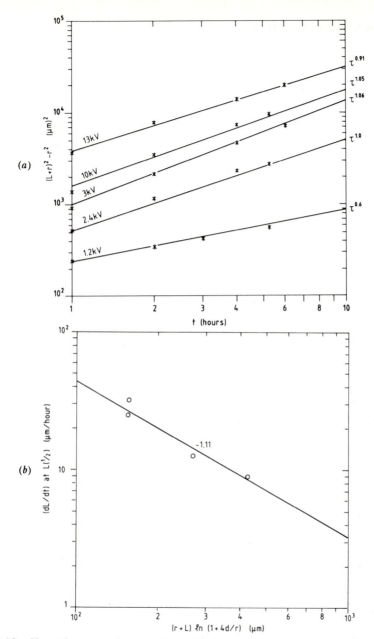

Fig. 4.12 Short time growth rate of individual water trees observed directly as they grow in a point-plane electrode configuration. Data taken from reference[246]. The growth rates dL/dt are linearised values for the first hour of growth, i.e. $dL/dt|_{t=0\cdot5\,\text{hour}}$

(a) Tree length as a function of time of growth for different voltages and constant radius of curvature, showing a fit to eqn. 4.8a with $p \approx 1$. The voltages indicated are peak values

(b) Growth rate at $t = 30$ min and constant voltage for needles with different radii of curvature

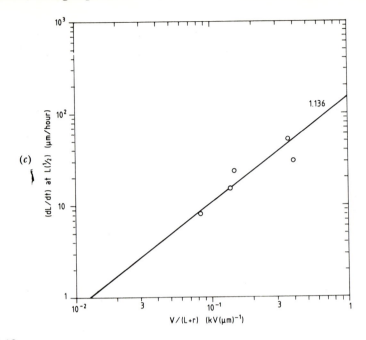

Fig. 4.12
(*c*)   Growth rate at $t = 30$ min and constant radius of curvature for different voltages. When combined, (*b*) and (*c*) show that the short time growth rate is a linear function of the field at the tree tip (eqn. 4.13)

Meyer[246], and contain limited numbers of data points with significant error in length. The data obtained from miniature cables[115] also rely on a few points and the cables were scratched. Although the results obtained for cable sections [201] were derived by means of a better analysis containing more data, the initiating sites were unusual artificial defects. Taken as a whole the balance of the evidence appears to lie in favour of a linear dependence upon a field which is modified in its length dependence in the manner suggested by Klinger (eqn. 4.10*b*), though this conclusion must be treated with some caution until confirmed unambiguously.

(c) Length dependence
The length dependence of the water tree growth rate is a central factor in the growth kinetics. Although numerical integration is required when exact model functions are adopted[228] for the field $E(L)$ in eqn. 4.7*a*, the approximate result of eqn. 4.8 shows that the behaviour can be obtained easily from the experimental data when $r < L \ll D$ which is often the case. Here a plot of $\log(L)$ as a function of $\log(\tau)$ will yield the inverse power of L in eqn. 4.7*b* via the gradient. The growth rate dependence upon controlled parameters can also be obtained by comparison of similar plots as the parameter is varied. In this way it has been found that the typical behaviour in the different geometries of point-plane, uniform field plaques[236] and cable sections[201] is an inverse square law, as shown in Fig. 4.13. Again the cable

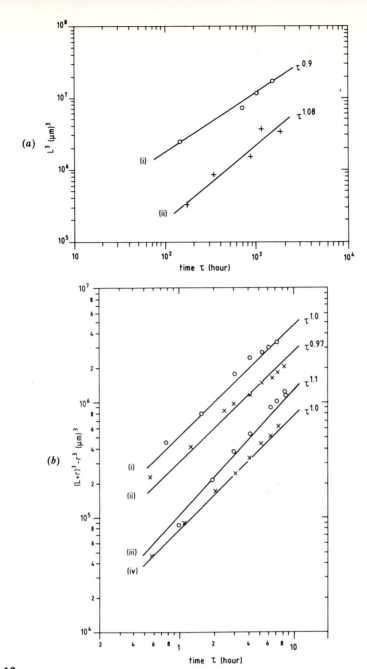

Fig. 4.13

(a) Length plotted against time for trees growing from the surface of flat plaques (i) and stripped cable (ii)[201]
     Copyright © 1988 The Institution of Electrical Engineers

(b) Tree data taken from Filippini *et al.*[247]. *In situ* measurements on unannealed samples of polyethylene with different molecular weights (i) MW = 48 000; (ii) MW = 81 000; (iii) MW = 161 000; (iv) MW = 119 000. *r* is taken as the radius of curvature of the needle point

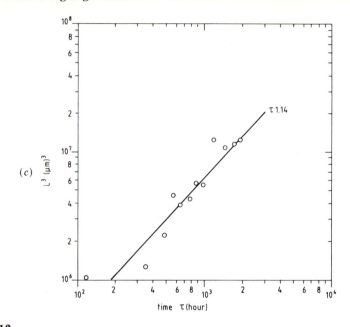

Fig. 4.13
(*c*)   Characteristic tree length observed in tree ensemble grown in Rogowski profiled plaques of polyethylene[236]
Copyright © 1986 The IEEE

sections with wedge shaped cuts give differing results most of which follow a L^{-1} law. Since some trees in this experiment do follow an inverse square law and both sets exhibit a linear voltage dependence it is possible to construe the results as support for Klinger's[243] field expression, however the tree lengths observed are much greater than the radius of the initiating cuts.

The investigation of the propagation of individual trees for which the growth site is known requires only a single measurement at the differing times and is not time consuming. In contrast investigation of the tree ensembles found in cables or plaques requires a laborious procedure of tree counting and measurement for large numbers of trees. It has been suggested[238] that the man-hour requirements may be reduced by the use of extreme value statistics. In this procedure only the largest tree, L_{max} is measured in each slice of insulation examined. The distribution of largest trees is determined and the time dependence of the characteristic longest tree obtained. When applied to bow-tie trees which are exponentially distributed in length the probability $P(L_{max} < x)$ that L_{max} is less than x, follows the first asymptotic extreme value distribution[248]

$$P(L_{max} < x) = \exp\{-\exp[-\alpha(x - \bar{L}_{max})]\} \qquad (4.14)$$

where $\bar{L}_{max}$ is the average longest tree and α a measure of the distribution width. This expression will also apply when the bow-tie length distribution is peaked as long as the long tree tail dies off no slower than exponential.

It should be noted that when the number density of trees is increasing its time dependence is incorporated into $\bar{L}_{max}$. This is clearly the case for bow-tie trees where a power law (t^b with $b > \sim 0.5$) behaviour is observed for $\bar{L}_{max}$ even though the lengths of individual trees are reported as saturating, as it is implied their number density does also. Thus $\bar{L}_{max}$ does not describe the behaviour of any one tree and should not be used for comparison with treeing theories. Instead it is a measure of the ensemble.

The length distribution of vented water trees is not exponential but peaked and has a shape close to that of a Weibull distribution[235,229,234]. However there must also be at any time a cut-off length, $w(t)$, beyond which no trees can exist (i.e. we do not have an infinite growth rate), such as is clearly shown in the data of Bahder *et al.*[235]. In this case the appropriate extreme value distribution is that of the third asymptote[248], (i.e. a Weibull distribution) rather than the first asymptote equation (4.14) that has been used without justification in some cases[238,249]. Thus

$$P(L_{max} < x) = \exp\left[-(\{w - x\}/\{w - \bar{L}_{max}\})^\beta\right] \qquad (4.15)$$

where $\bar{L}_{max}$ is now the characteristic longest tree, and β is the shape parameter of the tree distribution itself. Again $\bar{L}_{max}$ will incorporate time dependent changes in the number density and cannot be taken as a description of an individual tree.

Since the longest vented trees are the most damaging[175] $\bar{L}_{max}$ is a useful measure of the extent of deterioration but cannot be utilised for comparison with theoretical models of individual tree growth. In addition the distributions (eqns. 4.14 and 4.15) can be used to predict the maximum tree length expected in a given specimen size[248]. However the use of the extreme value analysis is restricted by the need to assume a tree length distribution of a particular class in order to choose the appropriate extreme value distribution[248]. It will also be difficult to determine an exact value for $w(t)$ where this is necessary. Furthermore it is always possible that the length distribution belongs to a class without a stable extreme value distribution. In view of these drawbacks, it is recommended that a full count and length measurement be carried out for comparison with theory. An extreme value analysis may be taken in addition as a measure of the deterioration[229].

4.2.3 Long time growth

At long times the initial power law propagation discussed in the previous section may cross-over to a different behaviour, with the cross-over time (k^{-1}) being 100 days or greater at conditions close to those of service (i.e. 3 kV, 50 Hz, 0·1 N NaCl electrolyte). This change is important because the reduction in breakdown strength associated with vented trees is determined by the longest such tree present in the ensemble[175]. The slowing down shown in Fig. 4.3 was identified from data obtained on an ensemble of trees grown in miniature cables, and was the result of incorporating the average tree into the tree propagation analysis. One possible explanation for this behaviour would be a continual initiation of slowly propagating trees.

Most individual trees do not show such a slowing down even when grown for 3000 hours, however there are some exceptions[228] whose growth was similar to that of Fig. 4.3, and in some cases the trees even stop growing[250] in the same way as bow-tie trees. It was suggested[228] that the reduction in growth rate experienced in these atypical cases was the result of in-built mechanical stresses in the material. If this is correct it is possible to see how such stresses which also occur in cables[222] may retard the growth of even individual trees. Another factor which will influence the growth rate of trees in an ensemble will be changes of shape, such as might occur through coalescence[119,250] or bifurcation[206]. In these cases the L dependence of the field at the tree tip will alter and the growth rate change. The propagation of coalescing trees will particularly resemble that of Fig. 4.3 since the joint tree boundary will be a lower stress enhancing region than either tree separately. This has occasionally been seen in direct observation experiments when trees which initiate at different times form a common growth surface[228]. In this case a change in the shape of the length distribution can be expected and the time of its occurrence may be identified with k^{-1} in the empirical growth behaviour, Fig. 4.3.

On the other hand bifurcation may well lead to an accelerating tree propagation. This effect occurs when the tree damage boundary is unstable to fluctuations[251] and each finger effectively grows as an independent entity. Because of their shape the fingers are likely to grow faster than the originating tree. Another possible origin for long time acceleration of tree growth is the increase in local field experienced by a tree growing from the outside in the radially non-uniform field of a co-axial cable. In this case the field increase appears to be insufficient to cause the acceleration sometimes observed[249] for vented trees at long times ($> \sim 0.7$ years) in power cable segments. A similar argument applies to the same type of behaviour observed[252] in the point-plane geometry, i.e. the trees have to be exceedingly close to the plane electrode before the field is large enough for such acceleration. Since this accelerating growth behaviour appears to be linked to a reduction of breakdown strength[253] its possible origins in vented tree bifurcation and the physical reasons for this behaviour need further investigation.

It is clear that a reliable prediction of the influence of water treeing on cable failure requires a description of the tree propagation, number density and length distribution at long times, since treeing deterioration and its effect upon breakdown strength is determined by the longest vented tree[175]. This stage of water tree growth therefore deserves more attention than it has been given to date.

4.3 Mechanisms of water tree growth

Before examining the wide range of mechanisms that have been postulated to explain water treeing it is useful to outline the conditions that must be satisfied by such a mechanism. These are:

(i)   to incorporate those forces known to be essential to water tree formation;

(ii)   to produce the known structure of a water tree with the available magnitude of forces;
(iii) to predict the same dependence upon controlled variables as observed in inception and propagation.

In determining what forces are essential to the water treeing process it is necessary to show that the process ceases when the force is removed. Thus, although many factors accelerate water treeing, only those which inhibit treeing when reduced from their normal level can be accepted as essential. The most obvious of these is an AC electrical stress, which must be applied to insulation in contact with an aqueous electrolyte. Both inception and propagation are accelerated by the frequency, f, of the AC field[206] and the near linear frequency dependence[228] of Q (see eqn. 4.7b and Fig. 4.14) implies that water trees should not grow under DC electrical fields. Although observations of DC water trees have been reported[254,255,256], attempts to verify their existence using point-plane geometries[232,257] and cable sections[249] have failed. It has been suggested that these trees are the result of an AC ripple superimposed on the DC stress[232], and it is known that in this case the DC component has no influence upon the propagation rate[257]. Thus it must be concluded that if DC water trees do occur their mechanism is different (and much slower[255,258], $\sim 0 \cdot 2 \; \mu m \cdot hour^{-1}$ at $22 \; kV \cdot mm^{-1}$) than those formed under AC conditions and the shape reported[255] (comet-like or filamentary) suggests that the type of damage will also be different.

4.3.1 General critique of proposed mechanisms
Since water trees consist of microvoids generated or enlarged by the treeing process it is natural to assume that mechanical forces are involved in treeing.

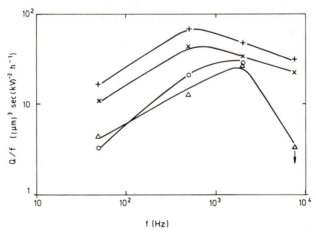

Fig. 4.14   The frequency dependence of the rate constant, Q, determined from observation of vented water tree growth *in situ*[228]. The needle radii of curvature are: $+3 \; \mu m$; $\times \; 10 \; \mu m$; $\bigcirc \; 40 \; \mu m$; and $\triangle \; 90 \; \mu m$; and the field is assumed to be determined by the electrode geometry
Copyright © 1988 The IEEE

This contention is substantiated by the accelerated propagation of trees parallel to a tensile mechanical stress[232,119,223] and the retardation of growth by isotropic compression[259]. The essential role of chemical reactions is perhaps not so clear cut, however it is known that anti-oxidants reduce the propagation rate[206] while oxidising agents increase it[223]. Furthermore it has been shown that treeing is inhibited by an inert (nitrogen) atmosphere[260,261]. When taken together with the evidence for oxidation in the treed region[213,219] it seems reasonable to conclude from these results that chemical oxidation[220,221,262] is also an essential feature of the water tree mechanism.

Over the years a large number of alternative mechanisms have been proposed to explain water treeing, and these have been summarised in Fig. 4.15. In this diagram the mechanisms have been divided horizontally in terms of their potential applicability to either inception or propagation, though in many cases the originating authors made no such distinction. A rough categorisation of the processes involved may be made although it is impossible to be precise with such diversity of form. Thus most mechanisms can be regarded as versions of an electro-mechanical process, however some (a, b, e, l, m, n) are essentially electrical, others electro-chemical (t, y, z) or electro-mechano-chemical (d), and a few rely upon non-local field driven thermodynamic forces (i, o). From this categorisation it is therefore clear that most authors identify either a mechanical or chemical component as essential to water treeing, and although one author[222,264] (g, s) regards these and other processes as alternatives, in only one case is the essential nature of both recognised through their utilisation in combination[116] (d). It is not surprising therefore that in spite of their diversity none of the postulated mechanisms are adequate to describe all the basic observations. The same conclusion has also been reached by Zeller[262] through an analysis of the conditions required for the effective operation of the electrical and electro-mechanical groups of mechanisms.

When the mechanisms are examined in greater detail some must be ruled out completely as being incompatible with experimental data, whereas others betray only an inadequacy to completely reproduce the observations and may be acceptable if modified. Among the former must be counted all mechanisms for which partial discharges[245,271] are essential. Since the original report[271] careful experiments with a sensitivity capable of detecting discharges smaller even than the magnitude suggested have failed to do so in connection with water treeing[275]. The electron bombardment theory[258] is another mechanism of this type. In essence this theory is derived from electrical treeing and would predict water filled channels, and a frequency independent propagation rate, both of which are contrary to experiment. Note that direct observation of tree propagation[228] determines Q to be approximately proportional to f (Fig. 4.14) and has also shown that AC water treeing suppresses electrical treeing and vice-versa[276]. This mechanism may possibly be the cause of DC rather than AC water treeing[258], however Fedor's theory[270] of electro-osmosis is also a likely candidate. As a theory of the AC process electro-osmosis has a number of drawbacks. It is for instance frequency independent and propagates via sequential cracking.

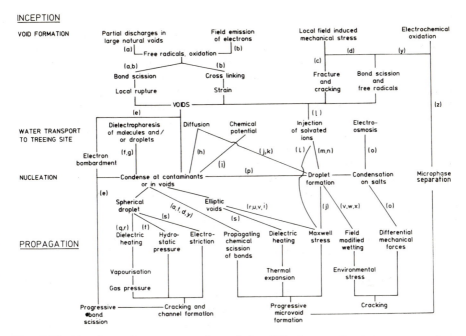

Fig. 4.15 A schematic representation of the different mechanisms proposed to explain water treeing. (a) Abdollal *et al.*[221]; (b) Bernstein[118]; (c) Morita *et al.*[240]; (d) Zhurkov *et al.*[116]; (e) Yamanouchi *et al.*[258]; (f) Patsch[263]; (g) Tanaka *et al.*[264]; (h) Isshiki *et al.*[265]; (i) Matsuba & Kawai[189]; (j) Mizukami *et al.*[266]; (k) Wilkens[267]; (l) Dissado & Wolfe[268]; (m) Morita *et al.*[269]; (n) Tu & Kao[259]; (o) Fedors[270]; (p) Katz & Bernstein[190]; (q) Nitta[271]; (r) Yoshimura *et al.*[245]; (s) Tanaka *et al.*[222]; (t) Yoshimitsu and Nakakita[196]; (u) Sletbak[272]; (v) Meyer *et al.*[273]; (w) Minnema *et al.*[274]; (x) Soma & Kuma[237]; (y) Henkel & Müller[220]; (z) Zeller[262]

Nor is it clear that the process necessarily requires an electrical field for its operation. However under DC conditions such a mechanism, together with field driven ion penetration of the polymer[277,278,279], could cause the comet shaped trail of damage observed for DC water trees[255,279]. Most other mechanisms fail to a lesser extent, though nonetheless sufficiently to rule them out as independent candidates for AC water treeing.

The fact that many routes lead to fracture or cracking of the insulation is a serious drawback to their acceptance in view of the experimental observations. In particular the theory of Patsch[263] which relies on molecular dielectrophoresis to maintain a steady condensation of water into voids with the resulting hydrostatic pressure causing rupture must also be criticised both because it is dubious as to whether dielectrophoresis can be operative on the molecular scale[231,264] and because of its lack of frequency dependence. In fact most water transport mechanisms (dielectrophoresis, chemical potential, diffusion, osmosis) are frequency independent as long as they are not required to overcome internal barriers in the polymer. For dielectrophoresis this will be the case[264] when the droplet size is less than 0·01 μm, and here as with chemical potential forces the frequency dependence of the dielectric constant may even cause a reduction in transport rate with increasing frequency. A frequency acceleration of the transport rate would however occur if the droplets were larger since transport then would require either local deformation, which is unlikely with the forces available[280], or the surmounting of a potential barrier, both of which involve repetitive motions.

Those mechanisms that transport water in its molecular form, such as diffusion and possibly the chemical potential process can be ruled out as sufficient since they are incapable of transferring electrolyte from the reservoir to the tree. However diffusion may be involved as a preliminary to capillary (radius <0·1 μm) condensation when trees are formed in a humid atmosphere[281]. The electro-osmosis model relies on molecular transfer to salt concentrations in the polymer and it is not clear whether the influence of an AC electrical field will introduce electrolyte transfer in this case either. Even dielectrophoresis which could be used to obtain an electrolyte concentration has been noted experimentally[264] to be insufficient to initiate a water tree by itself. It is also known that the rapid condensation of water in the presence of an AC field does not initiate water trees[237]. Furthermore there appears to be some doubt[262] as to whether the water concentration (2—10%) observed in water trees[211] could be achieved in this way at fields below those giving local breakdown. It is therefore clear that in general the proposed transport mechanisms cannot be responsible for the observed behaviour of the inception time[206], although they may contribute to it in specific cases as has been suggested for diffusion[196], and dielectrophoresis[264].

All mechanisms which rely for propagation on a force proportional to the square of the electric field will give a growth rate increasing with frequency if the force is used to produce insulation damage by means of a local fatiguing process. Rejecting those proposed models which lead inevitably to cracking the most promising group of mechanisms are those which rely on the progressive formation of microvoids. It has been suggested that for this to be possible a tensile component of the local oscillating stress is required[119,222], the appearance of which is usually ascribed to an elliptical void shape. As yet however no elliptically shaped voids have been detected in water trees[212], but it is possible that a combination of dielectric heating and Maxwell stress[238] will generate a tensile stress component from a spherical void. Alternatively compressive stresses combined with shear yielding could have the same effect.

The existence of an oscillating displacement confined to the treed region and initiating electrolyte electrode has been confirmed by means of a direct holographic examination of a growing tree[282]. These observations were consistent with a nearly radial pressure due to Maxwell stress, with the displacement amplitude proportional to the square of the voltage, and independent of frequency and electrolyte concentration. However Zeller[262] has pointed out that a local field sufficient to cause polymer deformation by these means will exceed the insulation strength and either lead to electrical trees, breakdown, or field moderation. Furthermore a DC field component will give an additive displacement and would accelerate tree propagation if the growth rate were proportional to the maximum pressure, a result contrary to experiment[257]. Thus even these electro-mechanical mechanisms appear to be incapable of generating a water tree without the assistance of some other contributory component such as an electrochemical process[262]. It seems likely therefore that the field drives a combined mechanical and chemical process during water treeing.

In most cases no distinction is made between vented and bow-tie trees, even though models based on local hydrostatic pressure may not apply if the tree were connected to a liquid reservoir through an open vent. A notable exception is the surface tension mechanism of Soma and Kuma[237] which has been applied specifically to bow-tie trees. Their results show that the number of bow-ties and the length grown in a fixed time were reduced when the aqueous reservoir was better able to wet the polymer. However no testable expression was suggested to relate the growth rate to the interfacial tension.

As a rule most quantitative theories give expressions for the force responsible for damage production and it is implied that tree growth is described by the advance of the damage boundary. The electron bombardment theory[258] is exceptional in quoting a formula for the tree growth rate,

$$\frac{dL}{dt} \propto E(L) \exp\left(-A/E(L)\right) \tag{4.16}$$

which however adds further weight to the rejection of this mechanism since it is demonstrably inapplicable to trees observed *in situ*[236]. An E^2 dependence is predicted for the damage producing force by most electro-mechanical models and also by the only quantitative expression for chemical damage[196]. In the latter case a chemical reaction generates an increase $d\varepsilon$ in the local permittivity. The progress of the reaction, $d\eta$, is driven by a field dependent component of the free energy and is expressed through a change in the equilibrium constant from $K(0)$ to $K(E)$, with

$$\log_e \{K(E)/K(0)\}\, d\eta = (E^2/8\pi)\, d\varepsilon \tag{4.17}$$

In any model the energy utilised for treeing originates either with the normally recoverable field energy stored during cycling, or with the component dissipated (as in the dielectric heating theory[271,245,222]). In the postulated models these energies are converted to a field dependent 'pressure' $\Pi(E)$ which it is implied produces the water tree.

By assuming that the polymer compression at the tree boundary stores the energy available for its advance by damage production, an energy balance equation will yield an expression for the propagation rate suitable for comparison with experiment. To this end g is defined as the volume fraction of the boundary layer of the tree occupied by the damage, with $V_T(\tau)$ being the tree volume at time τ and $dV_T(\tau)$ its incremental increase in the time interval $d\tau$. Denoting the tree surface area as $S_T(\tau)$, the damage produced in time $d\tau$, $(g\, dV_T(\tau))$ is given by

$$f\, d\tau\, \Pi(E) S_T(\tau) \propto g\, dV_T(\tau) \tag{4.18a}$$

where $\Pi(E)S_T(\tau)$ is proportional to the mechanical force produced during each of the $f\,d\tau$ applications of pressure in the time $d\tau$. Hence

$$[S_T(\tau)]^{-1}\,dV_T(\tau)/d\tau = dL/d\tau = Bf\Pi(E)/g \qquad (4.18b)$$

Here the proportionality constant B will be proportional to the ratio of the field dependent compressibility (electrostriction) coefficient[274] to bulk modulus if the damage is taken to require the generation or expansion of voids[283]. A propagation rate in the form of eqn. 4.7a is thus arrived at when $\Pi(E)$ is a power of the local field. Here Q_a contains all the details of a particular model as well as the factor Bf/g. However insufficient distinguishing predictions have been made to allow discrimination between the models on the basis of the parameter dependencies of $Q_a(\propto Q)$.

The previous discussion has indicated deficiencies in most models with respect to some aspects of the essential conditions (i) and (ii) of which the most notable are those concerning the roles of chemistry and ions. Under condition (iii) the only clear cut prediction made is that of a growth rate proportional to the square of the local field. However this behaviour has not yet been conclusively verified experimentally and it is possible that the propagation rate is a linear function of the field. Such an expression for $\Pi(E)$ can be expected to apply if ion injection into the polymer is a rate determining factor in water tree formation. In order to see how such a process may combine with others to cause water treeing, the experimentally known influences on the treeing process will now be examined in detail.

4.3.2 Experimentally observed features

It has already been pointed out that electrolyte material from the reservoir is carried into microvoids during the growth of water trees, and Meyer's[211] data showing that the volume fraction of water in vented trees is a linear function of the voltage suggests that an equilibrium density of solvated ions is maintained in the tree by the applied voltage. Some proposed mechanisms have made use of this idea to form the basis of an inception process in which solvated ions are driven into surface cracks opened up by Maxwell stress[268] or applied hydrostatic pressure[259]. The action of the environmental stress cracking agent Igepal in favouring inception[223] is consistent with this view, however its effect in reducing the growth rate indicates that the formation and penetration of field enhancing microcracks is not sufficient for rapid growth. With respect to this point it has been shown that the formation of vented trees correlates with decreasing electrolyte permittivity[284] and ion hydration entropy[269]. The former result implies that treeing is favoured by increased ion injection, and the latter that inception is favoured if the hydration shell is weakly bound to the ion.

Further light has been shed upon this stage of tree formation by the results of Meyer and Chamel[285], which show that ion permeability of the polymer is not increased by either DC or AC fields but rather by the formation of locally degraded regions, which the authors suggest were oxidised. Since this effect did not occur to an observable extent in de-ionised water, it seems likely that an explanation of the ion effect in inception must be sought in terms of a local chemical degradation brought about by dehydration of a layer of solvated ions adsorbed in the polymer-electrolyte interface[277,278]. This approach can be carried one step further by the aid of a recent examination of tree inception *in situ*[276]. Here it was observed that the first stage of tree growth was the rapid formation of short bouquets of microvoids running from the electrolyte needle along the field lines. This is then followed by a filling in of the global volume of the tree with further microvoids. The authors define an inception time, t_i, as the time required to grow a bouquet of the smallest detectable length and determine its field dependence from the position of inception on the needle surface. A re-plotting of this data, Fig. 4.16a, shows that t_i is proportional to $E^{-5.75}$, a dependence which is much stronger than that usually quoted for water trees[206] and near the weakest behaviour observed for electrical trees[122].

In their earliest stages then, water trees appear to produce microvoids in a similar way to those generated in some electrical trees. However the bond scission involved in microvoid formation[116] is for water trees most probably the result of an electro-chemical reaction[118,220,286] involving the injected

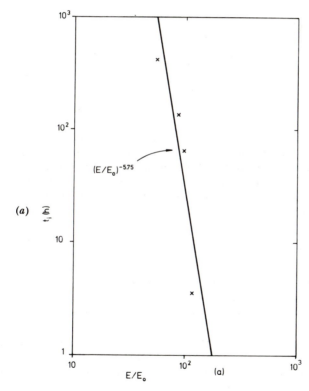

Fig. 4.16
(a) The time, t_i, required to initiate a bouquet of microvoids of length 10 μm during water treeing[276] as a function of the applied field. E/E_0 is the geometrical enhancement factor giving the local field E in terms of the nominally uniform applied field E_0

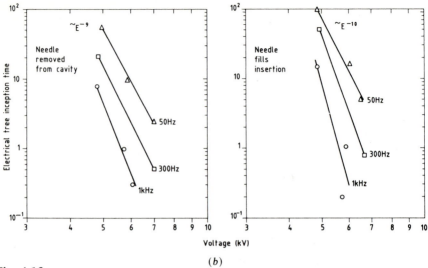

Fig. 4.16

(*b*)   Electrical tree inception time in XLPE as a function of voltage applied to a metal needle, with 5 μm radius of curvature[129]. Values for the case of a cavity introduced at the needle tip are compared with direct needle polymer contact. Both show essentially the same field dependence, indicating that the electrical tree inception mechanism is unchanged

electrolyte, rather than the field emission or electrode injection of electrons that is the case for electrical trees[287,288]. This explanation would be consistent with the failure[275] to observe light emission during water treeing. Unlike electrical trees however, the power available for local damage production will be insufficient for the expansion and coalescence of microvoids to form channels[189] which would yield a filamentary growth pattern[289]. Instead electrolyte is transported into the tree volume, which because of oxidation is now more permeable to solvated ions, and microvoid formation is continued within it up to an extent determined by the applied voltage[211]. It would be this period of consolidation therefore that would determine the inception time commonly noted[230] and give the weaker field dependence observed for t_d[206]. Subsequent growth could then be expected to maintain the equilibrium microvoid and water density as the damage boundary advances.

The most serious failure of the proposed propagation mechanisms is their inability to predict a dependence upon electrolyte concentration, which has the third strongest influence upon growth rate after voltage and frequency. Direct observations of tree growth have measured the effect of electrolyte concentration[290] over the range 1 N to 10^{-5} N (NaCl solution), and the values of Q deduced from this data have been plotted as a function of electrolyte concentration (C) in Fig. 4.17. Over this extended concentration range it can be seen that the square root dependence of Ashcraft[199] is too

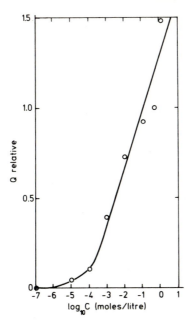

Fig. 4.17   The rate constant, Q, for water tree growth (see eqns. 4.7b and 4.8a) as a function of electrolyte (NaCl) concentration C (moles/litre) scaled to unity at 0.5 N NaCl. The data has been taken from Koo and Filippini[290]

limited, and the actual behaviour is

$$\frac{Q}{Q_s} = \frac{0.3}{z}\left(\log\frac{[C+A]}{[H^+]}\right)\qquad(4.19)$$

In this expression data from electrolytes with different valencies[209], z, and the pH dependence determined by Ashcraft[199] have been combined together, to give the rate constant, Q, scaled to that of a standard 0.5 N NaCl solution, Q_s. For large electrolyte concentrations ($> \sim 1.5$ N in the case of NaCl), the growth rate has been observed to drop sharply to the order exhibited by a 10^{-3} N solution[208,291,292] and eqn. 4.19 no longer applies. This form of concentration dependence is similar to that found for metal ion catalysis of polymeric oxidation[95] and implies a direct role for this type of chemistry in tree propagation[293,294]. In some investigations however it is possible that salt precipitation occurs at the polymer surface[115,208] giving a much lower effective electrolyte concentration.

Field driven oxidation has been proposed as an explanation of water treeing by two groups[221,286] whose mechanisms involve similar sequences of free radical reactions but differ in their initiation processes. The earlier model[221] relies upon partial discharges to produce monatomic oxygen, UV radiation and energetic electrons. The two former products generate polymer free radicals and polymer oxidation via peroxide intermediates. The electrons produce the same type of products by dissociating any water that is present into ions and radicals. Although partial discharges below the limit of detectability may occur in small voids ($< \sim 2\ \mu$m) the fields required ($> 5 \times 10^8$ V $\cdot$ m^{-1}) are of the order of the breakdown field and in excess of those involved in water treeing. Taken together with the failure to observe either partial discharges or luminescence during treeing[275] these considerations imply that this initiation mechanism is highly unlikely. The initiation step proposed in the later model[220,286] involves an electrochemical reaction with the water associated with a hydrophilic defect or contaminant[195]. Although such initiation sites are sometimes observed their resulting vented trees are easily distinguishable since they tend to initiate quickly and grow rapidly up to a saturation length. Most vented trees possess no such visible inception sites (see Fig. 4.7 for example), however the injection of solvated ions to a partially oxidised site which is itself hydrophilic will generate an equivalent inception site.

The sequence of reactions involved in the subsequent oxidation is given in References 221 and 286 to which the reader is referred for detail. Here only pertinent steps will be mentioned to illustrate the process. Firstly there is a discharge of dissociated water ions to form radicals:

$$OH^- \rightarrow \ ^\cdot OH + e \tag{4.20a}$$

$$H^+ + e \rightarrow H^\cdot \tag{4.20b}$$

which react to form polymer radicals during the time scale of field reversal, thereby insuring irreversibility for the dissociation electrochemistry, i.e.

$$^\cdot OH + RH \rightarrow R^\cdot + H_2O \tag{4.20c}$$

$$H^\cdot + RH \rightarrow R^\cdot + H_2 \tag{4.20d}$$

Liberated electrons from 4.20a may also react with oxygen to initiate a sequence leading to peroxide radicals

$$O_2 + e \rightarrow O_2^- \tag{4.20e}$$

$$O_2^- + H^+ \rightarrow HO_2^\cdot \tag{4.20f}$$

$$HO_2^\cdot + RH \rightarrow H_2O_2 + R^\cdot \tag{4.20g}$$

$$2H_2O_2 \xrightarrow[\text{catalysis}]{\text{metal ion}} HO_2^\cdot + HO^\cdot + H_2O \tag{4.20h}$$

or the polymer radicals may react with oxygen to generate polymer carbonyls, hydroxides and acids via peroxide intermediates.

$$R^\cdot + O_2 \rightarrow RO_2^\cdot \tag{4.20i}$$

$$RO_2^\cdot + RH \rightarrow R^\cdot + RO_2H \tag{4.20j}$$

$$2RO_2H \xrightarrow[\text{catalysis}]{\text{metal ion}} RO_2^\cdot + RO^\cdot + H_2O \tag{4.20k}$$

$$RO^\cdot + RH \rightarrow R^\cdot + ROH \tag{4.20l}$$

The initial step in this process has been assumed[286] to be the electrochemical dissociation of water

$$H_2O \rightleftharpoons OH^- + H^+ \tag{4.20m}$$

and it should be noted that the concentration of the hydroxyl ion, which is essential to the sequence, will be increased as the basicity of the electrolyte solution increases. Thus this electrochemical oxidation sequence is in accord with the specific form of eqn. 4.20 which implies a rate constant (Q) proportional to the electro-motive force (e.m.f.) of an electrochemical reaction which generates hydroxyl ions.

In the high pH conditions which accelerate treeing, reaction 4.20a, 4.20c and 4.20e will be favoured leading to polymer radical formation. Oxygen also plays the important rôle of initiating two main sequences via eqn. 4.20e and 4.20i. Use of a nitrogen atmosphere can be expected to inhibit these reactions and hence the subsequent radical formation which would lead to the reduction of treeing observed experimentally[260,261]. Indeed it is difficult to see how else nitrogen could affect treeing except through suppression of partial discharges. Both main sequences involve a metal ion catalysed reaction which would be catalysed by the ions injected from the electrolyte into the polymer[293]. The oxidation processes (eqn. 4.20k) suffer a sharp reduction in rate when the catalysing ions are present in high concentration due to their involvement in termination reactions[95]. The observed sharp drop in water tree growth rates for high electrolyte concentrations[292,195] is therefore consistent with a metal ion catalysed oxidation process.

In many ways this electrochemical oxidation mechanism is in accord with experimental observation, however no detailed suggestion has been made as to how the oxidation produces the known form of tree damage. Zeller[262] has proposed that a microphase separation occurs with solvated ionic clusters condensing on lamella surfaces now hydrophilic due to oxidation. In view of the free radical and carboxyl formation[295] and the observed[282] Maxwell stressing of the treed region it seems probable that the process involved is somewhat more complicated. Zhurkov *et al.*[116,95] have pointed out that free radical formation can involve chain scission. This will lead either to microvoid formation through crosslinking in the case of polyethylene (see the scenario of Fig. 4.18) or to crazing via chain unzipping as in poly(α-methylstyrene). The surfaces of these defects will contain carboxyl and hydroxyl groups which will bind metal ions such as alkaline and alkaline earths in the form of ion clusters[262,296]. In forming the counterion pairs fully hydrated cations must release some water molecules from their hydration shell[297], and thus the process is favoured by low hydration entropies in accord with Morita *et al.*'s[269] data on tree initiation. As the cluster develops ion bridges will be set up which stabilise the defect surface by means of surface crosslinks[296,220]. An equilibrium will also be established between bound and free water, bound and dissociated ions, and the polymer contractile pressure[297] resulting in an inverted micelle-like[298] microvoid containing mainly pure water.

Such centres will contain strong dipoles[298] bound to the polymer which will allow a low frequency AC electric field to drive the segmental motions. In a large magnitude field the mechanical force developed will be substantial[222] though below the polymer yield point[262]. As discussed earlier (Section 3.1) such a stress will both generate free radicals along the polymer chains, and possibly free volume in the intervening region. The net effect will be

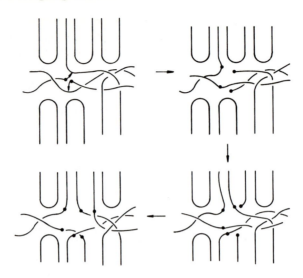

Fig. 4.18   Schematic representation of microvoid formation via the generation of free radicals through chain scission followed by a crosslinking chain reaction

the production of incipient oxidation sites in advance of the tree boundary together with a facilitation of the field driven transport of electrolyte to them via 'easy paths' opened up transiently in the amorphous region. Probably this transport mechanism will favour ions with smaller radii, although their greater tree propagation rates[217] may arise from other sources such as basicity or catalytic activity. Repetition of the overall process will allow the damage boundary to progress through the polymer.

One consequence of this mechanism is that the microvoids would contain water which has been de-ionised electrochemically with the metal ions of the electrolyte bound to the oxidised surface. While it is known from the DSC of frozen water trees[299], that the water can be strongly supercooled (freezing point = 230 K) indicating exceedingly small and pure water droplets, the state of the water in the microvoids can best be determined from the dielectric response[137,134]. Here a weak water sensitive loss peak (Fig. 4.19) has been observed at a frequency between 10^2 Hz and 10^3 Hz, which, it has been suggested[300], originates from the Maxwell-Wagner response of water-filled microvoids. Making use of Hanai's[301] model to relate the loss peak frequency[302] to the conductivity, σ_W, of the disperse phase, and given the dielectric permittivities of polyethylene and water together with the volume fraction, a value of $\sigma_W \simeq 2 \cdot 5 \times 10^{-6}$ S · m^{-1} was found. This exceptionally low value for water indicates the almost complete absence of an ionic component in the water phase consistent with the proposed mechanism. At frequencies above the peak the hydroxyl ion concentration available for electrochemistry at the surface will be reduced, and a decrease in Q/f can be expected such as is shown in Fig. 4.14. The same behaviour may however occur if the dielectric response originated with the polymer motions of polar regions, whose magnitude would increase on wetting and whose response time would relate to local packing and steric constraints.

At low frequencies ($f < 10^2$ Hz) the response defines the behaviour of the continuous (polymer) phase. Here the charge diffusion behaviour observed in unused polymeric insulation (see the $f^{-1/2}$ frequency dependent capacitance in Fig. 4.19a) becomes a transport process with a frequency dependent ($\sim f^{0.25}$) conductance in treed material (Fig. 4.19b). Such a dielectric response is typical of transport between clusters in a system lying below the

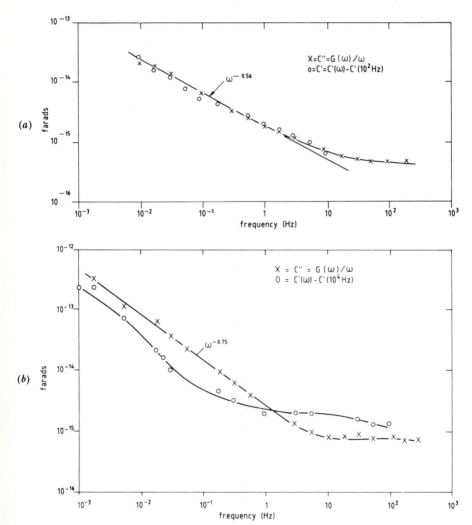

Fig. 4.19 Dielectric response of slices of cable insulation.
(a) Unaged insulation showing a response with a charge diffusion character.
(b) Insulation containing water trees. Response dominated by a dispersive transport process below ~ 10 Hz

percolation limit, see Reference 303 for details. The effective mobility observed for the treed regions is slightly greater than that of the diffusing entities and up to 10^4 times the DC conductivity expected for polyethylene. This is consistent with a percolation mechanism whereby electrolyte is supplied to the damage boundary via limited range easy routes rather than physical channels[262]. Such electrolyte supply is another essential component of tree growth in this picture, and may be the reason for the termination of the growth of bow-tie trees. In this case the supply of electrolyte is probably limited even though dielectrophoresis may ensure a continual water supply[218].

4.3.3 Water tree formation scheme

The present evidence is therefore in favour of a general scheme for water tree formation by means of the following combination of processes.

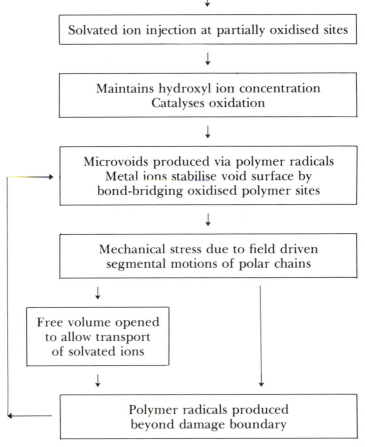

Throughout the preceding analysis stress has been laid upon the production of water tree damage. By defining the tree to consist of $N(\tau)$ microvoids with some characteristic volume V_v ($\propto a^3$) at time τ, the damage $V_v N(\tau)$

can be related to the tree volume, $V_T(\tau)$, through

$$V_T(\tau) = V_v g^{-1} N(\tau) \propto (L+r)^3 - r^3 \qquad (4.21a)$$

for a hemispherical tree. Here g is the volume fraction of microvoids within the tree (see eqn. 4.18a) and r is the effective radius of the initiating centre as in Section 4.2.2(a). In the following r may be taken to be the radius of curvature of a needle electrode, the radius of an initiating void, or that of a defect region (see Fig. 4.3c). A comparison of eqn. 4.21a to the growth equation (4.8b) with the typical value of 2 for p obtained when $L > r$, implies that the tree volume, $V_T(\tau)$, and hence $N(\tau)$ are proportional to τ with[304,252]

$$N(\tau) = \tau/(g^{-1} t_1 a^3 r^{-3}) = (\tau/t_v) \qquad (4.22)$$

defining a time t_v $(= g^{-1} t_1 a^3 r^{-3})$ which can be interpreted as the time required to form an average sized surface oxidised microvoid radius (a) at the tree boundary. If t_v is unaffected by the presence of other microvoids then the observable quantity t_1 will be proportional to g.

As long as the microvoid production rate remains constant when tree growth is fan-shaped or cylindrical, such as might occur through amalgamation around wedge-shaped defect electrodes, the growth law in this case will be given by

$$V_T(\tau) \propto r(L+r)^2 - r^3 \propto V_v g^{-1} N(\tau) = (\tau/t_1) r^3 \qquad (4.21b)$$

This expression has the form of eqn. 4.8b with $p = 1$ and could serve to explain the growth behaviour noted earlier for this type of defect.

When formulated in this manner water-tree growth is determined by the time, t_v, required to generate a microvoid, which is a function of g and t_1 both of which may vary with the growth conditions such as voltage, temperature etc. The voltage dependence of t_v can however be determined if g is taken to be proportional to the volume fraction of water in the body of the tree, since Meyer's data[211] shows that this is proportional to the voltage V. Thus

$$t_v = t_1 a^3/(A V r^3) \qquad (4.23)$$

$$= a^3/(3QAV^3) \quad \text{if } t_1 \propto V^{-2} \qquad (4.24a)$$

$$= a^3/(3QAV^2) \quad \text{if } t_1 \propto V^{-1} \qquad (4.24b)$$

where A is the proportionality constant relating g and the voltage, V. The expressions (4.24a) and (4.24b) for t_v are derived from the two possible voltage dependencies of t_1. A theoretical choice between these alternatives will require a detailed model for the way in which the e.m.f. of eqn. 4.19 drives the formation of microvoids.

Eqns. 4.24a and 4.24b also provide a means of relating propagation to the inception time since, if t_d is defined by extrapolating the growth curve to zero length, it may be regarded as the *apparent* time required to produce a single microvoid, i.e. t_v. In this case it is eqn. 4.24a which is most in accord with experiment[206], however it should be noted that the tree is not volume filling close to inception ($L < \sim r$) and the experimental value of τ/t_1 in this region is no longer proportional to the g^{-1} of the developed tree.

A recent investigation[276] of vented water trees has shown that the microvoid density (g/V_v) is essentially constant at the tree boundary but rises towards the base of tree. This result suggests that the tree advances by the production of a specific volume fraction of microvoids at its boundary, which is the sense of g in eqns. 4.18a and 4.21a, and that the same value will apply at the tree inception. However microvoids are also produced in the body of the tree contemporaneously with its advance and it is probable that it is their density that is proportional to the voltage. In this case the voltage dependence of t_v may be different to that of eqns. 4.24 and will depend on whether the time to produce a microvoid on the tree surface (t_v) remains unchanged or has to allow for the formation of all the extra microvoids in the tree body.

It would be reasonable to assume that the microvoid formation mechanism would be the same for bow-tie trees as for vented trees, and hence that t_v has the same form in the two cases. If this is so then the difference in growth behaviour[119] must arise from a time dependence in Q ($\propto g^{-1}$) which causes growth to saturate as the electrolyte supply of the initiating site is exhausted. Although the growth law of bow-tie trees is not at present established, the saturation length will be determined by the initial growth rate (i.e. the inception site conditions) when this is finite. Thus the saturation length of void initiated bow-tie trees (Fig. 4.3c) can be expected to depend upon the void size r, and Shaw and Shaw's[119] dimensionless growth curve implies that the dependence is linear up to[197] $r \approx 10 \ \mu$m. Data for void sizes up to 100 μm possibly indicate a sub-linear dependence[305], however the large scatter and the different fields experienced by the voids in the co-axial cable probably account for the deviation. The r dependence of the saturation length (and initial growth rate) provides a convenient physical origin for the length distribution observed[119]. An additional distribution of inception times cannot however be ruled out[197] and in some cases the number density has been observed to increase[119] continually up to[201] 1800 hours, even though the characteristic length had saturated much earlier.

The relationship between r and the saturation length does not necessarily mean that the tree length is controlled by the void contents and it has been suggested that access to the water content of the polymer influences both the number density and the length[306]. It is also known that the shape and hence length to volume relationship of bow-tie trees varies with frequency, and hence it is possible that the best description of bow-tie tree growth may be in terms of the total bow-tie volume as a function of time e.g. Fig. 4.20. At present however the analysis of bow-tie growth is insufficiently advanced to allow comparisons of t_v to be made between them and vented trees.

In the light of the evidence to date it appears that an electro-chemico-mechanical mechanism[294] is better able to describe the known facts of water tree inception and growth than any other yet proposed. However much more work is required to establish this contention and, in particular, the electrolyte dependence must be confirmed. Now the analytical description of water tree growth, centred on eqns. 4.21a and 4.21b makes it possible to relate the predictions of a given microscopic model to the observed macroscopic growth behaviour through t_v (eqn. 4.22), thereby allowing such predictions to be tested.

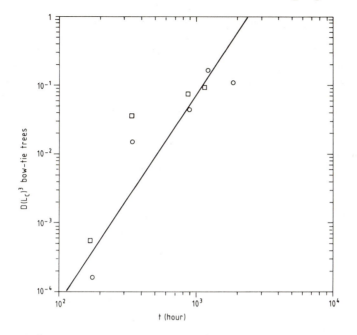

Fig. 4.20 $DL_c^3 \propto$ Bow-tie tree volume (mm³) per mm length of cable. L_c = characteristic length, D = tree density. □ Cable with outer semiconducting sheath; ○ Cable stripped of outer sheath

4.4 The gross morphology of water trees

The appearance of water trees can be quite varied and has in the past led to a range of epithets such as broccoli, bush, fan, etc. In many cases they resemble objects termed fractals[307] which are a subject of intense modern interest. Compare for example Fig. 4.7 with the illustrations of fractal forests given by Liu[307] and Meakin[308]. Fractals of this type are objects which are self-similar. This means that the relative geometrical arrangement of the complete object is reproduced on a smaller scale in its substructures, and hence magnification of a portion of the object up to the overall size will reproduce the original object (see Fig. 15.8). Such entities will have a fractional dimension (d_f), such that their volume L^{d_f} will be less than the space they occupy, L^3. Their density, L^{d_f-3}, will reduce, and consequently the unoccupied space will increase, with increasing size. In this case a process in which growth occurs by the addition of a succession of volume elements, one for each time element will lead to a growth law of the form

$$L^{d_f} \propto \tau \qquad (4.25)$$

In comparison eqn. 4.8b shows that within the limit of experimental error, vented water trees in the size range $r < L < D$ are compact space-filling entities, in contrast with the fractals which occupy a space without completely filling it. Eqn. 4.8b has been verified for uniform-field plaques and cables as well as divergent field conditions, and hence in most cases we must accept that water trees grow isotropically on a global scale in this region of size. It has also been suggested[309] that the microscopic morphology of water trees should exhibit a microvoid pattern[262] of fractional dimension such as is found in the channels of electrical trees[289,304]. No direct evidence for this contention has been obtained as yet and insofar as such a pattern would imply a structured form of growth, eqn. 4.8b is contrary to its existence.

The initial stage of growth ($L < \sim r$) however is in general not isotropic. Fig. 4.7 gives uniform field examples which may have a fractal form for which propagation is by a stochastic addition of damage at potential growth sites with a probability governed by the site conditions[251,310]. The bouquet of microvoids observed by Chen *et al.*[276] in spherically symmetric divergent fields is of a similar form. Among the factors giving rise to a spatial variation in the growth probability of sites are internal strains and applied mechanical forces which generate preferential anisotropic paths for electrolyte transport and progressive microvoid formation[119,223,232]. However the anisotropic propagation produced in these cases is eventually overcome and a typical compact shape formed[276,268] albeit with features such as striations overlaying it.

Large trees ($r \ll L < \sim D$) also have a tendency to develop a structure, particularly in cables where bifurcation occurs to give shapes like that of a hand, Fig. 4.21. Such shapes are visually similar to the viscous fingering which can occur when a fluid of low viscosity is forced into one of high viscosity[311,312] (μ). Here the viscosity difference causes a hemispherical boundary to be unstable to fluctuations, an effect opposed by the interfacial tension[311] (σ). The fluctuations therefore will only grow into fingers when the surface boundary ($\propto L$) is greater than a critical fluctuation wavelength[311], λ_c, with

$$\lambda_c \propto \left[\frac{M\sigma}{v}\right]^{1/2} = \left[\frac{k_p}{N_{ca}}\right]^{1/2} \tag{4.26}$$

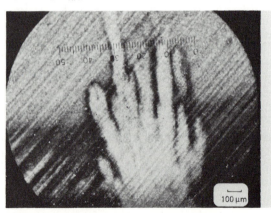

Fig. 4.21   Water tree observed in a cable exhibiting bifurcation as it grows into the water halo. Note that the abstraction of water from the halo gives dark regions surrounding the fingers of the tree
After Dissado *et al.*[206]. Copyright © 1983 IEEE

where σ is the interfacial tension, k_p the permeability and v the velocity of the boundary. The factors M $(= k_p/\mu)$ and N_{ca} $(= \mu v/\sigma)$ are the mobility of the displaced fluid[311] and the capillarity number[312] respectively. Essentially fingers grow faster than the median tree surface, because for $\lambda < \lambda_c < R$ the reduction in tip widths increases tip pressure sufficiently above the average value for the extra interfacial tension to be compensated.

Strictly speaking a water tree is not a liquid with a boundary determined by surface tension being forced into another viscous liquid. We may regard it however as a fluid phase of mobile solvated ions being driven into the amorphous component of the polymer, which it displaces to form microvoids, under the action of a pressure gradient as expressed in eqn. 4.18b. This picture possesses some of the features of models in which a low viscosity liquid displaces the high viscosity liquid from a porous medium[310,251,313,314]. In our case the displaced fluid is the amorphous polymer from the microvoids whose resistance to the deformations and chemical reactions involved in microvoid formation will be equivalent to a viscosity, μ. The mobility of the displaced component governs the advance of the tree boundary and may be determined by the permeability (k_p) of the polymer to the electrolyte as well as by the microvoid formation resistance. Thus the magnitude of t_v will be related to these two quantities. The interfacial tension of the tree boundary σ will depend upon the energy required to form the surfaces of the water-filled microvoids at the tree tip, and will thus depend upon the microvoid density, g. Local variations in any of the factors[312,313,314] will give fluctuations which can lead to fingering. Even when these are constant local differences in the driving force appearing as noise may give fractal structures[314], which appear at all size scales if the interfacial tension is zero.

On the basis of these equivalences it is not unreasonable to expect water trees to show a fractal aspect during their initial stages of growth for it is here that local variations can be expected to have the greatest influence. Firstly because only specific local sites allow inception, secondly because the next few microvoids are either chosen stochastically or as a result of favourable conditions, and thirdly because the material boundary region is probably not isotropic in material properties. During this early period the bigger trees will 'shield' the smaller ones, reducing $\Pi(E)$ at their tips and inhibiting further growth. However unlike typical fractal aggregations the tree structure does not continue to be determined by its initial form. This is probably because microvoid formation is not restricted to the tree tips and may continue in the body of the tree simultaneously with generation on the boundary. Given the slow advance of the damage boundary and the probably more uniform material properties, the tree is likely to approach an isotropic structure such as would occur for a low noise situation even when σ is zero[312].

In the next stage of growth the advance of the damage boundary occurs by the addition of volume elements of microvoids in each successive time interval. Incipient tips will either be unstable and smoothed out[312] or occur at many sites and grow at similar rates since t_v is a constant, thereby retaining a compact near homogeneous form for the tree. It should be noted that for an object grown in three-dimensional space a fractal aggregation would be expected to have a dimension of $\sim 2 \cdot 5$, denoting a rather compact figure whose dimension it would be difficult to establish experimentally either from visual inspection or from its expected growth law ($L \propto \tau^{0.4}$).

In our picture of water tree propagation therefore the advance of the tree will be through the formation of the minimum volume fraction of microvoids at the boundary such that the advance will be visible and access to the new boundary maintained. It is this meaning that should be ascribed to g (and $N(\tau)$) in eqns. 4.18b and 4.21a. During the stage of compact growth, g can be expected to be the same whatever the length of tree, even though the microvoid density behind the boundary will increase[211,276]. Those factors which can be expected to influence g will be the microvoid size and the requirement to maintain a percolation system in equilibrium with electrolyte flow.

Although incipient fingers will finally emerge in large growths under low noise (i.e. near spatially uniform) conditions[312], it is more likely that the bifurcations observed in large water trees are the result of local changes in the factors that govern t_v i.e. the equivalents of k_p, μ, and σ in eqn. 4.26, since such structures are rarely observed in plaques or point-plane geometry. In these latter laboratory samples care is taken to ensure uniformity of material morphology which may not be found in extruded cables. Local variations in microscopic permeability[312,314] and deformation resistance will occur under the influence of frozen stresses[264] and morphological changes, causing the electrolyte mobility to fluctuate. A non-uniform dispersion of antioxidant will cause variations in the resistance to electro-chemical reactions ($\equiv \mu$) and electric stress concentrations at defects or mechanical fluctuations will induce fluctuations in the 'pressure' $\Pi(E)$. Even an increased access to a local electrolyte supply may generate the fluctuations which result in bifurcation into fingers. This latter process is probably illustrated in Fig. 4.21 where the fingers have not amalgamated with the water halo but instead extracted water from it by dielectrophoresis leaving an untreed rim between themselves and the halo into which they have grown.

This stage of growth is of major importance with respect to the influence of water trees on dielectric failure. When fingers are formed they grow more rapidly than the median tree due to the concentration of growth favouring influences within them, and the data of Maloy et al.[313] gives an example of the acceleration effect than can be expected. Such accelerations have been observed for water trees both in the laboratory[252] and in cables[249,253], where it has been associated with a reduction in breakdown strength[253]. Simple calculation shows that the observed accelerations cannot be due to an increase in the tree tip field as the central conductor is approached and hence it is likely to be the result of fingering. Fig. 3.6 shows that the influence of such behaviour upon the breakdown strength is probably due to the generation of electrical trees by very long fingers as they approach the central conductor. It is therefore clear that any effort to reduce the deleterious effect of water trees upon polymeric insulation should include an investigation of the factors that cause water trees to bifurcate and produce fingers.

Chapter 5

Electrical tree degradation

5.1 General characteristics of electrical treeing

Electrical trees found in polymeric insulation grow in regions of high electrical stress, such as metallic asperities, conducting contaminants and structural irregularities[121]. Partial discharges in voids can also generate degradation structures from the void surface which are essentially electrical trees[126,127]. Those trees initiated at an electrode are termed vented trees, while trees generated in the body of the insulation have a branched channel structure roughly oriented along the field lines and are referred to as bow-tie trees[121,119]. In this case the term vented tree is an apt one as electrical trees are composed of inter-connected hollow tubules with the channel at the base forming a vent for the whole system which may in some cases allow access to the external environment[178,315], or to the atmosphere of the void for partial discharge initiated trees[129,178,316,126].

Not all vented trees however have the same shape. Three sub-categories can be roughly distinguished, which may be referred to as tree-like or branched, bushy, and bush-branch[129,178,317,318], and these shapes are illustrated in Fig. 5.1. Branched trees exhibit a multiply branched structure with the channel diameters ranging from tens of microns ($\sim 30\ \mu$m) in the trunk to around one micron ($\sim 1\ \mu$m) in fine filamentary channels at the tip. In bush-type trees the hollow tubules are so densely packed as to appear as a solid bush-shaped mass when viewed from the side. Bush-branch trees are essentially bush trees with one or more branches projecting from their boundary.

Noto and Yoshimura's data[129] indicate that for trees grown in polyethylene with AC electric fields at 50 Hz, branch-type shapes were formed below $\sim 5 \cdot 4 \times 10^8\ \mathrm{Vm}^{-1}$, bush-type trees between $5 \cdot 4 \times 10^8\ \mathrm{Vm}^{-1}$ and $6 \cdot 0 \times 10^8\ \mathrm{Vm}^{-1}$, and bush-branch trees above $\sim 6 \cdot 0 \times 10^8\ \mathrm{Vm}^{-1}$, with these values reducing as the frequency increased. Laurent and Mayoux's results[317] confirm the trend of a cross-over from branch to bushy type trees with increasing field, however bush-branch type trees were grown at a lower field ($\sim \frac{1}{4}$) than the other two. On the other hand these and other[318] sets of data coincide in their tree shape sequence if applied voltage rather than field is the determining parameter (i.e. branched $V < \sim 13$ kV; bushy ~ 13 kV $< V < \sim 20$ kV; bush-branch $V > \sim 20$ kV at 50 Hz).

Electrical trees have been studied in the laboratory using a variety of stress enhancing geometries which have been summarised by Mason[122]. The most typical construction is that of a point-plane geometry with metal needles[121], however flexible semi-conducting needles have sometimes been

(a)

(b)

(c)

Fig. 5.1 Illustration of different shapes of electrical trees (*a*, *b*, and *c*), and some of their features just prior to breakdown (*d* and *e*). The trees were grown in an epoxy resin and are reproduced by courtesy of J. Cooper, National Power TEC, Leatherhead, UK.

(*a*) Branch-type tree
(*b*) Bush-type tree
(*c*) Bush-branch tree

used[288,318,319] in order to avoid the formation of cracks, crazes, and mechanical stress during preparation of the system prior to testing. Using these techniques it has been shown that electrical trees grow not only in AC fields[199] but also under DC conditions where voltage ramps[287], short cir-

(d)

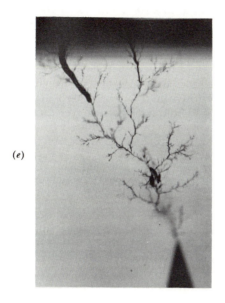

(e)

Fig. 5.1
(d) Bush-branch tree crosses the insulation just prior to catastrophic failure
(e) Beginning of catastrophic failure initiated by a branch-type tree. Note the widening of a branch on the left, the substantial widening of a branch on the right as the arc-type discharge begins to propagate, and the widening of a few channels in the body of the tree

cuit[287], polarity reversal[287], impulses[320], and constant DC voltages[316], have all been applied. Although some differences in inception[287,316,129], propagation[163,320], and tree shape[320,321] occur, electrical trees are essentially the same form of degradation under AC and DC conditions unlike water trees.

In common with water trees inception and propagation are identifiably separate stages[129] in the generation of electrical trees. When investigated under constant amplitude stresses, inception is characterised by an inception time, defined as the time required to generate an observable tree (usually about 10 μm in length). This time is strongly dependent upon the electric field strength[122], typically having a form E^{-a} with $5 < a < {\sim}20$, in contrast to the value of 3—4 observed for water trees[206]. An example for AC stresses taken from the data of Noto and Yoshimura[129] is given in Fig. 4.16, where it is compared to the weaker field dependence of the time required for the formation of a microvoid string at the start of water treeing. The inception time is also essentially inversely proportional to the frequency under AC stresses[129].

Data such as this implies that, though initiation is the result of a repetitive process in AC fields, it is generally much more field (voltage) dependent than microvoid formation in water treeing. With DC fields it is more common to define an inception field by ramping the voltage until a tree is initiated[287]. Under slow ramp rates ($< 5 \times 10^9$ V $\cdot$ m^{-1} s^{-1}) the inception field usually considerably exceeds the uniform DC breakdown strength (${\sim}5 \times 10^8$ Vm^{-1} in polyethylene), a fact ascribed to the moderating effect of injected space charge[287] upon the local field. At fast ramp rates ($\gg 5 \times 10^9$ V $\cdot$ m^{-1} s^{-1}) the space charge region is not allowed sufficient time to form completely and the inception field reduces to a constant value which is nonetheless often in excess of the breakdown field[133]. Although such a result may be taken as evidence of a threshold field for electrical tree inception under DC stresses it should be noted that trees can be initiated by AC stresses whose peak field is somewhat below the DC breakdown strength[129]. Furthermore if the tree is initiated by partial discharges in a void at the needle tip the inception field will depend upon the void size, ranging from greater than 10^9 Vm^{-1} for sub-micron sized voids down to around 3.5×10^6 Vm^{-1} for large voids[122] bigger than about 1 mm.

Trees grown from the larger voids can be clearly distinguished from electrical trees due to metal needles, however small voids of 2 μm or less are not easily detectable and would allow for tree inception fields below the DC breakdown value. The partial discharge inception field for large voids thus provides an absolute threshold field below which electrical trees cannot be initiated, in contrast to water trees where it is possible that it is only a matter of waiting sufficiently long to observe them at any level of stress[128]. A comparison between needles and needle cavities shows that the time to inception is reduced when the cavity is present[129], and this is probably also the case if cracks exist adjacent to the needle[322]. These observations almost certainly reflect the greater ease of generating a destructive discharge/avalanche in the gaseous contents of a void[129] or craze[322] than in the polymer[288,323].

In its first stage electrical tree propagation exhibits a decelerating growth rate[324,178,163], as do water trees. Eventually however an acceleration of the propagation rate occurs leading ultimately to failure[178,163,325]. A schematic representation of this form of behaviour is shown in Fig. 5.2. When tree propagation is examined in greater detail it is observed to be intermittent in form[121] with bursts of growth decaying to a standstill followed by further

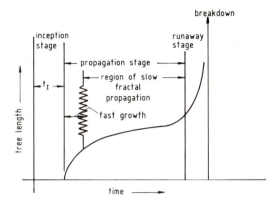

Fig. 5.2 Schematic representation of electrical tree growth

growth activity. This is particularly evident in the earliest stage of propagation under AC stresses where the generation of the first branch[326] is a
self-limiting process obeying the growth equation

$$L(t) = L_m(1 - \exp\{-(t - t_I)/\tau_e\}) \tag{5.1}$$

for the tree length $L(t)$ grown during the period $(t - t_I)$ following inception
at t_I, see Fig. 5.3(*a*). Here L_m is the final length of the first branch during
the initial burst of growth.

Subsequent to the formation of the first branch ($L_m = 11 \cdot 2\ \mu$m) tree
growth restarts as two new branches are added to the tip. Thus the inactive
periods can be associated with the initiation of tip splitting[326]. Extinction of
channel discharges is the reason usually advanced[121,129,178] for the cessation
of growth, however in this case discharges still continue on the positive
half-cycle. It is therefore possible that Wasilenko's proposal[325] that a threshold tip field is necessary for the continued propagation of a channel is
operative here, with the observed discharges originating from beyond the
channel tip. The fields determined by Wasilenko for crosslinked polyethyl-
ene with pulsed DC measurements (i.e. $\sim 8.5 \times 10^8\ \mathrm{Vm}^{-1}$ for positive polarity
or $\sim 10^9\ \mathrm{Vm}^{-1}$ for negative polarity) take no account of the space-charge
effects but are close to those often found for tree inception.

Once a number of branches have been formed it is possible for some to
be growing while others are inactive, and the individual growth curves of
each branch will be subsumed to give a propagation curve appropriate to
the overall tree structure. This behaviour may continue right up to the
runaway stage for both branched and bush type trees[163,317], however in
some cases an extended period of slow growth occurs prior to runaway. It
is possible to regard this stage as a large scale version of the initial process
with a bulk region in which channel formation has ceased initiating new
rapidly-growing branched structures from just a few of its tips. However
in most cases it is not possible to describe the pre-runaway propagation by
the self-limiting growth of eqn. 5.1, see Fig. 5.3(*b*). An inspection of the
trees appropriate to this data suggests that during the period in which the
overall tree length hardly changes, branches were initiated successively from

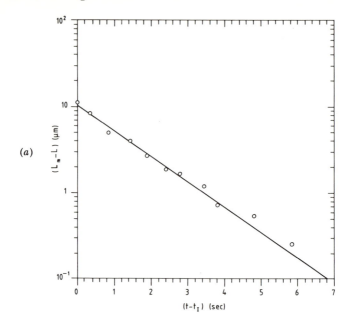

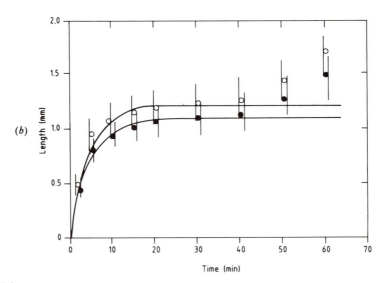

Fig. 5.3

(*a*)   Semi-log plot of the growth, L, of the first channel of an electrical tree, taken from the data of Reference 326. These results fit eqn. 5.1 with $t_I = 218 \cdot 17$ s, $L_m = 11 \cdot 2$ μm and $\tau_e = 1 \cdot 48$ s.

(*b*)   Electrical tree growth in XLPE[324] with the continuous lines given by eqn. 5.1 with zero for t_I. The experimental data is shown with their error bars as: XLPE 1: ○; XLPE 2: ●; and the fitted curves have values for τ_e and L_m of 4·79 mins and 1·23 mm for XLPE 1 and 4·53 mins and 1·1 mm for XLPE 2

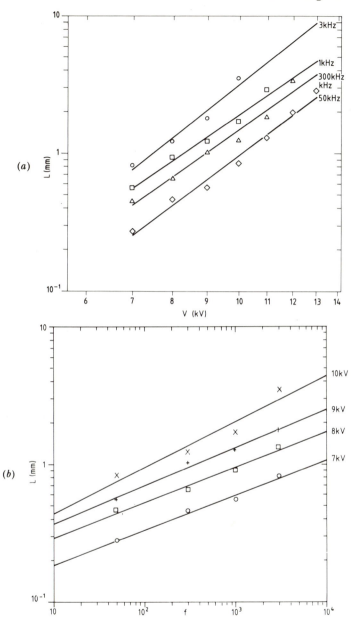

Fig. 5.4 Length of electrical trees after 60 minutes growth:
(*a*)   as a function of voltage; $L \propto V^{3 \cdot 8}$
(*b*)   as a function of frequency; $L \propto f^{1/3}$.
Data taken from Reference 129. Note that the results shown in (*b*) should not be
confused with those in Reference 329 where the length of the tree at inception is
given as a function of frequency. Because the time to inception and hence inception
voltage (in ramp experiments) decreases with increasing frequency, this 'initial' tree
length also reduces with increasing frequency

near the base of the tree and grew to similar lengths. It therefore seems important to measure separately the behaviour of sub-structures when investigating tree propagation dynamics[327].

One reason for the temporary cessation of tree growth may be the lack of a vent to release gaseous decomposition products[178] to the external environment[129]. This will be particularly the case for bow-tie electrical trees[121] and vented trees at a metallic asperity[129], rather than those initiated at voids[129] or at contaminants in the semiconducting screens of a cable[328]. Experiments with an introduced cavity as a vent can thus be expected to yield information on tree propagation for which only electrical effects are active in causing intermittency. This 'intrinsic' propagation behaviour has been studied by Noto and Yoshimura[129], and their data can be used to show that the tree lengths grown in the fixed time of one hour are voltage ($\propto V^4$) and frequency dependent ($\propto f^{1/3}$), see Fig. 5.4. A comparison with the results of Yoshimura *et al.*[324] and Ieda[163] shows that this region of time corresponds to tree propagation in the stage just prior to runaway. Hence the behaviour obtained refers to 'intrinsic' tree propagation at a stage when the build up of gaseous decomposition products might otherwise restrict growth.

The final stage of tree growth exhibits an accelerating propagation[178,317], corresponding to a runaway development of a tree branch from one of the existing tree tips. Very often this terminal growth is filamentary[317] (channel diameter ~ 1 μm), jagged but unbranched. Such features are similar to those of electrical trees grown in dielectric liquids[210,330,331] and which lead rapidly to failure. However in solid polymeric materials breakdown is not always an instantaneous (nano-second) process subsequent to the tree bridging the insulation, and can in some cases take up to one hour[178,332,333]. A more detailed discussion of the terminal process will be made in Section 6.2.

The study of electrical trees has two main advantages in comparison with water trees. In the first instance the trees are usually clearly visible and their growth can be observed *in situ* at the same time as other experimental investigations are carried out[288,318]. More importantly however the greater degree of degradation requires a larger amount of electrical power and gives rise to a wider range of observable physical effects. Thus the process of electrical treeing has been followed by such diverse techniques as: acoustic emission[334,335]; current pulses[288]; light emission (Bamji *et al.*[130,336] and references therein); electron spin resonance (ESR) and thermally stimulated discharge currents (TSD)[157]; thermally stimulated currents (TSC); thermoluminescence and electrode polarity effects[287,163]; FTIR[157]; and partial discharges and gas emission[315]. As a result of the data obtained by these investigations a broad outline for the mechanism of the formation of electrical trees has been established and, although further work is required to clarify some of the details, it now seems unlikely that the main features will be challenged.

5.2 Electrical tree inception

5.2.1 The role of space charge
The earliest theories of electrical treeing regarded inception as a local

breakdown process occurring as a result of the field enhancement of the needle electrode[199]. Because of the field non-uniformity the breakdown value would only be exceeded over a limited region around the needle tip and this would define the length of the initial tree channel. The requirement of inception fields in DC stresses substantially greater than breakdown values was explained by the presence of a homo-space-charge which moderated the field enhancement at the needle tip[287]. Under AC stresses initiation occurs below the breakdown value where it was considered that charge injection would no longer occur[129]. Instead it was proposed that either dielectric heating would generate a void at the tip[129], or Maxwell stressing would cause cracking around the needle[129,319,163]. In these circumstances the lower dielectric strength of the contents of the void or crack or craze[323] will allow the commencement of discharges and treeing.

It is clear that cracks or crack-susceptible regions may occur as a result of differential mechanical stresses arising from material preparation, particularly during the hardening of epoxy resins[322] and the cooling of polyethylene to low temperatures (~ 77 K)[319], and that these features will reduce inception fields and times to inception. However in their absence space-charge injection cannot be ruled out since this is now known to occur at fields of the order of one fifth to one third of the breakdown level[337,338,339] in homogeneous dielectrics, and about a tenth of this value in inhomogeneous materials. Estimation of the critical field for space charge injection (E_{sc}) yield values of $\sim 3 \times 10^8$ Vm^{-1} for epoxy resins[338] and $\sim 10^8$ Vm^{-1} for polyethylene[323,339,340]. This low value for polyethylene is consistent with the DC field dependence of the capacitance of thin polyethylene films[341]. Here it was found that below 3×10^7 Vm^{-1} the incremental capacitance was due to compressive Maxwell stresses, whilst between 3×10^7 and 9×10^7 Vm^{-1} polarisation of the trapped space charge contributed, and above 9×10^7 Vm^{-1} mobile injected charge carriers dominated producing irreversible effects. Thus AC stresses whose peak-to-peak value is greater than twice the critical level can inject and extract opposite sign charge carriers on alternative half-cycles[342,340]. Since some portion of the injected charge will be retained in deep traps[163] during the subsequent half-cycle[319] a heterocharge can be built up[342,343] which can cause the breakdown field to be exceeded[316] for a part of the cycle. Tree initiation can then proceed via a local breakdown[288] as in the DC case[287].

The most direct evidence for the existence of injected space charge is given by capacitance measurements[342] with a resolution of femtocoulombs[337,338,339] which have enabled the critical field for space-charge injection to be determined for a number of materials[340]. A different method is to measure the current flowing when the space charge is released to the electrode, and this has been carried out by laser stimulation[342] or by thermally stimulated discharge (TSD) currents[157,344]. This technique measures the charge trapped rather than that injected and yields some information as to the type of trap. It has also been shown that the injection of space charge coincides with the onset of luminescence[345]. Thus it is also possible to follow charge injection by monitoring the electroluminescence intensity[336,319].

According to the field-limited space-charge (FLSC) model[346,347] (Section 9.3.4), charge which can be injected from the electrode above a threshold field, can only move into the insulation once a higher critical field (E_{mc}) is exceeded. The subsequent penetration is limited by the requirement that the field at the space-charge boundary must be kept at E_{mc} if it is to advance. In equilibrium the space-charge field will prevent further injection at the electrode surface and have a boundary value less than E_{mc}. The size of the space-charge cloud will increase with the voltage applied to the needle[347]. At the fields typically used for treeing experiments, i.e. $< \sim 10^9$ Vm^{-1}, the penetration depth has been estimated[316,342,323] to be $< \sim 10$ μm. This figure is consistent with the length of trees formed by short-circuiting both electrodes, after DC stressing at these fields[287] (i.e. short-circuit trees).

The link between space charge and treeing can best be seen through DC pulse experiments where prestressing with the same polarity increases the inception voltage[348] in contrast to opposite polarity prestressing which reduces the inception voltage[348] and increases the tree length[287]. That negative space charge is injected from the needle is demonstrated by the dependence of the negative short-circuit tree length[287] and the tree inception voltage[342,349] upon metal work function and UV illumination[329]. These results cover a range of fields from 4×10^8 Vm^{-1} to 4×10^9 Vm^{-1}. In contrast the positive polarity short-circuit tree appears to be independent of work function, thus in the case of polyethylene 'holes' are not injected. Here the concept of a hole refers to the extraction of electrons from the ground state molecular orbitals of the polymer. However polarity differences in the electroluminescence intensity[343,319] imply that the positive space charge in polyethylene is not just bulk ions produced by impact ionisation. The most likely origin is thus the ionic dissociation of neutral centres and extraction of the electrons by the high convergent field of the positive electrode[329].

Short-circuit trees give an idea of the extent to which the space charge has penetrated the polymer on injection, and in polyethylene it is clear that electrons are more mobile than positive space charge[287]. Trees grown under an impulse however may not show such polarity differences for which the electron space charge will be spatially homogeneous until a channel starts to form[312], whereas the positive space charge is probably non-uniform and field enhancing. Examples of this behaviour are found for PMMA[321] and epoxy resins[320]. It should be noted that pulse sequence experiments[287,348] indicate the existence of two timescales in the dissipation of injected charge since a rest time of $\sim 10^2$ sec eliminates the pre-stressing effect on same polarity pulses whereas $\sim 5\cdot 10^3$ sec is required for alternate polarity pulses.

5.2.2 Alternating fields

Although trees can be formed by impulse injection of electrons it is more instructive to consider their inception by reversal from negative to positive polarity, since this is the manner in which trees are initiated under AC stresses. Here pulses of electroluminescence, which is the first observable process[130,319,350,351,352,343], occur either prior to tree inception[319], or at lower voltages in ramp experiments[130,336]. This process does not occur in DC fields[343,316] and at fixed AC stresses does not necessarily lead to electrical trees[336,316,322]. However a region of degeneration ($\sim 5~\mu$m) at the needle tip is irreversibly produced[322,319] during its action. The emission spectrum is weakly dependent upon polymer morphology[343] and differs from void discharges[275] in lying predominantly in the visible range[316,322,343]. It has therefore been attributed[343,319,336], to the recombination of injected space charge with the residual deep trapped charges retained from the previous half-cycle[353]. The onset-field for luminescence is thus determined by the barrier to charge injection which is reduced by the presence of electronegative gases[350] (such as air, O_2, or SF_6) which form acceptor states[336] at the interface, however the process can still occur in degassed specimens[353], albeit at higher applied voltages.

During injection a space charge is formed in shallow traps[343] ($n_{st} \approx 10^{14}$ mm^{-3} in PE[343,354]) within milliseconds[316] and penetration to the recombination centres ($n_r \approx 10^8$ mm^{-3} in PE[343]) takes place in the space-charge-modified field. Hence the apparent onset field increases as the range of space-charge injection increases (i.e. as the needle tip radius decreases[130,323,340]). It has also been shown[343] that the injection of homogeneous negative space charge is easier than positive charge[287,322, 329,340] in agreement with TSD data[342], which provides an explanation for the higher tree inception field in negative polarity DC experiments compared to those of positive polarity.

The injection current, which is proportional to the luminescence intensity, follows the Schottky emission law at the lowest field[336] but converts to Fowler–Nordheim tunnelling emission through the Schottky–Mott barrier at higher fields[336,343]. It has been suggested[70] that during injection electrons can reach a kinetic energy of 3—4 eV, which is sufficient to cause bond scission. Model calculations on divergent field injection of single electrons into the linear hydrocarbon n-$C_{37}H_{74}$, show that the kinetic energy achieved is strongly peaked at a distance of ~ 10 nm from the electrode tip[355], with the peak value exceeding 4 eV for fields in excess of $\sim 7 \times 10^8$ Vm^{-1}. Typical luminescence onset-fields lie below this value and thus it is likely that only a few bonds are broken during injection. The light emitted during recombination also has insufficient energy to break bonds, except for the small UV component[353], unless this is the residual fraction of a strongly absorbed emission.

In polyethylene the recombination centres are located in the amorphous-crystalline interface[163,106] and some structural modification such as free-volume generation[94,356] may take place there as a result of the electrostatic forces involved in trapping and recombination. In degassed (or N_2 filled) polyethylene therefore the deterioration zone[357] may be composed mostly of minor structural rearrangements on the amorphous-crystalline boundaries where the deep trapped charge is located[106].

The presence of oxygen (e.g. in air) complicates this picture. Here oxygen in the metal-polymer interface may be converted to an active[358] or excited state[130] (by the field, injected electrons, or UV quenching[358,130]) and hence cause bond scission[95] via oxidation[318,359]. At room temperature[318] where the excited molecules can diffuse and polymer segmental motion take place the deterioration may develop as far as the formation of microvoid strings close to the tip. In contrast at[316,319] 77 K the polymer motion is very restricted and some chain scission may take place without the opening up of microvoids[318,360]. Thus oxygen may cause tree-promoting damage particularly at room temperature. This is not always the case, for example in epoxy resins at room temperature (in which diffusion and segmental motions are severely restricted) the deterioration zone[351,322] may be due to enhanced curing, and could even be conductive enough to moderate the stress[322]. Only prolonged stressing will cause void formation[351] possibly by shrinkage or more likely as a result of local Joule heating.

In general the true onset of tree inception is the presence of current pulses rather than luminescence pulses. Such pulses have been observed in polyethylene prior to a visible tree channel, and in their earliest stage occur only on the positive half-cycle and with a very small amplitude[288,326,361] (0·04—0·3 pC). These current pulses must be associated with the extraction of the electrons which are retained into the succeeding half-cycle in deep traps. Their charge density has been estimated to be 10^{-3}—10^{-2} pC $\cdot$ μm^{-3} although the volume it occupies may vary[316,344,342] ($< \sim 10^3 \ \mu m^3$).

Extraction capable of giving current pulses will only occur when the local field due to the heterocharge has reached a magnitude sufficient to generate an avalanche[288], which will start to erode the polymer[163] in a manner first suggested by Ashcraft[199]. In this picture the electrons gain enough kinetic energy to ionise the polymer (> 10 eV), form cation radicals (< 10 eV) and break polymer bonds (~ 3.5 eV). Capture of thermal electrons by the cation radicals will lead to the formation of free radicals[362] which may then cause chain scission[320]. The type of degradation associated with these processes will depend upon the polymer as noted earlier[95], for example poly(α-methylstyrene) may unzip giving a craze or crack, PMMA may decompose and polyethylene is likely to form microvoids[357] by crosslinking[118], see Fig. 4.18. Because of the non-uniform extracting field the degradation will converge on the needle tip and a string of microvoids aligned along the field lines will be produced[276,318]. Maxwell stresses may be involved in converting the crosslinking to microvoids[163,324,363], which however will be too small ($< \sim 10$ nm) at this stage to sustain partial discharges[276,122]. Only under the severest cases of mechanical stressing, for example metal needles in samples of polyethylene under vacuum, will a sizeable ($\sim 3 \ \mu m$) void develop[318].

Repetition of the avalanche extraction in an AC field causes crazes[163,326] to develop via a microvoid coalescence[364] or extended unzipping. If electroluminescent deterioration in oxygen has previously occurred the subsequent treeing will start from this point. During this stage the current pulse magnitude will increase as the charge taking part is accumulated through cycling. Eventually the degradation and space-charge field will have accumulated sufficiently for current pulses to occur in both half-cycles[288,326], typically of a magnitude $< \sim 1$ pC in polyethylene. The discharge pulses on the negative half-cycle are probably the result of avalanches caused by injected electrons, however these avalanches will take place in the partially developed craze which contains considerable 'free volume'. They therefore have some of the characteristics of a gas discharge with their local onset field being dependent upon gas species and pressure[365,323]. At this stage burst-type acoustic emission can be detected in polyethylene and PVC[334], which most probably corresponds to the yielding of the polymer as the craze is converted into a microcrack[366]. The tree is now just within the limits of detectability in polyethylene[326], i.e. ~5 μm long and < ~1 μm in diameter, and thus the onset field for 1 pC pulses may be taken as the tree inception field. When the sample contains crazes prior to treeing the time required to reach this stage will be appropriately reduced[322].

5.2.3 Energy threshold approach

By assuming that the initial tree channel is produced by repetitive avalanching, Tanaka and Greenwood[342] have conjectured that the time to tree inception is given by the time required for a given amount of energy C_t to be transferred to the polymer, i.e.

$$ft_I[G_n - G_{th}] = C_t \tag{5.2}$$

Here G_n is the energy available from displacement of the injected charge in the local field, and G_{th} is its threshold value for damage producing avalanches, with ft_I, the number of cycles. C_t can be expected to be a material property, and its value is probably related to the tensile strength[367]. The energies G_n and G_{th} were taken to be determined by the current injected per cycle, which was regarded as obeying Fowler–Nordheim tunnelling through a Rose-type barrier[336], i.e.

$$G_n = A \exp\left(-B\phi^{3/2} V^{-1}\right)$$
$$G_{th} = A \exp\left(-B\phi^{3/2} V_0^{-1}\right) \tag{5.3}$$

where because of a fixed geometry the field variable has been replaced by the voltage with V its applied and V_0 its threshold values respectively. Eqn. 5.2 therefore provides an explanation as to why the inception voltage in ramp experiments is observed to approach a threshold value at high frequencies[329].

Substitution of eqns. 5.3 into 5.2 gives the relationship

$$\log(t_I) = \frac{a}{V} + b - \log\left\{D - \frac{1}{V}\right\} \tag{5.4}$$

Here a is a constant proportional to the $\frac{3}{2}$th power of the metal work function ϕ, b is another constant, and D^{-1} an effective threshold voltage close to V_0. Fig. 5.5 shows that the model expression is an extremely good fit to the data that the authors quote. Furthermore the ratio of constants a obtained for steel and for silver electrodes (1.19) is close to that expected from their work functions (1.18). The value of D^{-1} (4·975 kV) is independent of the metal as expected. The use of the Fowler–Nordheim law in this model should be valid since the voltage range in which Schottky emission dominates[336], will be in the region where the presence of a threshold determines the behaviour. Differences in the shape of the barrier, i.e. Rose or Mott–Schottky type, are not distinguishable over the limited voltage range available outside the threshold region.

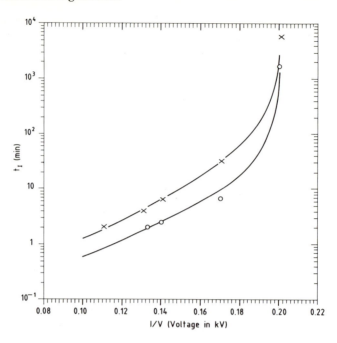

Fig. 5.5   Electrical tree inception times at 50% probability as a function of voltage,
× steel; ○ silver. Fit to eqn. 5.4 is shown by the continuous lines, with
   steel: $a = 12 \cdot 4$, $b = -2 \cdot 1334$, $D = 0 \cdot 201$
   silver: $a = 10 \cdot 4$, $b = -2 \cdot 352$, $D = 0 \cdot 201$
Data taken from Reference 342
Copyright © 1978 The IEEE

It should be noted that the model is not inconsistent with the empirically
determined power law dependence of t_I (i.e. Fig. 4.16). At high voltages
an approximate form V^{-6} is found, and when low voltage data is included
the inverse power may increase to any arbitrary value. Additionally the
inception times are distributed[129,342], and the average value is usually used.
As the voltage is reduced the distribution becomes wider as might be
expected from Fig. 5.5 if emission were governed by a small range of local
fields for a given voltage and needle geometry. This is precisely what would
be expected if emission takes place from localised areas on the tree tip[368]
such as occurs in the breakdown of liquids[369].
   Although luminescence does not occur in DC treeing and the repetitive
factor is missing, tree inception will nonetheless follow a similar path of
avalanche damage if the injected (or extracted) charge current is sufficiently
high in energy. Tree inception will therefore occur when the voltage is
rapidly changed[287], and hence there is no time for oxygen to have an
affect[360]. One extra factor that has an influence here is the spatial distribution
of the injected charge. While extraction of electrons can be expected[321,320]
to follow field lines in a filamentary manner, the injection of electrons may

be spatially homogeneous. In this latter case the field is required to be sufficiently large to cause the space-charge boundary to be unstable to fluctuations[327] which generate filaments capable of sustaining an avalanche[346]. However if electron injection only occurs from small local sites it will intrinsically lead to a filamentary space-charge distribution, which will only spread out uniformly if the field is insufficient to cause channel formation. Thus with (ft_I) replaced by the number of pulses, eqns. 5.2 and 5.4 should also be applicable to tree inception by repeated DC pulsing as long as sufficient time is allowed between pulses for the injected charge to dissipate[348].

5.3 Electrical tree propagation

5.3.1 Channel discharge

The first channel generated in AC treeing has a self-limiting growth behaviour[326], e.g. eqn. 5.1. Such behaviour may result from electro-fracture[366] depending upon the amount of strain energy available and its spatial distribution. However it is unlikely that this mechanism would give the observed[326,370,371,372] discharge behaviour, namely current pulses on *both* half cycles during channel formation followed by the disappearance of the negative half-cycle pulses and a reduction of those on the positive half-cycle to ~ 0.2 pC as the channel growth is terminated. These observations are consistent with a mechanism in which electrical discharges eroding the channel are temporarily extinguished. Although some gaseous decomposition products will have been formed thereby raising the gas pressure in the channel[373] this is not likely to increase the discharge inception voltage[374] of voids of this small size. However it is known that long cylindrical channels with diameters less than $\sim 100\ \mu$m can have discharge inception voltages close to the breakdown value of the polymer[375] and unusual discharge behaviour[376,377] due to the absorption onto the sidewalls of cations produced in the gas avalanches[376,375]. Tree channels are of this type with a length to diameter ratio of $\geqq 5$ and thus can be expected to behave in the same way. Because ions are produced by an avalanche in bands separated by an impact ionisation path length[378], positively charged bands will tend to be formed along the channel length. These bands will extinguish further discharges in the channel when their charge density and hence local field is sufficiently high to attract and neutralise the electrons.

It seems that behaviour of this type leads to a difference in the discharge-phase relationship between needle channels and disc voids where the inability to discharge the whole surface during the voltage rise time allows the greatest amount of discharging to occur as the voltage drops[379]. As a result discharge-phase relationships have been used to characterise the cross-over from void discharge to channel formation for partial discharge initiated trees in epoxy resins[126,380] (Section 19.2).

Although the extinction of forward-discharges in the first channel depends upon the detail of the fields caused by charge deposition on the sidewalls, it may be assumed that the channel continues to grow as long as the local field is greater than a threshold value E_{th} for discharge initiation[376,381]. The local field however will be determined by low mobility cations restricted to the sidewalls by absorption. Such ions can be expected to reduce the apparent field E_a by an amount proportional to the channel length if the maximum number of cations absorbed per unit length is a constant. Thus taking the growth rate as proportional to the over-field, $E_a - E_{th}$, (i.e. the discharge magnitude[381]) gives

$$\frac{dL}{d\tau} \propto E_a - E_{th} - bL \qquad (5.5)$$

where b is a material dependent constant, and $\tau \, (= t - t_I)$ is the time elapsed from initiation of the growth generating avalanches. Integration of eqn. 5.5 gives a form equivalent to that experimentally observed, i.e. eqn. 5.1, with the length at which the first channel ceases to grow, L_m, defined by

$$L_m = \frac{E_a - E_{th}}{b} \qquad (5.6)$$

which is a similar expression to that assumed by Bahder *et al.*[120] for partial discharge initiated trees.

At this point the first channel contains sufficient positive charge on its sidewalls to extinguish forward discharges, and only electron extraction avalanches on the positive half-cycle can occur[326]. Extraction from beyond the channel will initiate damage in a similar way to the initiation of the first channel and yield similar magnitude pulses. However since the extraction will take place to the sidewalls rather than the tip the zones of damage will potentially diverge and give rise to the possibility of branching[326]. Contemporaneously with this process some discharges will take take place to the sidewalls partially neutralising the cations there and causing erosion and widening of the channel[370]. The pulses associated with these latter processes will be exceedingly small and will probably only be detectable optically[127]. A schematic representation of tree formation up to this point is shown in Fig. 5.6.

5.3.2 Branch development
Continuation of the AC-treeing process will initially proceed via the formation of branches to the first channel. At some point neutralisation and diffusion[376] of the cations into the polymer will again allow both forward and backward discharges to occur within the tree[326]. When sufficient cations have accumulated on the sidewalls at the branch tips, back-avalanching of electrons deep trapped in the polymer allow the tree to advance via the formation of new branches, even though discharges in some of the existing channels may be temporarily extinguished[335,317].

This stage of propagation is characterised by the appearance of impulse acoustic emission[334] and light pulses[275,317], associated with the gas discharges which can also be observed as current pulses[288] in excess of 1 pC. The mechanical shock wave only contains 0·002% of the discharge energy and should not cause yielding in polyethylene though it may contribute to the

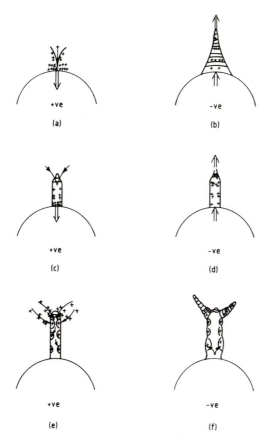

Fig. 5.6 Schematic representation of the initial stages of tree formation. (*a*) Back avalanche < ~0·1 pC; (*b*) Forward avalanche in low density region or craze = 0·5 pC; (*c*) and (*d*) Gas discharge in first channel ≃1 pC; (*e*) Back avalanche to channel sidewall ~0·2 pC; (*f*) Branching crazes and sidewall erosion

enhancement[382] of damage zones. In contrast the electron stream in such discharges can reach kinetic energies up to 10 eV at the discharge tip which is sufficient to extend the channel by bond breaking. However most degradation results from the cation stream[123,124] which can produce temperature increases of 100's K at the point of impact[383]. In air discharges uniform erosion (from oxygen) and pitting (from nitrogen) will widen the channels of trees in polyethylene[380] and produce bursts of gaseous decomposition products[315], among which acetylene has been identified by mass spectroscopy[315], and CO_2, H_2, and CO by their spectra in the light emission[384].

During the advance of the tree by the extension of existing channels and the addition of new branches the first channel will widen into a crater[120] and the main branches into channels with diameters >10 μm. The accompanying discharges will rapidly increase to[317] ~100 pC with a proportionally large acoustic emission[335], and stay at this level during the slow propagation region of Fig. 5.2. Acoustic emission pulses have also been measured in polyester resins[335] where it was shown that their phase distribution was similar to that of the current pulses, and their amplitude proportional to that of the discharge up to 300 pC. Since the acoustic pulses also have an impulse form it is reasonable to attribute them to the channel discharges rather than cracking in the polymer, though it is necessary to show that the two are in phase in order to establish this unambiguously. During this stage of branched tree propagation the overall tree length grows very slowly and the tree development is concentrated in the creation of a sub-branch system[370] which gives an apparent increase in the main branch diameters to $\sim 50 \mu$m.

5.3.3 Runaway propagation

The runaway stage of branched trees is typified by a cessation of the light emission and a reduction of the discharges[126,317] to below 5 pC. Although the discharge extinction has been attributed to a rise of pressure in the channels[178,129] because of gas production, the evidence appears to be more in favour of field moderation as a result of deposition of cations and semiconducting discharge products on the sidewalls[374,385,317]. For example hydrostatic compression has less influence on treeing if space charge has been introduced by prestressing[365]. Also the behaviour of treeing following the interruption of stress follows a pattern more appropriate to the dissipation of wall charges[386] rather than release of gas pressure by diffusion through the polymer[178], with small discharges forming new channels in some cases and the recommencement of large discharges in others[317]. The former behaviour retards the onset of runaway and failure, while the latter speeds it up[317].

In this stage of propagation the leading tree channels are highly filamentary with diameters of 2—3 μm and very little side branching[129,317]. The magnitude of the discharges and accompanying acoustic emission[335] observed at this stage in trees grown in uninterrupted stress indicate that they are of the same type as those occurring in the early stage of treeing[288], and hence are confined to the channels of the runaway branches. It seems likely that this form of growth initiates when sufficient cations have been accumulated to prevent discharges in the main body of the tree. Filament formation continues at a few sites on the tree surface with a favourable stress enhancement. Once a channel becomes wide enough charge repulsion will force cation wall charges into it from the main body of the tree. This channel thus acts like a conducting projection. The feedback of cations formed in the small discharges will reinforce the forward extension of the branch, whereas space charge shielding will rapidly extinguish the development of side branches[312].

5.3.4 Bush-type tree propagation

Bush-type trees propagate initially like branched trees[163], but then the formation of new branches from channels close to the electrode rather than the growth tip produces the denser bushy form[317,129]. During this process discharges continue within the tree eventually reaching magnitudes[317] of ~500 pC, with very large acoustic emission[335]. Once the tree has started to grow the local fields will be strongly dependent upon the space charge and its arrangement, rather than the field intensification of the original point-plane geometry. Possibly therefore the discharge and hence deposited wall charges will increase with applied voltage rather than the initial field. Increasing the frequency will reduce the time available for the wall charges to dissipate and hence allow a given surface charge to be maintained by a lower voltage[386]. Thus the most likely explanation of the cross-over to a bush shape is that sufficient sidewall charge is generated during the early formation of the tree to cause branch formation from the channel sides, possibly originating at surface pits. The removal or neutralisation of wall charge in this way will favour the continuation of discharging and the density of the tree channels will increase the magnitude.

Although it has been reported that discharges do not cease[317] during the growth of bush-type trees, this is probably because the electrode separation was too small in this case, and the tree could cross the gap before exhausting the possibility of wall charge removal via the formation of the bushy structure. However large gaps[317] and high voltages[129] will increase the wall charges to such an extent that discharging above 5 pC ceases[317], and the acoustic emission drops substantially[335]. At this point light emission from the bushy body ceases and only light originating in fine (diameter $< \sim 5\ \mu$m) channels around the surface of the bush can be observed[317]. If no interruption of the voltage occurs, discharges do not restart[335] in the bulk of the bush, but a few of the surface channels may acquire a large enough space-charge field to runaway as in the case of branched trees[317,335].

5.3.5 DC growth

Under DC stresses degradation can only occur by means of avalanches and discharges taking place in one direction. An impulse or ramp voltage will produce pulses of injected (or extracted) electrons[387] which initiate a sequence of avalanches sufficient to generate a first channel, the length of which will be determined by such factors as the impact ionisation path length and the polymer yield strength. At this point any heterocharge deposited on the sidewalls will be swept out and neutralised by the electrode, thereby transferring the maximum stress to a region around the channel tip. The tree will advance by repeating the avalanche erosion with each new channel tip acting as a potential branch point, and penetration will only cease when the space charge field at any one of the tips is insufficient to cause an avalanche[120,320,321]. On the other hand the length of short circuit trees is determined by the equilibrium penetration of homocharge[287,346], which is destructively extracted.

Experiment shows[287] that the negative space charge penetrates more deeply into polyethylene than does the positive space charge, and also that the depth increases strongly as the temperature is raised above 223 K. Such data may be applied to a determination of the shielding of the needle by homocharge in other types of situations[319,316], but will not have any bearing on the length of the tree in these cases. Only when the tree is produced by the thermal or mechanical effects of high density filamentary injection currents, such as might be produced in nanosecond high voltage impulses[388], will the high field drift mobility determine the tree penetration length.

The continued growth of DC trees under repeated impulses depends upon the time allowed between pulses for the disssipation of space charge from around the tree tips. If this is sufficient then all tips may act as potential growth sites under a new impulse. However the time required can be large[287] (~ hours) and it can be expected that repeat pulses of the same polarity will only cause growth from a few of the tips which are available[320,321]. On the other hand the space charge will activate all tips for growth when polarity reversal sequences are followed. Because of the heterocharge extraction during the impulse it is unlikely that bush-type trees can be formed during DC treeing, and the tree shape will usually resemble branch-type trees[320].

5.4 Fractal analysis of electrical trees

5.4.1 Fractal characteristics

The seminal work of Niemeyer *et al.*[289] showed that discharges onto a surface (Lichtenberg figures) had a fractal structure and presented a stochastic model capable of simulating them. Since then a number of attempts have been to apply the same type of description to electrical trees[389,338,390,391]. The aim of this work to date has been to analytically relate the mode of tree propagation to the form of structure generated. However it is possible to see it as the first step in the development of an analytical model of treeing breakdown. Here the fractal description provides the information on internal structure that is necessary, along with length at a given time[120], to assess the possibility for runaway to breakdown that is inherent in the situation.

The term fractal simply denotes a figure with a fractional dimension (see eqn. 5.7 below) and it seems to have been taken for granted that it applies to electrical trees as a glance at a branched tree appears to suggest. However this is not necessarily the case and the appellation must be justified by direct measurement[289]. A more precise definition relates the mass M to the length L through

$$M \propto L^{d_f} \tag{5.7}$$

Here M is the mass the figure would have if it could be made as a solid and weighed. It is determined by measuring the figure with small measuring blocks of unit size and mass. Each block is placed so as to cover a section of the figure, and the mass is the *minimum* number of such blocks required for its complete coverage. Similarly the length is the spatial separation of the extremities of the figure measured in terms of the unit of block size. In planar figures the block is replaced by a square or pixel whose size defines the length unit, and area the mass unit (see Reference 251). When the figure is composed of thin tubes it becomes essentially a line structure and the mass is then the total length of all its sections measured in the same units used to obtain its overall length.

Figures which possess the same structural density at all sizes have unit values of d_f equal to the dimensionality of the space in which they are formed (i.e. $d_f = 2$ for a planar figure, and $d_f = 3$ for Euclidean space), and are the familiar geometrical bodies such as square and cubic lattices. However where the dimension d_f is non-integer and less than that of the space in which it sits the figure is a fractal[392]. Thus the density of a fractal (i.e. its mass divided by the volume in which it sits) decreases as its size increases, that is the unoccupied space increases as L increases. In the case of trees the mass should be replaced by the sum of the arc-lengths of each branch (S_i) over all the branches within a length L of the initiating point, i.e.

$$S = \sum_i S_i \propto L^{d_f} \tag{5.8}$$

Electrical trees are grown in 3-dimensional space, though their own dimension d_f may not only take values below 3 but often even below 2. However photographs usually project the tree onto a plane, and, if the dimension is 2 or greater, the field of view will be uniformly filled as with bush-type trees. In order to determine the dimension in this case the tree must be sectioned. The figure that results will have a measurable dimension of $d_f - 1$, which can be used to determine d_f. This method can also be used to obtain the dimension of trees with $d_f < 2$, which will give a figure in the section with dimension less than unity. However in this case it is better to determine d_f from the projection which will have the same value of dimension as the tree itself[392]. Using these general rules it can be inferred that d_f for branch-type trees is less than 2, and for bush-type trees lies between 2 and 3.

Mitsui et al.[320] have published a set of tree structures produced in epoxy resins under a variety of conditions and a fractal analysis of some of these is presented in Fig. 5.7. The dimensions obtained in this way ($d_f \approx 1 \cdot 66$ for positive pulses; $d_f = 1 \cdot 3$ for negative pulses; $d_f \approx 1 \cdot 4$ for alternative polarity pulses) are significantly different to integer values, though less than those quoted for discharges[289] ($d_f = 1 \cdot 7$). Such trees are however stochastically formed and will have slightly different values for d_f each time they are grown. Their fractal properties are only present on average, and hence the value of d_f should be determined as an average over as many trees (structural realisations) as possible. The size of the tree used should also be as large as possible to allow the full range of statistical possibilities to develop[312]. The small value for the negative pulses may however be due to insufficient time being allowed between pulses for the dissipation of a field moderating space charge around the tips. This would inhibit further growth from all but the most advanced tips and constrain development around the lateral boundaries thereby reducing d_f, in a similar way to that observed in fluid displacements[251]. In contrast positive space charge need not be uniformly distributed and probably retains growth activity at each tip[321].

5.4.2 Tree dynamics

The time development of the tree may be derived by assuming that a fixed 'mass' (ΔS) or volume element of tree is added in equal time intervals, giving

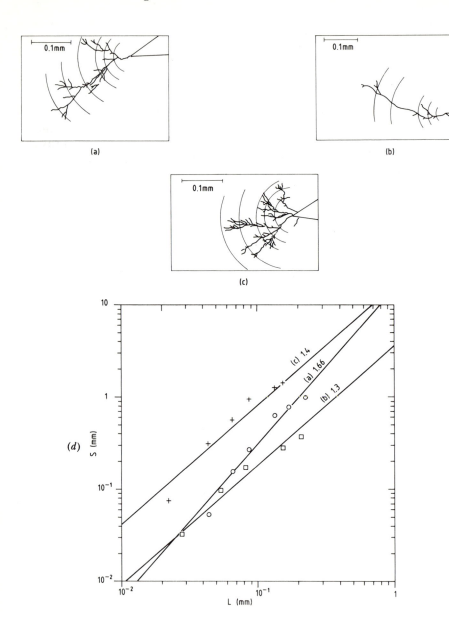

Fig. 5.7   Fractal analysis of electrical trees grown in an epoxy resin[320] by a sequence of DC pulses. (*a*) Positive pulses; (*b*) Negative pulses; (*c*) Alternate polarity pulses. Tree structures of different lengths (*L*) are defined by marking off the arcs shown from the inception point. The total length of all the branches (*S*) within an arc is then measured; (*d*) The fractal dimension, d_f, is found from the log-log plot of *S* as a function of *L*

as in eqn. 4.25

$$L^{d_f} \propto \tau \propto \text{number of pulses, } N_p(\tau) \qquad (5.9)$$

This behaviour may be expected as a consequence of the equal amounts of energy available for treeing during each pulse. When all tips are accessible to further growth as in alternate polarity pulses such a result is indeed found, Fig. 5.8, with $d_f = 1\cdot45$. Thus in this case the geometrical and time estimations of d_f are consistent. However when the polarity is kept constant the dimension of the time evolution is close to unity, which suggests that in this case each pulse (or time interval) advances the tree by a constant increment of length $((\Delta S)^{1/d_f})$ adding a variable number of volume elements according to the activity of the tips. The internal structure of each element added must however still be a fractal.

Trees grown under AC stresses would be expected to follow the same behaviour pattern as alternate polarity pulses, with $N_p(\tau)$ being replaced by $f\tau$. Although no such comprehensive data is available in this case, an evaluation of the time dependence of tree growth[163] in polyethylene shows that two classes of behaviour occur, namely one for which $d_f = 1\cdot6$—$1\cdot7$ (branched trees) and one with $d_f = 2\cdot4$—$2\cdot6$ (bush trees), see Fig. 5.9. This data is taken from the region of growth in between the initial fast propagation, and the rapid runaway, where the formation of multiple branches from increasingly

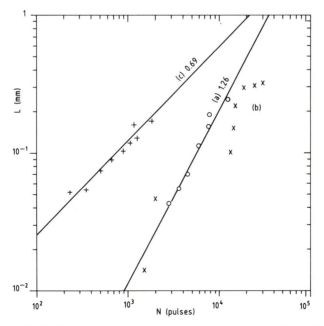

Fig. 5.8 Length of electrical tree in epoxy resin as a function of number of pulses. (*a*) ○, positive pulses, (*b*) ×, negative pulses, (*c*) +, alternate polarity pulses. Data taken from Reference 320

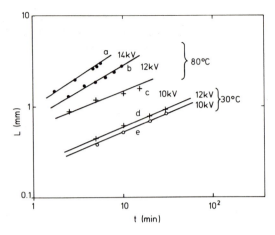

Fig. 5.9   Length of electrical trees in polyethylene as a function of time in the intermediate regime between initial growth and runaway. The data has been taken from Ieda[163] and exhibits the form of eqn. 5.9: $L \propto \tau^{1/d_f}$ with (a) $d_f = 1 \cdot 6$; (b) $d_f = 1 \cdot 7$; (c) $d_f = 2 \cdot 6$; (d) and (e) $d_f = 2 \cdot 4$

larger numbers of potential growth tips[370] would be expected to give a fractal structure. It should be noted that the dimension of the branched trees is close to that found in surface discharges[289], while that of the bush-type trees lies in the range found for diffusion limited aggregation[393] and dielectric breakdown simulation[394] on a 3-dimensional grid. A comparison with Fig. 4.13 will serve to illustrate the difference between this form of growth and the volume filling behaviour of water trees.

Eqn. 5.9 also predicts that if the proportionality factor is frequency independent, then the frequency dependence of tree growth should follow a f^{1/d_f} law. The cross-over in shape from branch-type to bush-type trees will however make it difficult to prove this relationship, and the observed exponent of $\sim \frac{1}{3}$ is roughly consistent with either of the two groups of behaviour given the uncertainty in estimating its value. The change in shape is also probably responsible for the observed reduction of the length grown in a fixed period as the voltage is increased[129,163]. Thus a reduction in the channel formation time at high voltages may, in a given voltage range, be overcome by the increased density of the bush-type tree as compared to the branch-type tree[386].

5.4.3 Tree simulation

Attempts to simulate electrical tree structures[389,390] are aimed at elucidating those factors in the propagation dynamics which determine the tree shape. Fujimori's model[389] is unique in allowing the tree channel to choose arbitrary directions and lengths of growth but unfortunately did not eliminate unrealistic choices (e.g. direction reversals and very long channels). All other model simulations start with a regular 2 (or 3) -dimensional grid one side of which is held at a higher potential than the other. Following the introduction of the first tree channel along a grid line parallel to the field at the electrode, adjacent grid lines are converted in turn to branches according to a set of rules applied to each grid point from which growth may develop.

The first such model[289], which has since become known in the literature as the dielectric breakdown model (DBM)[312], specified the random addition of conducting branches to a potential growth site chosen with a probability weighting which was a power of the local field at the site, i.e. E_{loc}^{n}. In order that the model would better reflect the physics of charge injection as expressed by the field-limited space-charge (FLSC) theory[346,347] (Section 9.3.4) Wiesmann and Zeller[390] modified it so that a potential difference (field) could exist within the tree itself, and branches could only be added if the local field at a branch tip exceeded a critical threshold value. In this way it was shown that when the channels are conducting a branch-like structure of dimension ~ 1.7 is obtained, see Fig. 5.10, but the structures become bushy and limited in growth as the field within them is increased and the potential at the tips reduced towards the threshold value for growth.

It was also shown that when the stochastic criterion for branch addition was replaced by an automatic branch formation along the grid line where the local field is in greatest excess over a spatially-distributed local threshold field, then a weakly branched structure grows along the line of least resistance. This result suggests that tree structures are the consequence of stochastic propagation dynamics rather than due to material inhomogeneities as sometimes suggested[326]. Thus this type of investigation can yield useful

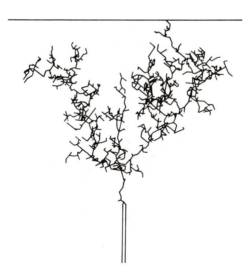

Fig. 5.10 Perspective view of a three-dimensional tree simulation[390]. The needle has a length of $20a$ and the tip-plate distance is $50a$. The tree structure was taken to be conducting (i.e. zero field along the branch channels) and a threshold field E_0 (equal to 0.5 in units of the initial field at the tip) was required at the end of a channel in order to add a new branch

insights into tree structure, for example the cross-over from branched to volume-filling structures may be related to the field within the tree due to wall charges.

Care must be taken in any simulation to see that its rules reflect the known physics of the process to be investigated. In this respect the DBM, even in its modified form, describes a discharge rather than a tree, i.e. a branch *must* be added at each step and the structure is nearly conducting. These features are what would be expected of an electron stream produced in a discharge. In this case we would expect a spatially uniform cloud of injected charge to be unstable with respect to charge density fluctuations for much the same reasons as occur in viscous fingering[311,312], i.e. large tip velocities (10^4 mm $\cdot$ s^{-1}) and low interfacial tension (due to charge repulsion) reduce the critical wavelength to such low values that the charge cloud divides and sub-divides into filaments[251] without any residual uniform region. However electrical treeing requires repetitive processes to erode the channels (~ 100 μm $\cdot$ hour^{-1}), and such a discharge model can only be expected to apply to impulse DC trees and possibly not even here since the channels are still probably formed by multiple avalanching rather than filamentary thermal or mechanical runaway[346].

In order to take account of these differences the simulation rules have been amended so that a branch can only be added when it has been chosen a fixed number[389,312] of times, m, rather than the once of the DBM. This restriction corresponds to the physical requirement that a number of damage forming events occur at a growth tip during the formation of a branch. Since each event (avalanche) at each potential growth tip produces irreversible damage that is accumulated the counters measuring the number of times a tip is chosen must retain their count even after one of the tips accumulates enough damage to grow a new branch. Only the tip of the new branch starts from zero. These simulation rules are known as the Tang[395] version of the diffusion limited aggregation (DLA) model and it appears[312] that in this case the dimension of the structure increases from unity to 1·7 (in planar simulations) as its size increases. This result is reasonably consistent with the observed structure of electrical trees. The inclusion of material inhomogeneity, which in viscous fingering is simulated by a site dependent permeability factor[312] alters the asymptotic dimension to 1·5, and thus does not appear to be a contributing factor except perhaps in epoxy resins. However more data will be necessary before any detailed comparison can be made.

A different DLA model[396] sets the counters to zero after the addition of each branch and may lead to confusion since in this case the dimension approaches unity for large structures (clusters) or large values of m. However it should be noted that such a model is not based on the physical nature of branch formation in electrical trees, with its accumulated damage, and hence should not be regarded as giving an electrical tree simulation. In simulation terms both these models can be regarded as equivalent to a much larger grid with m short steps along this grid in one direction appearing as a single branch on the scale of the actual grid used. Thus objections to the size of the grid used are overcome in either case.

All the models discussed so far add only one branch in succession, however failure in a random fuse system has been investigated by allowing all fuses carrying an above-threshold current to fail simultaneously[397]. The failure structure which has formal similarities to an electric tree[398] has in this case a dimension of 1·6, which lies within the range of the other applicable simulation models[289,390,312].

One of the difficulties in developing a simulation model of electrical trees is the fact that though discharges occurring inside the tree cause it to advance, the discharge itself is not the tree. A model under current investigation[399] recognises this problem and combines the two types of behaviour. In this case discharges simulated by the DBM are allowed to occur, but only within the already established tree structure. Where the discharge attempts to exit from the tree structure its charge is deposited in a counter tagged to the exit point, and a branch is added when the counter exceeds a specified value. This model allows a number of factors to be investigated for their effect on tree structures. For example the discharge can be varied from branched to linear by increasing η. In the branched case the charge must be sub-divided at each branch point. Discharges may be extinguished if a channel is used a specific number of times, and the effect of this on tree development evaluated. Furthermore it should be possible to allow discharges to restart, and with greater facility, in a channel following a quiescent period.

By using a model such as this which reflects the physics of tree propagation, the relative effects of the several factors determining tree structures can be evaluated. In particular the propensity of branches to grow rapidly for a short period of time can be checked, and the relationship of a runaway process to the charge density at the tips investigated. In the latter case it may be possible to determine the factors controlling the statistics of the process by performing a large number of simulations for each different condition.

Chapter 6
Tree-initiated breakdown

6.1 The water tree route to failure

6.1.1 Reduction of breakdown strength

A considerable amount of evidence has now been accumulated showing that water trees reduce the breakdown strength of polymer insulation (see Shaw and Shaw[119] and references therein). In this respect bow-tie trees perform a different role to that of vented water trees[175], causing a reduction of breakdown strength even when the bow-tie water tree length is small ($\sim 20\ \mu$m), in contrast to vented water trees[160] where a strong decrease in breakdown voltage only occurs when the trees become long[128,192,400], Fig. 6.1. This difference has been associated with the alternative manner in which insulation deterioration is accumulated in the two cases. Thus in the case of bow-tie water trees it is their number density[401,402] which is important in determining the average size of tree[238] sufficient to give the global deterioration[304,239] which will reduce the breakdown strength[403]. In contrast vented water trees rarely amalgamate in cables and the criterion for the deterioration they generate is the length of the individual tree compared to the insulation thickness[229]. Considered in this light the crucial factor in the reduction of breakdown strength is the extent to which water trees provide a degraded route across the insulation, which may be approximately measured by the degree of water treeing[200,222,237]. Measurements of the breakdown strength across tree sections[218] have shown that the treed region does not act as a conductor which decreases the dielectric thickness, but rather as a region with a much lower breakdown strength ($1 \cdot 5$—$3 \times 10^7\ \mathrm{V \cdot m^{-1}}$), the smallest value of which is found at the base[175,218,404] of the tree.

It has been estimated[306], using a reasonable value for the conductivity of the treed region ($< 10^{-5}\ \mathrm{S \cdot m^{-1}}$), that the field at the tree tip is only enhanced by a factor of $1 \cdot 74$, whereas inside the tree a value of $\sim 9 \cdot 5$ was obtained[262] due to water droplet polarisation. These enhancements are sufficient, at the apparent average breakdown stress, to produce discharges in air-filled voids of size greater than $10\ \mu$m at the tree tip and microvoids[122,405] ($< \sim 5\ \mu$m) inside the tree. Thus it is possible that the average (characteristic) breakdowns are the result of these discharges[405]. However the effect of water trees on breakdown tests is not just to shift the breakdown distribution to lower field strengths, but also to increase the probability of failure at low fields relative to the characteristic breakdown field[175,176], Fig. 6.2. Thus the most likely explanation of the water tree effect is that the enhanced field within the tree causes breakdown through[163] 'weak dielectric paths' between

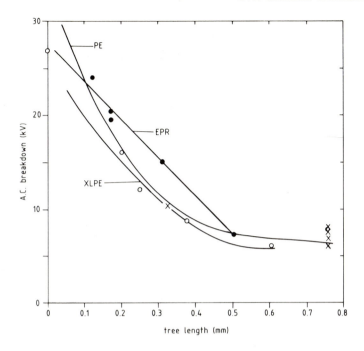

Fig. 6.1 Effect of water tree length on AC breakdown strength. After Bernstein, Srinivas and Lee[400]
Copyright © 1975 The IEEE

the water-filled microvoids. The weakness of these regions can probably be traced to the physical disruption of the lamella surfaces caused by the microvoid formation[18], free radicals and trapped ions in the amorphous-lamella interfaces, and oxidation of the microvoid surfaces[225].

Drying out the tree will remove the field enhancement, and collapse of the voids may relieve the strain in the inter-lamella regions but some damage together with local concentrations of contaminant salts[218] will remain. In this state the treed region will still initiate low-field breakdowns with a higher probability than was found prior to treeing[175]. This will be particularly the case for long trees (> ~1 mm) which have been shown to have reduced oxidative stability as a result of the removal of antioxidant[406]. The average breakdown strength however will correspond to breakdowns initiated outside the tree region and, in the absence of the field enhancement of the tree, will regain its original value. If partial discharges are responsible for this characteristic failure mechanism[405] it may even be possible that water vapour from the drying tree enters voids in the non-treed region and raises the breakdown stress[407] above its original value[218,212] either by increasing their threshold field for discharge or by the production of field relieving discharge products.

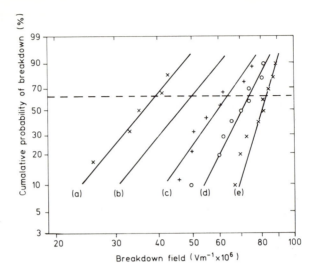

Fig. 6.2 The cumulative probability of breakdown of extruded polyethylene samples as a function of AC voltage measured using progressive-stress tests. The samples were 'aged' in electrolyte to form water trees for the following times: (*a*) 500 hours ($a + b = 4\cdot5$); (*b*) 340 hours ($a + b = 4.6$); (*c*) 195 hours ($a + b = 5\cdot3$); (*d*) 50 hours ($a + b = 7\cdot4$); (*e*) not aged ($a + b = 12$). Here ($a + b$) are the shape parameters of the Weibull distributions that best fit the experimental data, which are shown by the continuous straight lines in the diagram whose gradient is ($a + b$)
After Rynkowski[176]. Copyright © 1981 The IEEE

6.1.2 Initiation of electrical trees
Although the breakdown strength of polymer insulation will be reduced by water trees, the material can still be expected to withstand the service conditions even after the water tree has bridged the insulation[363,408]. Thus the deleterious effects of water trees upon cable insulation[121,128] are most likely to be related to their ability to initiate electrical trees[128,119,409]. It has even been suggested that, in the present generation of power cables, water trees are the precursors of most electrical trees[128,409,410] and hence the ultimate origin of cable failure[187].

A direct observation of a water tree initiated electrical tree was first made by Miyashita[411] as a cable sample was progressively taken to breakdown. In this case the electrical tree initiated opposite to the water tree and the breakdown channel started from a neighbouring imperfection, hence it is possible that the water tree was only marginally involved. Since then many examples of electrical trees initiated by or in conjunction with water trees have been noted[119], and direct observation of initiation has been made in at least two cases[276,408]. Thus it is known that water trees may generate electrical trees at their growth tip both from bow-tie water trees[134,409,412,413] and from vented water trees[276,409,412,414] as well as from the counter electrode[137,408,415]. This latter case has been observed both in service, Fig. 3.6 and in the laboratory[137], Fig. 6.3. Although the bifurcated trees of Fig.

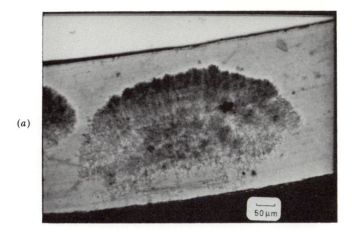

(a)

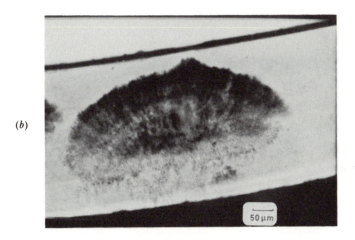

(b)

Fig. 6.3   Electrical tree initiated from the counter electrode by a water tree grown in a Rogowski plaque under laboratory conditions[137]. Here (a), (b) and (c) are sections through the water tree showing the protuberance which initiated the electrical tree indicated by an arrow in (c); (d) shows the disruption of the polymer surface accompanying the formation of the electrical tree

3.6 seem to be restricted to service conditions, in-service examples have also been found[416] which are similar to those obtained in the laboratory. Here the water tree has been sectioned in a number of places parallel to its direction of growth to show that the electrical tree apparently results from a local protuberance on the water tree tip which has crossed the short distance to the conductor.

(c)

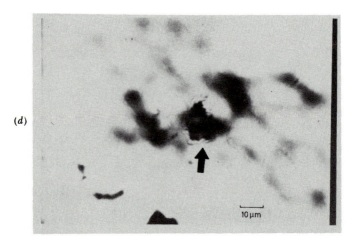

(d)

Fig. 6.3   Continued

The mechanism whereby water trees initiate electrical trees is not understood at present. Indeed the phenomenon is at first sight a little surprising
since the presence of water is mildly inhibitive to electrical tree generation
through a moderation of the local stress enhancement at asperities and a
reduction of the likelihood of partial discharges in voids[118]. This behaviour
is clearly illustrated in the experimental results of Chen *et al.*[276] where it
was found that water trees and electrical trees grow alternately from an
aqueous needle electrode, with the transport of water into the electrical tree
channels temporarily extinguishing the electrical tree in favour of the water
tree.

It has been suggested that voltage surges may be responsible for initiation with the water tree regions acting as nucleation sites because of their enhanced degradation[128,409]. However the observation of this process under controlled laboratory conditions rules out this possibility as an essential feature, although it may contribute in service. Another possibility is the reduction of the dielectric thickness by water tree growth leading to a field enhancement at the tree tip[128]. The drawback to this suggestion is the observation that the treed *region* has a dielectric permittivity close to that of the polymer[244] and is only weakly conducting[136,218]. Thus the value of 1·74 for the field enhancement at the tip[306], which is based on a conductivity appropriate to the water in the voids rather than the whole tree is likely to be an upper limit, and the effective reduction in dielectric thickness cannot be expected to enhance the field sufficiently for electrical tree inception. Furthermore electrical trees have been initiated from water trees in the bulk where a stress enhancement of this type would be weak, and at water trees grown from needle electrodes where the growth in fact serves to reduce the stress enhancement.

The data of Karner *et al.*[408] shows that there can be a delay before the initiation of an electrical tree even after the water tree has bridged the needle electrodes, during which time the local stress enhancement may be expected to drop as a result of the lower impedance of the water tree. It should be also noted that the results shown in Fig. 6.3 cannot be taken to support a field enhancement mechanism, since the broad tree front precludes any strong enhancement factor, and the electrical tree has been initiated by a bridging protuberance resulting from a local fluctuation on the water tree boundary.

We have seen that electrical trees may be initiated by the generation of microcracks or crazes containing space charge which is cyclically injected and extracted. Since a water tree cannot be regarded as a conducting electrode, it will not act as an electron injecting source, as is the case with metallic electrodes. Nor is its stress enhancement likely to be sufficient to cause partial discharges in neighbouring intrinsic voids[120] ($>10^7$ V · m^{-1}, void size $<100\,\mu$m), under the field and laboratory conditions ($\sim3\times 10^6$ V · m^{-1}, 50 Hz) known to lead to electrical tree conversion[137,231,414]. However if the water tree were to approach or even surround a conducting contaminant such as a small metal particle, the combined stress enhancement due to the polarised water filled microvoids[262,306] and the contaminant[417] may be sufficient for the particle to initiate an electrical tree.

In other cases an origin for the conversion of water trees to electrical trees may be found in their respective formation mechanisms. It is known that water tree propagation causes structural disruption of the amorphous (inter-lamella) regions. Electrochemical dissociation is also probably involved producing organic ions, and ultimately oxidation and microvoid formation via polymer radicals. Such electrochemistry is fully reversible in liquid electrolytes, however in the solid polymer the dissociated ions will be driven into the amorphous region where some will be trapped and take part in oxidation reactions. In an oscillating field therefore the injection and extraction of ions, some of them chemically bound to the polymer as for instance in the ion clusters of ionomers (i.e. polymers co-polymerised with organic acids containing neutralising metal ions which act as crosslinks[296]), will give rise to electromechanical fatiguing at the tree tip. In addition there will be a build-up of heterocharge due to trapped and bound ions in advance of the tree tip. Here we have the conditions necessary for the initiation of electrical trees, with a field driven chemical reaction taking the place of electron injection, by a repeated electrochemical ionic dissociation and recombination.

During normal water tree propagation the metal ion stabilisation of the microvoid surface and the condensation of water onto them[298,297] will prevent the process going as far as channel formation. Here the water will provide a liquid interface for the electrochemistry at the site and also act as a mild plasticiser. An advance of the damage boundary will therefore usually occur instead. If however the rate of electrochemistry and microvoid formation outstrips the supply of water for condensation, electromechanical fatiguing may cause longitudinal crazes and microvoid coalescence into cracks[38]. In this case ions trapped in the craze may produce high local fields[120] and initiate an electrical tree channel by invasive injections/extractions. The continued propagation of an electrical tree probably also relies on the absence of water, since water vapour in polyethylene voids extinguishes discharges by depositing semi-conducting oxalic acid crystals on the void surface[407].

The conditions for this type of initiation process will be favoured by the presence of ionisable organic contaminants (e.g. esters), and a low concentration of antioxidants[406]. This latter feature may also be a contributing factor to the growth of exceptionally long trees which can be expected to bifurcate on entering a region where the microvoid formation time (t_v) is reduced. The growth rate of the fingers of such trees can be expected to accelerate[311,249] and therefore they are the type of tree which is most at risk of outstripping their water supply. Other trees in this category are large bow-tie trees grown from contaminants which rely on dielectrophoresis for their supply of water[418]. Both types of tree may occur in cables, but it is the former which seems to be most dangerous.

When a tree approaches closely to the counter electrode its behaviour may be somewhat different. Here a random advance at a point on the tree surface will accelerate rapidly because of the strong increase of field over gaps of ~ 10 μm. Thus the electrolyte will be allowed access to the electrode while deterioration continues to accumulate in the rest of the water tree. At the point of contact the electrolyte will take part in a field driven electrode reaction, which can generate organic and metallic ions, precipitate salts, and decompose water to hydrogen and oxygen. Rapid chemical deterioration and stress cracking is thus likely in a region abutting an electrode which itself is subject to chemical and physical roughening and etching. The severe disruption of the electrode surface which can be clearly seen in Fig. 6.3, is much greater than usually occurs at the needle during electrical treeing[419]. In the combined field enhancement of tree microvoids, ionic space charge, small electrode gap, and sharp electrode asperities, partial discharges may occur in voids or channels formed by microvoid coalescence near to the electrode.

Chen *et al.*[276] have suggested that vapourisation of the water by dielectric heating may be instrumental in generating electrical trees which then propagate by partial discharges in the vapour-filled voids. While this is a possibility at the high frequencies (~10 kHz) used by this group to accelerate tree initiation, the mechanism seems more appropriate to subsequent electrical tree propagation rather than inception at the working frequency of 50 Hz.

In nearly every respect the electrical tree would be expected to follow the path of the water tree since this provides an 'easy' route of pre-existing deterioration marked by microvoids potentially available for rupture, together with a supply of electrolyte and vapourisable material. In nearly all the observations[119] this has been the case, with one exception however[420], where, after the insertion of a metal electrode into the water tree, the breakdown avoided the tree. This result is similar to breakdowns initiated from the anode by point-cathode streamers in liquid breakdown[421,422] and suggests that the water tree surface may be charged. It is also possible that the high field has initiated a separate electrical tree within the body of the water tree, as observed by Karner *et al.*[408], which repels the carriers injected from the metal.

Electrical trees generated by bow-tie water trees suffer from a number of restrictions to their formation. Firstly the amount of electrolyte available from the bow-tie water tree is limited and hence the chemical supply of charge necessary for electrical tree initiation can be expected to eventually cease. This factor may also inhibit the dissipation and neutralisation of wall charges produced by channel discharges. Since the electrical tree also has no vent for the release of gaseous decomposition products, propagation will terminate, at least transiently. All these factors act so as to place a limit on the extent of the electrical tree growth. It is clear that if bow-tie water trees are to contribute to cable failure in this way, their number density must be so large that the bow-tie electrical trees they generate can link up so as to be able to grow across the insulation, a rather improbable occurrence at working voltages. For this reason bow-tie water trees are not usually regarded as particularly damaging in cables and they rarely appear to be associated with cable failure[423]. On the other hand vented water trees can cross the insulation thereby providing a convenient route for a subsequent electrical tree, and are thus a major concern in cables[424].

6.2 Catastrophic failure

This section will be concerned with dielectric failure as the culminating process of treeing degradation rather than with the range of failure mechanisms that may be operative in polymers. These latter processes will be dealt with in Part 3.

In some ways it is possible to view electrical treeing as a deterioration mechanism lying just on the threshold for a runaway breakdown. Thus in its early stages electrical tree growth is almost self-limiting reflecting a near attainment of equilibrium. Some electrical trees of the bushy-type do in fact achieve an equilibrium and never progress further, in much the same way as some cathode streamers in liquids[425,419]. However in most cases the accumulation of damage in the early stages causes the system to become unstable leading to a runaway propagation, Fig. 5.2, which is the penultimate stage of treeing breakdown. This stage is characterized by a few rapidly propagating[317] (10 μm · min^{-1}) branches which initiate at the tips of the existing tree channels, and is similar in form to the secondary streamer stage observed in the breakdown of liquid dielectrics[421]. Such a parallel is most evident between the bush-branch trees of solid dielectrics and cathode streamer breakdown in liquids, where the bush tree appears to act as a secondary electrode which initiates a branch-type tree from selected tips (sites) on its surface rather as does a metal electrode[331,426]. At the time when runaway branching is initiated major discharges within the body of the tree will have ceased[317] and its branch tips will contain cation wall charges, thus the analogy with a structured or rough electrode is exceedingly apt.

When many trees (primary or secondary) are initiated around the same time they tend to grow only to short distances and then stop[427,419], probably because of their interactive electrical shielding. In other cases however there is a swift propagation of a branch which may slow and stop while another branch initiated at a different tip of the primary tree surpasses it in growth. This type of growth is not untypical of fractal objects produced by stochastic processes. Here the fractal dimension and time development are average properties both of single trees and of several realisations in identical conditions. Thus in the same tree there will be regions more heavily branched than the average and region less branched. These latter regions will be more susceptible to further growth than the former, because their space charge field intensification will be higher, their electrical shielding will be less, and possibly they will be more accessible to discharges in the body of the tree giving a larger deposit of space charge. Such regions experience a more rapid propagation than the average thereby allowing them to build up a greater than average shielding while the initial heavily shielded regions develop very little extra shielding and end up with less than average. In this way differently propagating regions in the fractal object are continually fluctuating about the average form[327].

In these simulations however the growing fractal is not usually allowed to approach the boundary of the grid, whereas this is often the case in the experimental situation where secondary tree branches once initiated will experience a significant increase of local field. This is due to a reduction in shielding of the cation space charge which is transferred from the primary tree (branched or bushy) into the growing channel by charge repulsion. The primary tree therefore acts as an injecting source, but of cations rather than electrons as in the liquid case[421]. Consequently the whole tree system behaves as a region of space charge around the initiating electrode with the secondary tree projecting towards the counter electrode. Extension of the projecting tip by channel formation involving small discharges[317] (<5 pC) can occur in a manner similar to that of the first tree channel. However cation transport into the projection will occur much faster than cation generation by discharges when the first channel is formed. Propagation of the projection tip will thus be the result of channel-forming avalanches to and from the advancing space-charge boundary which extinguishes channel discharges above[317] ~5 pC.

Because of the progressive enhancement of the tip field the progress of the secondary tree will tend to accelerate. Some branches may therefore approach closely enough to the electrode to experience the accelerating effect of the field increase and hence continue on to bridge the insulation[317]. In some cases though the field between branch tip and the electrode may even be sufficient to initiate a local breakdown between them via thermal runaway[428], electro-fracture[346], filamentary electromechanical break-down[429], or simply mechanical fracture[319,316], see Section 7.3.2. This run-away region will in general appear as a thin ($<1\ \mu$m) diameter unbranched channel following a line of least resistance between the tree tip and the electrode, an example of which can be found in Fujita *et al.*[335].

Although the screening of the leading channel in a secondary tree is weak it will possess some sub-branches ($d_f > 1$) and its propagation may not runaway. If the point-plane gap were sufficiently wide therefore it is possible that a secondary tree would slow down and not cross the insulation before other such trees had been initiated at the primary tree boundary by random stochastic processes. In this case the space charge will be shared between the secondary trees and the field at each one will be shielded by the others. As a result the structure will be converted to a larger scale fractal object as suggested by the simulations[327]. In some cases it appears that the secondary tree converts to a bush-shape at high voltages (>20 kV) and frequencies (>400 Hz) in thick ($\sim 6{\cdot}3$ mm) cable insulation[386]. However such experiments are difficult to perform, and realistic simulation models should be considered to check this contention[399], with particular attention being paid to the propagation of different sub-regions.

Once a tree has crossed the insulation an air channel exists connecting the electrodes. An instantaneous short circuit would thus be expected if the channel walls were conducting or semi-conducting at this stage. This however is not usually the case[178,317,386,430,431], with the tree able to withstand breakdown for an additional period of time which may last up to 100 hours[317]. If the stress along the air channel is not short-circuited by wall conduction it would normally be expected that a gas discharge would proceed rapidly[375] ($\leqq 0{\cdot}1\ \mu$s) to breakdown, especially since any excess gas pressure would be released on crossing the insulation. However spark discharges are inhibited in channels with very small diameters[376] by wall charges which may be either present initially[377] or introduced through gas avalanches which terminate on the walls[376,377]. Tree channels reaching the counter electrode typically have diameters less than 10 μm, often in the range 1—5 μm, and are thus unable to initiate a spark. Instead channel widening by eroding discharges precedes failure. One way this is achieved is via avalanches initiated in the gas which terminate on the channel walls. Such events may neutralise local concentrations of wall charges and cause erosion through the energetic oxygen cations[123,124,380] they produce from the gaseous contents of the channels[384]. The electrons may also be trapped in the polymer, produce anions on the walls particularly if electronegative gases are present, or even generate a positive wall charge via a small percentage of impact ionisations on the wall[386]. Surface tracking between oppositely charged surface regions will then also lead to channel erosion[165,432]. This process may also occur between any conducting patches on the channel surface, and will cause erosion and carbonisation in PVC and PMMA, but just erosion in PE. In brittle materials such as some epoxy resins electrostatic forces may also cause some local mechanical damage around the channels.

As the channel diameter approaches 10 μm only a few of the electrons in an avalanche are captured by the walls. Instead those electrons that do reach the wall generate secondary electrons[386] which together with those produced by the UV emission initiate a succession of avalanches converting the discharge to a streamer[381]. This type of discharge can reach magnitudes of ~5000 pC and occurs just prior to catastrophic failure.

The final event in treeing breakdown occurs when the local streamer gas discharge becomes self-sustaining. It originates at the ground electrode where sufficient wall charges have been removed to allow an initiating streamer to develop. When there are enough charge carriers of sufficiently high energy in the streamer colliding with the electrode to raise its temperature (and reduce the barrier to injection, see Section 9.2) to the point at which thermal injection of counter charges will occur, acceleration in the space charge field around the electrode will raise the current to a value at which the gas becomes substantially ionised. A self-sustaining flow of electrons and cations in opposite directions down the channel will then be formed with the temperature of the ionised gas plasma reaching[419] 4—7 × 10^3 K. At these temperatures the conducting plasma column is highly radiant in the visible region of the spectrum. Its tip will be charged and will advance at a speed[375] of ~10^5 m · s^{-1} initially following the tree channel.

The rise in temperature at the channel surface will be sufficient to cause melting and carbonisation, and possibly some radial cracking. Any wall charges will be neutralised by charges diffusing out of the plasma, which contains enough energy to create a path along the shortest distance between the electrodes, sometimes leaving the tree structure where this no longer fulfills the condition. It has been reported[317] that the field generated between the counter electrode and the plasma tip is sometimes sufficient to induce a counter propagating discharge column which advances to meet and join with the originating plasma finally culminating in an arc discharge between the two electrodes.

At the velocity of propagation of the plasma the arc will form within 10^{-8}—10^{-7} s and will continue until either the power supply is exhausted (laboratory experiments) or the discharge is short circuited by the formation of a conducting carbonised wall (service conditions). While the arc continues damage to the electrode surface through vaporisation and melting[256] may occur with ions from the metal electrodes accumulating on the wall of the discharge channel[433]. In capacitors with thin electrodes the vaporisation may be used to isolate the breakdown region through a complete removal of the electrode around the breakdown point[434,435]. This is not possible in most cases however and failure will be complete when the channel walls becomes sufficiently conducting to sustain a continuous current between the electrodes thereby short circuiting the capacitance formed by the residual insulation.

Chapter 7
Factors affecting treeing

7.1 Effect of morphology upon treeing

7.1.1 Water trees

As we have seen in Chapter 4, water trees are a very low level form of electrical degradation. Their principal identifying feature on the microscopic scale is the presence of microvoids (~ 1—$5\ \mu$m) marking their path, although there is circumstantial evidence for more weakly degraded regions which link the tree system together (Chapter 4 and Section 6.1). These latter regions are not physical channels and are probably best regarded as 'easy' transport routes[262] along paths containing increased free volume, some of which exists in the form of sub-microscopic voids[262] (~ 10 nm) which act as microvoid generating sites. It is probably these sites that are revealed as small voids ($< \sim 0.5\ \mu$m) on boiling in water[212]. Chemical changes within the tree take place at the limit of resolution and appear to be restricted to the surfaces of the microvoids[218,221].

It has been suggested in Section 4.3 that the kind of mechanism best able to explain water tree formation is an electrically-driven combined physical and chemical process. Under these circumstances it is rather difficult to separate the effect upon treeing of material morphology from that of chemical composition. This is even more the case when it is realised that the polymer morphology itself may be modified by a change in chemical composition. For these reasons this section will concentrate upon those effects which can be directly related to morphological changes, even when such changes originate with variations in chemical composition.

(a) Inception
(i) Vented trees
In Section 4.3 it is contended that the initiation of vented water trees relies upon the transfer of solvated ions from an external reservoir to inception sites in or very close to the polymer surface. Certainly inception will be favoured by such a transport process, while propagation will find it essential. Any means of facilitating the access of electrolyte to the polymer interior will therefore accelerate inception and reduce the inception time t_d. The most obvious route whereby this might occur is via surface scratches or microcracks[115,234]. Apart from externally-caused mechanical damage the incidence of surface crack formation will depend upon both the physical properties of the material and its preparation history.

In the specific case of extruded polyethylene it used to be thought that the surface region was composed of long crystals oriented transverse to the surface[170]. The inter-crystalline regions could thus be expected to function as access routes and potential cracks[268]. This picture however has since been shown to be false, and it is now known that the surface of extruded polyethylene appears to have the same type of lamella organisation as the bulk material[17,15]. This does not however preclude the possibility of a surface variation on a finer scale than that of the lamellae (i.e. $< \sim 0 \cdot 01 \ \mu$m). In particular it is possible that the surface of extruded material may contain uncrystallisable material in the inter-lamella space. This will give rise to frozen strains[222] and possibly extended free volume defects in the amorphous regions. Such defects at the surface will provide both easy routes for its penetration by solvated ions, and a means for their retention.

The morphological factors which favour vented tree initiation are thus the presence of molecular packing defects near the surface[16], which act as inception sites, and a flexibility in the surface chain structure necessary to allow solvated ions to penetrate. These two factors may not always be complementary. For example network polymers such as epoxy resins may have potential initiation sites near the surface, but are very rigid and hence difficult to penetrate. These polymers are only weakly susceptible to the formation of vented trees[436].

Semicrystalline polymers such as polyethylene do allow some surface penetration by solvated ions[437]. The more permeable the material the more rapid will be the penetration and the easier the tree inception[265]. For example the inception time is less[206] in PE than XLPE even though the crystallinity of polyethylene is the same or slightly higher (few percent) than XLPE in the commercial products normally used in water tree experiments. The sigmoidal curve obtained for the number density[119] (Fig. 4.6) in these materials implies that here the number of inception sites available under a given set of conditions is finite and a relatively small fraction of the surface. When extraneous factors such as surface flexural forces or the presence of environmental stress cracking agents[223] cause the extended packing defects to form surface microcracks an increase in the number of sites initiating trees will occur[223]. The increased access to an inception site and the local stress enhancement afforded by a crack or scratch will also reduce the inception time[206]. However the density of inception sites will also be increased if the surface penetration is increased regardless of the production of damage, as for example by the application of hydrostatic pressure[259].

Amorphous rubbery polymers such as EPR have considerable free volume and their surfaces are relatively permeable to moisture and ions[437]. They are therefore likely to be susceptible to water tree *initiation*. In its unfilled state EPR is rather brittle and tends to form cracks which acts as inception sites as expected. The commercial form of EPR however contains large amounts of a filler ($> \sim 40\%$) such as clay in order to give it flexibility and mechanical strength. The reduction in free volume and increase in stiffness can be expected to reduce its susceptibility to vented tree inception[219]. Until recently[119] it was difficult to see water trees clearly in filled EPR and it appeared that they initiated rather slowly[115,206]. However now it appears that they are initiated in considerable numbers[229,205], grow rapidly at first, but then only slowly to the size at which they could be resolved in the earlier experiments[115]. It is noticeable that the tree density continues to increase over the period of observation, Fig. 7.1, in contrast to the behaviour found for polyethylene. Probably the number of potential inception sites available is very large. It should be noted that the performance of EPR is critically dependent upon the type of filler[438] and this point will be returned to again in the following sections.

(ii) Bow-tie trees

Bow-tie water trees are initiated at a variety of imperfections in the bulk material. These include voids[197,305]; gel, amber and glass contaminants[190,403]; metal particles[190], and cellulose fibres[305,196], but not carbon particles[190]. Of all these potential inception sites it is the voids that are related to the material morphology[199,436] and will be influenced by any changes made to it. However some inception centres such as anti-oxidant clusters[118,190] are formed from additives which have reached a dangerous local concentration level, and their existence may be indirectly dependent upon the physical structure and its formation history.

Voids are an intrinsic feature of most polymers (see Chapter 1). In the commercially important case of polyethylene, crosslinking, which otherwise improves the quality of the insulation[437], can increase the void density considerably[199,436] (10×) when a steam curing process is used. If nitrogen or silane crosslinking is adopted instead the void density is not significantly different to that of the starting polymer. Radiation curing also leads to a low void density[439], however in this case care must be taken to avoid excessive doses which appear to facilitate tree propagation[400,115,206]. Thus it could be expected that the incidence of bow-tie trees would be low in clean XLPE insulation crosslinked in a suitable manner. However this may not always be so. Heat treating after crosslinking is relatively cheap and simple, and has also been used to remove voids in cables[440,306]. In this case the density of bow-tie trees increased more than twenty-fold[306] while the void density

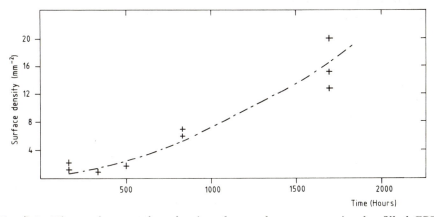

Fig. 7.1 The surface number density of vented water trees in clay-filled EPR plaques. The continually increasing number density implies that the initiation time distribution may be nearly constant over this time range after an initial period $t < 100$ hours where no initiation occurred, i.e. a step-up function

reduced from 10^6 mm^{-3} to 2×10^4 mm^{-3} after heating at 109°C for 17 hours. It has been suggested[441] that this increase can be associated with a structure giving a peak in the differential thermal analysis (DTA) melting curve, possibly related to secondary crystallisation in the amorphous regions. It is then argued[441] that microvoids are formed by this process thereby increasing the density of inception sites. However heat treating which includes a period *above* the melting temperature reduces the second peak to a shoulder but gives the same bow-tie densities. The most likely explanation of this behaviour has been provided by Namiki *et al.*[442] who showed (using scanning electron microscopy on freeze-fractured sections) that voids in XLPE cables are almost filled by low molecular weight polyethylene which exudes into the void in both crystalline and amorphous forms, when the material is heat treated above 60°C. In subsequent treeing experiments the bow-tie density was found to increase even though the apparent void density was reduced. Most probably the residual gap between the sides of the 'filled' void and its plug[190] favours a rapid condensation of electrolyte into a region whose shape leads to an increased stress enhancement. When combined with the mechanically weakened surface of the void it can be expected that large numbers of small bow-tie trees[190,442,306] will be formed in a relatively short time leading to the observed increase in number density.

A general feature of bow-tie inception sites appears to be that they act as centres at which water can collect, regardless of whether it is physical[190] (voids), chemical (hygroscopic contaminants[196]), or electrochemical (metals[190] and metal ions[188]), factors that allow them to function in this way. Consequently it is heterogeneous materials that are most susceptible to bow-tie initiation, if avoidable features such as contaminants and voids[443] are kept to a low level. For examples clay fillers in EPR can possess surfaces with up to one hydroxyl group per square micron and can thus adsorb water. Since the filler particles also allow free volume generation (reversal of physical aging) to occur under mechanical or electro-mechanical stresses even though the material is in a rubbery state (see Section 3.1), and trap charges on their surface[438], they will function as bow-tie inception sites. Hence clay-filled EPR can be expected to be very susceptible to bow-tie generation. A different type of heterogeneous material is found in blended polymers and random copolymers[444]. Here again boundaries between regions of different chemical composition and dielectric constant form structural interfaces at which water will collect when an electric field is applied, and hence initiate bow-tie trees. It is likely however that the structurally weaker boundaries provided by the blends will allow a more rapid initiation than in the co-polymers[444] where different regions are chemically bonded together.

The requirement of an access to water for bow-tie formation is probably revealed by the growth saturation observed for void-initiated trees[119,197]. Here the fact that the saturation length of the bow-tie tree is linear in the void diameter[197] probably indicates that growth will cease when the finite reservoir of the void is used up. This can be shown by taking the bow-tie to be a double-cone of length L, its volume is then proportional to L^3 and hence at saturation to the volume of the void. Other observations are consistent with the importance of access to water for bow-tie initiation. For example in steam-cured cables an annular region can be observed where water droplets collect. This 'halo' provides a source of condensed water from which bow-ties have been found to initiate very quickly[445] even though, in this particular instance, the void density was much less than in the rest of the cable where no bow-ties were observed. Pelissou[418] also found that the number of long (>80 μm) bow-ties was much larger in the halo region, although their total number density was almost the same throughout the cable. Clearly therefore the inception time for bow-tie tree initiation depends upon the access of a potential initiation site to water. If, for example, water-filled voids are present initially, the inception time will probably be too small to measure[238,191]. However if water has to condense at a site the inception time will be dominated by the transport rate[231], as suggested by Patsch[263] and Tanaka *et al.*[222]. In this latter case morphological features, which affect the water permeability of the polymer, will control the inception time for bow-tie trees. Here semicrystalline polymers can be expected to delay inception longer than amorphous polymers[400] having a similar chain chemical structure, though probably network polymers such as epoxy resins will be best of all.

The above considerations suggest that the controversy about whether or not bow-tie trees are instantly initiated (see Section 4.2 and Reference 119) may be due to differences in experimental practice. Thus if the sample is in contact with water prior to energising with an electric field, the inception time may be too small to measure[238,191]. Alternatively if the field is applied from the moment contact with a reservoir is established the inception time may become long enough to measure[218,190]. In fact the exponential inception rate[119,444] of eqns. 4.2a and 4.2b may be deduced for void-initiated bow-tie trees by assuming that the trees can only start growing when the void is filled with water and that this takes a time proportional to its diameter. Such a relationship would occur if the voids were filled by a water boundary advancing at a constant velocity through the material. This type of behaviour would not occur if diffusion under a concentration gradient were the only driving force, however penetration will also occur via percolation, osmosis, and the establishment of phase equilibrium between water vapour and condensed water. All these and other factors may lead to just such a limiting boundary velocity. The distribution of void sizes in polyethylene is usually taken to be exponential[119] (see Section 15.5) and hence the number of voids, $dN(r)$, in the size range $r \to r + dr$ will be

$$dN(r) \propto e^{-ar} dr \tag{7.1}$$

with r proportional to their inception time. Thus the number of trees initiated during the period $t \to t + dt$, i.e. $dN(t)$, will be

$$dN(t) \propto e^{-a't} dt \tag{7.2}$$

Integration of eqn. 7.2 over the range zero to t, then gives the number, $N(t)$, of bow-tie trees in the same form as eqn. 4.2a, i.e.

$$N(t) \propto N(\infty)[1 - e^{-a't}] \tag{7.3}$$

Note that if we take the saturation length to be proportional to r as suggested by experiment[197], eqn. 7.1 leads to an exponential distribution for such lengths, as is commonly observed[119]. The exponential number density assumed above for the void size formally yields a finite number of voids with zero radius and hence a zero inception time. Since however there must be a smallest size of void there must therefore be a shortest inception time. If the conditions of the experiment are such that the time required to condense water at these sites is appreciable, a shortest value for the inception time can be measured and the plot of $N(t)$ against t will have a sigmoidal form[237]. These considerations show that where bow-tie trees are concerned the state of the material prior to test is of major importance, and this should be noted when presenting experimental results. For example the length of bow-tie trees grown when there is no contact to a liquid reservoir increases with the relative humidity of the surrounding atmosphere[218] without reference to void size. It is also probable that the water content of a cable will have a strong bearing upon the formation of bow-tie trees during service[241]. In contrast vented trees rely on an external source for their inception, except when they grow from the central conductor[249,231], and the water content of the cable will only affect them once they have already been initiated[175]. Indeed water inside the cable may even reduce the density of vented trees[175]. A possible reason for this behaviour may be that some vented tree inception sites initiate rapidly through access to internal water and are thus able to compete more effectively with other potential sites for a continuation of the water supply from both the external and internal reservoirs.

Although essential a local water concentration alone is not sufficient to act as a tree inception site. For example in Reference 196 no trees were produced when a water concentration was established by dielectrophoresis. It is likely therefore that the tree inception sites require other attributes in order to be favourable to the initiation of the propagation process, such as adjacent regions of extended free volume or microvoid strings. Defects such as these could be formed by poor contact between contaminant and polymer[190], differential mechanical strains at a boundary, or even electric stresses. The ability of a site to initiate a bow-tie, and its inception time, will therefore depend to some extent on the degree to which its associated defect regions facilitate the first steps in the propagation process.

Since propagation requires the extension of free volume and the creation of microvoids at the boundary of the site by a repetitive (fatiguing) application of the field (see Section 4.3) a finite time will be required for inception. By assuming that water-filled voids initiate bow-ties when the electric field generated internal pressure exceeds a critical value independent of the void size, the radial stress applied to the polymer, and hence the inception time, will be proportional to the void radius[119]. This picture would recover the argument leading to eqn. 7.3 but would not give instantaneous initiation for all trees in water saturated samples, except as an apparent result when the time scale of observation was long compared to the longest inception time.

Inception must also be governed by the ability of the polymer to resist free-volume extension under the combined stresses (electromechanical, electrochemical, and mechanical-chemical) of the initial propagation steps. The mechanical component in this process will clearly be influenced by the local displacements available to the polymer chains. Therefore as the glass transition temperature, T_g, approaches the treeing temperature from below large scale chain displacements will become unavailable and when the material is in its glassy state only small scale intra-chain motions can contribute. Thus the number density of bow-ties should reduce with increasing T_g, just as found by Yoshimitsu and Nakakita[196] for a series of epoxy resins, Fig. 7.2.

(b) Propagation
Water trees grow from their inception sites by the generation of defects, probably in the form of free volume extensions, some of which develop into microvoids. The outline of the tree itself can be envisaged as a damage boundary which propagates through the insulation from an inception site which may, in the case of vented trees, be indistinguishable from any other region of the polymer. Although the means by which the tree causes the damage is not yet certain, mechanical and chemical changes have been identified, and it is suggested in Section 4.3 that the mechanism involved is a combination of electrically driven mechanical and chemical processes.

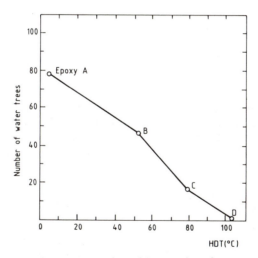

Fig. 7.2 The number of water trees found in a series of epoxy resin samples with different glass-transition temperatures. The resins were compounded in different proportions from two types (Shell: 'Epon 871' and 'Epon 828') which allowed the glass-transition temperature to be varied. This was characterised by the 'heat distortion temperature' (HDT) measured using standard ASTM D648-56. The glass-transition temperature in these materials was quoted[196] to be approximately 5 to 10°C below the HDT. The water trees were generated from cotton cellulose particles: a hygroscopic contaminant purposely added at a rated weight. The number of trees were measured after 63 hours at 50°C in a relative humidity of 80%. After Reference 196
Copyright © 1978 The IEEE

Apart from a limited access to water and inorganic ions bow-tie trees will propagate in a similar manner to vented trees, and will be influenced in the same way by material factors. Both types of tree will therefore be discussed together in this section, though attention will be drawn to any pertinent differences.

Because the damage involved in water tree formation occurs locally and on the small scale of the free volume (i.e. molecular packing order in the chains[16]) it is inevitable that variations in macroscopic properties such as density[265,232], crystallinity[444] and melt flow index[252,232] show little significant correlation with their growth. Instead the type of material properties that can be expected to influence water tree propagation will be those that alter the morphology on a local scale in the amorphous regions, and its ability to deform. We have already noted that the availability of mechanical relaxations favours the inception of bow-tie trees[446] and can expect them to favour propagation also. Conversely morphological modifications that inhibit such relaxations particularly in the amorphous regions of semicrystalline polymers will retard propagation. For example increases in the molecular weight of LDPE (from 48 000 to 161 000) increase the number of chain entanglements and tie-bonds between lamella and hence the elasticity limit. For this reason the resistance to free volume extension and microvoid formation is stronger in the samples with high molecular weight, and consequently their rate of water-tree propagation is slower[247].

In early work the relationship between water trees and material morphology was seen as a correlation between growth and softness[265], but now a much more precise picture can be formed. Transmission electron micro-

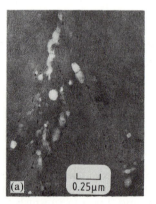

Fig. 7.3 Transmission electron micrographs showing water tree damage in materials with different morphologies. (*a*) Peroxide crosslinked LDPE; here voids formed in the amorphous inter-lamella regions coalesce into channels where the lamellae axes are favourably aligned, (*b*) Peroxide crosslinked LLDPE. This material possesses a spherulite (2–5 μm in size) organisation in contrast to that of (*a*). In this case chains of voids tend to develop in the interspherulitic regions around the spherulite boundary. Tree damage may also propagate through a spherulite along the axis of the lamella bundles. © IEEE 1987.

scopy (TEM), with a resolution of ~ 0.02 μm, has revealed that water tree damage is located in the inter-lamellae regions of semicrystalline polymers[18,16,38,447] in the form of voids (20 nm $\rightarrow$ 1 μm) and regions of chain disorder and strain (Fig. 7.3). The formation taken by this sub-microscopic damage was found to vary with the morphology[448,447] and the resistance of the material to fatigue environmental stress cracking[39,447]. The most facile material was LDPE where void coalescence could form channels up to 2 μm long and 40 nm wide, along stacks of lamellae possessing a common alignment. Extension of the damage in this way was also found to favour rapid water tree propagation. Localisation of the damage on the other hand reduced the growth rate and increased the resistance to fatigue failure, see Table 7.1 for some examples. This was found to occur when spherulites were present[39,447] for example in compression moulded PE, polypropylene, isotactic polybutene-1, and peroxide crosslinked LLDPE. Here the voids form preferentially around the spherulite surfaces where they sometimes coalesce into unconnected channels.

The spherulites (2—5 μm) therefore appear to restrict the rapid extension of aligned voids. A LDPE copolymer gave a different type of localisation with regions of severe damage and large voids but no coalescence into channels. This material also has a slower growth rate. The material with the slowest growth rate also had the greatest density (see also Reference 332) and largest tensile yield stress. In this case very few voids were found and the extent of the tree was essentially formed by the free volume extensions of the pre-void stage.

These results seem, qualitatively at least, to be in agreement with the analysis of Section 4.3 if the damage density g is taken to include free volume extension (molecular disorder) as well as void formation. There it was argued (eqn. 4.22) that for a given material the generation of a unit of damage required a certain amount of time t_v. Thus if the damage density required to advance the tree boundary is increased, for example by propagation around a spherulite boundary rather than by longitudinal growth, the

Table 7.1 Water tree growth rate constants, Q_a, for compression moulded samples compared to fatigue stress cracking time to rupture, T_r. Data taken from Reference 447

LDPE formulations	p	Q_a [μm^3 h^{-1} kV^{-2}]	T_r
A	2	3×10^3	5 min 41 s
B	2	2.6×10^3	7 min 32 s
C	2	1.6×10^3	11 min 46 s
Peroxide X-linked			
A'	2	6×10^3	15 min 28 s
B'	2	2.6×10^3	23 min 47 s

C' Damage density g increases with tree volume ($\propto V_T^{0.85}$) see eqn. 7.5

growth rate will be reduced. In this case however the greater damage density in the water tree may offset its small length and the tree may be as dangerous (see Section 15.5) to the insulation as much longer trees[175]. It would be interesting to see what effect the increase in damage density has on the dielectric strength and fatigue failure.

Microscopic analysis (optical and TEM) has been used to compare the damage structure of vented trees in both crosslinked and uncrosslinked versions of the same material[447]. From this it appears that the lower growth rate of XLPE, as compared to PE, is not due to a difference in the damage density, but rather to the ease with which the polymer forms the damage regions. Crosslinking can be expected to increase tie-chains between lamellae and hence the elasticity in the amorphous regions. Thus the time required to form a unit of damage, t_v, will be increased and the growth rate slower[115,206].

In filled materials the filler particles will behave rather like spherulites. It has already been pointed out that such particles cause the neighbouring polymer to be susceptible to physical aging (Section 3.1.1) giving locally modified regions which electro-mechanical stresses can be expected to easily open up into free volume or voids. Such centres are ideal for bow-tie inception (as we have noted), but when incorporated into vented tree growth will lead to a large damage density in the tree. It can therefore be expected that vented tree growth rates will be much less for filled material than an equivalent unfilled one. An example of this behaviour can be found in the growth of vented trees in filled EPR[229,205] shown in Fig. 4.10b. In this case the growth is not just slow, it in fact appears to stop, rather like bow-tie trees. Since access to an electrolyte reservoir is not impeded, this effect may be due to other factors associated with filler particles. For example the particle (clay) may absorb water and swell thereby acting both as a 'drain' for the electrolyte and a physical barrier to further penetration. When this occurs it appears that less favourable sites can then initiate a vented tree giving continual inception. Fillers can also produce a similar reduction in the growth rate of bow-tie trees[449], and it also appears that if suitably chosen they can inhibit inception as well.

The work of de Bellet et al.[447] also throws light upon the growth of water trees under mechanical stresses either applied or frozen in. In semicrystalline materials such stresses will tend to align the lamella, and hence the free volume defects in the amorphous regions, along their direction of application. It is therefore to be expected that water trees will grow preferentially and rapidly along these lines of defects and disordered surfaces[213] as observed[119,223,232] rather than breaking the polymer bonds that are weakened by tensile stress[119]. However it should be noted that over a long period of time the stress orientation may be overcome and the trees resume growth in the direction of the electric field[228]. Presumably the region of strain and defects becomes 'saturated' by the treeing damage after a while and it becomes more easy to open up regions transverse to the stress than extend the strain longitudinally.

Polymeric insulation sometimes undergoes prolonged heating at temperatures just below the melting point so as to remove dangerous by-product gases (e.g. peroxides) from the crosslinking process[247,252]. Under these conditions it is possible that the material inhomogeneities will be reduced and residual strains relieved, that is the material is annealed. Such temperatures (up to ~90°C in polyethylene) may also occur near the conductor of power cables during service. A number of investigations of their effect upon water trees have therefore been made[17,252,247,306]. In most cases the growth rate of the water trees was greater in the annealed specimens[17,252,306], and this was attributed to local changes of morphology. Heating semicrystalline polymers below their melting point causes longitudinal extension of the lamella, lamella thickening, and some secondary crystallisation in the amorphous regions. Since large scale segmental adjustments are not possible below the melting point these local changes in crystallinity will be accompanied by compensating strains and free volume increases in the amorphous regions. Such defect regions will be frozen in during rapid cooling, and will act as incipient centres for water tree damage[252]. They can thus be expected to accelerate water tree growth both by reducing the time required to generate water tree damage, t_v, and by facilitating the transport of water to the tree boundary. This later factor may be particularly important in the case of bow-tie trees[306]. Slow cooling from the annealing temperature will allow some re-arrangement of the chain segments in the amorphous regions and Gölz[252] has shown that in this case the growth rate reduces towards that of the un-annealed material. If structural memory is lost by heating above the melting point the water tree growth rate returns to that of the de-gassed but otherwise 'as-prepared' material[252].

In one case[247] annealing (24 hours at 85°C) caused a decrease in vented tree growth rates. It is noticeable that in this case the growth law has been modified from that of eqn. 4.21a to

$$(L+r)^3 - r^3 \propto \tau^{0.8} \tag{7.4}$$

see Fig. 7.4. If it is assumed that the time required to produce a unit of water tree damage, t_v, is constant throughout the period of tree growth, then eqn. 7.4 implies that the density of water tree damage, g, is given by:

$$g \propto (N_v(\tau) = \tau/t_v)/V_T(\tau)$$
$$\propto V_T^{0.25} \tag{7.5}$$

i.e. increases with the size of the tree. It therefore seems that in the point-plane geometry used here annealing has produced incipient defects which are not homogeneously distributed, leading to an increase in the density of tree generated damage as the tree grows.

7.1.2 Electrical trees

The degradation produced by electrical trees is much greater than that of water trees, being characterised by a connected system of hollow tubules (channels) whose diameters are 1 μm or greater. Although inception and growth still involve the scission of bonds in the polymer chains, rather than polymer ionisation, the power dissipated is much larger than in water tree

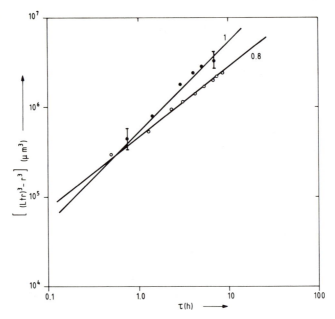

Fig. 7.4 Water tree volume $[(L+r)^3 - r^3]$ as a function of growth time τ (eqn. 4.21a) taken from the directly observed individual tree measurements in Reference 247. The material is polyethylene (containing antioxidant) of molecular weight 48 000. The symbols are: ● = unannealed material; ○ = annealed material. The numbers give the power-law relationship between tree volume and τ

formation. Channel formation is thus a rapid process which does not rely on a fatigue mechanism, even though channel enlargement and progressive extension require discharge or avalanche repetition (Chapter 5). In this case material morphology can be expected to have some effect, though it will not play the prominent rôle that it does in water treeing.

(a) Inception
The central feature of electrical tree inception is the formation of a channel by extraction and injection currents to and from the electrode (Chapter 5). The nature of the material will affect this process through the amount of energy required to produce a channel-forming rupture, i.e. C_t in eqn. 5.2. Thus the ease with which a tree is initiated and hence its initiation time can be expected to depend upon the yield stress of the material as reported[450,367]. Here trees will initiate faster the lower the tensile strength, Table 7.2. A similar trend has been found[121] for tree inception voltages in unfilled HDPE, LDPE, and EPR with the weakest material (EPR) initiating trees at the lowest voltage and the strongest (HDPE) at the highest voltage under 'identical' conditions, Table 7.3.

Nury[451] has also reported an increase in tree inception voltage (field) in high pressure polyethylene as the size of spherulites becomes larger. Since however LLDPE with larger spherulites had a smaller inception voltage[451]

Table 7.2   Electrical tree inception times, t_I, and times, t_g, for the tree to cross at least 90% of the insulation[367] (point-plane geometry, 7 kV r.m.s. applied at 50 Hz)

Material	Tensile strength (MPa)	Modulus of elasticity (MPa)	Inception time, t_I (hours)	'Growth' time, t_g (hours)
Elastomer PL-3	6·8—10·34	13·7	$10·75 \pm 2·55$	600 ± 50
Elastomer PL-2	58·5—75·8	207	$19·5 \pm 4·33$	450 ± 50
LDPE	75·8—103	344	$26·45 \pm 3·09$	290 ± 50
Polyester	137·8	3780	$36 \pm 5·3$	350 ± 50

it seems likely that the increased difficulty of tree inception is due to a higher yield stress in the amorphous regions. In the high pressure polyethylene these will be compressed by the bigger spherulites, and it is noticeable that the density increases as the inception voltage goes up for both materials. Therefore spherulites *per se* do not appear to inhibit tree inception by acting as a barrier, but rather through their effect upon the free volume in the adjacent amorphous regions. This dependence upon the availability of free volume and yield stress in amorphous regions can also be seen in the low inception voltage for XLPE when inhibiting chemical by-products have been removed[198]. Here the crosslinking is light and the material is non-spherulitic

Table 7.3   Electrical tree, 50%-inception voltage, V_I, in materials of different crystallinity using a double needle test geometry and AC, 50 Hz, field

Reference	Data		
121	Material	Crystallinity	V_I (kV)
	HDPE	High (large spherulites)	10·5
	LDPE	Medium (normal spherulites)	8·0
	LDPE	Low (smaller spherulites)	7·0
	Ethylene copolymer	Very low (very small spherulites)	5·5
	EPR	Amorphous	5·5
451	Material	Spherulite size (μm)	Density (Mg·m^{-3})	V_I (kV)
	High pressure LDPE	5	0·931	15
	High pressure LDPE	3	0·926	12
	High pressure LDPE	2	0·922	12·5
	High pressure LDPE	<1	0·922	11·5
	LLDPE	15	0·935	10·5
	LLDPE	11	0·920	8–9
	LLDPE	10	0·921	7

and thus its susceptibility to tree inception should be similar to LDPE, or perhaps worse if the de-gassing has introduced strains in the amorphous regions.

Since the primary step in electrical tree initiation is the establishment of a path consisting of extended free volume defects or crazes aligned on an injecting/extracting site on the electrode, the presence of such defects prior to the application of an electrical stress can be expected to reduce the inception time and fields. For example tree inception will become easier under the application of mechanical stresses that create this type of defect. This will typically be the case for applied tensile stresses[144] or Maxwell stressing by an AC field applied to the electrode[367]. Conversely hydrostatic compressive forces will tend to reduce the free volume and increase the inception times/fields[365]. Aligned free volume defects will also act as nucleation sites for microcracks which may be formed during material manufacture or aging if the electrical tree inception site is a metal asperity. For polyethylene this is most likely to occur during rapid cooling to low temperatures[319]. On the other hand polyesters and epoxy resins are prone to crack formation when shrinkage occurs during an inadequate curing or cooling stage[322]. In this case the earliest stage of the inception process is by-passed and inception times/fields will be lowered. Even where the aligned regions only represent an *incipient* microcrack, AC Maxwell stressing or mechanical stresses will induce crack formation and reduce inception times/fields[367]. However when mechanical and Maxwell stresses are combined so that there is no nett stress, cracks will not be formed, and the tree inception behaviour will be that of the original material. It has been shown[367] that under these circumstances tree inception in a polyester can be strongly delayed (up to 60 hours) at low voltages.

Polymer morphology may have a rather subtle effect upon tree inception in DC fields. Here injected space charge is trapped around the injection point and moderates the electric stress[287]. When the applied voltage is constant the field-limited space charge model predicts a fixed region of space charge[346,338]. However this model neglects the small but finite drift mobility of the space charge at low fields. The injected space charge density will therefore slowly drop and then be replenished by further injection which will likely take place from the same spot on the electrode. As a result a *slow* sequence of injection pulses[452] will occur which will eventually initiate a tree[387]. In this case tree inception will be favoured by a morphology with shallow traps and increased low-field mobilities, possibly a material with a low density of structural defects at the amorphous crystalline interface[106,107]. On the other hand if the injected charge is not dissipated in the bulk of the material, but is transported sufficiently so as to be redistributed about the injection region, an increased moderation of the local field may occur possibly reducing the likelihood of tree inception. Such an occurrence has been suggested to explain the behaviour of mica in EPR, with transport along the mica sheets redistributing the injected charge[452].

(b) Propagation
During electrical tree propagation discharges in existing channels extend or generate new channels at the tip. Clearly amorphous regions, with their

lower yield stress, will be strongly favoured during channel formation[453], even though subsequent erosion may widen them into regions originally occupied by the lamella. It has also been reported that electrical trees avoid certain filler particles[121] such as polystyrene when blended with polyethylene[454]. Probably this results from the high tree resistance of the filler[198], and a similar behaviour may sometimes be found for spherulites[453]. However the fractal shape of electrical trees is now known to be a property of their propagation dynamics (see Section 5.4) and not material inhomogeneity[346]. This latter feature will introduce an additional stochastic factor into tree growth but there is no evidence that it can alter the shape of an electrical tree from branch-type to bush-type for instance. Consequently it is unlikely that material inhomogeneity will substantially affect tree propagation rates, unless the barrier regions occupy a substantial fraction of the insulation, when the branching (and dimension, d_f) may be increased.

It has been suggested that electrical tree propagation rates are determined by mechanical factors[320,366] such as the ability of the material to sustain an elastic deformation[367]. The evidence however is not very convincing; for a range of elastic moduli running from 14—3800 MPa the average time required to grow a given distance ranged only between 600 and 300 hours with a variation of ±50 hours, Table 7.2. Indeed the times (350 ± 50 hours) for the material with the largest modulus were scarcely distinguishable from those (450 ± 50 hours) for the second smallest[367] (~200 MPa). In view of the variety of high energy bond breaking factors such as hot electrons, ion bombardment, chemical oxidation, thermal degradation, etc., occurring during discharges[123,124,383], such a result is hardly surprising. However it can be expected that bonds under tensile stress will be less able to resist the discharge than unstressed ones. Thus electrical tree propagation will be accelerated preferentially in a plane perpendicular to the direction of tensile stress. This results in a planar tree when the tree is of the branch type, and an elliptical structure if it is bush type[386], both with an increased propagation rate. These observations were made on mechanically stressed XLPE cables, but similar results were also found for polyethylene[455], epoxy resins[456], and

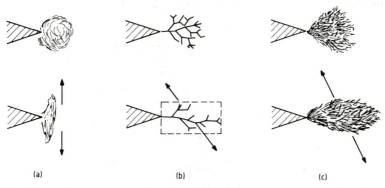

(a) (b) (c)

Fig. 7.5 Schematic representation of behaviour of trees in material under tensile stress (indicated by arrows). In each case the top drawing refers to the unstretched material: (*a*) water trees; (*b*) branch-type trees; (*c*) bush-type trees

polyesters[367]. The behaviour of electrical trees under tensile stress is in marked contrast to that of water trees which, as noted previously, follow the lines of extended free volume in the direction of the stress rather than attack the strained bonds, Fig. 7.5.

If the tensile stress is sufficiently high microcracks will be formed usually in the plane perpendicular to the stress direction, and electrical tree growth will be accelerated in the direction of crack propagation. This relationship between tree shape and microcrack alignment has been demonstrated using HDPE and LDPE samples[457,458], stretched in the direction of the field and perpendicular to it. The acceleration in propagation can be associated with the exposure of stretched bonds in the crack to discharges in the tree channels. For example it is known that corona and void discharges generate rapidly growing trees at cracks[459,460]. Electrostatic forces from the charges deposited may possibly lead to a runaway crack[144] propagation[127]. An electro-mechanical failure mechanism of this type may even be initiated by large electrical trees, and appears in some cases[335] to be the step whereby the tree finally crosses the insulation.

7.2 Influence of chemical composition upon treeing

7.2.1 Water trees

As we have stated previously (Chapter 1) the chemical composition of polymeric insulation is not simple. In addition to the host polymer chemical residues from the polymerisation and crosslinking processes are present, as well as additives deliberately introduced for commercial reasons (Section 1.6). Most investigations of the effects of additives are concerned with their tree inhibiting properties and these will be discussed in Chapter 8. Here we will be concerned only with those chemicals that appear either naturally in the insulation material or have been added for reasons other than tree inhibition.

(a) Polymeric composition
The most obvious way in which chemical composition can affect water treeing in hydrocarbon polymers is via the susceptibility of the material to oxidation. Oxidised regions in polymers are structurally weakening, hydrophilic and can bind solvated cations[118,221]. They can thus function as inception sites for vented trees when they occur at the polymer surface[285]. In addition oxidised and oxidisable regions in the bulk increase the ease with which solvated ions can penetrate the polymer and generate water tree damage, (see Section 4.3) thereby facilitating water tree growth[221].

Differences in water tree behaviour have been observed between homopolymers, copolymers, and polymer blends of different chemical composition and molecular weights. However such effects are not necessarily attributable directly to their chemical nature since the materials will also possess differences in morphology which themselves may be responsible for the alteration in water tree inception and growth. For example increasing the molecular weight of polyethylene increases its resistance to oxidation and hence should reduce water tree growth rates. Such a dependence has been observed[247] but since the presence of anti-oxidant did not significantly affect the trend (see Table 7.4), it was attributed instead to an increase in the resistance to local mechanical fatigue[247].

Annealing may also increase the susceptibility to local oxidation since contaminants and additives of all kinds will tend to become concentrated in the amorphous regions adjacent to the extended or newly crystallised lamellae. However, here too the change (increase) in growth rates can be taken to be the result of alterations in the morphology since slow cooling has a less severe effect then rapid cooling[252].

In the examples considered above the local oxidation resistance reduces as the local mechanical strength reduces, but this may not always be the case. A competition between chemical and morphological effects could occur in copolymers, blends and filled polymeric materials and is possibly the reason for the conflicting results reported for these materials[119,444,449]. We have already commented (Section 7.1.1) on the fact that an increase in bow-tie densities and growth rates can probably be correlated with increases in local strains and free volume due to the composition of random copolymers and polymer blends[444]. However in the random copolymer case the inception time, t_d, was also increased[444]. This result implies that the initiating step for the bow-tie trees was not just the accumulation of water and possibly reflects an increased resistance to oxidation of the copolymer inception sites. Later experiments with various fillers[449] indicate that their performance depended upon their chemical composition with oxidised or oxidisable fillers enhancing bow-tie trees and chemically inactive species (silicates, ethylene copolymers) inhibiting both their number density and propagation rates. These results highlight the need for more systematic studies in which the material morphology and chemical composition are characterised prior to treeing.

A recent investigation of vented and bow-tie trees in polystyrenes[461,443] has revealed a different way in which the chemical composition can influence water treeing. Here it was noted that the water trees consisted of very finely-branched channels, in contrast to the usual structure of isolated voids or disconnected channels. This difference may arise because an oxidative attack on polystyrenes leads to an un-zipping of the polymer which would tend to favour elongated defects, rather than the void-forming chain scission and crosslinking that occurs in polyethylene (see Section 3.1). Water trees

Table 7.4   Water tree growth rate constants, Q and Q_a, (defined in eqns. 4.8*a* and 4.7*a*, respectively) for different LDPE materials. Values obtained from the data of Reference 247, quoted in μm^3 hour^{-1} kV^{-2}

Mol. weight	Antioxidant	p	Q	Q_a
48 000	yes	2	3533	$1{\cdot}49 \times 10^5$
81 000	no	2	2100	$1{\cdot}07 \times 10^5$
119 000	no	2	553	$2{\cdot}81 \times 10^4$
161 000	yes	2	767	$3{\cdot}24 \times 10^4$

with this kind of structure can be expected to possess a different dimension to that usually found (i.e. 3), however the tree growth curves reported[461] show that the channel-like growth tends to saturate and be replaced by volume filling microvoid formation. Nonetheless it is likely that the elongated nature of the channel-like defects will cause a greater reduction in dielectric breakdown strength than that found for similar length trees in polyethylene, but here no data are yet available.

(b) Typical additives and residues

Antioxidants are essential additives to cable insulation since oxidation destroys its mechanical strength and flexibility as well as reducing the breakdown strength[103,41]. It would therefore be expected that their presence would inhibit water tree inception and growth, and in some cases this is what happens[115,206]. However staining antioxidants (amine) appear to accelerate tree growth[208,115,206,118] rather than retard it as does the non-staining (phenolic class). This is a rather puzzling result, and since the same effect is found in both inception and propagation[206] it does not appear that it can be attributed to a difference in the spatial dispersion of the two types of material. It is noticeable that fillers possessing amine groups also enhance bow-tie treeing[449] and thus the effect seems to be related to the amine group rather than any induced morphological factors. These groups are more polarisable than the phenolic systems, and can ionise in aqueous solution. They can also hydrogen-bond water molecules and so would behave similarly to an oxidised site. Probably therefore they form stress enhancement centres which are hydrophilic and can bind solvated ions, thereby aiding water tree damage generation.

Antioxidants are not very soluble in polyolefins[118] and when present in excess may migrate to the semiconductor shield interface[462] or condense into voids. In this way local concentrations of antioxidant may be formed which promote oxidation instead of inhibiting it[118,463]. Such centres will act as inception sites for water trees[190] and could also enhance their growth rates if they occur along the tree path. It is possible that all vented trees in service conditions are initiated in this way. A long inception time ($> \sim 4$ years[187]) may then be due to the migration of antioxidant[462] and oxidation of a local centre rather than ingress of solvated ions or water. It should however be noted that the field dependence of the inception stage in highly accelerated point-plane experiments is the same as that found in miniature cables, which indicates that the same inception process is being observed in the laboratory as in the field.

Acetophenone and cumylalcohol (eqn. 7.6) occur in crosslinked polyethylene as by-products of cross linking by dicumyl-peroxide[464] and strongly inhibit both the inception and growth rate of water trees[199,115,206].

$$
\begin{array}{cc}
\underset{\displaystyle\text{CH}_3}{\overset{\displaystyle\text{CH}_3}{\bigcirc\!\!-\!\!\text{C}\!\!-\!\!\text{OH}}}
& \underset{\displaystyle\text{CH}_3}{\bigcirc\!\!-\!\!\text{C}\!\!=\!\!\text{O}}
\end{array} \qquad (7.6)
$$

cumylalcohol (BP 215°C)   acetophenone (BP 202°C)

Indeed these two chemical residues are responsible for a substantial portion of the tree resistance of crosslinked polyethylene when compared to low-density polyethylene[199], and addition of acetophenone to low-density polyethylene markedly increases the inception time[115,206] and reduces the growth rate[199,115,206] (see Table 7.5).

Aromatic molecules, such as acetophenone and cumylalcohol, are known to trap free electrons by forming negative ions[198], and to moderate the stress at enhancement sites through a field-dependent increase in the conductivity[465,466]. Acetophenone and cumylalcohol are also relatively volatile, and may diffuse from the polymer over a period of time thereby causing it to become more susceptible to tree propagation, as can be seen in Table 7.5. Their volatility also enables them to migrate to and condense in the channels of electrical trees[464]. It is possible therefore that their effect on water trees is due to migration to potential inception sites which then become field graded. However the possibility that they fill mechanically-deformed regions and increase the local resistance of the polyethylene to yielding (i.e. act as a plasticiser) cannot be ruled out. Indeed they may just physically shield oxidisable sites from chemical attack by solvated ions. With a combined physical and chemical mechanism responsible for water trees it is possible that a number of features of acetophenone are active in inhibiting their formation.

7.2.2 Electrical trees

As we have seen in Chapter 5 the inception of electrical trees involves bond breaking by injected hot electrons. The chemical composition of the material may influence this process in a number of ways other than through the yield strength discussed in the previous section (7.1.2). For example differences in the metal-polymer interface can alter the barrier for charge injection[336]. Functional groups in the polymer may trap injected electrons without proceeding to bond scission in the backbone[198], or they may ionically dissociate in the field[198,465] and hence moderate the accelerating field experienced by the injected charges[466]. Such groups may also prevent bond scission by absorbing damaging UV radiation[130] produced by the trapping of accelerated electrons.

Table 7.5 Water tree growth rate constants, Q_a, relative to LDPE, taken from the data of Reference 199

XLPE		LDPE	
% acetophenone	Q_a/Q_a (LDPE)	% acetophenone	Q_a/Q_a (LDPE)
0 (vacuum treated)	0·8	0	1
~0·4 (aged 29 days)	0·32	0·2	0·74
~0·6 (aged 15 days)	0·12	0·64	0·31
~0·7 (fresh)	0·06	2	0·25

Propagation, on the other hand, proceeds through damage produced by discharges in the gas-filled tree channels. The contribution to the damage from the ballistic impact of ions and electrons will not be significantly affected by chemical composition, since the bond strengths of all polymeric materials are very similar (3.5—4 eV). However the resistance to chemical erosion[383] may vary with the composition and the type of by-product produced by the discharge. This may be particularly important in the case of amine-cross-linked epoxy resin insulation where acidic gases (NO_2 and N_2O), generated by a discharge[467], can chemically degrade the polymer. Absorbed gases and gaseous decomposition products also affect tree propagation by altering the inception field for discharges[384].

(a) Polymeric composition
Table 7.6 gives the double needle, 50%-inception voltage, V_I, for electrical trees in a number of different materials[198]. The values of V_I obtained for the polyethylenes and polyethylene copolymers have been taken to reflect the increase in inception resistance with increasing crystallinity (see Section 7.1.2 and Table 7.3). It is noticeable that the time for the tree to propagate across the insulation also follows this trend reflecting the confinement of the tree channels to the amorphous regions prior to widening by discharge erosion (Section 7.1.2), However the materials with the largest inception voltage are very different chemically to polyethylene, and their inception resistance cannot be attributed to either crystallinity or crosslinking. The

$$-\overset{\displaystyle O}{\underset{\displaystyle \|}{X}}-$$

side-groups of these polymers contain oxygen ($-X-$) or aromatic moieties which are relatively easily ionised and have high electron affinities[198]. It is likely therefore that their resistance to tree inception is due to a moderation

Table 7.6 Inception voltages (50% in double-needle geometry) and lifetimes (time for tree to cross insulation) for electrical trees, taken from Reference 198

Material	Inception voltage V_I (kV)	Lifetime (hour) at 15 kV
Polysulphone	27	—
Polyethylene terephthalate	25	—
Polycarbonate	19	—
Polystyrene	12·5	>120
HDPE	10·5	41·5
LDPE melt index 0·2	8·0	22
LDPE melt index 2·0	7·3	13·9
Polypropylene	7·0	—
XLPE (vacuum treated)	6·5	11·1
Ethylene copolymers { EVA 18% VA	5·5	6·5
EEA 18% EA	5·0	7·5
EPR	<5·0	6·5

of the field at the needle tip via a field-induced conductivity[466], although the non-destructive trapping of injected electrons[121,128] cannot be ruled out.

Chemical factors also play a rôle in the inception of electrical trees in epoxy resins. Here the application of a voltage to the needle electrode can enhance the curing process[320] producing discoloured regions around the point[322,351], and increasing the resistance to inception (i.e. the inception time, t_I, is increased). When a tree starts growing it does so in a region where the curing is less advanced[320], and hence least discoloured. The increase of inception time, compared to a ramped application of the voltage, is probably caused by an increase in the rigidity and yield strength of the regions of enhanced cure[320] (see eqn. 5.2), rather than field moderation through regions of conduction. Prolonged stressing can generate a void at the needle tip[351], either through differential mechanical stresses between regions of differing cure, or via joule heating from the injection currents.

(b) Chemicals normally present
All polymers absorb gases from the atmosphere and hence contain an equilibrium concentration of oxygen in the interface region. This electro-negative gas will produce acceptor surface states that lead to an inversion region after contact to a metal. Consequently the width of the barrier to electron injection will be reduced compared to a degassed material at the same voltage[336], and tunnelling injection will be facilitated, Fig. 7.6. Absorbed electronegative gases will thus decrease the inception voltage, whereas their replacement by neutral gases such as nitrogen, or removal by degassing will increase it[320,130].

In addition to forming a negative surface charge layer oxygen can also affect the tree inception process chemically[360,468,318]. During charge injection some oxygen molecules (O_2) in the polymer will be raised to an excited highly reactive state[469,130], both by direct interaction with the carriers and through quenching of the excitation of other molecules and their UV emission[130,336]. In this form oxygen will promote bond scission, see eqns. 4.20e—4.20l and thus facilitate channel formation[318]. It can therefore be expected that the absence of oxygen in degassed materials will increase the inception times and voltages for treeing[360], Fig. 7.7.

Because of the damaging effect of oxidation, polymeric insulation in the power cable industry routinely contains antioxidants. Here, as with water trees (Section 7.2.1), the phenolic and amine classes have a different effect upon inception, Table 7.7, only now it is the staining antioxidant (amine) which increases the resistance to tree inception, whereas the non-staining class appears to have little effect. It is noticeable that the addition of amines to LDPE strongly increases the tree inception voltage. Thus it seems likely that those factors which make amine antioxidants dangerous in water tree-ing, namely a high polarisability and capability of field dissociation, allow them to modify the local stress through an increase in conductivity, and thereby increase the electrical treeing resistance[466].

When polyethylene is crosslinked via the peroxide process acetophenone and cumylalcohol are produced as byproducts. These chemicals are volatile and are known to migrate into the channels of a propagating tree[464]. They

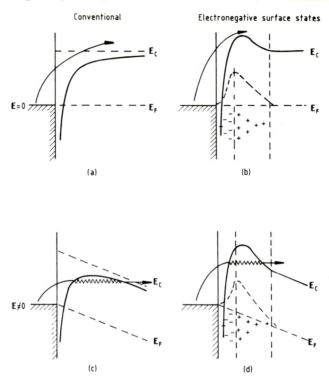

Fig. 7.6 Schematic representation of the metal–insulator contact: (*a*) conventional contact; (*b*) double-layer formed by presence of electronegative surface states, electron injection by thermally activated process; (*c*) and (*d*) same system in an applied field. Injection occurs via field-induced tunnelling; note the reduced barrier width and hence tunnelling distance in (*d*) which favours electron injection

also increase the resistance to tree inception and propagation, Table 7.8, an effect which has been attributed to their field dependent contribution to the conductivity[465]. In the region of a stress enhancement site this increase in conductivity will effectively grade the electrode boundary and reduce the local field[465,466]. Acetophenone appears to be particularly effective in this regard[466].

Discharges during both inception and propagation can be affected by the nature of the atmosphere that surrounds the insulation, since some gases will have higher discharge inception fields than others at STP. Interestingly water vapour in polyethylene insulation will also have a mildly inhibiting effect on discharge generation, because it interacts chemically with the CO and CO_2 produced by the decomposing polymer to produce a semiconducting oxalic acid deposit which reduces the potential drop along the channel. Although acidic gases (oxides of nitrogen and sulphur) may be produced when void discharges occur in some polymers[467] (including polyethylene) which then chemically degrade the insulation; oxygen is again the most

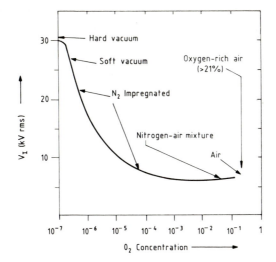

Fig. 7.7 Variation of electrical tree inception voltage V_I with oxygen content. Taken from Reference 360
Copyright © 1987 The IEEE

damaging species present. Here oxidation reactions with polymer radicals produced by the discharge will promote bond scission and channel formation.[384]

Since discharges also produce UV radiation, polymers that are susceptible to UV degradation will exhibit large propagation rates for electrical trees. In these cases however UV stabiliser chemicals are usually added, and since little of the discharge damage can be attributed to UV exposure[123,124], the risk of an increased tree propagation rate will be slight.

7.3 Effect of temperature

7.3.1 Water trees
Given the diverse processes that combine to produce water trees, it is most

Table 7.7 Effect of chemical additives on the tree inception voltage in LDPE (MI = 0·2). Data obtained in double-needle tests reported in Reference 198

Material	V_I (kV)
Nonstaining antioxidant + 0·5% 2-nitro-diphenylamine	23
0·1% diphenlylparaphenylene diamine	11
Staining antioxidant	11
Non-staining antioxidant	8
LDPE (MI = 0·2), no additive	8

Table 7.8 Effect of dicumyl peroxide ('DiCup') byproducts on tree inception voltages in the double-needle test. (*a*) Data from Reference 198, (*b*) Data from Reference 451

	Material	V_I (kV)
(*a*)	XLPE after curing (containing acetophenone and cumlyalcohol)	18
	LDPE (MI = 2·0) + 1·8% 'DiCup' R	13·5
	LDPE (MI = 2·0) no additives	7·3
	XLPE (vacuum treated)	6·5
(*b*)	XLPE crosslinked with DiCup at 200°C	
	—5 min curing time	11·5
	—10 min curing time	12·5
	—20 min curing time	15

unlikely that the effect of an increase in temperature will be the same in all of them. For example transport of aqueous electrolyte to the damage front may be thermally accelerated but so will diffusion of water vapour away from the treed region. The rate of chemical or electrochemical reactions may be thermally accelerated, but the equilibrium may be pushed towards the reactants at high temperature. Similarly the equilibrium between condensed water and water vapour may move towards the vapour component at high temperatures. Changes of phase such as in microphase separation are driven by free energy differences which increase with lower temperature, but often a potential barrier has to be overcome when forming a nucleus, and the rate will go through a maximum at some temperature[470]. Additionally morphological changes may occur when the temperature is raised which can either inhibit[247] or accelerate water treeing[252]. In view of these competing effects, it is hardly surprising that temperature dependencies ranging from positive to negative activation have been reported[119].

(a) Temperature gradients
During service the current carried by the conductor in power cables will tend to heat the neighbouring insulation, often up to 70°C, although temperatures of 90°C may be reached[471] for short periods of overload. The design criterion for cable insulation therefore includes the requirement that it retains mechanical, chemical and dielectric stability under these conditions[166]. Nevertheless the insulation close to the conductor will be more prone to structural and chemical modifications[102,112] than the rest of the polymer, particularly if such stabilisers as antioxidants are unevenly dispersed. During service aging therefore, sites susceptible to water tree inception may be generated near to the conductor[471]. However the high temperature at the conductor will increase the local water vapour pressure and hence cause an outflow of water in the vapour phase to the colder regions. Thus the probability of water tree inception may not be significantly altered when these conditions are continually maintained.

The thermal condition of power cables will depend upon both climatic and environmental conditions[472]. In particular where a cable is partially immersed in water an inversion of the temperature gradient may occur at the boundary[401,202]. Here the conductor is cooler than ambient just above the water boundary. In this case vapour pressure will cause water to be transported into the interior of the polymer, where it may find inception sites generated by thermal aging when the water boundary was higher. We may expect these regions to be particularly prone to water treeing[401], and this is borne out by the experience of the power cable users.

Both vented trees from the conductor, and bow-tie trees will be formed in the boundary regions[401], but most laboratory investigations have concentrated on bow-tie trees[238,191,202]. In this case inception sites may not require thermal aging for their formation, however an access to water will be essential to initiation and growth as previously noted, Section 7.1.1. For an initially 'dry' cable the application of different temperature gradients during treeing[202] has shown that the number density of bow-tie trees increases by an order of magnitude (or more) when the outer temperature was raised above that of the conductor. Additionally the radial penetration (outside to inside) of bow-tie tree formation was increased. When the outer temperature was below that of the conductor, both trends were reversed. All these results are consistent with water penetration from the external reservoir driven by the thermal gradient.

(b) Temperature cycling

This procedure attempts to reflect the effect of daily and seasonal changes in addition to power interruptions, and both the temperature of the water reservoir[249,238,202] and that at the conductor[238] have been varied. The conductor is dry and hot part of the time and therefore has the opportunity to form inception sites, while on alternate temperature cycles water from the reservoir is available for tree growth.

It has been found that temperature cycling increases both the length of bow-tie trees and their number density[249,202,238] (by an order of magnitude), compared to experiments carried out with the water reservoir held at a constant temperature. The results were also sensitive to pre-treatment of the cable. For example pre-soaking the cable so that the conductor region was always wet led to a further increase in bow-tie tree densities and lengths[238,191], with the trees located in the cable region of greatest water supersaturation. This behaviour illustrates the importance of water access to bow-tie tree formation (Section 7.1.1). On the other hand thermal aging (50°C for 120 hours) of a dry cable prior to water treeing increases the number density of large trees in the region close to the conductor. It would appear that in this case the aging produced favourable inception sites near to the conductor, possibly by morphological changes (Chapter 1), oxidation, or migration of chemicals such as antioxidants. Heating at higher temperatures (120°C) will introduce sites throughout the bulk by recrystallisation on cooling and leads to a uniformly high radial distribution of bow-tie trees.

Vented trees originating at the outer surface of the cable have continual access to a water reservoir and it has been noted that neither their number density nor length are particularly affected by temperature cycling[238,249]. It seems that either any new inception sites that may be produced by the higher temperatures are not developed fully or more probably that they cannot compete with those sites already present. Over an extended period of water treeing it can be expected that inception sites will be produced at the conductor leading to the formation of vented trees growing from inside outwards. This was not observed in some experiments[238] probably because the time-scale required was very large[249] (~1·7 years). In fact this very long inception time shows that it cannot be governed by water migration since bow-tie trees were found in the same region in the first cables to be examined (~1 month), and lends support to the contention that inception sites for vented trees can be produced by the aging process during operation. In well-manufactured cables with extruded inner and outer semiconductor screens there may be negligibly few inception sites[201] at installation, and the inception time for vented trees could therefore be dominated by thermal and chemical aging during service.

(c) Constant temperature

When polymeric materials are placed in contact with an electrolyte reservoir at a constant temperature, the vapour pressure of the water in the polymer will fairly rapidly come into equilibrium with the vapour pressure of the reservoir. As the temperature of the bath is increased, the vapour content of the material will rise[281,190] due to the increased ambient vapour pressure. As we have pointed out the inception of bow-tie trees is determined by access to a supply of water (Section 7.1.1), and will therefore be favoured by a temperature increase. The results of Matsubara and Yamanouchi[402] are consistent with this picture. Here the bow-tie number density was found to increase in proportion to the saturation water vapour pressure as the temperature was raised. If the polymer was dry prior to test an inception time for bow-tie trees would also be expected (see Section 7.1.1). Observations have shown that its value reduced with temperature[190] as the increased vapour pressure and polymer permeability reduces the time required for water to concentrate at an inception site. In general these results seem to be typical of bow-tie trees[240,119] although the fact that the initial state of the material is usually not reported makes it difficult to assess the experimental data.

Vented trees however usually have direct access to an electrolyte reservoir and are controlled less by water supply than by the time required to initiate and propagate damage formation. In one specific case however their inception is governed by water transport[231,281,196]. Here vented trees are grown from a copper wire conductor sandwiched between two plates of XLPE. Such a laboratory geometry may be regarded as a model for vented trees growing at the conductor in cable insulation. When the system is suspended in a water bath at a constant temperature, the inception of the vented trees at the conductor is controlled by the diffusion of water to the copper wire[231]. As the temperature increases the number density of vented trees increases and the minimum inception time decreases. In contrast when the trees are grown in a humid atmosphere of fixed relative humidity (96%) their number density *reduces* as the temperature increases[281]. Under these circumstances the water vapour pressure increases with temperature, and the results can be understood to indicate a shift towards the vapour phase in its equilibrium with the condensed liquid phase of the vented tree.

The effect of ambient conditions on both the sandwich geometry[281] and bow-tie tree experiments is a complex one involving the establishment of a dynamic equilibrium, between external liquid and vapour and internal water which is present both as a disperse 'vapour' phase and a condensed 'tree' phase. When the external reservoir is liquid (i.e. saturated) increases in vapour pressure with temperature will increase the rate at which water penetrates to set up the equilibrium with the internal condensed phase in a *dry* material. In this case the inception time will be reduced as the temperature increases. If the material is pre-soaked, however, no strong temperature effect may be found on inception although an immediate access to water may favour water tree propagation. When the external conditions are those of a humid atmosphere, increase of the relative humidity (vapour pressure) will increase the water content of an initially *dry* material and increase the number density of trees[218,281]. If however the material has been allowed to equilibrate with the environment, increase of temperature will increase the vapour pressure and hence reduce the proportion of the internal water present in the condensed (liquid) phase. In this case the extent of water treeing will decrease as the temperature is raised.

The results described above illustrate two competing effects of temperature that could be operative during the growth of vented trees from the outside of polymeric insulation where it is in contact with a liquid electrolyte. Transport of aqueous material to the damage boundary which can be expected to be accelerated by temperature will be opposed by vapourisation of the aqueous component at high temperatures[473]. The resulting vented tree propagation rate would exhibit a maximum between room temperature and 100°C. This sort of behaviour is a fairly common occurrence for vented trees[115,400,206,415] and an example is given in Fig. 7.8. Often though the results obtained from destructive experiments vary so much between the samples measured at different times that no discernible trend can be found. Recent results[474], including some continuously monitored individual trees, however confirm that the temperature effect in vented water tree propagation is caused by the ability of the system to access water in the liquid phase. Here the growth rate was observed to drop between 23°C and 65°C when water was available only at the high voltage needle point, and diffusion of vapour is a competing process. In contrast when water is also available at the ground plane vapour diffusion will not be effective since there is no concentration gradient, and it was found that the growth rate was larger, and although scattered, increased with temperature irrespective of material morphology.

Increase of temperature may influence the propagation of water trees in ways other than that of transport, some of which have been summarised at the beginning of this section. Of these the morphological changes that are known to occur in polyethylene[119] (> ~65°C) are the only factors that have been studied in relation to temperature effects. It has already been pointed out that prolonged heat treatment of polyethylene at 50°C (or greater) tends

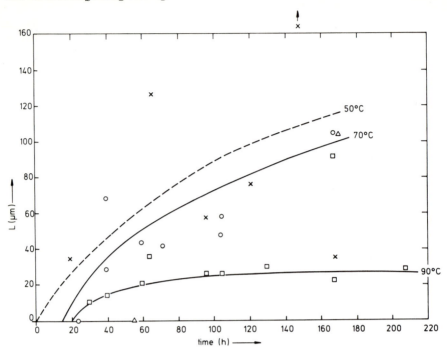

Fig. 7.8 Water tree growth in Rogowski plaques of XLPE at different temperatures (Dissado, Wolfe and Rowland — unpublished data) from 1 N NaCl solution under 6×10^6 V·m^{-1} r.m.s. at 50 Hz: $\triangle$ = room temperature; $\times$ = 50°C; $\bigcirc$ = 70°C; $\square$ = 90°C. Data obtained for room temperature, 70°C and 90°C have been shown to exhibit a reproducible trend through repeated experiments. The lines shown are aids to the eye only and so do not imply any fitted function

to plug existing voids with a low-molecular weight polymer, and increase the bow-tie number density[167,442] (Section 7.1.1). Sletbak and Ildstad[191] have also reported that temperature cycling introduces cracks around contaminant particles which facilitated bow-tie tree propagation. The temperature of the maximum propagation rate in Fig. 7.9 is suspiciously close to that at which morphological changes can be generated as the tree is being formed. Trees grown in the temperature range 50—75°C also show the greatest scatter in average lengths, whereas at 20°C and 90°C they are reasonably well behaved. It therefore seems possible that the reduction in growth rates observed in this and other experiments[115,400,119] at high temperatures (> ~70°C) may be due to a thermal annealing of the incipient damage on the water-tree boundary as the material softens and its structure relaxes.

The type of damage caused by water trees is also sensitive to temperature. This is particularly the case when the material is heated (>100°C) to drive off the water in preparation for a detailed examination. Here it has been noted that vented trees tend to coarsen in texture[306], and an increase in

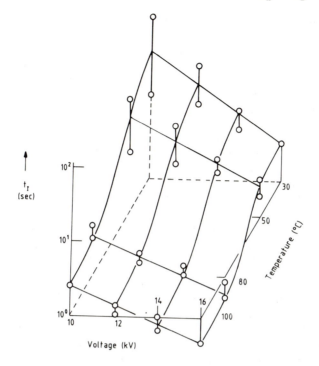

Fig. 7.9 Inception time for electrical trees, t_I, as a function of temperature and applied voltage. Taken from Reference 163
Copyright © 1980 The IEEE

the number of small microvoids[212] (~ 0.1 μm) is observed. Most probably this occurs because water first diffuses from the percolation or free volume system joining the voids, and then expands small microvoids as the liquid water inside vapourises prior to removal. It is therefore likely that power cable insulation containing water trees which has been temporarily raised to high temperatures ($\sim 100°$C) contains more severe damage within a given tree than that which existed prior to heating. This may serve to explain the greater breakdown risk arising from vented trees that approach the conductor where Joule heating is greatest.

7.3.2 Electrical trees
Temperature is not often used as a variable in electrical tree experiments, but from the limited range of data available it appears that between room temperature and 100°C the inception time decreases[163] and the propagation rate increases[163,386,475] (see Fig. 5.9) in polyethylene. It should be noted that these experiments were either carried out on LDPE or on XLPE, for which sufficient time had been allowed for volatile tree retarding chemicals such as acetophenone to diffuse out.

(a) Inception

It is known that charge injection from a needle electrode may occur via a Fowler–Nordheim tunnelling process[336,343] (Section 9.2.2), and we have shown in Sections 5.2 and 15.5 how this injection current may be related to the field dependence of the observed inception time (t_I) and its statistics. The temperature dependence of t_I cannot therefore be assigned to thermal activation of the injection process, nor do the observations follow an Arrhenius form, Fig. 7.9. In fact the behaviour[163] is remarkably similar to that of the characteristic breakdown strength over the same temperature range[163,476,166,356] (see Section 13.3, Fig. 13.10). It therefore seems likely that the reduction of t_I arises from a decrease in the local breakdown strength of polyethylene which has been attributed to an increase in the size of the free volume[476,356,477]. In this case injected electrons will travel over a longer path length[476,287], achieve a higher kinetic energy before collision, break more bonds, and thereby initiate an electrical tree channel sooner. In terms of eqn. 5.2 the energy C_t required for channel formation will be reduced as the temperature is increased and the free volume 'extends'.

At very low temperatures (77 K) polyethylene is below the glass transition. It is rigid, and the free volume is close to its value at[478,476] $\sim 10°C$. The tree inception voltages are however greater than those at room temperature even when these specimens are degassed[360]. An early report of this result is misleading since in that case a microvoid had been formed at the tree tip in the room temperature experiments[316]. In view of the fact that the same trend was observed regardless of oxygen content, the increase in resistance to inception must be attributed to an increase in C_t, i.e. the material has a higher tensile strength in the amorphous regions[316]. It is also possible that higher injection fields are necessary for the injected charges to acquire sufficient kinetic energy for bond breaking because the charge transport may be more restrained at these temperatures.

(b) Propagation

During propagation discharges erode and extend channels in the electrical tree system, and the tree propagation rate will be determined by a number of factors such as the magnitude of the discharge and the extent of the damage produced at the channel tip. An increase of temperature may affect both these factors in such a way that the propagation rate is increased. Firstly diffusion of adsorbed cations away from the channel walls and into the polymer will be facilitated. Thus the counter field to discharges within a channel will be reduced, the discharges will be bigger and the tree will propagate faster. Furthermore higher fields will be required to generate sufficient wall charges to produce bush-type trees (Chapter 5) and the lower dimension of the structure will also favour increased propagation rates (i.e. fewer channels must be formed to progress a given distance). A similar argument applies to the pressure of the gases within the channels. This change in shape has been noted by both Densley[386] and Ieda and Nawata[475] who found that for $T \gtreqqless 80°C$ only branch-type trees were grown no matter what the voltage stress whereas for $T < 70°C$ they only formed at voltages of 14 kV or less (see Fig. 5.9).

Increase of temperature is also likely to increase the extent of degradation produced at the channel tip by a given discharge, since the penetration of the polymer (polyethylene) by a charged pulse increases with temperature. This result follows from the temperature dependence of short circuit and impulse trees[287] and is probably caused by an increase in the size of free volume regions through which the charges are accelerated. This factor will also increase the propagation rate of AC electrical trees.

At temperatures below −50°C (223 K) the penetration range becomes small and independent of temperature[287]. Thus at liquid nitrogen temperatures (77 K) we can expect the trees to propagate slowly (as noted[319]) and to be rather compact. During treeing the gas pressure in the tree channels cannot be relieved by diffusion as at room temperature and is likely to rise substantially. It is probably this effect combined with the impact of a discharge that causes electrical trees at this temperature to fail the polymer by mechanical cracking[316] once they reach a given length[319]. Some contribution from electrostatic forces may also occur. These results also serve to illustrate the difference between electrical trees and electrofracture since it was observed that tree formation could take place giving a typical tree structure as long as the discharge level remained low, but cracking was almost instantaneous once large discharges occurred[316].

Chapter 8

Routes to tree inhibition

8.1 General aspects of material design and processing

As we have seen in Chapter 4, water tree formation can be thought of as due to an amalgamation of mechanical and chemical fatigue processes combined in a specific way through the agency of an electrical field. The degradation includes free volume extension, local strain and microvoid formation, and occurs as a form of in-service aging when polymeric insulation is in contact with a moist environment. Typically a time-scale of ~5 years may be required before the deterioration becomes dangerous[187]. Electrical trees are a more severe form of degradation which consists of a branched system of hollow tubules. Their initiation requires the injection of electrons from a stress-enhancing asperity or contaminant and occurs on a time-scale which depends on the magnitude of the stress enhancement (see Chapter 5 and Section 15.5). Laboratory time-scales are usually kept to 7 days or less[342,177] but the sites with lower stress enhancement that remain after proof testing may take years to initiate and propagate electrical trees across the insulation under service conditions[187]. Similar time-scales may be found for the initiation of electrical trees by void discharges[459].

All of these forms of electrical degradation will be enhanced and accelerated by aging processes[41,479] that may occur within polymeric insulation independently of an electric field. Of particular importance are chemical changes such as oxidation and free radical formation. Indeed when this type of deterioration occurs locally, rather than uniformly throughout the polymer[480], it may be responsible for creating the necessary conditions for tree initiation[134,285,336,118]. Such reactions are very slow at room temperature (years to decades) but can be strongly accelerated at the elevated temperatures[103] which may be produced near the conductor in power cables for example. Other forms of change may also occur as a result of heating, such as the disordering of crystal lamellae and changes in crystallinity both of which will affect the propagation of trees. Water halos[445,239], produced by the condensation of water droplets in field-aged cables, can also act as a water supply for water trees[39]. Other features intrinsic to the insulation which may influence tree growth are residual strains[222] and voids[263,481] both of which may be generated during aging[104].

The list of defects and aging processes given above is not exhaustive, but it serves to illustrate the range of processes that can be involved in degrading polymeric insulation during service, leading ultimately to a reduction of the working life of the insulation. Prevention, or at least suppression, to give a guaranteed lifetime of about 25 years is a major aim of manufacturers. To this end a number of steps have been taken to improve manufacturing practice in the cable industry. For example the triple-extrusion process which covers the conductor with a layer of semiconducting polymer[482] has effectively eliminated electrical tree-generating asperities at the conductor[121]. The size of particulate contaminants has been reduced by filtration[483] and better cleanliness standards, whilst improved semiconducting sheaths have been developed to prevent the leaching of atomic contaminants into the insulation[188]. Particular attention has been paid to the outer sheath with the aim of rendering it impermeable to water by blending the polymer (PVC) with hydrophobic additives[444]. However it is doubtful whether there is any means of permanently preventing the ingress of water other than by an expensive metal sheath[121]. Such metal sheaths are also not popular because they are easily split when the cable is laid. If this occurs, even in only a few places, water trees will grow at those places and the whole sheath is then effectively redundant.

Good thermal control during extrusion and curing also serves to prevent the formation of large voids, and the density of small voids has been reduced by the adoption of dry curing for crosslinked polyethylene in place of the steam technique[121,192]. Care should be taken however with radiative crosslinking to ensure that the dose is not excessive, as this can facilitate water treeing[115]. In some cases dry curing utilises inert gases such as N_2 or SF_6[121,178] which can migrate into voids and inhibit electrical treeing and partial discharges[130]. The insulation may also be deliberately impregnated with these gases, however it is to be expected that diffusion will reduce the effectiveness of this measure over a period of time.

The propensity for oxidation has been reduced by the addition of antioxidants. However it is essential to ensure a homogeneous distribution for otherwise deleterious antioxidant clusters may be formed[104,118]. Indeed even well-dispersed antioxidants may diffuse to the semiconducting sheaths over a period of years[484] where they can be absorbed by carbon black particles[462]. Inception sites for vented water trees may be produced in this way[285,118,104] and therefore the antioxidant material should preferably be as soluble as possible in the polyethylene insulation. If possible the antioxidant should be of the non-staining (phenolic) class since it suppresses water tree growth and is neutral with regard to electrical trees (Section 7.2). The alternative staining (amine) class is deleterious as far as water trees are concerned, and although it inhibits electrical tree initiation, alternative chemicals are available for this purpose[198].

Material alternatives to LDPE have also been considered, and a major improvement in degradation resistance has been brought about by the change over to crosslinked polyethylene[128] (XLPE). This material not only has superior mechanical properties, longer oxidation induction times and higher breakdown strengths than LDPE, but is more resistive to water tree inception and propagation. Its resistance to electrical tree inception and growth may however be only marginally superior to that of LDPE once the crosslinking by-product acetophenone has diffused out. Another material that has received some attention is EPR. It has been reported that most of its insulation properties, i.e. breakdown strength, dissipation factor, and mechanical toughness are inferior when compared to XLPE[437]. However the properties of EPR are extremely sensitive to the type of filler used and it is possible that many of them may be brought into an acceptable range by a suitable choice. A feature that makes EPR a potentially attractive material is its resistance to water tree propagation. In fact there is even some indication that the breakdown strength may increase when small vented trees are present[205], probably because the water causes field moderation at stress-enhancement sites. However as yet no long term water treeing data appears to be available, so it is not clear whether the trend to growth saturation will continue through a service lifetime. In view of its low inception field for electrical trees this material is only viable at low electrical stresses unless tree inhibiting chemicals have been added[333,198].

8.2 Suppression of electrical trees

The most effective step to be taken towards eliminating electrical trees is the adoption of manufacturing designs and processes which reduce metallic contaminants and asperities. The measures described in this section should be seen therefore as a means of ensuring that those inception sites which still exist following proof tests do not cause a treeing breakdown during the projected service lifetime of the insulation system.

8.2.1 Inception

It has been known for some time[121], that the incorporation of some chemicals into polymeric (and paper/oil) insulation substantially reduces the inception of electrical tree breakdown. Many of these additives contain aromatic moieties[121,464,198] some of which have electron withdrawing side-groups, and it was thought that they acted by scavenging energetic electrons to form aromatic anions[121,332] thereby preventing bond scission in the polymer. However electron affinities are proportional to ionisation potentials in aromatic molecules and the different effects of these molecular properties can only be distinguished when non-alternate hydrocarbons (non-aromatic) are included in the assessment. It was then found that the stabilisation of the inception voltage was correlated with their ionisation potential rather than electron affinity[198], Fig. 8.1. Unfortunately only a limited number of non-aromatic additives were considered and there may even be some error in the electron affinities of these. It is therefore clearly desirable that the correlation with ionisation potential be checked with a wide range of non-aromatic additives. Nonetheless it was suggested[198] on the basis of this analysis that impact ionisation of the additive was easier than that of the polymer and gave less energetic secondary electrons. Recombination of the resulting cation with thermalised electrons would regenerate the aromatic in preference to the production of free radicals. Any polymer cation radicals produced by energetic electrons would also be quenched, e.g.

Recently it has become clear that electrical tree inception relies on electrons whose kinetic energy is well below that required for ionisation (~8 eV) but sufficient to cause bond scission (~3·5—4 eV). Thus although the above mechanism may retard inception by removing electrons from the high energy tail of the distribution and hence suppress avalanches, it is probable that a different effect is partly responsible for the substantial increase in inception voltages observed[198], (Table 8.1). Rzad and Devins have proposed[485] that the inhibitor additives act as weak electrolytes. In a high field they ionically dissociate giving a field-dependent contribution to the conductivity[465,466], which will moderate the local electric stress in the region of a field enhancement site[466]. The voltage required to initiate an electrical tree is therefore increased. It should be noted however that an additive of this type which is not uniformly dispersed in the polymer may serve to enhance breakdown in nominally uniform fields through the provision of stress-enhancing filamentary tracks[486]. This may be particularly the case with semicrystalline polymers where the additive would be confined to the amorphous regions.

A major factor in the choice of additive is its retention by the polymeric insulation during the service lifetime. We have already seen that acetophenone though very effective in inhibiting electrical trees, is easily lost by volatilisation[464], (~30 days)[199] and other small molecules are also apt to migrate through the insulation. For this reason grafted materials containing effective groups, such as polystyrene[198,454] and polysulphone, have also been considered (see Table 8.1), with some success[454].

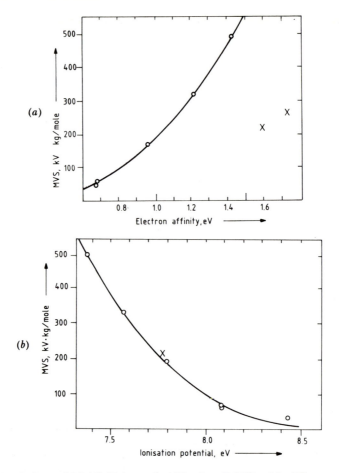

Fig. 8.1   Variation of Molal Voltage Stabilisation (MVS) with different additives. MVS is defined as the increase in inception voltage with additive divided by its molal concentration: ○ = aromatic hydrocarbons; × = non-aromatic hydrocarbons. After Reference 198
(*a*)   Correlation with electron affinity of additive
(*b*)   Correlation with ionisation potential of additive
Copyright © 1976 The IEEE

8.2.2 Propagation

Although field-grading additives[465,466] retard electrical tree inception their field-dependent conductivity may lead to filamentary thermal degradation at low onset stresses[337,338] and hence an acceleration of tree propagation once the tree starts to grow. This form of behaviour has been observed for voltage stabiliser additives of this type by Wagner and Gölz[333]. These authors also identified two other types of tree-retardant additive, namely ones which temporarily inhibited inception and ones whose effect was permanent ($>10^4$ min $\approx$ 1 week). Both these types appear to retard the onset of a

Table 8.1 Double-needle tree-inception voltages for LDPE (MI = melt index) containing a variety of additives, reported in Reference 198

Material	V_I (kV)
LDPE (MI = 2)	
+ 1·8% 'DiCup' R	13·5
+ 30% grafted polystyrene	10·5
no additive	7·3
LDPE (MI = 0·2)	
+ 2·0% α-Bromo-Naphthalene	19·5
+ 0·5% o-Nitroanisole	19·5
+ 0·1% Diphenylparaphenylenediamine	11·0
no additive	8·0

runaway tree branch (Sections 5.3 and 6.2), possibly by forcing the tree growth to be of a compact bushy form[487].

Additives intended specifically to retard propagation have usually been chosen for their ability to suppress partial discharges in the developing tree channels. They therefore have the added advantage of retarding tree initiation from void discharges. Leaving aside gases such as SF_6, and oils, the selection has fallen onto semiconducting organic liquids[121,122] which migrate into the channels[464,416] and form a semiconducting coating upon the channel walls. As a result the field along the channel (void) is short-circuited by the walls and discharges do not take place. Most of the additives investigated are aromatic derivatives, and the most effective of these appear to be the phenols (see Table 8.2). The alkyl alcohol dodecanol has also been shown to be effective in this regard.

In general these materials do not increase the tree inception voltage, but rather inhibit inception by preventing the discharges that produce the first

Table 8.2 Electrical tree inception and growth parameters for different additives taken from data given in Reference 487. Voltage 18·8 kV r.m.s. 60 Hz, needle radius 3 μm, point-plane separation 2·68 mm. 't(2·68 mm)' is the minimum time taken for a tree to grow across the insulation

Material	t_I (min)	t (2·68 mm)	Electron affinity (eV)
LDPE	~0	~2·5	
LDPE + benzene	~0	~4·5	~0
LDPE + α-naphthylamide	~13	~9	1·12
LDPE + aniline	~24·5	~2·5	1·12
LDPE + α-naphthol	~23·5	~6	1·83
LDPE + m-cresol	~28	~21	1·83

channel. When the liquid has to migrate to the tree inception site it suppresses growth at the stage reached when the liquid enters the channel[416]. Following discharge suppression in a channel it is likely that new channels will be formed from the base of the tree. Thus these additives will have the further retarding effect of forcing the tree to take the compact bush-form (Section 5.4). Kim and Lim[487] have attributed this effect to charges bound to the additive on the channel walls and related the influence of the additive to its electron affinity. However in aromatic materials many properties are correlated with the electron affinity[198] (this also implies that they are correlated with the ionisation potential for example), thus this contention cannot be regarded as substantiated.

The main difficulty with molecular additives is that, if they are mobile enough to migrate to a tree inception site, they will probably diffuse out of the insulation during service as for example acetophenone and cumyl alcohol. Although McMahon[416] has claimed that dodecanol stabilises its concentration in polyethylene at an effective level after about 50 days of accelerated aging, it is still not generally accepted that this will remain the case under service conditions. For this reason attention is being paid to materials which contain grafted inhibitors[488,454]. In one case a flame retardant material made by mixing chlorinated polyethylene with polyethylene was found to have an improved resistance to internal discharges when compared to the unmixed polyethylene[488]. Here the discharge products of the chlorinated material suppress oxidation, restrain the discharge, and also act as electron acceptors and radical terminators. However since an excess of chlorinated additive will degrade the polymer it is necessary to determine the required concentration with some care.

Blends and copolymers of polyethylene with polystyrene have recently been proposed[454] as tree retardant materials. In the blends the trees avoid the polystyrene particles ($0\cdot6$—20 μm) during propagation probably because the field required for tree inception in polystyrene is higher than in polyethylene, Table 7.6. When the polystyrene filler particles are well dispersed, the fractal dimension of the tree will increase and so will the time required to propagate a given distance. Random copolymers give a better dispersion than the blends, however the aromatic side-groups cannot be retained in the crystalline lamellae. Instead they project into the amorphous region and there is an optimum concentration ($\sim1\%$ by weight) beyond which the increase in free volume caused by these bulky projecting groups outweighs their tree inhibiting capabilities, and treeing degradation increases. Grafted copolymers which are revealed by TEM[454] to have well dispersed regions of polyethylene crystal lamellae and polystyrene particles ($0\cdot1$—$0\cdot3$ μm) up to a concentration of $\sim40\%$ (by weight) therefore appear to be the best of this range of materials.

8.3 Inhibition of water trees

Although the formulations suggested as rendering polyethylene insulation water tree retardant are not quite as varied as the mechanisms proposed for the process itself, they nonetheless cover quite a range of additive or material modification. Until the late 1970s most additives were chosen on a 'try it and see' basis (see Shaw and Shaw[119] and references therein). Now, however, the material formations available appear to have been developed as a response to what is seen by the group responsible as *the* water tree mechanism. Thus for example, modifications to the polymer morphology[449,489] are advocated by groups who favour an electro-mechanical mechanism, whereas those who believe in an electro-chemical mechanism suggest the use of reaction suppressing additives[220,286]. It is perhaps an indication of the composite nature of the process that both groups are able to claim some success. A more 'neutral' approach has been the use of additives aimed at reducing the condensation of water droplets[196,490].

In general little distinction is made as to whether the formulation is intended to prevent inception or propagation. This is a quite different situation to that for electrical trees where a tree retardant material is assessed in terms of its relative tree inception voltage[198] (or inception time), and the tree propagation curves[333,487]. Ideally water tree retardant material should be evaluated in terms of: (i) the mean initiation time t_d; (ii) the number density of trees; (iii) the growth rate constant Q, eqn. 4.7b, and growth law; and (iv) the damage density. Instead ill-defined and composite factors are often quoted, such as the 'treed area' and 'length grown in a fixed time'. The reader should be warned that this latter factor includes both the effect of the inception time and growth law, which is not linear in time, and is often erroneously termed the water tree growth rate[128,447,489] (WTGR). In view of the imprecise nature of these terms, it is rarely possible to determine which of the factors (i)–(iv) have been modified by the 'tree retardant' formulation. A comparison between formulations will only be possible if they have been examined under the same experimental conditions, and is then only valid in that context. Given these limitations some general statements can be made regarding the type of formulation proposed, its manner of operation, and its effectiveness.

8.3.1 Material modifications

It has been claimed[128,489], with some justification that suitable alterations to the chemical composition of polyethylene can significantly increase its water tree resistance, and two brands of material are now commercially available. Tree retardant polyethylene (TRPE) is a blended polymer containing low molecular weight polymers of a tree retardant functionality (oligomers), whereas in tree-retardant crosslinked polyethylene (TRXL) a different chemical functionality is grafted onto the main chain of the polyethylene molecule.

The effectiveness of these formulations in retarding water tree growth was demonstrated through a comparison of their WTGR (under unspecified conditions) with that of polyethylene. A reduction of about an order of magnitude was reported[128,489], which was retained in spite of pre-heating to temperatures (75°C for TRPE and 90°C for TRXL) at which many chemical additives lose their effectiveness by diffusing from the system. It is impossible to determine from the results reported by this group what features of water tree growth are affected in order to reduce the so-called WTGR, however later work[491] has shown that the number density of vented trees ($\geqq 50$ μm) and their maximum lengths[491,492] are reduced. It was also noted that the amount of moisture absorbed by TRXL was twice that of standard XLPE, and the number density and maximum length of bow-tie trees were both increased (see Section 7.1.1).

The increase in bow-tie deterioration does not seem to affect the dielectric strength which decreases less with aging time in the tree-retardant material[491,492], than standard XLPE. This result may be expected on the basis of the smaller vented tree growth found in TRXL. An interesting feature of TRXL and TRPE is that the characteristic breakdown strength for a given maximum length of vented tree, is greater than that of PE and XLPE, see Fig. 8.2. This behaviour suggests that the tree resistant materials function through an increase in the difficulty of generating free volume and void damage. In the terminology of Section 4.3 this implies an increase in the time required to generate a typical unit of damage (t_v). The decrease in breakdown voltage is then less severe with TRXL and TRPE because the damage density is less, as discussed in Section 7.1.1 for HDPE. It would be interesting to know whether the distribution of damage severity within the trees were significantly different in the two materials, but the failure distributions[492] are too imprecise to allow this kind of assessment to be made.

The results of de Bellet *et al.*[447] suggest that high molecular weight polyethylene may also act in the same way as TRXL in inhibiting vented tree growth, and blends of high and low density polyethylene have also been proposed as a tree retardant material[493]. The materials most similar in composition to TRXL are however the XLPE blends with styrene and ethylene copolymers proposed by Nagasaki *et al.*[449]. Here the growth of

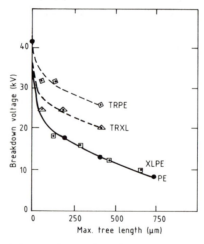

Fig. 8.2   The effect of the maximum length of vented water trees upon the characteristic breakdown strength of full size 25 kV cables: TRPE = tree retardant PE; TRXL = tree retardant XLPE. Taken from Reference 128

vented trees in the field direction appears to be inhibited in favour of a lateral spreading out. Such an alteration in geometry ought to be reflected by a change in the growth law, eqn. 4.5, unfortunately no data is available, as is also the case for TRXL. Unlike TRXL, whose functional groups appear to be hydrophilic, these materials reduce the number density and maximum length of bow-tie trees, with the ethylene copolymer (EVA) being the most effective.

It is suggested that the copolymers act as fillers which provide a barrier to water tree propagation around which the damage must move[447] (see Section 7.1.1). In view of the previously reported deleterious behaviour of polymer blends[444] it seems that the effectiveness of these formulations is dependent upon the ability of the filler to 'wet' the polymer so that there are no gaps or regions of strain at the interface[442,190]. These materials have also been observed to give an increase of about an order of magnitude to the dielectric life (aged in water) when compared to XLPE, which has been associated with the inhibition of vented and bow-tie water trees. As yet however no relationship such as that of Fig. 8.2 has been presented to indicate whether the degree of damage has also been affected along with the tree lengths.

A different approach to bow-tie tree inhibition has been an attempt to modify the void density in the insulation by adding a nucleation agent and void filler[494]. This method seems to have failed however since no significant change in structure or mechanical properties resulted[495].

8.3.2 Additives

Among the additives known to reduce both the inception and growth[220,199,444] of vented water trees are some such as acetophenone[115,190] and antioxidants[208,118] which also inhibit electrical treeing. In principal their effectiveness can be quantified by the increase they cause in the inception time t_d and in the time required for microvoid or typical damage formation, t_v, which influences the growth rate by reducing Q_a. A rough categorisation has been given by Dissado *et al.*[206] based on the experimental data of Reference 115. This analysis suggests that antioxidant is consumed at the damage boundary, thereby generating low density regions and voids at sites of high stress which are susceptible to oxidation (Section 6.1). The removal of antioxidant may therefore be a contributory factor to the conversion of water trees to electrical trees at the damage boundary.

The aromatic additives which are most effective in suppressing electrical trees have little effect on water trees. This reflects the fact that energetic electrons are not involved in water treeing, and indicates that the field grading ability of additives is of little consequence. Thus dodecanol which suppresses the propagation of both electrical and water trees[416], without affecting their inception, is unlikely to function as a field grader in respect of water trees. In action it migrates to the growing tree, and thus its influence may just lie in a moderation of local strains (i.e. as a mild plasticiser). Other possible explanations would see the dodecanol function as a barrier to ions and water which was unable to support electrochemistry. The fact that a similar chemical ethylene glycol was unable to initiate *vented* trees when used as a liquid electrode[115] supports this latter contention. Bow-tie water trees which rely upon the internal water reservoir for initiation were not affected[115].

It is possible that those additives which inhibit both electrical and water trees[119] do so by suppressing oxidation which has an accelerating effect on both phenomena. Henkel and Müller[220] have carried this supposition a step further in the case of water trees by contending that additives which stabilise organic hydroperoxides (and hydrogen peroxide) will interrupt the electrochemical reaction scheme (eqns. $4.20a$—$4.20m$) and hence retard both tree inception and propagation. On the basis of this argument up to 50 chemicals have been found which retard water tree growth[220], however many are not suitable for use, with additive retention being a major problem. Of the additives investigated, pyrogenic silica was found to be a useful tree retardant which was technically compatible with polyethylene cable insulation. Of those additives identified in this manner barbituric acid[220] and some of its derivatives[286] were found to be particularly effective.

X_1	X_2	
O	O	barbituric acid
S	O	
O	NH	} useful derivatives
S	NH	

These chemicals are able to scavenge peroxide free radicals because of their easily dissociated H atoms, and therefore inhibit oxidation of the polymer. It is noticeable that barbituric acid is able to function as a retardant even though it is poorly wetted by the polymer and crystallises[286] out leaving gaps and voids which would usually act as bow-tie tree inception sites. This fact points to the chemical rather than physical nature of their retarding mechanism. Among the derivatives investigated some have been reported to be both well wetted by the polymer and effective in retarding tree growth, and these form a possible basis for a tree inhibiting formulation. This may be particularly effective because it is possible for the inhibitor to be regenerated in an AC field[286].

Copper is a common metal for use as a conductor in cables, and thus is a typical metallic contaminant in the insulation. Copper is also notoriously susceptible to aqueous electrochemistry which produces ionic salts and it has been shown that Cu^{2+} ions are very effective in generating bow-tie water trees[195]. Copper has also been identified in both atomic[219,433] and in ionised form[215] in vented trees. As pointed out in Section 3.1 copper ions will catalyse the thermal oxidation of polyethylene[496] and it is likely that it is through this mechanism that they accelerate inception and propagation of water trees. Although semiconductor sheaths around the conductor will protect it to some extent from electrochemical attack by liquid water condensed in the region, they themselves sometimes act as a source of damaging contaminants[188], even in some cases titanium used as a white pigment in the tape.

In order to reduce the danger from metal ions it has been suggested that additives which either yield insoluble salts with heavy metal ions or form a colloidal sheath around the active ion centres should be used[220]. Kato and Fujita[195] demonstrated that 8-hydroxyquinoline (HOQ), which was part of a proprietary additive containing ferrocene and a silane oligomer as well, did in fact suppress the activity of copper ions in inception and growth[232] rather as suggested. Here the HOQ deactivated the copper by forming a complex with the ions which effectively surrounds them by an organic sheath (somewhat as the protein does to iron in haemoglobin). It was also shown that other ions such as Fe^{3+} and Zn^{2+} were also inhibited by HOQ in the same way. Recently a series of papers[497,498,499] have shown that the Cu^{2+} deactivating agent[500] ethanedioic acid bis[(phenylmethylene)hydrazide], (OAHB):

$$\langle\bigcirc\rangle - CH=N-NH-CO-CO-NH-N=CH-\langle\bigcirc\rangle$$

forms complexes with Cu^{2+} in organic solvents and probably functions in the same way in polyethylene although this additive is known to separate as micro-crystals in the polymer[497]. Possibly the complexes form on the micro-crystal surfaces. A commercial deactivator ('CUNOX AX' from Hitachi) which forms a $1:2$ complex with Cu^{2+} has also been assessed by this group.

This organic ligand binds with more than one group (chelates) to form a complex organic sheath around two Cu^{2+} ions, and should therefore be very effective in inhibiting water trees caused or accelerated by this type of contaminant.

A number of developments have been aimed specifically at the inhibition of bow-tie trees. The most simple of these is the addition of a 'surface active agent' which is intended to convert hydrophilic centres to a hydrophobic nature[196], with the aim of preventing water condensation at contaminants and discontinuities. This method has been shown to work with hygroscopic fibres in epoxy resins[196], but does not appear to be effective in polyethylene[220,449] (LDPE and XLPE). Voltage stabilising agents have also been suggested[240,332] possibly with the aim of field grading at the site of a contaminant or the prevention of oxidation, and these appear to reduce the number density of bow-tie trees. Soma and Kuma[237] have recommended stearates as bow-tie inhibitors. These additives favour the wetting of poly-ethylene by water and reduce the contact angle to around 62°. The aim here is to reduce the mechanical force experienced by polyethylene when water droplets flex periodically in an AC field. It was shown that the number density of bow-tie trees, and hence the number with a given length were reduced. Thus it appears that a number of potential sites (voids) were unable to initiate a tree at the given stress, due to the treatment.

A different approach has been adopted by Matsuba *et al.*[490] who suggest the addition of salts so as to equalise the chemical activities inside and outside of a bow-tie initiating void, with the intention of preventing damaging water condensation into the void. In this case the salts are meant to increase the conductivity and lower the permittivity of the water in the void, but it is not clear how they are meant to function in an initially dry cable where they are likely to act as condensation sites for any water entering the system. Nevertheless these authors report[232,490] that when a concentration of $\sim 10^{-3} \, \text{mol} \cdot \text{kg}^{-1}$ of inorganic salt (e.g. Na_2SO_4, $CaCl_2$, $CuSO_4$, etc.) was dispersed in polyethylene slabs no trees were produced in up to 455 hours as compared to an initiation time of 272 hours when no additive was present. However similar experiments carried out by Sletbak and Ildstad[191] showed considerable bow-tie treeing ($\sim 1000 \, \text{tree} \cdot \text{mm}^{-3}$) when salts ($\sim 1\%$ by weight) were present even when aged in humid air. The bow-tie number density was found to increase with the equilibrium humidity of the saturated salt solution. Possibly Matsuba's treatment[490] may lead to a competition for the available water and reduce the length of bow-tie trees for a given water content. Alternatively the wettability may be altered as suggested by Soma and Kuma[237]. Certainly though inorganic salts cannot be accepted as bow-tie inhibitors without favourable results in long term aging experiments including breakdown studies.

Among the tree inhibitors discussed in this section, antioxidants and Cu^{2+} deactivators would be routinely added to cable insulation. Acetophenone may also be present following crosslinking to give a temporary protection. The diversity of form of the other additives for which success has been reported vividly illustrates the complex combinative nature of the water tree mechanism. Attempts to inhibit water condensation, to render droplets less mechanically dangerous, to prevent electrochemistry and block the passage of solvated ions, all appear to work to some extent. It seems that each of these processes are essential but not sufficient for the generation of water trees (see Section 4.3). For such additives to be useful however it is necessary for them to be retained during service, to be compatible with the polymer, and not to interfere with any of the other beneficial mechanical and dielectric properties of the polymer[437]. For these reasons it is necessary to know whether inhibition is confined to inception or extends to propagation, does it alter the damage density and the growth law and what is the effect on long-time aging? In very few cases[492,416] is sufficient data presented in the literature to enable an accurate assessment of these qualities to be made, and it must be concluded that no inhibitor has as yet been shown to be fully effective.

PART 3

DETERMINISTIC MECHANISMS OF BREAKDOWN

'Breakdown theory faces a dilemma. The relatively general models can be stated exactly with specific parameters and solutions computed in a straightforward manner. However, the results lend themselves only to order of magnitude estimates on real substances, and do not reflect the complexity of experimental data. Alternatively, one could propose a different model for every different dielectric (and possibly for different thicknesses and methods of preparation of the same dielectric) leading to complex computing that offers little insight into the physical processes involved.'

O'Dwyer[501]

© IEEE 1984

Introduction

Even good insulators such as polymers conduct to some extent (see Chapter 2) and the useful range of materials is restricted to those whose conductivities are so low that their impedance is dominated by their capacitive capabilities under service conditions. Thus at working stresses the DC leakage current is negligible and is expected to remain stable as the stress is raised even though it may exhibit a non-linear field dependence[502,503]. At some high value of the field however an unstable situation will develop and the current will rise by many orders of magnitude without any further increase of the field. During this process the power dissipated in the insulator will melt and probably vapourise it[150] forming a narrow conducting channel between the electrodes unless the power supply is limited in some way, as for example in reverse-biased zener diodes. In this latter case the situation is reversible and the system will revert back to an insulating state on reducing the applied field. Whilst a similar situation could occur in polymers, perhaps under trap-filled space-charge limited current conditions[147], or due to the Schottky[148] or Poole–Frenkel[149] mechanisms (see Chapter 9), the term breakdown is here reserved for the destructive generation of a conducting path. Under such circumstances the insulator will have been irreversibly converted to a conducting state.

In the above description we have reduced breakdown to its fundamentals, namely a cross-over in the current from stability to instability at some field, with consequent material modification. In order for the current to behave in this manner it is essential that a positive feedback mechanism exists, i.e. an increase in the current changes some material property in such a way that the current is enhanced. The means by which this occurs depends upon the particular breakdown mechanism, the most important of which for polymers are discussed in this part of the book. Typical feedback processes are local heating[162], and impact ionisation[378,146]. The current reinforcing effect of these processes will be opposed by a 'dissipative' mechanism such as thermal conduction[162], or the counter field of the ions[378,146] (in the case of avalanches), and at fields below breakdown the current will in time reach a stable equilibrium value. The breakdown field for a particular model is then defined as the maximum field for which a stable equilibrium can exist. A schematic representation of the current-time characteristics in both stable and unstable regimes is given in Fig. III.1a. An interesting comparison can be made with Figs. 4.3 and 5.2 from which it can be seen that water tree growth usually corresponds to a nearly stable change, whereas the propagation of most electrical trees represents an unstable process. These considerations clearly reflect the different degree of degradation that is represented by the two processes.

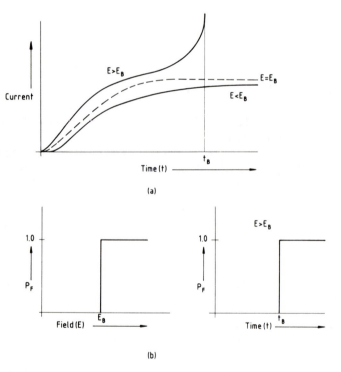

Fig. III.1 (*a*) Schematic representation of current-time characteristics for a deterministic breakdown model, where t is the time of application of the field. Examples are shown for fields below the breakdown value E_B and fields above E_B. (*b*) The cumulative failure probability, P_F, for a deterministic model shown both as a function of the applied field, and as a function of the time to breakdown, t_B, at fields in excess of E_B

Models of breakdown which attempt to define balance equations for the current (or energy density) in terms of inter-linked causal mechanisms are denoted here as deterministic[504]. In this class of model each step in the current response to an applied field is completely determined by previous conditions and itself completely determines future events. A model-dependent breakdown field, E_b, can be determined for which the system response to the applied field just becomes unstable. At fields below E_b the system is stable and breakdown will not occur. Above E_b a dynamic process responding to the applied field becomes unstable, thereby 'switching' from being self-limiting to self-enhancing, leading to breakdown. This process may be a single event or a sequence of 'cause and effects' which finally establish the instability rather like an electrical tree. In either case however the causal chain of events uniquely determines the time to breakdown for a given applied field. These results are summarised in Fig. III.1b. In practice breakdown is a statistically-distributed process (see Chapter 14) and thus to be genuinely applicable to real systems the deterministic models must be modified to allow for the possibility of stochastic features in the development of the instability (see Chapter 15). (By stochastic we mean that there is a choice of alternative possibilities for development at each step in the process.) Because of this limitation a comparison between predicted breakdown fields (or times) and experimental values can only be made for measurements accurately defined as the 'mean', 'minimum', or another characteristic value of the observed distribution. Thus unless the whole distribution follows a predicted dependence a given deterministic model can only be demonstrated to be the underlying breakdown mechanism for a small fraction of the observed breakdowns. In spite of these limitations and problems of evaluation the deterministic models represent a useful basic framework in which to examine the types of process that may lead to breakdown.

The catastrophic process whereby a conducting path across the system is finally formed will be electrically power driven and ultimately thermal in the sense that the discharge track in solids involves at least the melting and probably the vapourisation of the insulator[150] (see Section 6.2). Many features of this final stage are likely to be common to all solids and all mechanisms[161], and hence deterministic models are usually categorised according to the process(es) that are postulated to lead up to it. Workers often subdivide breakdown models into various groups dependent on experimental parameters rather than physical models. We have found it useful to consider four basic categories for polymeric insulating systems: *thermal* ($E_b \sim 10^5$—10^9 V·m^{-1}), in which the applied power causes (local) heating and a concurrent increase in conductivity; *electro-mechanical* ($E_b >$ $\sim 10^8$ V·m^{-1}), in which the combination of electrostatic attraction of the electrodes and (local) softening (or cracking) causes the dielectric to collapse (or fracture); *electronic* ($E_b \sim 10^7$—10^9 V·m^{-1}), in which breakdown initiation is caused by (local) high electric fields; and *partial discharge breakdown* ($E_b > \sim 10^6$ V·m^{-1} perhaps), in which voids in the dielectric ionise and thereby cause progressive deterioration.

It is unlikely that any single mechanism can be responsible for the many diverse phenomena observed in the breakdown of polymers, and hence in the following chapters the principles of each breakdown mechanism will be qualitatively described; this will be followed by a more quantitative treatment and examples of polymer breakdown attributed to this mechanism will be given. An overview of these mechanisms has already been given in Section 3.2. The reader is also referred to the many works on breakdown in solids that already exist (e.g. References 82, 133, 150, 159, 161, 163, 165, 378, 503, 505, 506, 507). Much of the experimental work using highly viscous liquids is also capable of aiding a better understanding of breakdown in solids and will be of interest (e.g. References 210, 330, 427, 426).

Breakdown processes (including degradation processes which ultimately lead to breakdown) are generally thought of as being in competition such that the 'winning' process is that which leads to breakdown quickest under the service conditions of the insulating system. For example see Vermeer's results for different times and temperatures reported by Klein[161] in a systematic way in which it was clearly shown that thermal and electrical breakdown mechanisms were both operative but which one dominated depended on both these factors. The results of Park *et al.*[508] on PVDF at low thicknesses and different temperatures indicate a similar electrical breakdown mechanism dominating at low temperatures and a thermal breakdown mechanism operating at high temperatures. Most deterministic models are based on ideal insulating systems and the instability criterion is chosen such that breakdown quickly ensues after this criterion is satisfied. Thus breakdowns attributed to deterministic breakdown processes usually occur in a very small time scale (typically 10^{-9} to 1 second) at a well-defined high field. Under service conditions the fields would be much lower (typically by a factor of at least 100) to avoid such breakdown. However degradation processes may occur at these fields leading to a weakening of the insulation over a relatively-long period of time and ultimately culminating in breakdown. The 'winning' breakdown mechanism is therefore highly dependent upon field and operating conditions (e.g. temperature) and history and one mechanism may take over from another as time progresses.

As well as competing, breakdown mechanisms may also cooperate or accumulate. These 'cumulative' models of breakdown/degradation are described in Section 15.4 and take account of the variety of breakdown mechanisms by considering the breakdown process to occur in several separate stages during which the dielectric is progressively degraded, each stage establishing the conditions in the dielectric necessary for the initiation of the subsequent stage. In this case breakdown generally ensues at fields lower than that predicted by an individual breakdown mechanism. Whilst the model appears to be essentially deterministic, the state of deterioration is subject to considerable local and temporal variation because of the cooperation required between local sites to produce the necessary and sufficient conditions required for the subsequent stage in the overall breakdown process. Potentially therefore it embodies considerable stochastic possibilities, and hence is included in Chapter 15.

It is unclear which category of breakdown/degradation partial discharge breakdown should fall into. The onset of ionisation in voids can be well-defined (and is well characterised) in terms of field and other parameters so it could be thought of as a deterministic mechanism. However the degradation it causes is cumulative, usually changing over to an electrical treeing mechanism, and stochastic depending on the shape of the void and the local morphology. We have included partial discharges in this part of the book as their description follows on naturally from electrical avalanche breakdown.

In practice breakdown fields may be lowered (and the degradation processes enhanced) by imperfections in the insulating system. In fact breakdown is usually initiated at an inhomogeneity occurring either in the bulk of the insulator or at the insulator/electrode interface. Typical inhomogeneities include:

- electrode aberrations such as 'splinters' of conductor penetrating the insulation or surface scratches, asperities and depressions;
- regions of free volume in the polymer due to morphological irregularities with sizes typically less than ~10 nm (the so-called Matsuoka voids[168]), small (sub-micron) voids due to movement of additives etc. which may only affect very-long term degradation processes, and larger (super-micron) voids generally due to imperfections in the manufacturing technique such as insufficient pressure during high-temperature cross-linking;
- impurities in the insulation, including impurity particles (inclusions), moisture and sites of imperfect mixing and coagulation of fillers.

All such imperfections will give rise to local changes in electrode geometry or changes in dielectric, mechanical, and/or thermal properties resulting in locally non-uniform electrical and mechanical stresses and temperature gradients thereby leading to an increased probability of breakdown either in the inhomogeneity or in the surrounding insulation. Such 'physical' inhomogeneities may also give rise to space charge clouds which may be immobile relative to the charge carriers; these then form a type of electrical inhomogeneity as they cause field distortion. Furthermore the thermally-activated movement of polymer chains within the free volume gives rise to temporal fluctuations of mechanical and electrical stresses at any point within the bulk (Section 15.2). This is one of the factors contributing to the stochastic behaviour of breakdown (Part 4 of the book).

Much of the recent work on breakdown mechanisms has been on 'filamentary' variations of established models. In such cases breakdown is precipitated from a weak point or develops in a filamentary manner. In order to avoid the effect of weak points it is possible to use self-healing electrodes in which the heat produced by a localised breakdown, usually in a thin film, leads to evaporation of the thin electrodes and isolates the region of breakdown from the remaining intact insulation. Klein has advocated this technique for investigating deterministic breakdown in which the samples were conditioned by an electric field until all the weak spots had been isolated. Using this technique breakdown could be produced on a broad front across the sample rather than simply at an unrepresentative weak spot. Self-healing electrodes are also commonly used to increase the reliability of capacitors following the suggestion of Strab[509]. However the self-healing process is not feasible in most insulation systems either because they do not have thin electrodes or because the breakdown would be too extensive; the first breakdown therefore usually leads to catastrophic failure.

Whilst terms such as 'breakdown field/stress/strength' are widely used and usually defined as the applied voltage at breakdown divided by the dielectric thickness, it should be noted that:

(i)   the electric field is likely to vary considerably through the thickness of the dielectric because of electrical and physical inhomogeneities which may also be fluctuating in time and space;

(ii) in imperfect materials there is always a distribution of breakdown strengths, not a single breakdown strength.

There is usually very little direct evidence to confirm that a particular breakdown mechanism is operative in a particular situation. For example although there have been some observations of the spatial evolution of a temperature rise prior to thermal breakdown, these have been generally lacking. Indeed O'Dwyer[82] notes that whilst the best currently available concept of purely electrical breakdown has, as its precursor, collision ionisation there is in fact no direct and convincing evidence of collision ionisation; it merely seems reasonable as it is known to happen in semiconductors and reasonable alternatives are difficult to visualise. Coupled with the problems of measuring 'a specific breakdown strength' and the difficulty of estimating absolute values of breakdown strength from theoretical considerations, the applicability of most theories has been tested by comparing the theoretically and experimentally determined dependence of breakdown strength on thickness, time, and temperature.

In spite of these problems of evaluation, definition, and attribution, there are some general features of breakdown in polymers which are worth mentioning in an introductory manner. It is generally found that breakdown field decreases with increasing time of application of the electric field and with increasing sample thickness[161]. This is shown schematically in Fig. III. 2. The limiting value of electric field at very short times and very thin samples is now known as the 'intrinsic' breakdown strength and is typically about 10^9 V $\cdot$ m^{-1} for polymers[158,159,160]. This term was originally used to imply a specific breakdown mechanism which was independent of specimen shape or volume (e.g. Whitehead[510]) rather than the maximum strength available. This usage is now generally out of favour (e.g. Cooper in Bradwell[511]) since it seems likely that under such conditions breakdown will be electrode dominated and is not therefore intrinsically related to the material. O'Dwyer[82] however defines the intrinsic critical field strength as 'the more or less abrupt transition from circumstances in which collision ionisation is negligible to those in which it is not'. Under the technologically-important conditions of working lifetimes and thick dielectrics, the breakdown field may be two or three orders of magnitude lower than this intrinsic value.

The effect of temperature on the breakdown strength of polymers has been reviewed by Cooper[512] and Ieda[163] (most of their results are based on the original work of Ball[513] and Oakes[514]). It is usually found that the breakdown strength is reasonably constant ($\pm \sim 20\%$) from very low temperatures up to a critical temperature, often around room temperature, at

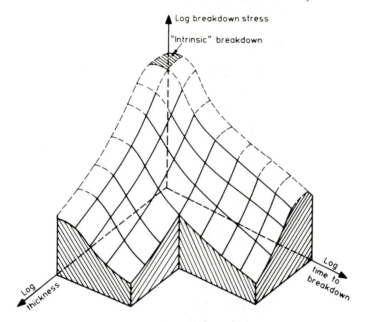

Fig. III.2 A schematic representation of the relationship between the breakdown field, the time to breakdown and the thickness of the insulation. The shaded area marked 'intrinsic breakdown' corresponds to thin samples and short times and represents the highest breakdown strength attainable. At the technologically important longer times and greater thicknesses the breakdown strength may be two or three orders of magnitude lower than this intrinsic value

which it suddenly starts to decrease, Fig. III.3. This can usually be associated with the softening of the polymer although not necessarily with its glass-transition temperature[133,512]. This effect may also be dependent on the type of test used, in particular the sample thickness and the electrode thermal conductivity and rigidity[161].

At low temperatures breakdown in polymers is likely to be of the avalanche type which may be exacerbated by carrier injection from electrodes. Just above the softening point various mechanisms could be responsible such as electromechanical breakdown, scattering of hot electrons with consequent lattice damage (the Fröhlich amorphous-solid model), or thermal breakdown. Thermal breakdown is likely to dominate over electromechanical breakdown at higher temperatures especially in lower-resistivity and dielectrically lossy materials. Notice the 'competitive' nature of the breakdown mechanisms are exemplified here as different mechanisms dominate (rather than operate) at different temperatures. In thicker insulation over a service lifetime the breakdown is likely to be due to the progressive deleterious effects of thermal aging, partial discharges and electrical (and possibly water) trees.

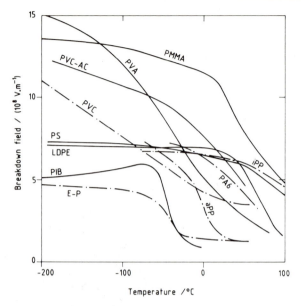

Fig. III.3   Breakdown field as a function of temperature for various polymers. The solid lines are taken from Ball[513] who reported work by W. G. Oakes and the Electrical Research Association. These tests were carried out on recessed specimens using impulses at the higher temperatures and progressive-stress tests at low temperatures to prevent thermal breakdown occurring and give the highest possible results. The broken lines are reported by Ieda[163] for DC breakdown in films and will tend to be lower than the other data by as much as 50%. Notice that for most cases, and especially the non-polar polymers, the breakdown strength drops rapidly at a critical temperature. (PVA = polyvinyl alcohol; PMMA = polymethyl methacrylate; PVC-AC = polyvinyl chloride-acetate; PS = polystyrene; LDPE = low-density polyethylene; PIB = polyisobutylene; iPP/aPP = isoactic/atactic polypropylene; E-P = ethylene-propylene copolymer, PVC = polyvinyl chloride, PA6 = polyamide 6)

Many of the breakdown mechanisms to be described require an understanding of the way in which carriers may be injected into insulators from electrodes and subsequently move through the bulk of the material under high-field conditions. In particular it is important to realise how external factors such as electric field and temperature influence charge carrier injection and movement. A detailed discussion of this subject is worthy of a book in its own right and there have been many reviews to which the reader is referred (60, 82, 515, 516, 517). For the sake of completeness however a brief overview of the factors affecting charge carrier injection and transport is given in chapter nine.

The references quoted in this part of the book represent only a small selection of those published on the subject. For example between 1970–1990 there were over ten thousand papers published on electrical breakdown over a thousand of which related to polymers. We hope to have included major works but the selection of other papers is to some extent arbitrary depending largely on the authors' experiences.

Chapter 9
Charge injection and transport in insulating polymers

9.1 Low-field conduction mechanisms

A description of energy band theory and an overview of the movement of electrons, holes and ions through polymers has already been given in Chapter 2. The main thrust of this chapter is to consider charge movement in a more quantitative manner. In particular this chapter examines the effects of high electric fields and other factors such as temperature on both charge transport through the bulk and the way in which charge may be injected into insulators from the electrodes.

At low electric fields the voltage–current relation tends to follow Ohm's law, although this is difficult to verify in insulators as currents are usually very small at low electric fields and there may be relatively large transients due to the dielectric relaxation effects. This is usually explained in terms of 'Ohmic' conduction in which charge carriers, usually electrons or holes, acquire an average velocity proportional to the field. Their average velocity is limited by collisions with the lattice, and, since these are more likely at higher temperatures, the conductivity has a negative temperature coefficient. The Ohmic conductivity model always predicts reasonably high mobilities (of the order of magnitude of those found in metals and heavily-doped semiconductors) which are not found in insulators. A modification to the model includes traps in the forbidden band gap in which carriers may be resident. This effectively reduces the number of carriers available or, equivalently, lowers their effective mobility. Both trap-free and trap-limited Ohmic conduction models will be discussed in Section 9.1.

Ionic conductivity, briefly outlined in Section 9.1.2, is usually discussed in terms of the movement of a vacancy through an ionic lattice and equations may be derived for both low and high field conditions. At low fields a linear voltage–current relation is found, but at high fields the behaviour becomes 'super Ohmic', i.e. the current increases faster with voltage than that predicted by Ohm's law. At high fields however a different mechanism usually dominates and this behaviour is unlikely to be observed. Ionic conductivity displays an Arrhenius-type conductivity–temperature relationship, i.e. $\log(\text{conductivity}) \propto 1/\text{temperature}$ under both low and high field conditions. Whilst the theory does not appear to be directly applicable to polymers since there is no regular polar lattice through which the ions could move, the basis of the theory is the hopping of carriers over potential barriers which can occur in polymers[71,518,519]. Under an electric field the relative heights of the potential barriers either side of the carrier may be changed and hopping becomes more likely in the field direction. This is developed further in Section 9.3.

Having considered the two mechanisms which may operate at low fields, the effects of high fields are described next. These fall into two categories: electrode-controlled effects and bulk-controlled effects. It will be shown that there exists a potential barrier between electrodes and insulators and that this normally restricts the transfer of electrons from one to the other. (Notice that ions necessarily build up at metal electrode-insulator interfaces under DC conditions unless some electrolytic reaction occurs. Hole transfer is possible however since electron–hole recombination would then occur at the electrode.) High fields can reduce both the height and width of this potential barrier. The reduction in thickness enables electrons to 'tunnel' through the barrier, which is a quantum-mechanical effect without analogy in our directly observable world, and is described as Fowler–Nordheim injection. For very thin films ($< \sim 10$ nm) it is sometimes possible for electrons to tunnel right through the film to the counter electrode by this mechanism. The other effect of high fields mentioned, the reduction in barrier height, increases the probability that there will be electrons with sufficient energy to overcome it; this is known as Schottky injection. These two mechanisms are described in Section 9.2. The nature of the electrode–insulator interface, and in particular the energy band structure of this interface has been greatly idealised for the sake of clarity as a detailed description is outside the scope of this book. The reader is referred to References 520, 521, 522.

In Section 9.3 various mechanisms are described which occur at high-fields:

(i)   Once charge has been injected into the bulk from the electrodes under high fields it may be that it is difficult for the charge to move further through the bulk. Thus a space charge may be formed at the electrodes and throughout the bulk which limits the further movement of the charge. This is known as space-charge limited conduction and the theory is generally developed in terms of an energy band in which there are traps.

(ii)   It has also been argued that the drift velocity of injected carriers is a non-linear function of high fields giving rise to a negative differential resistance phase for individual charge carriers. This may give rise to a maximum possible field within a space-charge region; and is known as the field-limited space-charge (FLSC) model.

(iii)   The band structure may be less well defined, for example like the one shown in Fig. 2.6 in which states are scarce around the Fermi level and gradually increase in concentration towards the 'mobility edges'. In this case there are very few carriers in the higher-energy states which are sufficiently close for carriers to easily move from one to another and there are many carriers trapped in the lower-energy states which are too far apart for carriers to easily move between them. Conduction may however take place by thermally-activated hopping between states over potential barriers.

(iv) At high fields the electronic carrier density may be enhanced and there are two important models for this: the Poole–Frenkel mechanism in which potential barriers are lowered by the field and thus release carriers which would otherwise be localised (this will assist the thermally-activated hopping process), and what is often known as the Fröhlich amorphous-solid model. The latter however may also be considered as an electrical breakdown mechanism as it predicts a critical field above which the rate of energy gained by electrons would be greater than that at which they could lose it. We have therefore included this model in the chapter on electronic breakdown in this part of the book (Section 12.1.3).

Whilst these theories of conduction are to be found in various standard texts, none of these appear to discuss them all in a reasonably complete manner. We hope that our descriptions, whilst in some cases being simplified and in many cases being capable of further development, are reasonably complete within the scope of these limitations.

9.1.1 Ohmic conductivity

The model for Ohmic conduction which is generally cited (e.g. Shockley[523]) describes the movement of an electron in a crystal lattice such as is found in a metal or semiconductor. We will consider this case first and then introduce trap states into the energy-band diagram such as would be found in amorphous solids.

(a) Trap-free ohmic conduction

In the absence of an electric field conduction electrons move around freely in a solid due to their thermal energy and 'collide' with thermal vibrations of the lattice and crystal imperfections such as those due to impurity atoms, and voids. These collisions ensure that at any one time the electrons move in all directions, and with different speeds, such that their net *velocity* is zero. Their mean kinetic energy however will be dependent upon the lattice temperature and their average speed is of the order $(3k_B T/m^*)^{1/2}$ which is $\sim 10^5 \text{ m} \cdot \text{s}^{-1}$ at room temperature assuming $m^* \sim m_e$, i.e. the 'effective mass', m^*, of the electron is close to the true or *free electron* mass.

If an electric field, E, is now imposed the electrons will experience a force equal to $-e \times E$ and will therefore accelerate in the opposite direction to the field between collisions (since their charge, $-e$, is negative). Thus the trajectories of their position between collisions will be parabolas instead of straight lines as they 'fall' through the electric field. There will therefore be a net movement of electrons, whose net mean velocity is known as the drift velocity, v_d, in the opposite direction to the electric field and this movement of charge will constitute an electric current. It was shown in Section 2.1 that this current due to electrons is given by $n \cdot e \cdot \mu \cdot E$ where n is the number of electrons per unit volume and μ is their mobility which is defined as their drift velocity per unit electric field. In polymers mean mobilities greater than $10^{-10} \text{ m}^2 \text{ V}^{-1} \text{ s}^{-1}$ are rarely encountered even at high fields, and maximum fields before breakdown are typically $10^9 \text{ V} \cdot \text{m}^{-1}$. Thus drift velocities greater than $0 \cdot 1 \text{ m} \cdot \text{s}^{-1}$ are unlikely to be encountered.

This is five to six orders of magnitude smaller than average thermal velocities so the distribution of electron *velocities* is only slightly shifted by the application of even high electric fields. The *mean free time* between collisions and hence the mean free path is therefore unlikely to be substantially changed by the electric field.

Let us define the mean free time between collisions, τ_c, such that the probability that an electron has a collision in any small interval of time, dt, is dt/τ_c, where τ_c is a constant. (Notice this is not the same as saying that $v_d = l_c/\tau_c$ where l_c is the mean distance between collisions in the direction of the electric field.) Without the application of an electric field the total electron momentum in the system must be zero at any particular time since the thermal velocities are randomly orientated. Under the influence of an electric field this total momentum has a finite value and direction. Considering the total momentum of the n electrons in a unit volume we have the vector sum:

$$\boldsymbol{p}_{\text{tot}} = \boldsymbol{p}_1 + \boldsymbol{p}_2 + \boldsymbol{p}_3 + \cdots + \boldsymbol{p}_n \tag{9.1}$$

giving the average momentum of an electron as $\boldsymbol{p}_{\text{tot}}/n$ so that the drift velocity is:

$$\boldsymbol{v}_d = \frac{\boldsymbol{v}_1 + \boldsymbol{v}_2 + \boldsymbol{v}_3 + \cdots + \boldsymbol{v}_n}{n} = \frac{\boldsymbol{p}_{\text{tot}}}{nm^*} \tag{9.2}$$

The electric field acts to accelerate the electrons, $\dot{\boldsymbol{v}} = -e \cdot \boldsymbol{E}/m^*$ which would cause the momentum to increase, (i.e. $\dot{\boldsymbol{p}} = -e \cdot \boldsymbol{E}$) if it were not for the effect of the collisions. A dynamic equilibrium is established in which the effect of the electric field is to set up a stable change in the momentum. The increase of (the magnitude of) the total momentum due to the electric field in time dt is:

$$d\boldsymbol{p}_{\text{tot}} = -ne\boldsymbol{E}\, dt \tag{9.3}$$

and is counteracted by collisions. Since in any period of time, dt, the proportion of electrons that will collide is dt/τ_c (assuming a large number) the decrease of momentum due to collisions in time dt is:

$$d\boldsymbol{p}_{\text{tot}} = -\boldsymbol{p}_{\text{tot}}\, dt/\tau_c \tag{9.4}$$

In dynamic equilibrium these two are equal so that:

$$\boldsymbol{p}_{\text{tot}} = -ne\boldsymbol{E}\tau_c \tag{9.5}$$

Substituting for $\boldsymbol{p}_{\text{tot}}$ from eqn. 9.2 gives:

$$\boldsymbol{v}_d = -e\tau_c \boldsymbol{E}/m^* \tag{9.6}$$

(Some texts incorrectly state $\boldsymbol{v}_d = -e\tau_c \boldsymbol{E}/(2m^*)$ because they use an incorrect averaging procedure for finding $\boldsymbol{v}_d$.) From the definition of mobility, μ, we have:

$$\mu = e\tau_c/m^* \tag{9.7}$$

(Mobilities are always taken to be positive. Since an electric field always cause a current to flow in the same direction irrespective of the sign of the charge carriers this is reasonable, i.e. the eqn. 2.2, $\boldsymbol{J} = ne\mu\boldsymbol{E}$, is correct provided e is taken to be the magnitude of the charge on the charge carrier.) This constant mobility implies that the current density is linearly related to field and so Ohm's law holds. In a sense this is simply a result of the assumption of the above dynamic equilibrium which is based on the oservation of Ohm's law so perhaps the argument is rather circuitous!

Whilst band conduction has been thought to occur in some polymers[55,518,524–530] in general the situation is much more complicated. Electrons are unlikely to be free to move at all and those that do have sufficient energy to be in the conduction band within polymer molecules are likely to be scattered very quickly and fall back into a trapped state. The theory of trap-limited mobility will be developed in the next sub-section and the above concepts of collisional scattering and mean free time will be useful. The effect of increasing temperature on this true Ohmic conduction should be to decrease the conductivity since the increase in lattice vibrations (or phonon density) would make scattering more likely and decrease the mean free time. Whilst this is observed in metals and intrinsic semiconductors, this temperature dependence is not generally observed in polymers even at low fields. So even if a linear low-field relation between E and J is observed in polymers, true 'Ohmic' conduction is unlikely to be responsible.

(b) Trap-limited mobility

Given the formula for mobility derived in the last section (eqn. 9.7) it is possible to calculate a typical minimum value for mobility by noting that $\tau_c = l_c/s_{th}$ where s_{th} is the thermal speed of the electrons which was shown to be of the order of 10^5 m $\cdot$ s^{-1} at room temperature. The minimum value for l_c is likely to be of the order of 0.2 nm which results in a minimum mobility value of $\sim 10^{-4}$—10^{-3} m^2 V^{-1} s^{-1}. Typical values[531–528] for polymers are 10^{-15}—10^{-8} m^2 V^{-1} s^{-1}. A trap-limited mobility model may be used to account for this large discrepancy.

Consider a dielectric in which the band gap contains trap states below the conduction band. The traps are considered to all be at the same level, $\mathbb{E}_t$, an energy $\Delta \mathbb{E}_t$ below the bottom of the conduction band (i.e. $\mathbb{E}_c - \mathbb{E}_t = \Delta \mathbb{E}_t$), and of number density, N_t, Fig. 9.1. The number density of conduction

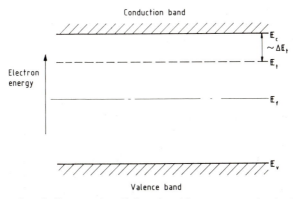

Fig. 9.1 Energy-band diagram for dielectric with trap states in the band gap

band electrons, n_c, may be calculated using the Fermi–Dirac distribution (eqn. 2.5) and since the band gap is wide ($\mathbb{E}_c - \mathbb{E}_f \gg k_B T$) the approximation given in Section 9.2.1 may be used. That is:

$$n_c \simeq N_{\text{eff}} \exp \left\{ -\frac{\mathbb{E}_c - \mathbb{E}_f}{k_B T} \right\} \tag{9.8}$$

where N_{eff} is the effective density of states for the conduction band. Similarly the number density of *occupied* trap states, n_t, is given by:

$$n_t = N_t \left(1 + \exp \left\{ \frac{\mathbb{E}_t - \mathbb{E}_f}{k_B T} \right\} \right)^{-1} \tag{9.9a}$$

which for shallow traps, i.e. ones well above the Fermi level, approximates to:

$$n_t \simeq N_t \exp \left\{ -\frac{\mathbb{E}_t - \mathbb{E}_f}{k_B T} \right\} \tag{9.9b}$$

Combining eqns. 9.8 and 9.9b and remembering $\mathbb{E}_c - \mathbb{E}_t = \Delta \mathbb{E}_t$ gives the ratio of free charge carriers to those trapped:

$$\frac{n_c}{n_t} = \frac{N_{\text{eff}}}{N_t} \exp \left\{ -\frac{\Delta \mathbb{E}_t}{k_B T} \right\} \tag{9.10}$$

The number of carriers available for conduction is therefore reduced from n to $\theta \cdot n$ where $\theta = n_c/n_t$ given by eqn. 9.10. Historically this has been known as 'trap-limited mobility' rather than 'trap-limited carrier concentration' since θ is also approximately the ratio of the time carriers spend in the conduction band to that spent in traps. Thus in the presence of traps the carriers could be said to take longer to drift the same distance as without traps so their mobility is apparently reduced whilst their number density can be said to be constant. Either way the conductivity $\sigma = ne\mu$ becomes multiplied by the factor θ which may be as small as 10^{-10}—10^{-6}. Apparent mobilities of $10^{-14} \, \text{m}^2 \, \text{V}^{-1} \, \text{s}^{-1}$ are therefore reasonable in the light of this model.

9.1.2 Ionic conductivity

Low-field conduction in polymers is frequently found to obey the Arrhenius-type relationship (e.g. References 44, 539–547):

$$\sigma(T) = \sigma_0 \exp \left\{ -\frac{\phi}{k_B T} \right\} \tag{9.11}$$

where σ_0 and ϕ are experimentally-determined values which are found to be constant over a given range of temperature. This is usually attributed to an ionic conduction mechanism. The identification of carrier type is very difficult. Seanor[44,71] has used mass transport methods in polyimide 6,6 but the volume in such cases is very small. The effect of an increased pressure may be to make electron orbital overlap increase thereby increasing electronic conductivity but making ionic percolation more difficult[81,548–551]. Ionic conductivity may also be expected to increase and electronic conduction to decrease as the temperature is raised through the glass-transition[45] and melting temperatures[92]. Electrochemical reactions at the electrodes have been observed by Sawa *et al.*[552]. Ionic conductivity has been reported for a number of polymers[553–559] whereas electronic mechanisms have been proposed for others[546,560,561]. In some polymers there are disagreements (e.g. PET[561,547,562,563]).

The theory of ionic conduction expounded here was developed originally for the movement of ions or vacancies through an ionic crystal and considers both the concentration of charge carriers and their mobility as functions of temperature. A more sophisticated treatment (oustide the scope of this book) based on this approach has been developed by Seanor[71] and Barker[564]; Belmont[519] has dealt with the high-field case. The carriers are considered to be thermally activated and to be located in potential wells and so their mobility is limited by the ease with which they can escape from these wells. Both these concepts are reasonably applicable to polymers under certain circumstances, and, although the scaling factors may be expected to be different, the dependencies on temperature and field may be expected to be not too dissimilar.

Consider the cubic crystal structure shown in Fig. 9.2. If the central cation is missing then any of the other twelve cations could fill its place provided they had sufficient thermal energy to overcome the coulombic forces which hold them in place. From classical statistical mechanics the probability that this nearest-neighbour transition takes place in time dt is given by:

$$P(\text{transition}) = \nu_0 \exp\left\{-\frac{\Delta G}{k_B T}\right\} dt \qquad (9.12)$$

where ΔG is the Gibb's free energy of activation and ν_0 is known as the

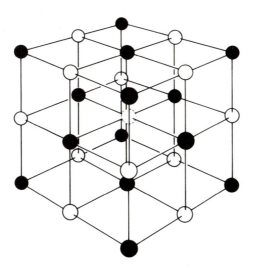

Fig. 9.2 A cubic crystal structure containing anions (●) and cations (○) with a vacancy in the central position instead of a cation

'attempt to escape' frequency and is approximately given by:

$$\nu_0 = \frac{k_B^3 T^3}{dh^3 \nu^2} \qquad (9.13)$$

Here d is the number of directions in which the defect can move and ν is the vibrational frequency around the defect site in a plane orthogonal to the direction of movement. The latter is typically about 10^{12} Hz so that ν_0 is $\sim 2 \times 10^{13}$ Hz at room temperature ($\nu_0(T = -100°C) \simeq 4 \times 10^{12}$ Hz, $\nu_0(T = 100°C) \simeq 4 \times 10^{13}$ Hz). Whilst ν_0 is usually assumed to be independent of temperature in comparison with the exponential term in eqn. 9.12 significant errors can occur around room temperature for values of $\Delta G < \sim 0.2$ eV. At temperatures above some Gibb's-free energy-dependent value, T^*, an Arrhenius-type plot of $\log \{P(\text{transition})\}$ versus 1/temperature starts to deviate above the straight line which would be expected if ν_0 were temperature independent. The temperature at which this causes an inaccuracy of a factor of 10 in P(transition) is shown for various values of Gibbs energies in Fig. 9.3. It can be seen that there is a sudden increase in this temperature (T^*) between 0.1 and 0.2 eV indicating possible inaccuracies below this value.

Consider now the simple case where an electric field of magnitude, E, is applied along an axis containing a vacancy and two possible sites, Fig. 9.4a. The potential barrier inhibiting the movement of the vacancy in the field direction is thereby reduced in height (energy) by $\frac{1}{2}Eea$ where a is the spacing between cations, Fig. 9.4b. Similarly the potential barrier inhibiting its movement in the reverse direction will be increased by the same amount.

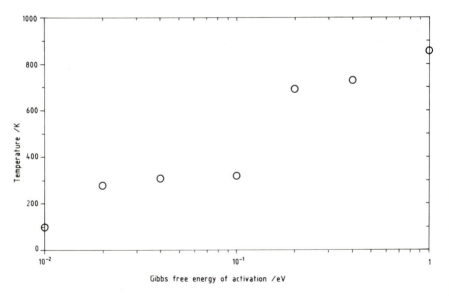

Fig. 9.3 The temperature at which significant deviations from an Arrhenius behaviour would be expected for ionic conductivity as a function of band gap

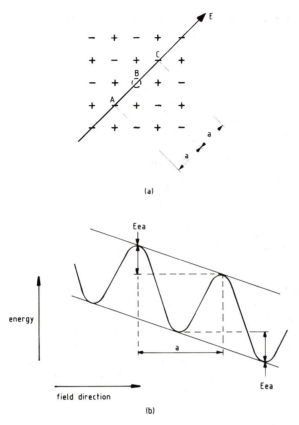

(a)

(b)

Fig. 9.4 An electric field will increase the potential barrier in one direction and decrease it in the other. (*a*) Direction of the electric field through the vacancy. (*b*) The effect of the electric field on the potential barrier

The probability that the vacancy moves in the field direction is thereby increased from its values given in eqn. 9.12 to:

$$P(+a) = \nu_0 \exp\left\{ -\frac{\Delta G - \frac{1}{2}Eea}{k_B T} \right\} dt \qquad (9.14a)$$

and the probability that it moves in the reverse direction is decreased to:

$$P(-a) = \nu_0 \exp\left\{ -\frac{\Delta G + \frac{1}{2}Eea}{k_B T} \right\} dt \qquad (9.14b)$$

The net probability of vacancy movement in the direction of the field is therefore given by:

$$P_T = P(+a) - P(-a)$$

$$= \nu_0 \exp\left\{ -\frac{\Delta G}{k_B T} \right\} dt \times 2 \sinh\left\{ \frac{Eea}{2k_B T} \right\} \qquad (9.15)$$

Since this defines the proportion of the (large) number of carriers that move a distance a in time dt, the magnitude of the drift velocity in the direction of the field is:

$$v_d = a\nu_0 \exp\left\{-\frac{\Delta G}{k_B T}\right\} \times 2 \sinh\left\{\frac{Eea}{2k_B T}\right\} \tag{9.16}$$

Since for $x < 0.8$ radians $\sinh(x) \simeq x \pm 10\%$, at low and moderate fields $(E < \sim 10^8\ \text{V} \cdot \text{m}^{-1})$ this approximates to:

$$v_d = \frac{Eea^2\nu_0}{k_B T} \exp\left\{-\frac{\Delta G}{k_B T}\right\} \tag{9.17}$$

yielding a field-independent mobility, $\mu = v_d/E$:

$$\mu = \frac{ea^2\nu_0(T)}{k_B T} \exp\left\{-\frac{\Delta G}{k_B T}\right\} \tag{9.18}$$

At high fields $P(+a) \gg P(-a)$ so $P_T \simeq P(+a)$ and corresponding expressions can be evaluated. However this high-field behaviour is rarely observed as other high-field effects generally dominate.

The number density of carriers, n, is also likely to be temperature dependent. O'Dwyer[82] treats this in detail for ionic crystals but such an approach cannot be justified for the very different morphologies found in polymers. However it is reasonable to assume, at least to a first approximation, that the carriers are likely to be generated thermally so that their number density is of the form:

$$n(T) = N \exp\left\{-\frac{\mathbb{E}_A}{k_B T}\right\} \tag{9.19}$$

where $\mathbb{E}_A$ is an energy of formation and N is a concentration of possible formation sites. Since $\sigma = ne\mu$ (eqn. 2.2) we have for lower-field ionic conductivity:

$$\sigma = \frac{e^2 a^2 \nu_0(T) N}{k_B T} \exp\left\{-\frac{\mathbb{E}_A + \Delta G}{k_B T}\right\} \tag{9.20}$$

Two characteristics of low-field ionic conduction are:

(i)   *Very low mobility*
       Very low mobility is usually encountered. For example taking values of $a = 0.2$ nm, $\nu_0 = 2 \times 10^{13}\ \text{s}^{-1}$, $T = 293$ K, and ΔG values of 0.2, 0.3, 0.5, and 1.0 eV gives, from eqn. 9.18, mobilities of 4.9×10^{-9}, 5.7×10^{-11}, 7.9×10^{-15}, and $1.8 \times 10^{-24}\ \text{m}^2\ \text{V}^{-1}\ \text{s}^{-1}$.

(ii)   *Current after application of a step voltage*
       If a step voltage is applied to an ionically-conducting insulator then an initial current decay is observed which is due to the buildup of cations and anions at the electrodes.

9.2 Charge injection from electrodes

The nature of the electrode–insulator interface is complex. In particular the energy-band diagram is complicated by the presence of physical, chemical and electrical defects such as: surface roughness; imperfect contact, chemical impurities including oxidation and moisture, dangling bonds, local polarisation resulting in band bending, donor–acceptor states, and trap states. This has been the subject of reviews by Lewis[520,521,522].

The two electron-injection mechanisms described here, Fowler–Nordheim injection and the Schottky effect, are based on the very simple electron energy-band diagram shown in Fig. 9.5a. In this picture electrons must overcome a potential barrier to leave the metal and enter the insulation. The height of this barrier depends on the interface. The theories were initially developed for electron injection from a metal into a vacuum in which case the barrier height was equal to the work function of the metal, ϕ_m. In the case of the metal–polymer interface however this barrier height and shape is modified by the local conditions obtaining. Taylor and Lewis[43] have considered both types of electron injection in PET and PE. In an ideal case the electrons would have to be excited into a conduction band in the

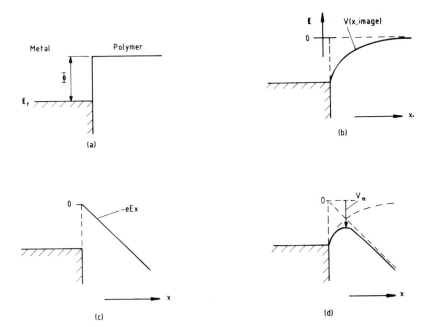

Fig. 9.5 Modification of the coulombic potential barrier at a metal–polymer interface by an applied field
(a)   Total barrier height
(b)   Shape of barrier including effect of coulombic image force
(c)   Potential energy due to applied electric field
(d)   Total barrier shape

polymer in which case the barrier height would be decreased by the electron affinity, χ, of the polymer (Fig. 2.3). Note that in some polymers (e.g. polyethylene[189,68]) the electron affinity is negative (i.e. only hot electrons with energies greater than the vacuum level are delocalised) and so the barrier height is increased. This is because electrons with energies greater than the electron affinity are naturally emitted from the insulator. In practice however the situation is much more complicated for the reasons given earlier and, since both models of injection make many implicit assumptions and we are mainly interested in dependencies on field and external parameters only, we will assume the simple picture given in Fig. 9.5a to be reasonable. Ieda has reviewed interfacial phenomena in polymer materials[59]. The barrier height, Φ, is unspecified in physical terms and depends on the metal–insulator work function difference and local conditions of electrical polarisation.

9.2.1 Schottky injection

In the simple band picture of Fig. 9.5a the barrier was assumed to be abrupt between the metal and insulator. In the Schottky injection model the picture is refined by assuming the barrier is due to the electrostatic attraction between the electron and the metal, the latter being positively charged since the electron has left it. This attraction gives rise to a gradually changing barrier due to the potential energy of the electron. Fig. 9.6a shows the lines of flux between a point charge (cf. the electron) and an infinite conducting flat sheet (cf. the electrode). In order to calculate the attraction between the point charge and the sheet we assume, from symmetry, that the sheet can be replaced by an equal and oppositely charged point an equal distance behind the surface of the sheet, Fig. 9.6b. This is known as the image charge theorem (e.g. Reference 565). If the electron is distance, x, from the interface then the distance separating electron and image is $2x$ and the force of attraction is given from Coulomb's law as:

$$F(x) = \frac{e^2}{4\pi\varepsilon_0\varepsilon_r(2x)^2} \tag{9.21}$$

The potential energy of the electron $V(x)$ may be defined as the energy required to take the electron from position x to infinity and is found by integration of eqn. 9.21:

$$V(x, \text{image}) = \int_x^\infty F(x)\,dx = -\frac{e^2}{16\pi\varepsilon_0\varepsilon_r x} \tag{9.22}$$

The equation is not valid as the electron approaches the surface closely and the surface can no longer be assumed to be smooth, i.e. $x \gg$ local features such as the metal lattice spacing, a. Furthermore the potential energy must approach the metal work function, ϕ_m, and not $-\infty$, as $x \to 0$. This is shown in Fig. 9.5b. Applying this to eqn. 9.22 yields:

$$V(x, \text{image}) = -\frac{e^2}{16\pi\varepsilon_0\varepsilon_r x + (e^2/\phi_m)} \tag{9.23}$$

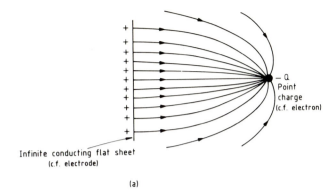

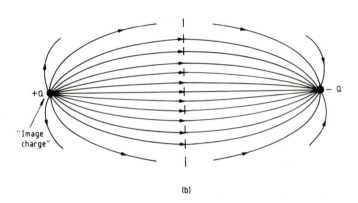

Fig. 9.6   Illustration of the classical law of images. (*a*) A point charge placed in front of an infinitely conducting, infinitely large, flat sheet coulombically attracts opposite charges to the surface. The lines of force between the sheet and the charge are the same as if the sheet were replaced by an equal and opposite charge placed equidistant behind the sheet as shown in (*b*)

If an electric field is now applied and we assume that this is constant in the region of the interface and of magnitude E in the x-direction then the potential is modified by (Fig. 9.5*c*):

$$V(x, \text{field}) = -eEx \tag{9.24}$$

resulting in a reduced potential barrier of the form (Fig. 9.5*d*):

$$V(x) = -\frac{e^2}{16\pi\varepsilon_0\varepsilon_r x + (e^2/\phi_m)} - eEx \qquad x \geqq 0 \tag{9.25}$$

The height of the potential barrier, $V_m = V(x = x_m)$ may be found by differentiating this potential function and setting the differential to zero to indicate a maxima. Furthermore eqn. 9.25 may be simplified by assuming

$16\pi\varepsilon_0\varepsilon_r x \gg e^2/\phi_m$. If this were not true the theory would breakdown anyway as $x \to a$, the interatomic spacing, and the influence of particular lattice atoms would predominate. Thus:

$$\frac{dV(x)}{dx} = \frac{e^2}{16\pi\varepsilon_0\varepsilon_r x^2} - eE = 0 \qquad x = x_m \qquad (9.26)$$

so that

$$x_m = \left(\frac{e}{16\pi\varepsilon_0\varepsilon_r E}\right)^{1/2} \qquad (9.27)$$

and

$$V_m = -\frac{e}{2}\left(\frac{eE}{\pi\varepsilon_0\varepsilon_r}\right)^{1/2} \qquad (9.28)$$

For very high values of field x_m does approach typical interatomic spacings. For example if $E = 10^9$ V · m^{-1} and $\varepsilon_r = 3$ then $x_m = 0.35$ nm and the theory, if not the insulation, may start to breakdown. Above these values Fowler–Nordheim injection is more likely (Section 9.2.2). The barrier height is now reduced by V_m (Fig. 9.5d) so that the effective barrier height is:

$$\phi_{\text{eff}} = \Phi - \frac{e}{2}\left(\frac{eE}{\pi\varepsilon_0\varepsilon_r}\right)^{1/2} \qquad (9.29)$$

In order to find the current density, the rate at which electrons arrive at the barrier with enough energy to clear it must be found. An energy distribution is therefore required. In the metal $N(\mathbb{E})\,d\mathbb{E}$ is the number density of electrons with energy in the range $\mathbb{E}$ to $\mathbb{E} + d\mathbb{E}$ (Section 2.2). The effective density of states of conduction electrons in a metal is (e.g. Allison[566]):

$$N_{\text{eff}} = \frac{2^{7/2}\pi m^{3/2}\mathbb{E}^{1/2}}{h^3} \qquad (9.30)$$

so that

$$N(\mathbb{E})\,d\mathbb{E} = \left(\frac{2^{7/2}\pi m^{3/2}\mathbb{E}^{1/2}}{h^3}\right) P(\mathbb{E},\,T)\,d\mathbb{E} \qquad (9.31)$$

where $P(\mathbb{E},\,T)$ is the Fermi–Dirac distribution (eqn. 2.5), which is reproduced here for clarity:

$$P(\mathbb{E},\,T) = \left[1 + \exp\left\{\frac{\mathbb{E} - \mathbb{E}_f}{k_B T}\right\}\right]^{-1} \qquad (9.32)$$

For these conduction electrons $\mathbb{E} = \frac{1}{2}m^*S^2$, where S is their speed. Assuming $m^* \simeq m$ then $\mathbb{E} = \frac{1}{2}mS^2$ and $d\mathbb{E} = mS\,dS$ so that:

$$N(S)\,dS = \frac{8\pi m^3}{h^3} \cdot \frac{S^2\,dS}{1 + \exp\left\{(\mathbb{E} - \mathbb{E}_f)/(k_B T)\right\}} \qquad (9.33)$$

However the velocity distribution rather than the speed distribution of the electrons is required since only electrons with a component of velocity in the x-direction will leave the metal surface. Let us define $N(v)\,dv$ as the number density of electrons with velocity in the range v to $v+dv$, or, considering the directional components, $N(v_x, v_y, v_z)\,dv$ as the density of electrons with velocity in the range (v_x, v_y, v_z) to $(v_x+dv_x, v_y+dv_y, v_z+dv_z)$. Referring to Fig. 9.7, all electrons contained within the shell have *speeds* between S and $S+dS$, and the incremental cube contains $N(v_x, v_y, v_z)\,dv_x\,dv_y\,dv_z$ electrons. Thus:

$$N(v)\,dv = N(S)\,dS \times \frac{\text{volume in velocity space}}{\text{volume in speed space}}$$

$$= N(S)\,dS \times \frac{dv_x\,dv_y\,dv_z}{4\pi S^2\,dS}$$

$$= \frac{2m^2}{h^3} \cdot \frac{dv_x\,dv_y\,dv_z}{1+\exp\{(\mathbb{E}-\mathbb{E}_f)/(k_B T)\}} \tag{9.34}$$

and

$$\mathbb{E} = \tfrac{1}{2}mS^2 = \tfrac{1}{2}m(v_x^2 + v_y^2 + v_z^2) \tag{9.35}$$

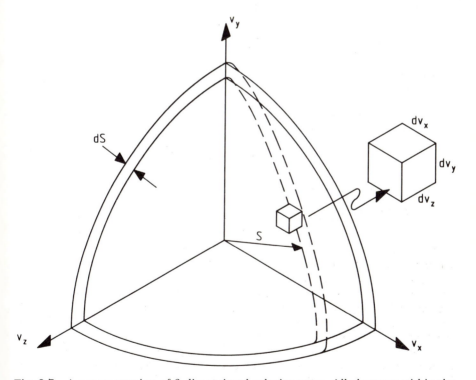

Fig. 9.7 A representation of 3-dimensional velocity space. All electrons within the spherical shell of inner radius S and thickness dS have speeds within the range S to $S+dS$

The electrons which cross the barrier must have an x component of velocity, v'_x, such that $\frac{1}{2}m\,(v'_x)^2$ is the height of the barrier and they must not be reflected by the surface of the metal. If the proportion of electrons which are reflected is denoted by R (the reflection coefficient $R \leqq 1$), the current density is:

$$J = \frac{2(1-R)em^3}{h^3} \int_{v'_x}^{\infty} \int_{-\infty}^{\infty} \int_{-\infty}^{\infty} \frac{v_x\,dv_x\,dv_y\,dv_z}{1+\exp\{(\mathbb{E}-\mathbb{E}_f)/(k_BT)\}} \tag{9.36}$$

Since insulators must have a wide band gap $\mathbb{E}-\mathbb{E}_f \gg k_BT$ for those electrons with a large enough $\mathbb{E}$ for injection, and the term

$$[1+\exp\{(\mathbb{E}-\mathbb{E}_f)/(k_BT)\}]^{-1} \simeq \exp\{(\mathbb{E}_f-\mathbb{E})/(k_BT)\}.$$

Using this approximation and the relation 9.35 gives:

$$J = \frac{2(1-R)em^3}{h^3} \exp\left\{\frac{\mathbb{E}_f}{k_BT}\right\} \int_{v'_x}^{\infty} v_x \exp\left\{\frac{-mv_x^2}{2k_BT}\right\} dv_x$$

$$\cdot \int_{-\infty}^{\infty} \exp\left\{\frac{-mv_y^2}{2k_BT}\right\} dv_y \cdot \int_{-\infty}^{\infty} \exp\left\{\frac{-mv_z^2}{2k_BT}\right\} dv_z \tag{9.37}$$

Now substituting $u = v(m/(2k_BT))^{1/2}$ gives:

$$\int_{-\infty}^{\infty} \exp\left\{\frac{-mv^2}{2k_BT}\right\} dv = \left(\frac{2\pi k_BT}{m}\right)^{1/2}$$

so that the product of the last two integrals in eqn. 9.37 is $(2\pi k_BT/m)$. The first integral may be calculated by substituting $u = mv_x^2/(2k_BT)$ so that:

$$\int_{v'_x}^{\infty} v_x \exp\left\{\frac{-mv_x^2}{2k_BT}\right\} dv_x = -\frac{k_BT}{m}\left[\exp\left\{\frac{-mv_x^2}{2k_BT}\right\}\right]_{v'_x}^{\infty}$$

$$= \frac{k_BT}{m}\exp\left\{-\frac{(\phi_{\mathrm{eff}}+\mathbb{E}_f)}{k_BT}\right\} \tag{9.38}$$

This yields the expression for Schottky injection as:

$$J = \frac{4\pi emk_B^2(1-R)T^2}{h^3}\exp\left\{\frac{-\Phi}{k_BT}\right\}\exp\left\{\frac{e}{2k_BT}\left(\frac{eE}{\pi\varepsilon_0\varepsilon_r}\right)^{1/2}\right\} \tag{9.39}$$

A constant term $A = 4\pi emk_B^2/h^3 = 1\cdot20\times10^6\,A\cdot m^{-2}\,K^{-1}$ is often identified. In order to verify Schottky injection, 'Schottky plots' of $\log_e(J/T^2)$ versus $E^{1/2}$ are often constructed. These yield straight lines although at low fields space charges and surface inhomogeneities tend to cause deviations. The intercept giving the pre-exponential terms in eqn. 9.39, $A(1-R)T^2$ $\exp\{-\Phi/(k_BT)\}$, and the slope giving $(e/(2k_BT))\cdot(e/(\pi\varepsilon_0\varepsilon_r))^{1/2}$ both tend to be low. The pre-exponential term may be six or seven orders of magnitude less than the theoretical value. Lewis[567] has suggested that this may be caused by a metal oxide layer about 2 nm thick. A consequence of the resulting barrier, which is not field limited, is that the right-hand side of eqn. 9.39 becomes multiplied by the probability of the electron crossing the barrier. Note that in eqn. 9.39 the high frequency value of the permittivity should be employed as the distance from x_m is very small and so the time to cross it is correspondingly small. A depletion or accumulation layer may also form at the surface.

9.2.2 Fowler–Nordheim injection

At very-high fields in the insulator ($\sim 10^9$ V $\cdot$ m^{-1}) next to the contact the potential barrier becomes very thin and a different mechanism comes into play. In classical mechanics a particle cannot enter a region in which its total energy is less than the potential energy required. However particles which exhibit particle–wave duality (Section 2.2) such as electrons and photons do have a finite probability of existing in such regions. This is observed in classical wave-optics in which light waves may pass through regions in which, because of their refractive index, their velocity of propagation becomes imaginary (e.g. Reference 568). Similarly it is found that electrons may pass through thin potential barriers despite having insufficient energy to surmount them; this is known as tunnelling. When this occurs at the contact barrier it is known as Fowler–Nordheim injection[569].

It is convenient to ignore the image forces in this analysis, their effects will be mentioned later, and consider the potential barrier to be the triangular shape shown in Fig. 9.8. If the surface defines the origin at which $x = 0$ then the potential energy $V(x)$ of the barrier is given by:

$$
\left.
\begin{aligned}
V(x) &= 0; & x &< 0 \\
V(x) &= \Phi - eEx & x &> 0
\end{aligned}
\right\}
\tag{9.40}
$$

In order to estimate the transmission coefficient of the barrier, that is the probability that an electron incident upon the barrier will be transmitted through it, it is necessary to use quantum mechanical theory and this is briefly outlined here.

Since we are only interested in the progress of the electron in the x-direction, we start by writing the general one-dimensional formula for any time-independent wave phenomenon:

$$
\frac{d^2\psi}{dx^2} + \left(\frac{2\pi}{\lambda}\right)^2 \psi = 0
\tag{9.41}
$$

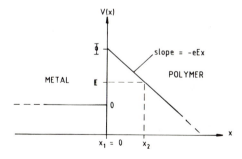

Fig. 9.8 The triangular potential barrier shape used in calculating Fowler–Nordheim emission (cf. Figs. 9.5c and 9.5d)

where $\psi(x)$ is the *wave function* describing the wave as a function of position. For an electron this is a complex function which can be interpreted such that $\psi^2 \, dx \, (= \psi \cdot \psi^* \, dx$, where the asterisk indicates the complex conjugate) is the probability that the electron exists between x and $x + dx$. The relationship between wave and particle characteristics is given by

$$h = \lambda \cdot p \qquad (9.42)$$

(Section 2.2.2, known as de Broglie's hypothesis) where λ is the wavelength and p the momentum. We can re-write this in terms of the particle's total energy, $\mathbb{E}$; potential energy, $V(x)$; and kinetic energy $\mathbb{E} - V(x) = \frac{1}{2}mv^2$ thus:

$$\lambda = \frac{h}{(2m(\mathbb{E} - V))^{1/2}} \qquad (9.43)$$

The one-dimensional, time independent, wave equation for a particle is therefore:

$$\frac{\hbar^2}{2m} \frac{d^2\psi}{dx^2} - (V(x) - \mathbb{E})\psi = 0 \qquad (9.44)$$

where $\hbar = h/(2\pi)$. This is a one-dimensional form of the Schrödinger equation. It is necessary to find the form of the wave function $\psi(x)$ which will satisfy the tunnelling problem. Let us assume that the barrier exists between two values of x: x_1 and x_2. We have already defined this region in terms of energy to be such that the particle's total energy, $\mathbb{E}$, is less than the potential energy, $V(x)$, required to enter it. Thus:

$$(V(x) - \mathbb{E}) > 0 \qquad x_1 < x < x_2 \qquad (9.45)$$

An approximate form of $\psi(x)$ which satisfies the Schrödinger equation with this boundary condition is:

$$\psi(x) = \exp\{\alpha(x)\} \qquad (9.46)$$

This is known as the WKB approximation as it was proposed (separately) by Wentzel[570], Kramers[571] and Brillouin[572]. Substituting this into eqn. 9.44 yields:

$$\frac{\hbar^2}{2m}\left(\left[\frac{d\alpha}{dx}\right]^2 + \frac{d^2\alpha}{dx^2}\right) - (V(x) - \mathbb{E}) = 0 \qquad (9.47)$$

In the WKB approximation $\alpha(x)$ is assumed to be a slowly varying function of x such that $d^2\alpha/dx^2 \ll (d\alpha/dx)^2$. Eqn. 9.47 therefore has the approximate solution:

$$\alpha = \pm \int_{x_1}^{x_2} \left[\frac{2m}{\hbar^2}(V(x) - \mathbb{E})\right]^{1/2} dx \qquad (9.48)$$

which can be substituted into eqn. 9.46 as the required solution of $\psi(x)$. The probability that the particle is transmitted through the barrier is given by the probability that it exists at position x_2: $\psi(x_2) \cdot \psi^*(x_2)$, i.e. the transmission coefficient is:

$$T = \psi(x_2) \cdot \psi^*(x_2)$$

$$= \exp\left\{-2\left(\frac{2m}{\hbar^2}\right)^{1/2} \int_{x_1}^{x_2} (V(x) - \mathbb{E})^{1/2}\, dx\right\} \tag{9.49}$$

Having established an expression for the transmission coefficient we can now apply it to the particular barrier shape in question defined by eqns. 9.40 with:

$$x_1 = 0$$

and

$$x_2 = \frac{\Phi - \mathbb{E}}{eE} \tag{9.50}$$

(see Fig. 9.8) to calculate the tunnelling probability for a given energy. This can then be multiplied by the number of electrons with this energy arriving at the barrier in order to calculate the tunnelling current density. Inserting eqns. 9.40 and 9.50 into 9.49 and noting:

$$\int_{x_1}^{x_2} (\Phi - \mathbb{E} - eEx)^{1/2}\, dx = \frac{2}{3} \frac{(\Phi - \mathbb{E})^{3/2}}{eE}$$

gives the transmission coefficient of the barrier as:

$$T = \exp\left\{-\frac{4}{3}\left(\frac{2m}{\hbar^2}\right)^{1/2} \frac{(\Phi - \mathbb{E})^{3/2}}{eE}\right\} \tag{9.51}$$

By using a similar argument to that used to arrive at eqn. 9.34 from the effective density of states (eqn. 9.30) it is straightforward to show that the number density of electrons with momentum in the range (p_x, p_y, p_z) to $(p_x + dp_x, p_y + dp_y, p_z + dp_z)$ is $(2/h^3) \cdot P(\mathbb{E}, T) \cdot dp_x\, dp_y\, dp_z$. Each of these has a velocity in the x-direction, v_x to $v_x + dv_x \simeq v_x$ so that the number within these momentum constraints arriving at the surface per square metre per second is:

$$v_x \left(\frac{2}{h^3}\right) P(\mathbb{E}, T)\, dp_x\, dp_y\, dp_z$$

and the number arriving with x-momentum in the range p_x to $p_x + dp_x$ (without constraint on the momentum components in the plane of the surface) is:

$$N(p_x)\, dp_x = \int_{-\infty}^{\infty} \int_{-\infty}^{\infty} \frac{p_x}{m} \frac{2}{h^3} P(\mathbb{E}, T)\, dp_x\, dp_y\, dp_z \tag{9.52}$$

This is difficult to solve (see O'Dwyer[82] for a fuller solution) and a common approximation is to assume that the temperature is absolute zero since, as is shown later, the tunnelling current is only weakly temperature dependent. In this case:

$$\left.\begin{array}{ll} P(\mathbb{E}, T = 0) = 1 & \mathbb{E} \leq \mathbb{E}_f \\ P(\mathbb{E}, T = 0) = 0 & \mathbb{E} > \mathbb{E}_f \end{array}\right\} \tag{9.53}$$

and the maximum momentum is for electrons at the Fermi level:

$$p_f^2 = 2m\mathbb{E}_f \geq p_x^2 + p_y^2 + p_z^2 \tag{9.54}$$

Carrying out the integration of eqn. 9.52 with respect to y and z using these simplifications and multiplying by the electronic charge, e, and the transmission coefficient $T(p_x)$ gives the current density:

$$J = \frac{2\pi e}{mh^3} \int_0^{p_f} T(p_x) p_x (p_f^2 - p_x^2) \, dp_x \tag{9.55}$$

The integration may be performed using the dummy variable $\Theta = p_f - p_x$. Since $T(p_f - \Theta)$ decreases rapidly with Θ, use of the following approximate relations (van der Zeil[517]):

$$p_f^2 - p_x^2 \approx 2 p_f \Theta$$

$$p_x \approx p_f$$

and

$$(\Phi - \mathbb{E})^{3/2} = \left(\phi + \frac{p_f^2 - p_x^2}{2m} \right)^{3/2}$$

$$= \phi^{3/2} + \frac{3}{2} \left(\frac{\phi^{1/2}}{m} \right) p_f \Theta$$

where $\phi = \Phi - \mathbb{E}_f$, and extending the upper limit of integration to infinity gives:

$$J = \frac{4\pi e p_f^2}{mh^3} \exp \left\{ -\frac{4}{3} \left(\frac{2m}{\hbar^2} \right)^{1/2} \frac{\phi^{3/2}}{eE} \right\}$$

$$\times \int_0^{\infty} \exp \left\{ -2 \left(\frac{2m}{\hbar^2} \right)^{1/2} \frac{\phi^{1/2}}{meE} p_f \Theta \right\} \Theta \, d\Theta$$

$$= \frac{e^3 E^2}{8\pi h \phi} \exp \left\{ -\frac{4}{3} \left(\frac{2m}{\hbar^2} \right)^{1/2} \frac{\phi^{3/2}}{eE} \right\} \tag{9.56}$$

A more accurate analysis[82,1] will result in replacing m by m^*, multiplying the pre-exponential factor by (m/m^*), and replacing $\phi^{3/2}$ with $\phi_{\mathrm{eff}}^{1/2} \cdot \phi$ in this equation where ϕ_{eff} is the field-reduced barrier height (eqn. 9.29).

If the image force is to be taken into account eqn. 9.40b is replaced by (cf. eqn. 9.22):

$$\left. \begin{aligned} V(x) &= 0 & x < 0 \\ V(x) &= \Phi - eEx - \frac{e^2}{16\pi\varepsilon_0\varepsilon_r x} & x > 0 \end{aligned} \right\} \tag{9.57}$$

The effect of this is twofold[573]. First the pre-exponential term is multiplied by a function of $(\Delta\phi/\phi)$ where $\Delta\phi$ is the barrier height reduction due to the field (i.e. $\Delta\phi = V_m$, eqn. 9.28). However this function is always of order unity and makes virtually no difference to the value of current density. Secondly the exponent is multiplied by the function $v(\Delta\phi/\phi)$ which is

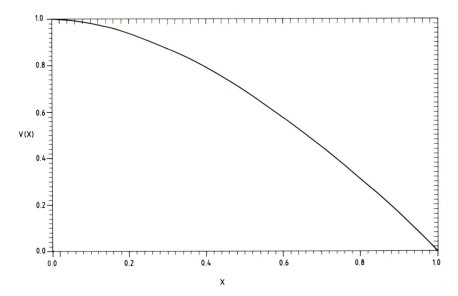

Fig. 9.9 The form of $v(\Delta\phi/\phi)$ from Good and Müller[573] for modifying the Fowler–Nordheim emission equation

described by Good and Müller[573] and shown in graphical form in Fig. 9.9. This may be responsible for a change of many orders of magnitude in the current density.

The effect of temperature is to multiply the current density by a factor $f(\phi)T/\sin[f(\phi)T]$ where:

$$f(\phi) \simeq \frac{\pi k_B 2(2m^*\phi)^{1/2}}{\hbar eE}$$

The multiplying factor is generally small for typical values of ϕ and T, some representative values are tabulated in Table 9.1.

Table 9.1 Temperature correction factors for Fowler–Nordheim injection current densities

	Multiplying factor			
T (K)	$\phi = 1\,eV$	$\phi = 2\,eV$	$\phi = 3\,eV$	$\phi = 4\,eV$
100	1·013	1·026	1·040	1·067
200	1·053	1·111	1·172	1·312
300	1·126	1·275	1·454	1·943
400	1·239	1·570	2·048	4·050
500	1·411	2·123	3·570	79·21

Fowler–Nordheim injection followed by avalanche breakdown (Section 12.2) has been observed[46] in very thin (10·2 nm) films of poly-*p*-xylene films in which the conductivity was almost independent of temperature from $<-150°C$ to $>50°C$ and in PET and polyethylene naphthalate films in which the conductivity suddenly increases very rapidly with field above $10^8 \text{ V} \cdot \text{m}^{-1}$.

9.3 High-field conduction mechanisms

9.3.1 Space charge limited conduction

Space charge limited current is found generally in insulating thin films and is found to be highly dependent upon thickness. In a thin film which is insulating but has good contacts, then, although there may only be a very small amount of free charge in the material at first, more free charge may be injected. If the dielectric constant is high this may lead to a high charge concentration build up within the material. Thus, at least initially, more charge may enter the dielectric than leave it.

It will be assumed that this charge injected into the material is uniformly distributed. Although this may be unreasonable towards the electrodes (where much charge is built up) in the bulk the assumption usually holds reasonably well. Whether space charge limited current becomes the 'rate determining step' controlling current between the electrodes will depend on the charge concentration, the type of charge, its mobility, how good the electrodes are at injecting charge, and the charge trapping characteristics. The latter may be critically dependent upon experimental conditions, for example Noel *et al.*[358] have shown that space-charge polarisation in LDPE is critically dependent upon the adsorption of atmospheric gases. (Although charged traps may not move in the lattice, their charge will still contribute to the space charge.)

The ideal insulator with good injecting electrodes will be considered initially and the interaction of injected charge with traps and thermally-generated carriers will be introduced later. Holes will not be discussed as this makes the treatment much more complicated. If there are two space charges then as the bias across the electrodes increases the space charges may close together and eventually overlap. If this happens then the two layers may cancel each other out given sufficient time and a negative resistance region may form (see Lamb[575] for example). The discussion will be in two parts; first the case of a dielectric without traps will be considered, and secondly the case with traps.

(a) Trap-free dielectric

Consider an ideal dielectric in which there are:

- no thermally generated carriers;
- no traps;
- Ohmic contacts implying good injection;
- only negatively-charged carriers (i.e. electrons).

The analysis is then similar to that from a thermionic cathode into a vacuum since once electrons travel from the metal into the conduction band of the dielectric they will form a space-charge cloud similar to that in a vacuum diode tube. Consider the dielectric to be infinite in area and sandwiched between two parallel-plate electrodes. The x-direction will be orthogonal to the electrodes, i.e. straight through the dielectric, and measured from the cathode surface as its origin. The thickness of the film will be s, i.e. the anode is at $x = s$ and we can define the field and charge-carrier number density to be functions of x, $E(x)$ and $n(x)$ respectively.

The total current density in the dielectric is made up of the three components of drift, diffusion and displacement:

$$J = ne\mu E - eD_n \frac{dn}{dx} + \varepsilon_0 \varepsilon_r \frac{dE}{dt} \tag{9.58}$$

where D_n is the Fick's diffusion coefficient for electrons and the other symbols have their usual notation. Assuming steady-state conditions then $dE/dt = 0$ and this becomes:

$$J = ne\mu E - eD_n \frac{dn}{dx} \qquad \frac{dE}{dt} = 0 \tag{9.59}$$

Now from Poisson's equation (eqn. 2.10):

$$\frac{dE}{dx} = \frac{ne}{\varepsilon_0 \varepsilon_r}$$

so that:

$$J = \varepsilon_0 \varepsilon_r \mu E \frac{dE}{dx} - \varepsilon_0 \varepsilon_r D_n \frac{d^2 E}{dx^2} \tag{9.60}$$

The space charge build up near the injecting electrode will have the greatest effect on diffusion, i.e. the field will be zero there and the current will be a pure diffusion current. However we are, for now, ignoring this effect and assuming the space charge concentration to be constant throughout. We may therefore neglect the diffusion term and assume a constant electric field, across the bulk of the dielectric thus:

$$J \simeq \varepsilon_0 \varepsilon_r \mu E \frac{dE}{dx} \tag{9.61}$$

Rearranging this expression and integrating both sides gives:

$$E = \left(\frac{2J}{\varepsilon_0 \varepsilon_r \mu} (x + x_0) \right)^{1/2} \tag{9.62}$$

where x_0 is a constant of integration. This may be estimated using the boundary condition that at $x = 0$, n is given by the concentration of electrons injected over the electrode–insulator barrier (Lamb[575]), $N_0 = n(x = 0)$ which at low fields (i.e. much less than those at which the Schottky or Fowler–Nordheim processes would have a role) is approximately given by:

$$N_0 = 2 \left(\frac{2\pi m^* k_B T}{h^2} \right)^{3/2} \exp \left\{ \frac{\chi_i - \phi_m}{k_B T} \right\} \tag{9.63}$$

where χ_i is the electron affinity of the insulator and ϕ_m is the work function of the metal. From eqns. 9.62 and 9.59 (ignoring the diffusion term) x_0 is given as

$$x_0 = \frac{\varepsilon_0 \varepsilon_r J}{2 N_0^2 e^2 \mu} \tag{9.64}$$

The current density to voltage relationship may be calculated from eqn. 9.62 using the fact that:

$$V = \int_0^s E \, dx$$

$$= \int_0^s \left(\frac{2J}{\varepsilon_0 \varepsilon_r \mu} (x + x_0) \right)^{1/2} dx$$

$$= \frac{2}{3} \left(\frac{2J}{\varepsilon_0 \varepsilon_r \mu} \right)^{1/2} ((s + x_0)^{3/2} - x_0^{3/2}) \tag{9.65}$$

It is usual to assume that $x_0 \ll s$ so that the current density is given by:

$$J = \frac{9 \varepsilon_0 \varepsilon_r \mu V^2}{8 s^3} \tag{9.66}$$

and the current is proportional to the square of the voltage. This is known as the Mott & Gurney[576] square law (or sometimes Child's law for solids, note that Child's law for a vacuum predicts $J \propto V^{3/2}$). The inequality constraint $x_0 \ll s$ is only true for electrodes with good injection properties as can be seen by calculations using eqns. 9.63 and 9.64. Note, however, that at high fields, the injection processes described in Section 9.2 will substantially reduce the barrier thereby increasing N_0 and reducing x_0 significantly. The charge density in the dielectric, n, is made up of two components, n_0 and n_1 where n_1 is an injection term. The current density can therefore be thought of as being made up of two parts:

$$J = n_0 e \mu \frac{V}{s} + \frac{9 \varepsilon_0 \varepsilon_r \mu V^2}{8 s^3}$$

(Ohmic)   (Space charge limited)

$$\tag{9.67}$$

and if $n_1 \gg n_0$ then space charge limited current dominates. A transition voltage, V_{tr}, may be defined such that above this voltage the space charge limited current dominates over the Ohmic component. This is easily found experimentally by plotting $\log (J)$ versus $\log (V)$ and observing the voltage at which the slope changes from unity to two (Fig. 9.10). Equating the Ohmic and space charge limited currents gives:

$$V_{tr} = \frac{8 e n_0 s^2}{9 \varepsilon_0 \varepsilon_r} \tag{9.68}$$

From this the space charge in the material may be found. Note that the space charge is made up of localised and delocalised charges as trapped charges also contribute to the total space charge. This may give rise to a problem in evaluating the current density in the material.

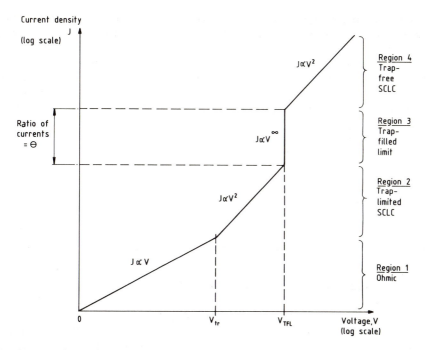

Fig. 9.10 Schematic graph showing current density versus voltage for an ideal case of space-charge limited current

(b) Dielectric with traps

We have already considered the effect of traps on Ohmic conduction (Section 9.1.1) and eqn. 9.10 gave the proportion of electrons $\theta = (n_c/n_t)$ available for conduction. If electrons are injected into the dielectric then, although they will all contribute to the space charge, only this proportion θ will contribute to the current. This fraction may be very small, ($\sim 10^{-6}$—10^{-10}) and considerably affects the Mott and Gurney square law and the transition voltage:

$$J = \theta \frac{9\varepsilon_0\varepsilon_r\mu V^2}{8s^3} \qquad (9.69)$$

$$V_{tr} = \frac{1}{\theta} \frac{8en_0s^2}{9\varepsilon_0\varepsilon_r} \qquad (9.70)$$

The above expressions are for the cases of shallow traps, i.e. where $\mathbb{E}_t - \mathbb{E}_f \gg k_B T$ (see Fig. 9.1) and it is possible to approximate eqn. 9.9*a* to 9.9*b*. However, at a sufficiently high voltage, the charge-carrier number density being injected is approximately equal to the number density of traps ($n \rightarrow N_t$) and the Fermi energy level will rise above the trap energy level. In this case the traps must be considered deep instead of shallow and eqn. 9.9*a* approximates to $n_t \simeq N_t$, i.e. all the traps are filled. As this so-called trapped-filled limit, V_{TFL}, is reached the value of θ quickly changes from its small value to unity and the current density rapidly rises (eqn. 9.69). For the hypothetical case considered of traps all at a single energy level this current increase with voltage would be infinitely fast and once the traps were filled and θ was unity the Mott and Gurney square law, $J \propto V^2$ (eqn. 9.66) would again resume. The voltage, V_{TFL}, can be calculated using Poisson's equation (eqn. 2.10) since at the trap-filled limit the space charge is virtually all due to traps and so $\rho_c = e \cdot N_t$. Thus:

$$\frac{d^2 V}{dx^2} = \frac{e \cdot N_t}{\varepsilon_0 \varepsilon_r} \bigg|_{V = V_{TFL}}$$

so that

$$V_{TFL} = \frac{eN_t s^2}{2\varepsilon_0 \varepsilon_r} \tag{9.71}$$

There are therefore four relations between voltage and current density shown as 1 to 4 on the schematic log (J) versus log (V) plot in Fig. 9.10:

- *Region 1:* $0 < V < V_{tr}$ (V_{tr} defined by eqn. 9.70), Ohmic conduction due to thermally-generated carriers, slope = 1.
- *Region 2:* $V_{tr} < V < V_{TFL}$ (V_{TFL} defined by eqn. 9.71), trap-limited space charge limited conduction, $J \propto V^2$ (eqn. 9.69), slope = 2.
- *Region 3:* $V = V_{TFL}$, all traps at energy level, $\mathbb{E}_t$, fill up, slope = ∞.
- *Region 4:* $V > V_{TFL}$, trap-free space charge limited conduction, $J \propto V^2$ (eqn. 9.66, Mott and Gurney square law), slope = 2.

In practice with a large distribution in energy of traps, the trapped-filled limit is not well defined. As the voltage is increased and more charge is injected the Fermi level moves up as more and more traps are filled. The distinction between regions 2 and 3 is therefore normally blurred. Since the current increases rapidly before region 4 is encountered, breakdown usually takes place before the last region is reached. An excellent demonstration of all four regions however has been shown for $1 \cdot 2 \; \mu$m thick films of Al/PVAc/Al by Chutia and Barua[577].

9.3.2 Hopping conduction

In Section 2.2.2 we briefly described the concept of a mobility gap instead of an energy band gap. In the energy-band picture the density of states, $N(\mathbb{E})$, is zero in the band gap, $\mathbb{E}_v < \mathbb{E} < \mathbb{E}_c$, and increases quickly at the band edges to give delocalised states in the conduction and valence bands. In the mobility-gap picture the band tails are not so abrupt and the density of states gradually increases, the band edge being roughly defined as the energy at which states are sufficiently close that the potential barriers inhibiting carrier movement between them are only of the order of $k_B T$ or less. In insulators the Fermi energy is less than the mobility edge, $\mathbb{E}_c$, and conduction is either by excitation to the mobility edge (cf. trap-limited mobility and space-charge limited mechanisms) or by thermally-activated transfer between localised sites often termed 'hopping'. Excitation to the mobility edge is unlikely in insulators with large mobility gaps in which states are likely to be filled to around the Fermi level but mostly empty a few $k_B T$ above it. Conduction is therefore more likely to be via a thermally-activated local transfer mechanism (provided $N(\mathbb{E}_f)$ is finite) in which an electron in an occupied state near the Fermi level acquires sufficient energy from the thermal vibrations of the lattice to move to a nearby unoccupied state of similar energy.

The process by which an electron may move from one site to another can either be by thermal excitation over a potential barrier, tunnelling through it, or a combination of thermal excitation and tunnelling. Activation *over* a barrier is not likely to be a competitive process here because the barrier will be $\sim \mathbb{E}_c - \mathbb{E}_f$. A pure tunnelling process will also be uncompetitive except at $T = 0$ K; firstly because the exponential reduction of the electron wavefunction with distance (see eqns. 9.46 and 9.48) limits the range over which it can be effective, and secondly because the principle of energy conservation restricts the participating sites to those having the same energy level. The dominant process at $T > 0$ K is thus one in which thermal promotion at one site raises the electron to a level which is equi-energetic with that of an empty neighbouring site (at $T > 0$ the energy levels will be smeared by $\sim k_B T$). Tunnelling between the two sites is allowed at this energy level and there will be a finite probability that the electron resides at the previously empty site. The overall probability of transfer in this mechanism is the joint probability for the two processes. This is given[60] by the product of the thermal probability that the electron is in an excited eigenstate that is equi-energetic with a neighbour separated by a distance R, i.e.:

$$P(\text{therm}) \propto \exp \left\{ -\frac{\Delta \mathbb{E}(R)}{k_B T} \right\} \qquad (9.72)$$

and the probability with which it resides on the neighbouring site in this eigenstate, i.e. the tunnelling probability:

$$P(\text{tun}) \propto \exp \left\{ -2\alpha R \right\} \qquad (9.73)$$

The promotion energy $\Delta \mathbb{E}(R)$ in eqn. 9.72 is the difference between the energy levels of the two sites involved (if Coulombic interactions are ignored), and can be determined as a function of R by means of the density of states, $N(\mathbb{E}_f)$, at the Fermi level[60]. Here the number of sites in the energy interval $\mathbb{E}_f$ to $\mathbb{E}_f + d\mathbb{E}$ that lie within a range R of an occupied site at $\mathbb{E}_f$ is $d\mathbb{E}(4\pi/3)R^3 N(\mathbb{E}_f)$. Thus, on average, the range R over which an electron has to tunnel to find *one* unoccupied site equi-energetic with its excited state at $\mathbb{E}_f + \Delta \mathbb{E}(R)$, is given by

$$\Delta \mathbb{E}(R)(4\pi/3)R^3 N(\mathbb{E}_f) = 1 \qquad (9.74)$$

which thus defines the R dependence of $\Delta \mathbb{E}(R)$ in eqn. 9.72.

From eqn. 9.74 we see that on average the promotion energy $\Delta \mathbb{E}(R)$ that is required for the electron to tunnel over a distance R reduces as R increases. The thermal probability factor for the composite process (eqn. 9.72) therefore increases as R increases in contrast to the tunnelling probability factor (eqn. 9.73). At a given temperature there will thus be an optimum range for which the product of the probabilities in eqn. 9.72 and 9.73 is a maximum, i.e. when the exponent:

$$2\alpha R + \left(\frac{4\pi}{3} R^3 N(\mathbb{E}_f) k_B T\right)^{-1} \qquad (9.75)$$

has a minimum. This occurs for

$$R = R_{opt} = \left(\frac{8\pi N(\mathbb{E}_f)\alpha k_B T}{9}\right)^{-1/4} \qquad (9.76)$$

and substituting this into eqn. 9.75 gives the *dominant* contribution to the probability of the combined transport mechanism, for which the conductivity is:

$$\sigma = A \exp\left\{-\frac{B}{T^n}\right\} \qquad (9.77)$$

where A is a constant of proportionality, $n = \frac{1}{4}$ and

$$B = 4\left(\frac{2\alpha^3}{9\pi k_B N(\mathbb{E}_f)}\right)^{1/4} \qquad (9.78)$$

Mott refers to this mechanism as 'variable-range hopping'. However the term 'hopping' is something of a misnomer since what actually takes place is a thermally-assisted tunnelling from site to site, and in no sense does the electron pass over the top of a potential barrier. Less simplified analyses give slightly different expressions for B and values of n between $\frac{1}{4}$ and $\frac{1}{2}$ which have been experimentally verified for various amorphous materials. The literature discussing the value of A has been reviewed by Emin[578] and Summerfield and Butcher[579]. Theories of hopping conduction in non-crystalline solids have been presented in depth by Böttger and Bryksin[516] and stochastic effects have been discussed by Brechmer *et al.*[574].

9.3.3 The Poole–Frenkel mechanism

The Poole–Frenkel mechanism[149] is the bulk limited analogue of the Schottky effect. In this case however barriers localising carriers *within* the dielectric (or semiconductor) are lowered by the field in contrast to the Schottky effect where the electrode–insulator barrier is lowered. For the Poole–Frenkel mechanism to occur the insulator must have a wide band gap and must contain donors or acceptors. Unlike the situation in *n*-type or *p*-type semiconductors used for making electronic devices however, these donors or acceptors require more energy than is generally available in order to be ionised. At room temperature therefore they do not generally donate electrons to the conduction band (creating free electrons) or accept electrons from the valence band (creating free holes). In terms of the energy band diagram the donor states are located at many $k_B T$ below the conduction band and the acceptor states many $k_B T$ above the valence band. In this analysis we will consider the case of the insulator containing only donors as this simplifies the treatment. The analysis is similar for acceptors or both donors and acceptors.

Since the band gap is wide we may assume that there are no thermally-generated free electrons and that the only electrons in the conduction band are those excited from donor states. The number density of conduction band electrons is therefore:

$$n_c = N_D - n_D \qquad (9.79)$$

where N_D is the number density of donors and n_D is the number density of 'occupied' donors, i.e. those which have not donated their electrons. Donors are essentially neutral when they are not ionised. However as an electron is removed, a Coulombic force exists between the electron and ionised donor which is given by:

$$F(r) = \frac{-e^2}{4\pi\varepsilon_0\varepsilon_r r^2} \qquad (9.80)$$

where r is the distance of separation. The potential energy, $V(r)$, associated with this Coulombic interaction (Fig. 9.11a) is given by:

$$V(r) = \frac{-e^2}{4\pi\varepsilon_0\varepsilon_r r} \qquad (9.81)$$

As $r \to 0$, $V(r) \to -\infty$ and the 'width' or 'dimension' of the ionised donor becomes important. With the application of an external field, E, and neglecting any local variations in the applied field, the potential energy becomes:

$$V(r) = \frac{-e^2}{4\pi\varepsilon_0\varepsilon_r r} - eEr \qquad (9.82)$$

Adopting the same procedure as in Section 9.2.1, the potential energy is differentiated and set to zero:

$$\frac{dV(r)}{dr} = 0 = \frac{e^2}{4\pi\varepsilon_0\varepsilon_r r^2} - eE \qquad (9.83)$$

so that the position of the maximum in the potential surface after application of an electric field (see Fig. 9.11b) is:

$$r_m = \left(\frac{e}{4\pi\varepsilon_0\varepsilon_r E}\right)^{1/2} \qquad (9.84)$$

Thus the maximum height of the barrier in the field direction is changed by:

$$\Delta V_m = -2\left(\frac{e^3 E}{4\pi\varepsilon_0\varepsilon_r}\right)^{1/2} \qquad (9.85)$$

If S is the capture cross-sectional area of donors for electrons then an expression may be derived for the concentration of donated electrons. Since the interchange of electrons between the conduction band and donor states is in a dynamic equilibrium n_c and n_D are constant. Thus the rate of thermal excitation of electrons from the donor to the conduction band is equal to the rate of capture.

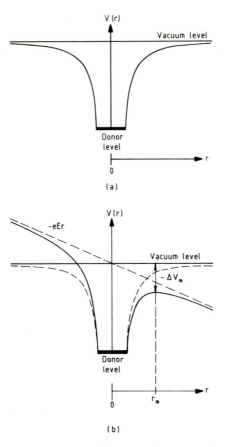

Fig. 9.11 Diagrams of electron potential energy due to coulombic and electric field near an ionised donor site. (*a*) no (or low) field. (*b*) high electric field

The rate of thermal excitation of electrons from the donor to the conduction band is given by:

$$n_D \nu_0 \exp\left\{-\frac{\phi_{\text{eff}}}{k_B T}\right\} \tag{9.86}$$

where ν_0 is the attempt to escape frequency (Section 9.1.2) and ϕ_{eff} is the reduced barrier height:

$$\phi_{\text{eff}} = \Delta\mathbb{E}_d + \Delta V_m \tag{9.87}$$

The rate of capture by donors is proportional to the number in the conduction band and the number of unoccupied donor sites and is therefore given by:

$$n_c(N_D - n_D)vS = n_c^2 vS \tag{9.88}$$

where v is the thermal velocity of electrons in the conduction band. Equating the rates 9·86 and 9·88 to satisfy the requirement of dynamic equilibrium, using eqn. 9.79, and noting that for most situations $v_0 = N_{eff} vS$, where N_{eff} is the effective density of states in the conduction band, gives:

$$n_c^2 vS = N_{eff} vS (N_D - n_c) \exp\left\{-\frac{\phi_{eff}}{k_B T}\right\}$$ (9.89)

For deep donors only a small proportion will be ionised so that $N_D - n_c \approx N_D$ and the number density of free carriers in the conduction band is thus given by:

$$n_c = N_{eff}^{1/2} N_D^{1/2} \exp\left\{-\frac{\phi_{eff}}{2k_B T}\right\}$$ (9.90)

Remembering that $\sigma = n_c e\mu$ and substituting for ϕ_{eff} using eqns. 9.85 and 9.87 gives:

$$\sigma = N_{eff}^{1/2} N_D^{1/2} e\mu \exp\left\{\frac{-\Delta\mathbb{E}_d}{2k_B T}\right\} \exp\left\{\frac{e^{3/2} E^{1/2}}{(4\pi\varepsilon_0\varepsilon_r)^{1/2} k_B T}\right\}$$ (9.91)

which is the Poole–Frenkel expression. A plot of log (J) or log (σ) versus $E^{1/2}$ yields a straight line giving an intercept of log $(N_{eff}^{1/2} N_D^{1/2} e\mu) - \Delta\mathbb{E}_d/(2k_B T)$ and a slope of $e^{3/2}/[(4\pi\varepsilon_0\varepsilon_r)^{1/2} k_B T]$.

The physical basis of the Poole–Frenkel model is very simplified and it is clear that the analysis must be treated as an order-of-magnitude calculation. It is generally difficult to distinguish the Schottky and Poole–Frenkel mechanisms by plotting the same data on the two different types of plot; generally data will fit both types of plot equally well. Various empirical and theoretical refinements have been suggested for the Poole–Frenkel model. There was a lot of discussion in the literature on this subject at the end of the 1960s and this has been reviewed in depth by O'Dwyer[82].

9.3.4 The field-limiting space-charge model

It has been understood for a long time that if an electrode injects charge into an insulator then a space charge cloud may form around the electrode and reduce the field locally. For example this occurs if a high negative voltage is put on a pin electrode in a vacuum. Electrons are injected into the vacuum and locally reduce the field preventing breakdown. There are also many reasons why the mobility of charge carriers should increase with field; some of these have been discussed in this chapter, for example the Poole–Frenkel mechanism.

Zeller[346] and Schneider[141] have proposed the concept of a field-limiting space charge (FLSC) which is a natural progression from these ideas. They propose that above a critical field, E_{mc}, the mobility of carriers in polymers is increased sharply from typical low values ($< \sim 10^{-10} \text{ m}^2 \text{ V}^{-1} \text{ s}^{-1}$) to those encountered in band transport ($> \sim 10^{-4} \text{ m}^2 \text{ V}^{-1} \text{ s}^{-1}$) and that this comes about as electrons acquire sufficient kinetic energy to remain above the mobility edge of the σ-electron conduction band. This change of mobility is so great that they propose that a reasonable approximation is:

$$\mu = 0 \qquad E < E_{mc}$$

$$\mu = \mu_\infty \qquad E > E_{mc}$$

(9.92)

where μ_∞ is the high-field limiting value of μ. In some ways this is analogous to the space charge limited conduction model at the trap-filled limit where the current suddenly rises with voltage. In practice neither of these effects are seen due to distributions of physical parameters but nonetheless one might expect to see a strong increase in mobility at E_{mc}. The situation is rather more complicated since the σ-conduction band would not exist in a well-defined way in amorphous regions of the polymer and may exhibit anisotropy in the crystalline regions. That is the longitudinal direction of the σ-bands may not be orientated with the field direction. Furthermore E_{mc} may well be in excess of the breakdown strength.

The FLSC model is usually discussed in terms of a point cathode in an insulator which is capable of injecting charge if the applied voltage is such that the field E is greater than E_{inj}. In the absence of a critical field for mobility, space charge is injected when $E > E_{inj}$ and diffuses throughout the insulation until at equilibrium the field at the cathode is reduced to E_{inj}. The inclusion of a mobility transition (eqn. 9.92) introduces a number of possibilities depending upon the relative magnitudes of E, E_{mc}, and E_{inj}. For example if $E_{inj} > E > E_{mc}$ any negative space charge, which may be intrinsic to the insulator or due to contact, will rapidly move away from the cathode leaving behind a space charge free region. The field in this region will be reduced to a low value ($< E_{mc}$) by the shielding effect of the displaced space charge. Beyond some point within the insulation the local field will initially lie below E_{mc} and any space charge in this region will not be able to move at first. The space charge will thus be formed into a thin radial shell which will continue to advance as a solitary wave as long as the field it generates on its boundary is in excess of E_{mc}. Alternatively, if traps are filled *en-route*, the wave will become dispersed as it propagates, i.e. it will become wider but of a smaller magnitude[580].

More typical of polymeric insulation however is the case in which $E > E_{mc} > E_{inj}$. Here the dynamic situation is very complicated. Initially injected space charge moves rapidly away from the cathode under the influence of the high ($E > E_{mc}$) local field, while simultaneously reducing the field in the region behind the space charge boundary, Fig. 9.12.

Zeller and Schneider's[141] picture of this situation seems to treat the space charge cloud as a resistor with a parallel shunt and residual insulation as a series resistance, Fig. 9.13. Initially the current is carried in the low-resistance branch of the circuit which supports only a small potential difference. As the space charge region extends the field (potential gradient) drops in this branch until it reaches E_{mc} when the current is switched to the high resistance branch. Now the potential drop across the space charge cloud increases, the field rises and the current is switched back to the low-resistance branch (i.e. $E \gtrsim E_{mc}$). In this way the field within the space charge region would be self-limited to $E = E_{mc}$.

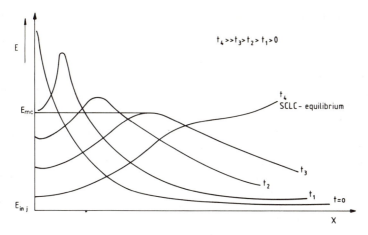

Fig. 9.12 Schematic diagram showing evolution of field as a function of distance from an injecting point

For this approach to be valid however there must be no current gradient or space charge between the electrodes, neither of which requirements are satisfied in the transient conditions of the situation we are considering. Instead we must expect a peak in the field at the space charge boundary with a field gradient on either side. In a region between the cathode and the space charge boundary the field will fall below E_{mc} and the charge here will become immobile. Radial extension of the charge boundary will continue to reduce the value of the boundary field simultaneously with a reduction of the field at the cathode tip as charge injection increases the quantity of low mobility space charge in its vicinity.

When the boundary field drops below E_{mc} the model goes over to the space charge limited case with the space charge spreading throughout the insulation due to a non-zero low field drift mobility and diffusion, as has recently been graphically demonstrated in a computer calculation by Hare[580]. The substitution of condition 9·92 by a negative differential resistance (NDR), which gives a hysteresis in the $I - V$ characteristic, does not

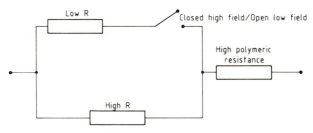

Fig. 9.13 Zeller and Schneider[141] treat a space charge cloud as the circuit shown whereby the change in resistance upon closing the switch is due to the increase in mobility when $E \geqq E_{mc}$

significantly affect this picture, except insofar as the minimum field at which the high mobility state can be *maintained* (E_h) may be small enough that the space charge boundary can advance across the insulation prior to space charge control of the current. Subsequent events depend upon whether the counter electrode is ohmic or blocking for the carriers. If the ohmic case is considered a high current will continue to flow, whereas when the electrode is blocking a high counter-field will build up which eventually reduces the injection current from the cathode to zero.

The concept of a FLSC is simple and makes possible computations of space charge without knowing any parameters such as mobilities and trap concentrations, however at the expense of a number of field parameters, i.e. E_{mc}, E_{inj}, and E_h. This is somewhat advantageous as generally the calculation of the space-charge distribution in a real situation is difficult if not impossible. The concept must not be taken too literally though. The mobility is unlikely to change very suddenly at a critical voltage, and E_{mc} is likely to be a function of position in inhomogeneous dielectrics such as polymers.

Zeller and Schneider[141] have also suggested that while the field at the space charge boundary is greater than E_{mc} (or E_h), a radially symmetric boundary is unstable with respect to filament formation. In contrast to a radial advance the field at the tip of a space charge filament can be expected to increase with its length, thereby allowing the high mobility condition to be maintained[338] in the absence of branching. As in the radial case we can expect a field gradient to be developed along the filament although now it is unlikely to be sufficient to reduce the field to below E_h. Thus when the material exhibits a NDR (see Fig. 9.14) a *single* filament will be reinforcing and carry a high mobility current. In general however single filaments will themselves be unstable with respect to branching, and branches will moderate the field around the tips thereby slowing down their advance and even perhaps reducing the current in some filaments to the low mobility state. It has been suggested[141,346] that this is the means whereby electrical trees are formed, however the mechanism seems more applicable to impulse breakdown.

It should be noted that the theory has as yet only been developed for DC fields and it is an open question as to the modifications required for the AC case. If we consider the radial space charge cloud case (without NDR), then it seems likely that the space charge boundary will be reduced to a diffusive (or drift) advance before the applied voltage starts to reduce (except at very high frequencies). We can thus expect a fuzzy-edged radially-symmetric space-charge cloud whose spatial extent will be determined by the diffusion coefficient or low-field mobility. This will appear and disappear on each cycle. On the other hand in the NDR case, which we assume will always lead to filamentary development, the filamentary system will be sharply defined, with a length determined by the high field mobility and the frequency. In this case the growth that is generated around the 'peak' of one half-cycle will first be frozen and then removed around the 'trough' of the next half-cycle. Since the filaments carry large current densities (i.e. high-mobility states) such repetitive displacements of charge may well indeed

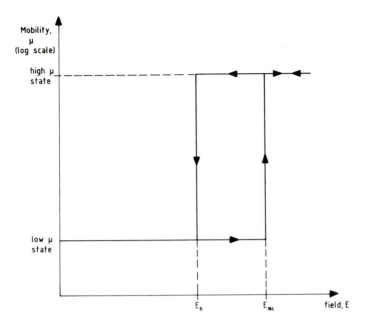

Fig. 9.14 According to the negative differential resistance (NDR) model, when the field is lowered from E_{mc}, the critical field required to produce a high mobility state, the mobility remains in the high mobility state provided the field remains above some holding value, E_h

contribute to the initiation of electrical trees as proposed by Zeller and Schneider[141].

It is generally accepted that it is difficult to incontrovertibly demonstrate that a particular high-field conduction mechanism is operative. A typical recent example of this is provided by Das–Gupta and Doughty[46] in their work on conduction in poly(ether ether ketone) (PEEK). Amongst various other measurements they plotted for different temperatures bilogarithmic current–field plots which showed the classic change from Ohmic conduction to space-charge limited; log(current) versus (field)$^{1/2}$ plots which gave good straight lines indicative of Schottky injection; Poole–Frenkel plots of log(conductivity) versus (field)$^{1/2}$ also gave reasonable straight lines as did log (current) versus linear field plots which tend to support an ionic conduction hypothesis.

Chapter 10

Thermal breakdown

In thermal breakdown electrical power dissipation causes heating of at least part of the insulation to above a critical temperature which results directly or indirectly in catastrophic failure.

When a voltage is applied to an insulator some current will flow, electrical power will be dissipated, and the temperature will increase. If breakdown does not occur then the temperature will continue to increase until the cooling of the insulator is equal to the electrical power dissipation and a steady-state heat flow is set up. If breakdown does occur this may be because the temperature of at least part of the insulation is such that either (i) the insulation is physically changed so that its breakdown strength is lowered to below the voltage applied (for example it may melt); or (ii) the conductivity of the insulation and hence its electrical power dissipation is increased causing a further increase in temperature and 'thermal runaway'.

Klein[161] has called the first case 'destructive breakdown' and the second 'thermal instability'. This is illustrated in Fig. 10.1a as a graph of power versus temperature for three different voltages $V_1 < V_2 < V_3$. The straight line on the graph shows the rate of heat loss from the insulating system, (i.e. cooling) as a function of temperature assuming Newton's law of cooling. If a voltage, V_1, is applied to the system then at the ambient temperature, T_0, there is no cooling and the electrical power dissipation causes the system's temperature to increase. At temperature, T_1, a steady state is reached where the electrical power dissipation is equal to the rate of cooling. This steady-state condition is stable since if the temperature were just above T_1 then the cooling rate would exceed the electrical power dissipation and the system would cool down to T_1 to achieve a steady state. If however the critical (breakdown) temperature were less than T_1 then the sample would breakdown after the application of voltage, V_1, before reaching this steady state; this corresponds to the first scenario described above (destructive breakdown). If a breakdown did not occur and a voltage, V_2, were applied then the temperature would stabilise at a higher temperature, T_2.

This temperature is metastable since if it were increased slightly, the power dissipation would increase faster than the rate of cooling and thermal runaway would ensue. The voltage, V_2, corresponds to the maximum value attainable with respect to thermal breakdown because, even if T_2 is well below the temperature at which irreversible damage occurs, then, provided the voltage is maintained, the temperature will continue to increase until breakdown does occur. This corresponds to the second scenario above (thermal instability). Temperature T_2 is often referred to as the highest stationary temperature. The voltage V_3 is always unstable and the temperature increases until breakdown occurs.

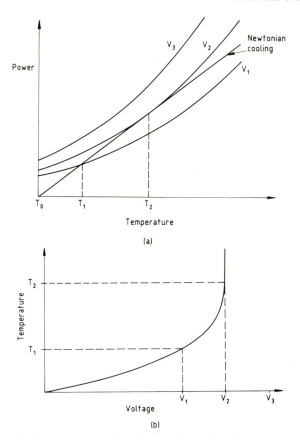

Fig. 10.1
(a)   A schematic graph of rate of heat input versus temperature for three different voltages: $V_1 < V_2 < V_3$; the straight line represents rate of heat loss due to thermal conduction (assuming Newton's law of cooling).
(b)   Variation of steady-state temperature of the hottest part of the insulation with voltage

Fig. 10.1b shows how the steady-state temperature of the hottest part of the insulation varies with voltage; V_1, V_2, V_3, T_1, and T_2 correspond to Fig. 10.1a. At voltage V_1 the temperature has increased to T_1; if T_1 is greater than the critical temperature then destructive breakdown occurs. If breakdown does not occur and the voltage is increased to V_2 then the temperature increases to T_2. If the voltage increases to $V_2 + \delta V$ then the temperature increases until breakdown does occur (thermal instability) and it is obvious why V_2 was termed the maximum thermal voltage by White-head[505]. (One could argue it is the minimum voltage for thermal runaway to occur.) There is no steady-state temperature corresponding to V_3.

Thermal breakdown is perhaps the most obvious form of breakdown mechanism but the concept of an instability occurring at a critical voltage for which the electrical power transferred to the system cannot be matched by a power loss is common to all the breakdown mechanisms in one form or another. The thermal breakdown has been discussed since the turn of the century. Whitehead[505] (1951) notes that 'there is little recondite about thermal breakdown ... Skinner mentioned it in 1904' and he continues, 'Everest in a discussion at the IEE in 1913, described a test based upon thermal instability, which he called the highest maintained electric strength test, ... which found its way, after 1918, into specifications'. The theory of thermal breakdown in solid dielectrics as a process of thermal instability was first suggested by Wagner[581] (1922) and later modified by von Kármán[582] (1924), Rogowski[583] (1924) and Dreyfus[584] (1924). Fock[585] (1927) gave a complete solution for whole sample breakdown. Moon [586] (1931) extended Fock's analysis to cover all thicknesses and a wider range of temperatures, a systematic numerical technique for calculating breakdown voltages (including extremes of thickness) and gave expressions for temperature and current. In 1939 Copple et al.[587] carried out numerical integration, using a mechanical integrating machine, and showed that, whilst any field in excess of the critical field would lead to breakdown, the time taken for breakdown to be initiated (the 'formative time lag') decreases from infinity as the breakdown field is increased above its critical value, Fig. 10.2. (Hikita et al.[938] have also presented a general numerical solution for the steady-state case.) Whitehead[505] (1951) showed for a limiting case of thick insulation that there was a maximum thermal voltage which was independent of thickness. O'Dwyer[82] (1973) and Klein[161] (1969) extended the analysis to include the effects of the thermal resistance of the electrodes although Fock[585] (1927) had already indicated how this could be done. The analyses up to this point considered samples of infinite area with uniform heating of any layer in the plane of sample. Klein[161] (1969) however was particular in noting that this rarely occurred except in specimens in which weak spots had been electrically isolated using self-healing electrodes. Filamentary thermal breakdown, in which breakdown takes place at localised spots, has been the subject of relatively recent experiments and modelling by Japanese workers including Hikita and Mizutani[428,588–594] (1987).

In the discussion here we will derive the general equations governing thermal breakdown. These can only be solved numerically in general however various limitations can be imposed which, if realistic, is useful as dependencies upon experimental conditions can be predicted. We will first of all consider the dielectric of infinite area and the conditions for breakdown on a broad front across the whole area. Further limitations include: (i) considering either steady-state or impulse voltages since thermal capacity can be ignored in the former case and thermal conduction ignored in the latter; (ii) considering thick specimens with constant temperature electrodes in which case the temperature varies through the thickness or thin specimens with finite thermal-resistance electrodes in which the temperature is constant throughout the sample; and (iii) considering either a low field case in which the electrical conductivity is only a function of temperature or a high field case in which it is also a function of field. Finally we will review work on filamentary thermal breakdown in which breakdown occurs at a spot rather than across a broad front.

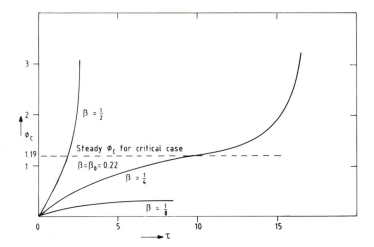

Fig. 10.2 Time variation of central temperature of a dielectric slab of infinite area assuming rate of heat production per unit volume at temperature T is $\alpha E^2 \exp(\gamma T)$. The plot uses non-dimensional reduced variables:

$$\phi_c = \gamma(T_c - T_0) \propto \text{rise in central temperature}$$

where T_c is the central temperature and T_0 is the ambient temperature;

$$\beta = \gamma \alpha \exp\{\gamma T_0\} E^2 d^2/(4\kappa) \propto \text{field squared}$$

and

$$\tau = \frac{4\kappa}{C_p D\, d^2}\, t \propto \text{time}$$

It is found that there is a critical field corresponding to $\beta = 0 \cdot 220$ and a temperature rise $\phi_c = 1 \cdot 19$. Above this field no steady-state condition is possible and breakdown always eventually results. After Copple *et al.*[587]

The electrical power dissipation in an elemental cube of sides δx, δy, and δz is given by the product of current density, electric field, and volume:

$$\text{electrical power dissipation} = \mathbf{J} \cdot \mathbf{E}\, \delta x\, \delta y\, \delta z$$

and since $\mathbf{J} = \sigma \mathbf{E}$ this can be written in terms of field as:

$$\text{electrical power dissipation} = \sigma \mathbf{E}^2\, \delta x\, \delta y\, \delta z \qquad (10.1)$$

For simplicity we will only consider a DC field and simply note here that for an AC field of angular frequency, $\omega = 2\pi f$, where the thermal response time of the insulation is much less than ω, the conductivity term would be modified to $(\sigma + \omega \cdot \varepsilon_0 \varepsilon_r'')$ where ε_r'' is the imaginary part of the complex permittivity and the field is the RMS value. The latter may be important in practice. For example Sletbak *et al.*[595] consider that in the bisphenol-A cast-epoxy resin system at high temperatures or/and at high water contents there may be thermal breakdown at even modest electric fields due to dielectric losses.

The heat flow into the cube in the x-direction through the face of area $\delta y \cdot \delta z$ is proportional to the rate of change of temperature, T, in that direction:

$$-\kappa \frac{\partial T}{\partial x} \delta y \, \delta z$$

where κ is the thermal conductivity. If $\partial T/\partial x$ is not constant, then on the opposite face of the cube, it will have increased by $(\partial^2 T/\partial x^2) \cdot \delta x$ so that the rate of heat flow out of that face of the cube in the x-direction is:

$$-\kappa \left(\frac{\partial T}{\partial x} + \frac{\partial^2 T}{\partial x^2} \delta x \right) \delta y \, \delta z$$

The net rate of heat flow into the cube in the x-direction is then given by that into the first face less that out of the opposite face:

$$\kappa \frac{\partial^2 T}{\partial x^2} \delta x \, \delta y \, \delta z$$

Summing up the contributions from the other faces gives:

$$\text{conductive heat flow in} = \kappa \left(\frac{\partial^2 T}{\partial x^2} + \frac{\partial^2 T}{\partial y^2} + \frac{\partial^2 T}{\partial z^2} \right) \delta x \, \delta y \, \delta z$$

$$= \kappa \nabla^2 T \, \delta x \, \delta y \, \delta z \tag{10.2}$$

The heat flow into the sample causes the temperature to increase:

$$\text{heat flow in} = C_p D \frac{dT}{dt} \delta x \, \delta y \, \delta z \tag{10.3}$$

where C_p is the specific heat (assuming the system has time to thermally expand under constant pressure) and D the density. The equation assumes no phase changes. The total heat into the elemental volume of the sample is given by the sum of eqns. 10.1 and 10.2 and equating this with eqn. 10.3 gives:

$$\frac{dT}{dt} = \frac{1}{C_p D} (\sigma E^2 + \kappa \nabla^2 T) \tag{10.4}$$

This is the general power balance equation. Provided there is a solution such that $dT/dt = 0$ at $T < T_c$, the critical temperature, then thermal breakdown will not take place. The breakdown field may be determined by setting either $T = T_c$ (destructive breakdown) or either $dT/dt = \infty$ or $T = \infty$ (thermal instability). Note however that the electrical and thermal conductivities, the specific heat and the density are all functions of temperature. Most analyses assume however κ, C_p, and D to be independent of temperature. The reasons for this are that: the temperature dependence of σ normally dominates and it may be known; the T_c at which thermal runaway occurs is usually found to be only a few tens of degrees above ambient, thus κ, C_p, and D do not normally change much whereas $\sigma(T)$ does; and it simplifies the analysis. At higher fields especially, σ is also a function of field, and it may also be time and position dependent (for example from transient dielectric responses and space charge movement). The dependence on field is often taken into account but the dependencies on time and position are usually ignored. The equation is usually solved in conjunction with an electrical boundary condition such as homogeneous current density:

$$\nabla J = \nabla(\sigma E) = 0 \tag{10.5}$$

We will first consider the case of steady-state thermal breakdown in which the voltage is changed slowly such that the insulation is always in thermal equilibrium, i.e. the electrical power dissipation is equal to the conductive cooling and eqn. 10.3 can be ignored.

Next we will consider the other limiting case of impulse thermal breakdown in which it is assumed that breakdown occurs before any heat has been conducted away, i.e. eqn. 10.2 can be ignored. The general case of breakdown at arbitrary times can only be solved numerically (e.g. References 938, 587, Fig. 10.2). For both steady-state and impulse breakdown we will consider an insulator in the form of a slab of infinite area sandwiched between parallel-plate electrodes. Both the electrical and heat flux is in the direction orthogonal to the electrodes, i.e. straight through the dielectric, and conditions at a given depth into the slab are the same at any position on the area chosen. This implies that breakdown will take place on a broad front across the dielectric. Later we will consider breakdown in a filament through the dielectric.

10.1 Steady-state thermal breakdown

Since thermal equilibrium is assumed, $dT/dt = 0$, and eqn. 10.4 simplifies to:

$$\sigma(T, E)E^2 + \kappa\nabla^2 T = 0 \tag{10.6}$$

It will be convenient to consider the low-field case of field-independent electrical conductivity first since this greatly simplifies the analysis; field-dependent conductivity will be left until Section 10.1.4. The temperature, by symmetry, is a maximum in the centre layer of the insulation. A limiting case is when the electrodes are assumed to be at a known temperature, T_0, which may be ambient or the operating temperature of the system. This is the thick slab approximation and gives the interesting result that the thermal breakdown voltage is independent of thickness. This was noted by Whitehead[505] who termed it the 'maximum thermal voltage'.

10.1.1 The maximum thermal voltage (the thick-slab, low-field approximation)

(a) The general form of the maximum thermal voltage (uniform field conditions)
In order to calculate the maximum thermal voltage it is necessary to find an expression relating the applied voltage, V_{app}, to the temperature at the

hottest part of the insulation, that is, in the central layer. We can then substitute this for the critical temperature, T_c, to find the breakdown voltage. According to the arguments above we can substitute $T_c = \infty$ for the thermal runaway case.

Since all flux is straight through the insulation, let us define this as the x-direction and use a 1-dimensional analysis for simplicity. Furthermore we will assume that the electrical conductivity, σ, is only a function of temperature (i.e. it is field independent) at this stage and that the thermal conductivity is constant, i.e. $\kappa(T) = \kappa_0$. Noting that the component of field in the x-direction is $-\partial V/\partial x$, eqn. 10.6 can be rearranged to:

$$\sigma(T)\left(-\frac{\partial V}{\partial x}\right)^2 + \kappa_0 \frac{\partial^2 T}{\partial x^2} = 0 \tag{10.7}$$

Assuming uniform field conditions (i.e. $d^2 V/dx^2 = 0$) re-arranging and integrating both sides gives:

$$-\sigma(T)\frac{\partial V}{\partial x}\int \frac{\partial V}{\partial x}\, dx = \int \kappa_0 \frac{\partial^2 T}{\partial x^2}\, dx$$

so that:

$$-\sigma(T)\frac{\partial V}{\partial x} V = \kappa_0 \frac{\partial T}{\partial x} + \text{constant} \tag{10.8}$$

where the constant is the constant of integration. Notice that the derivation of this equation required the assumption of uniform-field conditions; this equation or subsequent equations based on it are frequently incorrectly used in the absence of such conditions. In order to simplify the analysis by making this constant equal to zero, the origin, at $x = 0$, can be defined to be in the centre of the slab and the voltage $V(x)$ can be defined as zero at this position. Thus the voltages on the electrodes can be defined as $\pm\frac{1}{2}V_{\text{app}}$. Since the temperature is a maximum, T_{max}, at $x = 0$, $\partial T/\partial x = 0$ at this position and the constant of integration is zero. Separating the variables (this is possible since σ is not field dependent), rearranging and integrating from the centre of the insulation to an arbitrary point within the insulation gives:

$$\int_0^V V\, dV = -\int_{T_{\text{max}}}^T \frac{\kappa_0}{\sigma(T)}\, dT \tag{10.9}$$

If the integration is carried out from the centre to the electrode then $V = V_{\text{app}}/2$ and $T = T_0$ so that:

$$V_{\text{app}}^2 = 8 \int_{T_0}^{T_{\text{max}}} \frac{\kappa_0}{\sigma(T)}\, dT \tag{10.10}$$

The applied voltage reaches the thermal breakdown voltage, V_{th}, when T_{max} reaches T_c, the critical temperature. For the unstable case of thermal runaway $T_c \to \infty$ as $V_{\text{app}} \to V_{th}$ so that the *maximum thermal voltage*[505] is given by:

$$V_{th} = \left(8 \int_{T_0}^{T_c = \infty} \frac{\kappa_0}{\sigma(T)}\, dT\right)^{1/2} \tag{10.11}$$

Values of maximum thermal voltage calculated for various polymers by Whitehead[505] are given in Table 10.1. In order to calculate values of V_{th} an expression for $\sigma(T)$ is required. Since we are assuming that σ is not a function of E, then the obvious conduction mechanisms are the low-fields ones: ohmic and ionic. However for ohmic conduction the conductivity *decreases* with temperature so that thermal runaway cannot occur. (Putting $\sigma(T) \propto 1/T$ in eqn. 10.11 would result in $V_{th} = \infty$.) The usual form of $\sigma(T)$ considered is the Arrhenius form for ionic conductivity, eqn. 9.11. We will consider this and another phenomenological equation used by Klein[161].

(b) The maximum thermal voltage for ionically-conducting insulators
Substituting eqn. 9.11 into 10.11 gives:

$$V_{th}^2 = \frac{8\kappa_0}{\sigma_0} \int_{T_0}^{\infty} \exp\left\{\frac{\phi}{k_B T}\right\} dT \tag{10.12}$$

which is approximately:

$$V_{th}^2 = \frac{8\kappa_0}{\sigma_0} \left[-\frac{k_B T_0^2}{\phi} \exp\left\{\frac{\phi}{k_B T}\right\} \right]_{T_0}^{\infty}$$

for $\phi \gg 2k_B T$. Although this is obviously not true as $T \to \infty$, in practice the temperature at which runaway starts (T_2 on Fig. 10.1) is usually only a few tens of degrees above T_0 and the approximation is good. Thus for ionically-conducting insulators the maximum thermal voltage has the form:

$$V_{th} = T_0 \left(\frac{8\kappa_0 k_B}{\sigma_0 \phi}\right)^{1/2} \exp\left\{\frac{\phi}{2k_B T_0}\right\} \tag{10.13}$$

This gives the thermal breakdown voltage for thick slabs as being thickness-independent with an Arrhenius-type dependence on ambient temperature (i.e. $\propto \exp\{\text{constant}/T_0\}$).

Table 10.1 The maximum thermal voltage calculated by Whitehead[505] for various polymers

Polymer	Maximum thermal voltage (MV RMS at 50 Hz)
Phenol formaldehyde	0·2—1·4
Aniline formaldehyde	0·15
PVC (plasticized)	0·1—0·2
PAV	0·9
PE (commercial)	3—5
	(0·05 at 10^6 Hz)
Polystyrene	5
	(0·05 at 10^6 Hz)
Acrylic resins	0·3—1

(c) The maximum thermal voltage for Klein's conductivity dependence

Klein[161] has suggested that a mathematically-simple purely-empirical functional form for $\sigma(T, E)$ which fits data well for many insulators is given by:

$$\sigma(T, E) = \sigma_K \exp\{a(T - T_0) + bE\} \tag{10.14}$$

with $\sigma_K = \sigma(T = T_0, E = 0)$ (the subscript K indicates Klein). At low fields such that $bE \ll a(T - T_0)$ this approximates to:

$$\sigma(T) = \sigma_K \exp\{a(T - T_0)\} \tag{10.15}$$

with $\sigma_K \simeq \sigma_0 \exp(aT_0)$. Substitution into eqn. 10.11 then gives:

$$V_{th} = \left(\frac{8\kappa_0}{a\sigma_0}\right)^{1/2} \exp\left\{-\frac{aT_0}{2}\right\} \tag{10.16}$$

Again this is thickness independent but with $V_{th} \propto \exp\{-\text{constant} \times T_0\}$.

10.1.2 The thin-slab low-field approximation

The 'thick slab' approximation given in Section 10.1.1 assumed that the temperature distribution was contained within the insulator so that the electrodes remain at T_0. In this section we assume the converse; the dielectric is so thin that the temperature is uniform throughout it and the temperature increase is due solely to the thermal resistance of the electrodes. The electrode–insulation interface therefore rises to a higher temperature, T_1, than the electrode–environment interface which is assumed to remain at ambient, T_0.

(a) General solution for thin slabs

In order to proceed we must define the relationship between the rate of heat flow through the electrodes and $(T_1 - T_0)$. Assuming a temperature-independent thermal resistance between the electrode–insulation interface and the environment we can then define an electrode thermal conductance per unit electrode area, γ. In the infinite-area parallel-electrode insulator system discussed above half the heat generated flows out of each electrode so that:

$$\tfrac{1}{2}JV_{app} = \gamma(T_1 - T_0) \tag{10.17}$$

Replacing J by $\sigma V_{app}/d$ gives:

$$\frac{\sigma(T)V_{app}^2}{2d} = \gamma(T - T_0) \tag{10.18}$$

where T is the uniform temperature of the thin film of insulation. Given a functional dependence of $\sigma(T)$ the thermal breakdown voltage may be found by using the criteria from Fig. 10.1a that both the electrical power dissipation and cooling curves are equal as are their derivatives with respect to temperature at the point of thermal runaway.

(b) Thermal breakdown in thin slabs of ionically-conducting insulators

Again using eqn. 9.11 for $\sigma(T)$ and assuming no dependence of conductivity on field, the power balance eqn. (10.18) becomes:

$$\frac{\sigma_0 V_{app}^2}{2d} \exp\left\{-\frac{\phi}{k_B T}\right\} = \gamma(T - T_0) \tag{10.19}$$

At breakdown $T = T_c$ and the temperature derivatives of each side of eqn. 10.19 are equal:

$$\frac{\sigma_0 V_{app}^2 \phi}{2 d k_B T_c^2} \exp\left\{-\frac{\phi}{k_B T_c}\right\} = \gamma \qquad (10.20)$$

Substituting $T = T_c$ in eqn. 10.19 and dividing by eqn. 10.20 gives

$$\frac{k_B T_c^2}{\phi} = T_c - T_0$$

which has the approximate (second-order) solution:

$$T_c \simeq T_0\left(1 + \frac{k_B T_0}{\phi}\right) \qquad (10.21)$$

Substituting $T \simeq T_c$ in eqn. 10.19 (noting $(1 + k_B T_0/\phi)^{-1} \simeq 1 - k_B T_0/\phi$) gives:

$$V_{th} = T_0\left(\frac{2\gamma k_B d}{\sigma_0 \phi \exp(1)}\right)^{1/2} \exp\left\{\frac{\phi}{2 k_B T_0}\right\} \qquad (10.22)$$

Note that the breakdown voltage for thin slabs is proportional to the square root of thickness and that the ambient temperature dependence is the same as for thick slabs.

(c) Thermal breakdown in thin insulating slabs using Klein's $\sigma(T)$ dependence
In this case Klein's conductivity dependence[161], eqn. 10.15, must be substituted into the power balance equation (eqn. 10.18):

$$\frac{\sigma_K V_{app}^2}{2d} \exp\{a(T - T_0)\} = \gamma(T - T_0) \qquad (10.23)$$

Following the same procedure as before (equating temperature derivatives and dividing) it is easily shown that:

$$T_c = T_0 + 1/a \qquad (10.24)$$

Substituting $T = T_c$ into eqn. 10.23 gives the breakdown strength:

$$V_{th} = \left(\frac{2 d \gamma}{a \sigma_K \exp(1)}\right)^{1/2} \qquad (10.25)$$

Again noting $\sigma_K \simeq \sigma_0 \exp(a T_0)$ for the low-field approximation of Klein's conductivity–temperature dependence, this gives:

$$V_{th} = \left(\frac{2 d \gamma}{a \sigma_0 \exp(1)}\right)^{1/2} \exp\left\{-\frac{a T_0}{2}\right\} \qquad (10.26)$$

Again this has a square-root dependence on thickness and, like the previous case of Klein's conductivity, a dependence on ambient temperature of the form $\exp\{-\text{constant} \times T_0\}$.

10.1.3 The low-field thickness-dependent solution

The more general case discussed in this section can be obtained when the electrodes have a finite thermal resistance and there is a distribution of temperature throughout the slab; first analysed by Moon[586]. To be completely general the solution for steady-state breakdown must be the solution of eqn. 10.6; the case cited in this section for example would not apply to a coaxial power cable where all the heat is dissipated through the outer electrode and the inner electrode is therefore at the maximum temperature.

(a) General solution
In this case eqn. 10.17 still governs the electrode temperature, T_1, but the temperature within the dielectric is higher. Since $\boldsymbol{J} = \sigma \cdot \boldsymbol{E} = \sigma(-\partial V/\partial x)$ eqn. 10.7 can be re-written:

$$\kappa \frac{\partial^2 T}{\partial x^2} - J \frac{\partial V}{\partial x} = 0$$

and substituting for J from eqn. 10.17 gives:

$$\kappa \frac{\partial^2 T}{\partial x^2} - \frac{2\gamma(T_1 - T_0)}{V_{app}} \frac{\partial V}{\partial x} = 0$$

Integrating both sides with respect to x gives:

$$\kappa \frac{\partial T}{\partial x} - \frac{2\gamma(T_1 - T_0)}{V_{app}} V = 0 \qquad (10.27)$$

(Again defining $V = 0$ at $x = 0$ in the centre of the slab where $\partial T/\partial x = 0$ makes the constant of integration zero.) Substituting for V from eqn. 10.9 and integrating from centre to electrode gives:

$$\frac{\gamma(T_1 - T_0)d}{V_{app}} = \int_{T_1}^{T_{max}} \kappa \left(2 \int_{T}^{T_{max}} (\kappa/\sigma)\, dT \right)^{-1/2} dT \qquad (10.28)$$

Because the electrode temperature is T_1 instead of T_0 eqn. 10.10 becomes modified to:

$$V_{app}^2 = 8 \int_{T_1}^{T_{max}} (\kappa/\sigma)\, dT \qquad (10.29)$$

These equations (i.e. 10.28 and 10.29) are the general expressions for this case and may be solved to eliminate T_1 given suitable forms for $\kappa(T)$ and $\sigma(T, E)$.

(b) Solution for ionically-conducting insulators
The mathematical ingenuity of Moon[586] provided a numerical solution of eqns. 10.28 and 10.29 for the Arrhenius dependence of $\sigma(T)$ appropriate to ionically-conducting insulators given by eqn. 9.11. He used the dimensionless variables:

$$v = V_{th} \left(\frac{k_B \sigma_0}{8\phi\kappa_0} \right)^{1/2} \propto \text{ breakdown voltage, } V_{th};$$

$$c = \frac{\gamma d}{2\kappa_0} \propto \text{ thickness, } d, \quad \text{and}$$

$$w = \frac{\phi}{k_B T} \propto \text{ 1/temperature, } T.$$

His results of v versus c (effectively breakdown voltage versus thickness) for different values of w_0 ($\propto 1/T_0$, i.e. 1/ambient temperature) are given in Fig. 10.3. Paticularly noticeable is the thickness dependence. For thin slabs ($d < \sim 0.2\kappa_0/\gamma$) the breakdown voltage is proportional to the square root of thickness in agreement with eqns. 10.22 and 10.26. For thick slabs ($d > \sim 200\kappa_0/\gamma$) the breakdown voltage is independent of thickness in agreement with eqns. 10.13 and 10.16. The temperature dependence is in agreement with the ionically-conducting insulators (eqns. 10.13 and 10.22) since this was a premise of Moon's analysis. Moon did, in fact, attempt to analyse a $\sigma(T) = \sigma_0 \exp\{\alpha T\}$ dependence like that proposed by Klein but the analysis was mathematically intractable.

10.1.4 High-field (field-dependent) conductivity

General expressions for thermal breakdown involving field-dependent conductivity are not generally feasible because they become extremely complicated very quickly and there are a large number of possible expressions for $\sigma(E)$. This means that they are not easily tested experimentally. Individual insulating systems are best modelled numerically in this case. In this section we briefly investigate the dependency of common field-dependent (high-field) conductivities in thin films since, being able to assume a uniform temperature throughout the film greatly simplifies the analysis. In each case the analysis follows the principles described in Section 10.1.2.

(a) The high-field Klein model
Using Klein's conductivity dependence[161] at high fields, eqn. 10.14 and

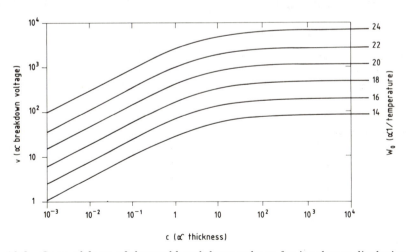

Fig. 10.3 General form of thermal breakdown voltage for insulators displaying an Arrhenius-type conductivity dependence on temperature (e.g. ionically conducting) in dimensionless units. After Moon[586]

simplifying it to:

$$\sigma(T, E) = \sigma_0 \exp\{aT + bE\} \qquad (10.30)$$

yields the following equation for breakdown strength after substitution into 10.18, equating derivatives etc. as before:

$$V_{th} = \frac{d}{b}\left(\log_e\left\{\frac{2\gamma d}{a\sigma_0 V_{th}^2}\right\} - aT_0 - 1\right) \qquad (10.31)$$

with the critical temperature for thermal runaway again being given by eqn. 10.24. Assuming that the logarithmic term is relatively slowly varying, the breakdown voltage is now approximately proportional to thickness with a linear negative temperature dependence.

(b) Schottky and Fowler-Nordheim emission
The equation for Schottky emission (9.39) can be written in a simpler form as a function of temperature and field namely:

$$J = a_S T^2 \exp\left\{-\frac{\phi - b_S E^{1/2}}{k_B T}\right\} \qquad (10.32a)$$

In order to determine the critical temperature for thermal runaway this can be further simplified to:

$$J = a_S T^2 \exp\left\{-\frac{H_S}{k_B T}\right\} \qquad (10.32b)$$

Substituting this into eqn. 10.17 to form the power balance equation and following the same procedure as before of equating the derivatives gives the following quadratic for the critical temperature:

$$T_c = 2T_0 - \frac{H_S}{k_B} + \frac{H_S T_0}{k_B T_c}$$

Solving this and assuming $H_S/(k_B T_0) > 1$ which is likely gives the approximate critical temperature:

$$T_c \simeq T_0\left(1 + \frac{k_B T_0}{H_S}\right) \qquad (10.33a)$$

This critical temperature is only a few degrees above ambient except for exceptionally high fields (i.e. $E^2 \sim \phi$) just as in the cases for ionically-conducting insulators and those with the Klein[161] conductivity dependence (eqn. 10.21 and 10.24). Substituting this into eqns. 10.17 and 10.32b gives:

$$\exp\left\{-\frac{H_S}{k_B T_0}\right\} = \frac{2\gamma}{a_S}\frac{k_B}{H_S \exp(1)}$$

Substituting for H_S and taking logarithms gives:

$$-\frac{\phi}{k_B T_0} + \frac{b_S E^{1/2}}{k_B T_0} = \log_e\left\{\frac{2\gamma k_B}{a_S(\phi - b_S E^{1/2})\exp(1)}\right\}$$

This can be rearranged to give the breakdown voltage as:

$$V_{th} = \frac{d}{b_S^2}\left(\phi + k_B T_0 \log_e\left\{\frac{2\gamma k_B}{a_S(\phi - b_S E_{th}^{1/2})}\right\} - k_B T_0\right)^2 \tag{10.33b}$$

Just as in the Klein model the breakdown voltage in this case would be proportional to thickness (which is interesting as the Schottky effect is due to electrode barrier lowering). As far as we are aware no attempts have been made to test this equation for polymers.

The other high field electrode effect is Fowler–Nordheim emission. This cannot be responsible for thermal breakdown since it is temperature independent.

(c) Space-charge limited current
The equation derived in Section 9.3 governing space-charge limited current flow does not contain any explicit temperature-dependent terms and one would therefore not expect this to lead to thermal breakdown. However breakdown would normally be observed around the trap-filled limit as the current increases very rapidly with voltage. This is likely to be a thermal breakdown but not necessarily a thermal runaway; i.e. a destructive break-down upon reaching a critical temperature, T_c, is likely. Making the common assumption:

$$J = a_{sc} V^{b_{sc}} \tag{10.34}$$

where a_{sc} is a constant of proportionality and b_{sc} is an exponent of order 5–10 and substituting into eqn. 10.17 and using $V_{app} = V_{th}$ at $T_1 = T_c$ gives:

$$V_{th} = \left(\frac{2\gamma(T_c - T_0)}{a_{sc}}\right)^{1/(b_{sc}+1)} \tag{10.35}$$

So V_{th} is not likely to be strongly dependent upon ambient temperature. This is not surprising as an increase in voltage above V_{TFL} would increase the temperature very rapidly.

(d) Poole–Frenkel conductivity
Poole–Frenkel conductivity can be written in the form

$$\sigma = \sigma_0 \exp\left\{-\frac{\phi - \beta E^{1/2}}{k_B T}\right\} \tag{10.36}$$

which gives a breakdown *field* of:

$$E_{th} = \beta^{-2}\left(\phi - k_B T_0 \log_e\left\{\frac{\sigma_0 E_{th}^2 d(\phi - \beta E_{th}^{1/2})}{2\exp(1)\gamma k_B T_0^2}\right\}\right)^2 \tag{10.37}$$

The formula for the critical temperature is complicated but the temperature rise is less than $\sim 100°C$ for $(\phi - \beta E^{1/2})/k_B < 10^7$ K for normal values of T_0.

10.2 Impulse breakdown

Steady-state breakdown was a limiting case of the general power balance eqn. (10.4) in which the temperature rise was so slow that the thermal capacity term could be ignored ($\nabla^2 T \gg C_p D dT/dt$). We now consider the opposite extreme of impulse breakdown in which the temperature rise can be considered so fast that thermal conduction may be ignored. This simplifies the analysis as the temperature of the whole slab can always be considered uniform. In this case the insulator is considered to breakdown at the end of the impulse, i.e. the time to breakdown is the length of the impulse. Eqn. 10.4 therefore becomes:

$$C_v D \frac{dT}{dt} = \sigma(T, E)E^2 \qquad (10.38)$$

(C_v is the specific heat at constant volume: the impulse would allow insufficient time for thermal expansion to occur and instead the pressure rises in the sample.) In the analysis here we will consider the breakdown strength of an infinite slab with parallel plate electrodes which is subjected to a square pulse of length, τ. Thus for $t < 0$, $V = 0$; for $0 < t < \tau$; $V = V_{th}$; and at $t = \tau$ the sample breaks down. We will assume a temperature-independent thermal capacity and summarise the results of low-field ($\sigma(T)$) Klein[161] and Arrhenius conductivity relations and high-field ($\sigma(T, E)$) Klein[161] and Poole–Frenkel conductivities (we noted in Section 10.1 that it is difficult to be quantitative about the other high-field mechanisms of conduction in this type of analysis). Finally we will compare impulse and steady-state breakdown values. We will assume thermal runaway so that the breakdown criterion is $T = \infty$.

10.2.1 Low-field impulse breakdown
(a) Klein conductivity relation
Using the Klein[161] low-field (i.e. field-independent) conductivity relation, eqn. 10.15, in the power balance equation for impulse thermal breakdown, eqn. 10.38, we have:

$$C_v D \frac{dT}{dt} = \sigma_K E^2 \exp\{a(T - T_0)\}$$

which may be integrated from $t = 0$, $T = T_0$ to $t = \tau$, $T = \infty$:

$$C_v D \int_{T_0}^{\infty} \exp\{-a(T - T_0)\}\, dT = \sigma_K E^2 \int_0^{\tau} dt$$

to give:

$$V_{th} = d\left(\frac{C_v D}{a \sigma_K \tau}\right)^{1/2} \qquad (10.39)$$

or, writing $\sigma_K = \sigma_0 \exp\{a T_0\}$, in the usual way:

$$V_{th} = d\left(\frac{C_v D}{a \sigma_0 \tau}\right)^{1/2} \exp\left\{-\frac{a T_0}{2}\right\} \qquad (10.40)$$

Similar equations are derived by Klein and Burstein[596]. This gives the voltage inversely proportional to the square root of the length of the impulse (i.e. the time to breakdown). Unlike the steady-state cases (eqns. 10.16, 10.26) the breakdown *field* is thickness independent (i.e. the breakdown *voltage* is proportional to the thickness) but the relation to temperature is the same.

(b) Arrhenius conductivity
Using the Arrhenius (ionic) conductivity relation (eqn. 9.11) and following the same procedure as above gives:

$$V_{th} \simeq dT_0 \left(\frac{C_v D k_B}{\phi \sigma_0 \tau} \right)^{1/2} \exp \left\{ \frac{\phi}{2 k_B T_0} \right\} \tag{10.41}$$

Again $V_{th} \propto d\tau^{-1/2}$. At typical ambient temperatures $\phi \gg k_B T_o$ so the exponential temperature term dominates and $V_{th} \propto \exp \{ \phi/(2 k_B T_o) \}$ which is identical to both the steady-state cases found (e.g. eqns. 10.22 and 10.13).

10.2.2 High-field impulse breakdown
(a) Klein conductivity relation
Using Klein's expression for high-field (i.e. field dependent) conductivity, eqn. 10.14, and the same method as above we find:

$$V_{th} = \frac{d}{b} \log_e \left\{ \frac{C_v D d^2}{a \sigma_K V_{th}^2 \tau} \right\} \tag{10.42a}$$

and again substituting $\sigma_K = \sigma_0 \exp \{ a T_0 \}$ gives:

$$V_{th} = \frac{d}{b} \left(\log_e \left\{ \frac{C_v D d^2}{a \sigma_0 V_{th}^2 \tau} \right\} - a T_0 \right) \tag{10.42b}$$

Like the high-field Klein steady-state equation this gives an approximate negative-linear relation between V_{th} and T_0 and an approximately thickness-independent breakdown *field*. The dependence upon the time of the pulse is also less pronounced than that at low fields with $V_{th} \propto \log (\text{constant}/\tau)$.

(b) Poole–Frenkel conductivity
Using the simplified Poole–Frenkel expression for conductivity given in eqn. 10.36, we find that the impulse breakdown *field* is of the form:

$$E_{th} = \left(\frac{1}{\beta} \left(\phi - k_B T \log_e \left\{ \frac{\sigma_0 E_{th}^2 \tau (\phi - \beta E_{th}^{1/2})}{C_v D k_B T_0^2} \right\} \right) \right)^2 \tag{10.43}$$

10.2.3 Comparison of steady-state and impulse breakdown
In all cases, as one would expect, the impulse breakdown strength is higher than that for steady state conditions.

For the Klein [161] conductivity model the ratio of impulse to steady-state breakdown fields is inversely proportional to the square root of the impulse time. At high fields and short pulses (O'Dwyer[82]) $(\tau \kappa d^2/(C_v D) \ll 1)$ a plot of the ratio versus $\log (\tau)$ is linear with a negative slope.

For the Arrhenius (low-field) and the Poole–Frenkel (high-field) cases the impulse time dependence is found to be the same as for the respective low-field and high-field Klein conductivity relations.

The expressions derived for V_{th} have been found useful in distinguishing thermal from other types of breakdown (e.g. References 596, 597).

10.3 Filamentary thermal breakdown

Breakdown does not usually occur on a broad front across the insulation area but at weak spots. This is in the competitive nature of breakdown: the temperature of a weak spot (with higher local electrical conductivity, lower thermal conductivity, field enhancing inclusion or the like) reaches the critical temperature before the rest of the insulation. Such behaviour is difficult to analyse in a general manner as different assumptions give rise to a wide variety of boundary conditions. We will therefore illustrate filamentary thermal breakdown with reference to specific experimental results.

Two experimental methods have been used to investigate filamentary breakdown: prebreakdown current measurements and direct observation of the spatial and temporal evolution of specimen temperature. It should be noted that both types of observation confirm *filamentary* breakdown but it is only their interpretation that suggests filamentary *thermal* breakdown. Other types of filamentary breakdown may be responsible for similar observations.

10.3.1 Prebreakdown currents

The electrical engineering group of Nagoya University, Japan, have interpreted their observations of prebreakdown current in various polymer films (including polyimide, polyamideimide, non-stretched polyvinylidene fluoride, and polyethylene) as evidence for filamentary breakdown (e.g. References 428, 588–598). Many of their conclusions are summarised in the paper by Mizutani *et al.*[594]. We discuss their experiments on polyimide films presented in this paper; our analysis is slightly simpler but our conclusions are much the same. We will carry out a simple analysis first, the simplifying assumptions are justified by Mizutani *et al.*'s[594] numerical analysis described later.

Mizutani *et al.*[594] gradually increased the voltage across thin ($\sim 25\ \mu$m thick) polyimide films at 150°C such that breakdown took place about five minutes after the start of the application of the voltage ramp. The workers monitored the current through specimens of different areas. For small area ($0\cdot28$ cm^2) specimens a rapid rise of current from $\sim 1\ \mu$A to $>10\ \mu$A was noted during the millisecond before breakdown despite the voltage increase being very small ($< \sim 4$ p.p.m.) over this small period of time. Larger area specimens ($1\cdot1$ cm^2) did not display any sudden increase in prebreakdown current but, being less resistive, conducted more current ($\sim 10\ \mu$A) at the breakdown voltage. If breakdown on a broad front was being experienced then one would expect the current in both cases to display similar characteristics such as a sudden increase in prebreakdown current. It is concluded that the sudden increase in current is due to the breakdown of a filament through the sample; in the smaller area specimen this is observable whilst in the larger area specimen the conductive current dominates until times closer to breakdown[590].

Their results for the prebreakdown current on several small-area specimens are shown in Fig. 10.4 as a function of time before breakdown, $(t_b - t)$; the theoretical curves will be described later. Since breakdown appears to have taken place over about a millisecond it is reasonable to assume that thermal conduction was insignificant and that the breakdown approximated to the case of impulse breakdown, eqn. 10.38. An Arrhenius (ionic) form for $\sigma(T)$ has been shown to apply for these polyimide films so that eqn. 9.11 is applicable. Substituting the form of $\sigma(T)$ given by eqn. 9.11 into the impulse power balance eqn. 10.38, rearranging and integrating gives:

$$\frac{1}{\sigma_0} \int_T^\infty \exp\left\{\frac{\phi}{k_B T}\right\} dT = \frac{E^2}{C_v D} \int_t^{t_b} dt \qquad (10.44)$$

These limits of integration have been chosen such that at the time of breakdown, $t = t_b$, the temperature is assumed to have 'run away' to infinity. The lower limit is taken to be the time t at which runaway starts to occur. The temperature, T, at time, t, is required. Since the breakdown time is known precisely this is a better method than integrating from $t = 0$ which is badly defined in this experiment. Performing the integration using the same approximation as in eqn. 10.12 gives:

$$\frac{k_B T^2}{\sigma_0 \phi} \exp\left\{\frac{\phi}{k_B T}\right\} \simeq \frac{E^2}{C_v D} (t_b - t)$$

Here the temperature dependence of σ_0 will be very weak and the T^2 term can be taken as T_0^2. (If $\phi = 1$ eV then $\exp\{\phi/(k_B T)\}$ doubles every 11°C at

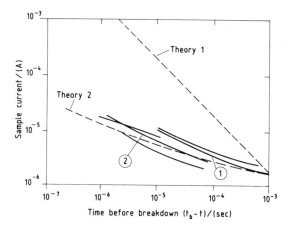

Fig. 10.4 A bi-logarithmic plot of current through several samples as a function of time before breakdown $(t_b - t)$. The samples were thin (~ 25 μm) films of polyimide at 150°C. The data marked '1' and '2' correspond to those used in Fig. 10.5. The broken line 'theory 1' assumes a constant radius of breakdown filament whereas that for 'theory 2' is the result of a numerical calculation in which the radius is found to change with time. After Mizutani *et al.*[594]

$T_0 = 150°C$ whereas T^2 only increases by 15%. Furthermore the analysis is invalid for $T > \sim 300°C$, the softening point of polyimide, so that the error cannot be greater than a factor of 2.) Noting $J = E\sigma = E\sigma_0 \exp\{-\phi/(k_B T)\}$ and that $E = E_b$ the breakdown field gives:

$$J = \frac{k_B T_0^2 C_v D}{\phi E_b (t_b - t)} \tag{10.45}$$

It is again argued that the heating takes place too fast to allow thermal expansion hence the specific heat should be evaluated at constant volume. Note that this implies an increase in the pressure in the filament which causes the filament to exert a mechanical force on its surroundings. From eqn. 10.45 it can be seen that in the region of thermal runaway, i.e. in the filament, the average current density, J, is inversely proportional to the time before breakdown. If we make the simplifying assumption that the observed increasing prebreakdown current is solely due to the enhanced current density in the filament and that the filament can be represented by a cylinder of radius, r_f, then the filamentary current, I_f, is given by:

$$I_f = \pi r_f^2 J = \frac{\pi r_f^2 k_B T_0^2 C_v D}{\phi E_b (t_b - t)} \tag{10.46}$$

which again appears to give $I_f \propto (t_b - t)^{-1}$. This is shown as line 'theory 1' on Fig. 10.4 and is clearly not in agreement with the data which appears to show $I_f \approx 1 \cdot 7 \times 10^{-7} (t_b - t)^{-0 \cdot 31}$ although the parameters cannot be determined very accurately. Whilst the assumption that the filament is similar to a cylinder is probably a reasonable assumption, the implicit assumption that r_f is constant is not justified. One would certainly expect the higher temperature at the axis of the filament to cause the thermal runaway to be much faster near the axis than the surface of the 'cylinder'. This is equivalent to saying that the effective radius of the cylinder decreases with time so that:

$$\begin{aligned} r_f &= \left(\frac{\phi E_b I_f (t_b - t)}{\pi k_B T_0^2 C_v D} \right)^{1/2} \\ &= \left(\frac{\phi E_b 1 \cdot 7 \times 10^{-7} \text{ amps } (t_b - t)^{0 \cdot 69}}{\pi k_B T_0^2 C_v D} \right)^{1/2} \end{aligned} \tag{10.47}$$

This is shown in Fig. 10.5 together with two curves estimated from prebreakdown currents indicated as 1 and 2 in Fig. 10.4. The parameters used were those of Mizutani et al.[594]: $\phi = 1$ eV, $E_b = 3 \cdot 1 \times 10^8$ V · m^{-1}, $T_0 = 150°C$, and $C_v \times D = 1 \cdot 6 \times 10^6$ J · K^{-1} m^{-3}.

Mizutani et al.[594] also carried out a numerical analysis using the general thermal power balance eqn. 10.4, and an activated conductivity–temperature dependence, e.g. eqn. 9.11. They assumed that thermal conduction only took place in the radial direction and that the temperature fluctuation when the filament was initiated had the form:

$$T(t = 0, r) = T_0 + \frac{\Delta T_0}{1 + (r/r_0)^2} \tag{10.48}$$

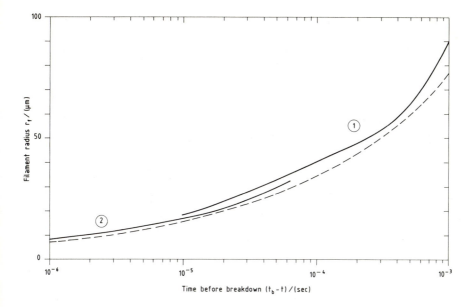

Fig. 10.5 Effective radius of the filamentary path as a function of time before breakdown after Mizutani *et al.*[594]. Experimental curves '1' and '2' were estimated by Mizutani *et al.* from currents '1' and '2' in Fig. 10.4. The broken line shows the solution of eqn. 10.47

where r is the radial distance from the filament axis, ΔT_0 is the temperature above T_0 at the filament axis, and r_0 is the initial filament radius defined as the radius at which the temperature drops to $T_0 + \Delta T_0/2$. They calculated the current density as a function of both r and t; this is plotted in their original paper. They also calculated the filament current using

$$I_f(t) = \int_0^{r_c} 2\pi r J(t, r)\, dr$$

with r_c = electrode contact radius. This is plotted as 'theory 2' on Fig. 10.4 and, whilst realising that their calculations are semi-empirical, it can be seen to be an excellent fit to their data. The line is close to the straight line predicted by eqn. 10.47 thereby justifying the earlier assumptions. Indeed although Mizutani *et al.*[594] used a value of $\kappa = 0.19\ \mathrm{W \cdot m^{-1}\, K^{-1}}$ they found almost identical results using $\kappa = 0\ \mathrm{W \cdot m^{-1}\, K^{-1}}$, which suggests that the thermal conductivity plays almost no role as long as thermal fluctuations can establish a temperature gradient across a filament (i.e. eqn. 10.48).

10.3.2 Direct observations of localised heating

Various attempts have been made to monitor the spatial and temporal evolution of the temperature of thin polymer films after the application of an electric field. It should be noted that although these have indicated the existence of hot spots, this does not conclusively demonstrate that the initiating breakdown mechanism is thermal.

Winkeinkemper and Kalkner[599,600] have monitored the temperature of circular $\frac{1}{4}$ mm thick disks of various polymers (PVC, PMMA, polyoxymethylene (POM), PET, and PS) under 50 Hz AC fields. The circumference of the disks were at a defined ambient temperature so that the centre of the disks heated up more quickly than the circumference. They measured the spatial and temporal evolution of temperature and correlated this with theory assuming all the heat flow was in a radial direction and steady-state conditions obtained. In some samples thermal runaway occurred upon reaching a highest stationary temperature and their theoretical and experimental results agreed well. The heating in these samples under the AC field was due to dielectric loss and it was found, as would be expected, that only those materials whose loss, $\varepsilon_r \cdot \tan \delta$, increased with temperature (PVC, POM, cellulose paper, and PET) gave rise to thermal runaway. Others (PMMA, PS) did not display these characteristics. In these experiments the theory and data agree well and there is no dispute that a thermal mechanism was responsible for the breakdown. However this was partly due to the construction of the samples; the edge cooling, radial heat flow, and lack of cooling at the centre meant that the temperature was bound to increase mainly in the central region.

Nagao *et al.*[601] have monitored the spatial and temporal evolution of temperature in thin (25 μm thick) films of polyethylene as a DC field was applied; no circumferential cooling was used. In this case, as the field was increased, hot spots appeared on the film and gradually one of these dominated and led to breakdown. Although this was clearly Joule heating it is not clear that the breakdown mechanism was purely thermal. The analysis of this experiment is much more difficult because the hot spots occurred naturally, and apparently randomly, presumably at weak points of the film and local electronic breakdown, for example, would also lead to hot spots as the local current density increased. In fact initial measurements of breakdown strength as a function of ambient temperature and prebreakdown current as a function of time and hot spot evolution appear to suggest that an electromechanical breakdown mechanism following the production of a local hot spot due to an inhomogeneity was the most likely breakdown process.

Chapter 11

Electromechanical breakdown

11.1 The Stark and Garton mechanism

Stark and Garton[602] noticed that the breakdown strength of many thermoplastics dropped when their temperature was increased such that they started
to soften. They speculated that this was due to a mechanism which is often
termed electromechanical breakdown. Electromechanical breakdown occurs
when the mechanical compressive stress on the dielectric caused by the
electrostatic attraction of the electrodes (or, more accurately, by electrostriction) exceeds a critical value which cannot be balanced by the dielectric's
elasticity. The electromechanical breakdown voltage can be evaluated by
equating these two stresses for the equilibrium situation before breakdown
in a parallel-plate dielectric slab:

$$\frac{\varepsilon_0 \varepsilon_r}{2} \left(\frac{V}{d} \right)^2 = Y \log_e \left\{ \frac{d_0}{d} \right\}$$

electrostatic   opposing

compressive =   elastic   (11.1)

stress   stress

where Y is the Young's modulus of elasticity, d_0 is the initial dielectric
thickness, and d $(=d(V))$ is the reduced thickness after the application of
voltage, V. The logarithmic form of the mechanical stress-strain relation is
required as the strains may be large in thermoplastics. Rearranging eqn.
11.1 gives:

$$V = d \left(\frac{2Y}{\varepsilon_0 \varepsilon_r} \log_e \left\{ \frac{d_0}{d} \right\} \right)^{1/2}$$ (11.2)

A plot of thickness as a function of voltage is given in Fig. 11.1 on normalised
scales and it can be seen that there is a critical voltage above which the thickness goes to zero. This occurs when $d[d(V)]/dV = -\infty$ or $dV/d[d(V)] = 0$.
From eqn. 11.2:

$$\frac{dV}{d[d(V)]} = \left(\frac{2Y}{\varepsilon_0 \varepsilon_r} \right)^{1/2} \left[\left(\log_e \left\{ \frac{d_0}{d} \right\} \right)^{1/2} - \frac{1}{2} \left(\log_e \left\{ \frac{d_0}{d} \right\} \right)^{-1/2} \right]$$

which is zero when $d/d_0 = \exp\left(-\frac{1}{2}\right)$ ($\simeq 0 \cdot 6$). Thus an electromechanical breakdown voltage, V_{em}, may be defined as

$$V_{em} = d_0 \left(\frac{Y}{\varepsilon_0 \varepsilon_r \exp(1)} \right)^{1/2}$$ (11.3)

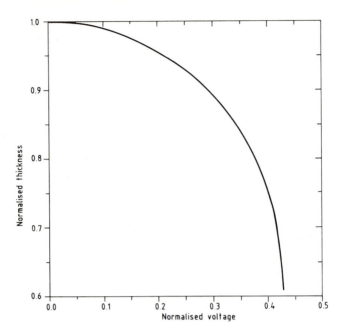

Fig. 11.1   A plot of thickness as a function of voltage for electromechanical breakdown according to eqn. 11.2

although the reduced thickness and consequent increased field may cause
another breakdown mechanism to become operative prior to the thickness
reaching zero, and hence lower the breakdown voltage.

Stark and Garton realised that their model of breakdown implicit to this
analysis was rather unrealistic as it assumed the dielectric material somehow
disappeared to an infinitesimal thickness at $V \gtrsim V_{em}$, and it ignored plastic
flow and the dependence of Young's modulus on time and stress. Their
experiments were carried out on 1 mm thick disks of polyethylene (both
uncrosslinked and radiation crosslinked) which were recessed with spherical
electrodes to about 50 μm thick; they found general agreement with eqn.
11.3. Since both electrical and mechanical stresses would have been concentrated at the narrowest point of the recession it is likely that the plastic flow
would have been small and the material was easily displaced. They also
noted that similar breakdown[603] and Young's modulus[604] measurements on
polyisobutylene agreed with the electromechanical breakdown theory. Following their paper other workers also attributed the breakdown of various
polymers to this mechanism[605,606,607].

11.2 Filamentary theories of electromechanical breakdown

Stark and Garton[602] also pointed out that at microscopic areas of stress
concentration an instability may occur giving rise to a localised thinning

and breakdown. This was considered by Blok and LeGrand[608] who considered that local regions subjected to higher-than-average electric fields experienced a shear stress which tended to form an indentation. The indentation produces an even more inhomogeneous field producing a sharp depression causing the material to flow radially away. With sufficient voltage this progresses right through the material. They supported this model by experiments in which polarised light was transmitted through a recessed polymer disk (minimum thickness 10—15 μm) with a high field applied. Since polymers are not generally optically anisotropic normal to the plane of a sample undergoing uniaxial compression but are if they undergo shearing they were able to observe plastic deformation at weak spots prior to electromechanical breakdown. The electromechanical breakdown processes in polyethylene, a copolymer of polycarbonate and polydimethyl siloxane (35/65 wt/wt ratio) and polyurethane was followed at room temperature using this technique. An attempt was also made to prevent electromechanical breakdown in polyethylene by reducing its temperature to 77 K (in liquid nitrogen), well below its glass transition temperature, however this breakdown mechanism was still observable. Only by restraining plastic flow in the polyethylene by encapsulating it in a crosslinked silicone rubber which was brittle at 77 K were they able to prevent electromechanical breakdown; in this case the polyethylene fractured instead.

Zeller *et al.*[366,141] modified and quantified the Blok and LeGrand process in a mechanism they termed *electrofracture* which is the growth of a filamentary (tubular) crack through a dielectric due to electrostatic forces. They proposed this as a model for the initiation and growth of partial discharge channels based on the concepts of fracture mechanics in which the change in electrostatic energy at growth is equated to the mechanical energy required for channel formation. Although their model appears feasible as an aging mechanism, it is difficult to quantitatively show that it is in fact responsible for aging. Subsequently Fothergill[429] proposed a similar mechanism which he termed filamentary electromechanical breakdown which is operative at higher (local) fields and is a *breakdown* rather than an *aging* mechanism. In contrast to the electrofracture mechanism quantitative experimental evidence is available to support filamentary electromechanical breakdown. We will present both analyses together here and point out the differences between them at appropriate junctures.

Both mechanisms have an analogy with fracture mechanics in which a crack spontaneously propagates when the energy required to create the crack is less than the strain energy liberated by the cracked material (the Griffith's criterion[609]). When mechanically stressed a unit cube of material has a strain energy $\frac{1}{2}\sigma_m\varepsilon_m = \sigma_m^2/(2Y)$ where σ_m and ε_m are the mechanical stress and strain respectively, so that for each unit volume of crack this energy is released. In order to break the bonds to make the crack an energy per unit area of crack surface is required. This is a macroscopic parameter known to engineers as the *toughness.*

In the electrofracture and filamentary electromechanical breakdown mechanisms a conducting tubular filamentary crack, which may be taken to be a cylinder with a hemispheric end, is considered to propagate. This gives rise to an enhanced field, E, at the tip resulting in an electric flux density $D = \varepsilon_0 \varepsilon_r E$ and a consequent electrostatic energy density (W_{es} per unit volume) $= \frac{1}{2}\varepsilon_0 \varepsilon_r E^2$. In the filamentary electromechanical breakdown model the effect of the electric field in inducing mechanical stress is also considered. The mechanical compressive stress induced by an electric field is $\sigma_m = \frac{1}{2}\varepsilon_0 \varepsilon_r E^2$ resulting in a strain energy density (W_{em} per unit volume) of $\sigma_m^2/(2Y) = \varepsilon_0^2 \varepsilon_r^2 E^4/(8Y)$. If a tubular crack of radius r_f propagates by a length δl then the volume displaced is $\pi r_f^2 \delta l$ and the energy released is:

strain

energy $= W_{es} + W_{em}$

released

$$= \left(\frac{1}{2}\varepsilon_0 \varepsilon_r E^2 + \frac{\sigma_m^2}{2Y} \right) \pi r_f^2 \, \delta l$$

$$= \left(\frac{1}{2}\varepsilon_0 \varepsilon_r E^2 + \frac{\varepsilon_0^2 \varepsilon_r^2 E^4}{8Y} \right) \pi r_f^2 \, \delta l \tag{11.4}$$

The electrostatic (E^2) component therefore dominates over the electromechanical (E^4) if $E < 2[Y/(\varepsilon_0 \varepsilon_r)]^{1/2}$ which, in polyethylene, is $\approx 2 \cdot 5 \times 10^9$ V $\cdot$ m^{-1}. Zeller *et al.* argue that a linear space-charge (FLSC) cloud would form very quickly ($< \sim 10^{-9}$ s) around the crack point giving rise to a field-limiting structure with a maximum field, E_{mc}, where $E_{mc} \approx 10^9$ V $\cdot$ m^{-1} in polyethylene. As we discussed in Section 9.3.4 however the field required for injection of carriers E_{inj} may be greater than E_{mc} and so the field at the surface of the crack may be greater than E_{mc}. It is also possible that the space-charge cloud may take longer to form than originally supposed[610]. Furthermore the FLSC model assumes a sharp rise in mobility at E_{mc} due to promotion of electrons into a continuous σ conduction band. Such a conduction band may not exist in the amorphous regions of a polymer where breakdown has been observed to take place preferentially in polyethylene films[611,612,613,434] and where the toughness may be lower.

In order to form a filamentary crack *two* energy components need to be overcome, the surface energy required to overcome the toughness, W_s, (as in the conventional brittle mechanical fracture) and also the plastic deformation energy required to push the material aside to form the crack. Considering the surface energy component, the extra area created by advancing the cylindrical crack by a length δl is $2 \pi r_f \, \delta l$ so that the surface energy required to overcome the toughness is:

surface

energy $= W_s = G 2 \pi r_f \, \delta l \tag{11.5}$

required

There is a major difference between the mechanisms here. Zeller *et al.*[366,141] (electrofracture) use γ, the surface tension, instead of the fracture toughness which is generally used to characterise crack propagation and is used in the filamentary electromechanical breakdown mechanism. These are quite different, for example[366,141] in polyethylene $\gamma \simeq 0.1\,\mathrm{J}\cdot\mathrm{m}^{-2}$ whilst $G \simeq$ $6500\,\mathrm{J}\cdot\mathrm{m}^{-2}$ in tension[8] and would be even higher in compression; a value of $G = G_{IC} \simeq 2 \times 10^4\,\mathrm{J}\cdot\mathrm{m}^{-2}$ may be more appropriate. Physically the electrofracture approach refers to a crack formed by displacing polymer chains without breaking bonds (analogous to crazing) in contrast to the filamentary electromechanical model.

As the tubular crack is formed the surrounding polymer becomes deformed and compressed. By considering the formation of a narrow tube of diameter very much less than length and insulation dimensions, it can be shown that the plastic deformation energy, W_p, required is:

$$\text{plastic deformation energy} = W_p = Y \pi r_f^2\, \delta l \tag{11.6}$$

Here Zeller *et al.*[366,141] use the yield strength rather than the Young's modulus. The use of the yield strength implies a permanently-deformed tube and this is appropriate for electrofracture which is considered to be an aging mechanism. For the filamentary electromechanical breakdown mechanism the Young's modulus is more appropriate. For most polymers the Young's modulus and yield strength are of similar magnitude and so it makes little difference in practice.

The general criterion for the crack to propagate is therefore:

$$\text{general criterion for tubular crack propagation}: W_{es} + W_{em} > W_s + W_p \tag{11.7}$$

In the electrofracture mechanism the local field at the crack tip is considered to be insufficiently high ($E < 2[Y/(\varepsilon_0\varepsilon_r)]^{1/2}$) for the W_{em} contribution to be significant, i.e. $W_{es} \gg W_{em}$. This is because of the FLSC condition and because electrofracture is an aging mechanism operative at lower fields than filamentary electromechanical breakdown. Furthermore, because Zeller *et al.*[366,141] (possibly incorrectly) used γ instead of G in eqn. 11.5 they considered $W_p \gg W_s$ so that for electrofracture the criterion is:

$$\text{electrofracture criterion:} \quad W_{es} > W_p \tag{11.8}$$

so that the electrical field required at the end of the crack tip E_{ef} is given by:

$$E_{ef} = \left(\frac{2Y}{\varepsilon_0\varepsilon_r}\right)^{1/2} \tag{11.9}$$

It is interesting to note that even this mechanism predicts a breakdown field which is comparable to E_{mc} ($E_{ef} \sim 2 \times 10^9\,\mathrm{V}\cdot\mathrm{m}^{-1}$ in polyethylene). This field has the same dependence on Young's modulus as the conventional Stark and Garton[602] electromechanical breakdown model, i.e. $E \propto Y^{1/2}$; indeed the values only differ by $E_{ef}/E_{em} = [2 \cdot \exp(1)]^{1/2} \simeq 2.33$. However note that the electromechanical breakdown field is the 'applied' field whereas the electrofracture field is the local field at the tip of the crack.

The filamentary electromechanical breakdown model assumes fields at the filament tip that are sufficiently high $(E > 2[Y/(\varepsilon_0\varepsilon_r)]^{1/2})$ for $W_{em} > W_{es}$ so that the appropriate criterion is:

filamentary
electromechanical $: W_{em} > W_s + W_p$ (11.10)
breakdown criterion

so that a filamentary electromechanical breakdown field can be defined as:

$$E_{fem} = \left(\frac{8Y(2G+Yr_f)}{\varepsilon_0^2\varepsilon_r^2 r_f}\right)^{1/4}$$ (11.11)

This is in agreement with the results of Hikita *et al.*[428] who measured the breakdown strength and Young's modulus of their 25 μm films of polyethylene at different temperatures. For polyethylene $W_s \gg W_p$ for the following reasons. The value of r_f depends on what initiates the filament. For example this may be a microvoid, an inclusion, an electrode aberration, an electrical tree, or a feature of the microstructure such as the morphology. It seems unlikely that r would exceed ~ 10 μm in these 25 μm thick polyethylene films. Since $G > 6500$ J · m^{-2} and $Y \approx 3 \times 10^7$ Pa we find that $2G \gg Y \cdot r$ in eqn. 11.11 so that

$$E_{fem} \approx \left(\frac{16GY}{\varepsilon_0^2\varepsilon_r^2 r_f}\right)^{1/4}$$ (11.12)

and a plot of $\log(E_b)$ versus $\log(Y)$ should have a slope of $\frac{1}{4}$. This assumes both that the filament diameter is constant from specimen to specimen (which is possible if it is a function of morphology or production technique) and that the toughness is not related to the Young's modulus which may be true especially in the amorphous regions. Hikita *et al.*[428] altered the temperature to change Y and V_b but did not increase it sufficiently to cause severe softening so it is reasonable to treat the value of G as constant at this stage. Their results, plotted on a log–log plot, are shown in Fig. 11.2 and the best straight line fitted has the formula:

$$V_b = 149\, Y^{0.250} \quad \text{(in M.K.S. units)}$$ (11.13)

If it is assumed that the enhanced electric field at the crack tip is proportional to V_b then the dependence of E_{fem} on Y is exactly as predicted by eqn. 11.12. For example using $V = E \cdot r_f$ at the end of the filament gives:

$$V_{fem} = \left(\frac{16\,G}{\varepsilon_0^2\varepsilon_r^2}\right)^{1/4} r_f^{3/4} Y^{1/4}$$ (11.14)

which yields a value of $r_f = 0.84$ μm using $G = 20\,000$ J · m^{-2}, $\varepsilon_r = 2.2$. This value is comparable to the gross morphological features of polyethylene[611] and is at the low end for channel radii in electrical trees in polyethylene[335,317]. One might expect this to be the case since partial discharge activity in electrical trees erodes and makes them wider.

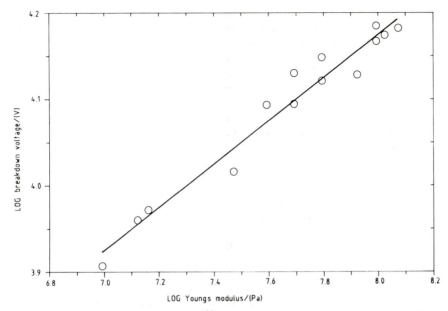

Fig. 11.2   The results of Hikita *et al.*[428] showing the (1/4)th power law dependence of breakdown voltage on Young's modulus. The results give support to the filamentary electromechanical breakdown mechanism according to eqn. 11.12

A more sophisticated relationship between applied voltage and electric field at the tip of a filament has been given by Eichhorn[121] as:

$$E = \frac{V(1 + r_f/d_{fe})^{1/2}}{r \, \text{arc tanh} \, ((1 + r_f/d_{fe})^{-1/2})} \qquad (11.15)$$

where d_{fe} is the distance from the filament to the counter electrode and is assumed to be the full thickness of the specimen in this case (i.e. a worst-case estimate). Solving eqn. 11.15 numerically gives $r_f = 0.18 \, \mu m$ which is also quite reasonable.

Both the electrofracture and filamentary electromechanical breakdown mechanisms assume a crack propagation process. The speed of such a propagation is limited to that of a longitudinal elastic wave (i.e. the speed of sound) in the material and is given by $(Y/D)^{1/2}$ where D is the density. In polyethylene $D \simeq 930 \, \text{kg} \cdot \text{m}^{-3}$ which results in a maximum speed of $180 \, \text{m} \cdot \text{s}^{-1}$. This is likely to be the final speed of the propagation at high fields since the crack tends to reinforce the local enhanced field and ensure its own rapid propagation. The time for such a crack to cross the $25 \, \mu m$ film is therefore $\geqq 0.14 \, \mu s$. Since the crack is not likely to cause a large current rise until it approaches the counter electrode this calculation is consistent with the observation by Hikita *et al.*[593] of increases in prebreakdown current in such films approximately 10^{-7} s before breakdown. Note that this would not allow time for a thermal breakdown process to occur and is rather long for an electronic breakdown process (see Chapter 12).

Interestingly the *limiting* high voltage propagation velocity of impulse electrical trees in polyethylene has been measured[614] to be somewhat higher than the sound velocity estimated here, i.e. $380 \, \text{m} \cdot \text{s}^{-1}$, negative point; $1700 \, \text{m} \cdot \text{s}^{-1}$ positive point. This implies that electromechanical processes only play a role in such electrical treeing once a faster, presumably electronic, mechanism has generated low-density regions.

11.3 Discussion

Although electromechanical breakdown has not received much attention in many reviews (for example in O'Dwyer's excellent book[82] it is not considered in the main section on breakdown) it is likely to be a common breakdown mechanism in polymers. In certain polymers, such as polyethylene, in which hot spot formation through Joule heating has been observed (Nagao *et al.*[601]) but thermal runaway does not appear to take place, it is likely that the consequent reduction in Young's modulus may often give rise to electromechanical breakdown. The effects may be frequently combined, the Stark and Garton dielectric thinning and consequent rise in local field giving rise to a local temperature rise and decreases in Young's modulus. This breakdown mechanism is more likely to occur in thin films in which a small deformation or indentation gives rise to a sharp increase in electric field although electromechanical breakdown may also play its part in the transformation of partial discharges into propagating electrical trees in thicker insulation systems. The common observation of a reduction of the breakdown strength of many polymers with increasing temperature during softening but below the melting point gives further support for electromechanical breakdown, Part 3, Fig. III.3. It is usually assumed that at low temperatures, i.e. below the polymer's glass transition temperature, T_g, the breakdown mechanism is electronic. However the results, discussed earlier, of Blok and LeGrand[608] demonstrating electromechanical breakdown at 77 K in polyethylene even cast some doubt on this. Below T_g the value of Young's modulus would be high and constant and it is possible to conjecture that the high and constant value of breakdown strength may therefore also be consistent with an electromechanical breakdown mechanism.

Chapter 12
Electronic breakdown

In electronic breakdown the field causes either the number or the energy of the electrons to reach unstable magnitudes such that they rise catastrophically. Ultimately this causes destruction of the lattice at least locally.

We can consider electronic breakdown to be essentially of two types in polymers:

(i)   *So-called 'intrinsic' breakdown.* It is found that there is, in terms of electron energy, a maximum rate at which the free-electron system can lose energy to the lattice by electron–phonon scattering but that the rate at which it can acquire energy from the field is a monotonically-increasing function. There must, therefore, be a critical field and corresponding electron energy above which the electrons indefinitely acquire energy faster than they can lose it thereby leading to breakdown.

Although this appears to be a mechanism which is truly intrinsic to the material, in fact high-energy electrons may also lose energy by mechanisms other than electron–phonon scattering such as inelastic collisions with defects or scattering with other electrons be they free or trapped. This mechanism may therefore only lead directly to lattice disruption if the lattice is very weak, perhaps after being weakened through earlier local energy-dissipating phenomena (see Section 15.4). In the case of collisions between the high-energy electrons and other free electrons the energy distribution will tend to stabilise until the overall rate of increase of electron energy cannot be matched by losses to the lattice. Collisions between the high-energy electrons and *trapped* electrons however may lead to 'avalanche' or 'impact ionisation' breakdown.

(ii) *Impact Ionisation or Avalanche Breakdown.* In this case electrons with a high energy, either as a result of acceleration in the field, or hot injection from the electrode, or purely from chance fluctuations, collide with trapped or bound electrons imparting sufficient energy for both electrons to be free after the collision. Given a sufficiently high field both electrons rapidly gain enough energy to each cause a second generation of collisions resulting in four free electrons. If this chain-reaction continues the local concentration of high-energy electrons builds up to such an extent that local destruction of the lattice ensues. However if the sample is too thin the electrons may reach the anode before the avalanche has built up to a destructive size. The avalanche may also be quenched if the liberation of localised electrons results in the production of relatively-immobile positive species, i.e. holes or ions. It should be noted that, contrary to much of the literature, this is not necessarily the case in amorphous materials such as polymers in which free electrons may have been captured in local potential wells which had no previous net charge. In this case *filled* traps are essentially negatively ionised and an avalanche-type collision results in their deionisation (neutralisation). Here it is assumed that the charges ensuring global neutrality are random in space giving a uniform background field. If, however, a region of positive space charge does appear to the rear as the avalanche progresses this may serve to moderate the field in the region of the avalanche thereby quenching it. (Indeed one would expect that high-energy electrons producing avalanches which are quickly quenched by the resulting space charge would be naturally found as an intrinsic process in many materials to which moderately high fields were applied.) In the case of a large avalanche a large positive space charge cloud may therefore almost nullify the field in the avalanche region between it and the anode. However this will increase the field between the cathode and the space-charge cloud which may result in a greatly enhanced field emission of electrons from the cathode, with these high-energy electrons causing immediate catastrophic damage around the cathode region. This is sometimes referred to as space-charge breakdown (O'Dwyer[82]).

Another type of electronic breakdown is 'Zener' or 'field-emission' breakdown in which electrons are excited at a high rate from the valence to conduction band thereby greatly enhancing the conductivity. Although this mechanism (and avalanche breakdown) are observed in semiconductor p–n junctions, it requires very high fields ($> \sim 10^{10} \ \text{V} \cdot \text{m}^{-1}$) and has been thought not to occur in polymers which generally breakdown at lower fields due to other causes.

The fundamental models of 'intrinsic' and avalanche breakdown are reasonably straightforward if rather simplistic. The models have been refined since their original proposals to the point at which their more-accurate mathematical descriptions tend to obscure the physical processes. We say 'more-accurate' since the descriptions are still very much approximations for polymers. Interestingly these refinements have made relatively little difference to the predictions of the models and, since breakdown measurements on polymers always result in a wide spread of results, the agreement between theory and data has not generally been significantly improved. For these reasons we have decided not to describe the mechanisms in as much detail as elsewhere but simply to point out the major simplifications and refer the reader to other reviews for more detail. The most authoritative of these is probably by Stratton[615], which is to be particularly recommended since he prefaces his more detailed analysis with an 'elementary description' which gives considerable insight into the physical processes under consideration.

12.1 'Intrinsic' breakdown

In 'intrinsic' breakdown, as in thermal and electromechanical breakdown, there is a power balance equation which can only be satisfied below a critical value of field or voltage. For intrinsic breakdown the power balance relates to the rate of gain and loss of electron energy. Free electrons on average gain energy from the field at a rate:

$$A = \frac{JE}{n} \tag{12.1}$$

In this equation both J and n may depend on the temperature, T, and other parameters, α, describing the electron energy bands, trap levels, etc. The rate of energy loss is unlikely to be explicitly dependent upon field except inasmuch as it may result in heating. We can therefore define an electron power loss $B(T, \alpha)$ such that in equilibrium:

$$A(E, T, \alpha) = B(T, \alpha) \tag{12.2}$$

One of the first attempts to describe intrinsic breakdown was made by von Hippel[616] who considered a single 'average' electron in the conduction band. This is a useful starting point as, although the complete distribution of electrons should be considered, the one-electron approximation is simple to visualise and there are many analogies with more complicated electron-distribution models.

12.1.1 The von Hippel, single electron, model (the low-energy criterion)

If we consider one electron then, following a similar argument to Section 9.1.1 it will be accelerated by the field at a rate $e \cdot E/m^*$ and assuming a scattering rate $1/\tau_c$ (cf. eqn. 9.4) we have that its rate of increase of energy is given by:

$$A(E, \mathbb{E}, T_0) = \frac{e^2 E^2 \tau_c(\mathbb{E})}{m^*} \tag{12.3}$$

The electron may lose (or gain) energy by scattering with lattice vibrations of energy $h\nu$; Goodman, Lawson and Schiff[617] have shown that the probability of collisions involving the emission or absorption of two quanta to be insignificant. The ratio of the probabilities of emission and absorption of a quanta is $(n_\nu + 1)/n_\nu$ where n_ν is the average number of quanta of energy $h\nu$, per site, i.e. the Bose distribution function:

$$n_\nu(T_0) = \left(\exp\left\{ \frac{h\nu}{k_B T_0} \right\} - 1 \right)^{-1} \tag{12.4}$$

Thus the net loss of the electron's kinetic energy to the lattice (emission − absorption) per collision is $h\nu(n_\nu + 1 - n_\nu)/(n_\nu + 1 + n_\nu) = h\nu/(2n_\nu + 1)$. If the scattering rate is again $1/\tau_c$ (assuming isotropic scattering) then the rate of energy loss becomes[615]:

$$B(\mathbb{E}, T_0) \simeq \frac{h\nu}{\tau_c(\mathbb{E})} \frac{1}{2n_\nu + 1} \left(1 - \frac{k_B T_0}{\mathbb{E}} \right) \tag{12.5}$$

where the term $(1 - k_B T_0/\mathbb{E})$ ensures that there is no energy loss for electrons of energy of order $k_B T_0$.

The functions $A(E, \mathbb{E}, T_0)$ and $B(\mathbb{E}, T_0)$ are plotted schematically in Fig. 12.1 as a function of electron energy, $\mathbb{E}$, for different fields, E. At $E = 0$, A and B are zero and the electron energy is $k_B T_0$. For field E_1 there are two possible steady-state solutions at which $A = B$ shown as energies $\mathbb{E}_1$ and $\mathbb{E}_2$. As the field is raised to E_{vH} (von Hippel) there is only one solution and this represents the maximum field at which the electron energy increase can be matched by an equal and opposite decrease B. This was termed the breakdown strength by von Hippel and has subsequently been termed the low-energy criterion for reasons which will become apparent in the next section (12.1.2). The value of E_{vH} can be found by assuming a form for $\tau(\mathbb{E})$ and by noting that at E_{vH} both A and B and their derivatives with respect to $\mathbb{E}$ are equal. This von Hippel criterion defines the minimum field at which *every* electron will continually increase its energy under the influence of the field until breakdown occurs.

12.1.2 Fröhlich's high-energy criterion
Fröhlich[618] noted that breakdown may occur at fields lower than that defined by von Hippel since electrons with energies greater than the higher of the two stable energies (Fig. 12.1) would continue to acquire energy at a rate

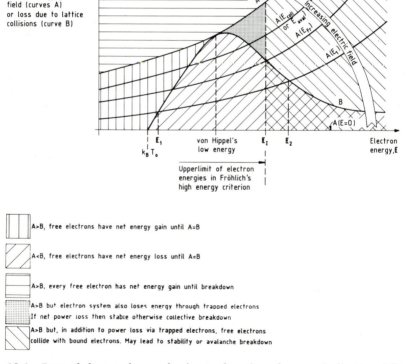

Fig. 12.1   Rate of electron loss and gain as a function of energy indicating different regimes in which different loss processes dominate

greater than they could lose it. Thus, for field E_1, electrons in the distribution with energies greater than $\mathbb{E}_2$ would continue to gain energy until breakdown occurred unless their energy was dissipated by mechanisms other than electron–phonon scattering. Fröhlich considered the case where the concentration of free electrons was sufficiently low to be able to ignore energy dissipation by scattering between electrons (this will be considered in the next section, 12.1.3).

During an ionising collision a high energy electron ($\mathbb{E} > \mathbb{E}_2$) collides with a lattice atom and liberates another electron. After the collision both electrons usually have energies less than $\mathbb{E}_2$ and therefore continue to lose energy until they reach the stable energy $\mathbb{E}_1$. The distribution of electron energies therefore normally has an upper limit of $\mathbb{E}_i$. Fröhlich defined the critical field for breakdown such that the higher (meta)stable point was equal to the ionisation energy, this is shown as E_{Fr} in Fig. 12.1. Equating A and B and setting $\mathbb{E} = \mathbb{E}_i$, the ionisation energy, it can be seen that the Fröhlich high-energy criterion is given by:

$$E_{Fr}^2 = \frac{m^*}{e^2} \frac{B(\mathbb{E}_i, T_0)}{\tau_c(\mathbb{E}_i)} \simeq \frac{m^*}{e^2} \frac{h\nu}{\tau_c^2(\mathbb{E}_i)(2n_\nu + 1)} \tag{12.6}$$

We will later consider the case in which ionisation produces a self-sustaining avalanche and it is notable that the Fröhlich high-energy field defines the limit at which significant ionisation results. The actual breakdown field for the avalanche process must thus be higher than this value. In contrast the von Hippel criterion defines the field at which all electrons continue to gain energy indefinitely, and hence places an upper limit on the breakdown field.

It is clear that the value of B will tend to decrease slowly with increasing temperature as the probability of energy absorption from the lattice approaches that of emission (i.e. n_ν increases). The breakdown field will therefore also slowly decrease with increasing temperature. However at a certain temperature the number of electrons in the conduction band will be sufficient for electron–electron interactions to dominate over electron–phonon interactions. Fröhlich showed that in impure and inhomogeneous materials this results in a strong negative temperature coefficient (see following section, 12.1.3). In practice it is to be expected that this latter process will dominate in polymers at and above room temperature.

Fröhlich's model is supported by experimental evidence[514,619,593] in polymers in which polar groups and defects increase the low temperature breakdown strength presumably by acting as extra scattering centres. For example[514] poly(methyl methacrylate); poly(vinyl acetate); and poly(vinyl alcohol) all have extremely high breakdown strengths (Part 3, Fig. III.3) and the addition of chlorine or carbonyl groups to PE also increases its breakdown strength. Ieda[620] has noted the increase in electric strength of polyethylene when co-polymerised with PBPM, TBPM or DBNM. In this case the halogen co-monomers are assumed to act as electron traps; this is supported by the observations of the reduction in high-field conductivity[620] and TSC measurements[621]. However these results may also be consistent with a thermal breakdown mechanism.

12.1.3 The Fröhlich amorphous solid model

Fröhlich[622,623] subsequently considered the effects of electron–electron interactions in an amorphous or impure solid. Although such interactions are not likely to dominate in a crystalline insulator (in which the concentration of free electrons is too low) it is possible in an insulator which has traps below the conduction band. In this case electron–electron interactions may take place between conduction-band and trapped electrons so that they form part of the same energy distribution defined by an electron temperature, T_e. The electron system will gain energy from the field through the acquisition of kinetic energy by the free electrons and lose it by lattice scattering primarily via the far more numerous trapped electrons. For this energy loss to take place T_e must be greater than T_0 the lattice temperature.

The energy band diagram Fröhlich considered is shown in Fig. 12.2 in which there is a band of width ΔE of shallow trap states situated below the bottom edge of the conduction band. Electrons in the trap states are assumed to be capable of absorbing or emitting quanta of energy $h\nu$ by making transitions within the band of traps from E to $E \pm h\nu$. The trap levels must therefore be sufficiently dense and $\Delta E \gg h\nu$. A conventional energy band gap is considered to lie below the band of traps separating it from ground-state levels. Fröhlich's analysis does not appear to have been modified in the light of mobility-gap theory but by assuming his conduction band edge is analogous with a mobility shoulder the band of shallow traps would be a natural consequence and the two interpretations would almost certainly lead to very similar results.

If $\Delta E \ll$ band gap then the Fermi level will be located approximately half way between the bottom of the trap band, E_t, and E_v, i.e. an energy E_h below E_c using the nomenclature of Fig. 12.2. Again we will write a balance equation between electron energy gain and loss but, since we are not dealing with a

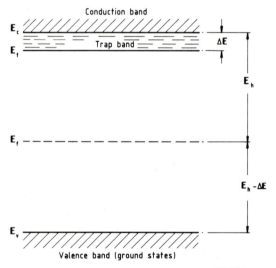

Fig. 12.2   The energy band model used by Fröhlich[622,623] assumes the existence of a band of shallow traps immediately below the conduction band

system of individual electrons, it is more convenient to define these as per unit volume rather than per electron as was the case in Section 12.1.1. The power provided by the field is therefore:

$$A(E, T, \alpha) = JE$$

and from eqn. 2.2:

$$A(E, T, \alpha) = n_c e \mu E^2$$

and using eqn. 9.7:

$$A(E, T, \alpha) = \frac{e^2 \tau_c n_c}{m^*} E^2 \qquad (12.7)$$

where n_c is the concentration of (free) electrons in the conduction band. Since energy is gained by the free electrons and lost by both free and trapped electrons it is necessary to calculate their relative concentrations. The number of electrons in the conduction band is given by:

$$n_c = \int_{\mathbb{E}_c}^{\infty} N(\mathbb{E}) P(\mathbb{E}) \, d\mathbb{E} \qquad (12.8)$$

(cf. eqn. 2.7) where $N(\mathbb{E})$ is the number density of states and $P(\mathbb{E})$ is the Fermi–Dirac distribution (eqn. 2.5) describing the probability of their occupancy. The density of states in the conduction band is given by (Allison[566]):

$$N(\mathbb{E}) = \frac{2^{7/2} \pi (m^*)^{3/2}}{h^3} (\mathbb{E} - \mathbb{E}_c)^{1/2} \qquad (12.9)$$

Substituting for $N(\mathbb{E})$ and $P(\mathbb{E})$ in eqn. 12.8 gives:

$$n_c = C_1(T_e) \exp \left\{ \frac{-\mathbb{E}_h}{k_B T_e} \right\} \qquad (12.10a)$$

with

$$C_1(T_e) = 2 \left(\frac{2 \pi m^* k_B T_e}{h^2} \right)^{3/2} \qquad (12.10b)$$

so that $C_1(T_e)$ is an effective density of states and is slowly varying with temperature in comparison with the exponential. At $T_e = 300$ K, C_1 is typically $\sim 10^{25}$ m^{-3}. The equation for n_c can be substituted into eqn. 12.7 to give the rate of electron energy gain as:

$$A(E, T_e, \alpha) = \frac{e^2 \tau_c C_1}{m^*} E^2 \exp \left\{ \frac{-\mathbb{E}_h}{k_B T_e} \right\} \qquad (12.11)$$

The concentration of electrons in the trap band, n_t, depends on the density of states in the band, $N(\mathbb{E}_t < \mathbb{E} < \mathbb{E}_c)$ so that:

$$n_t = \int_{\mathbb{E}_t}^{\mathbb{E}_c} N(\mathbb{E}) P(\mathbb{E}) \, d\mathbb{E} \qquad (12.12)$$

where $P(\mathbb{E})$ is the Fermi–Dirac probability distribution (eqn. 2.5). Assuming for simplicity that the concentration of traps is constant in the band, i.e. that $N(\mathbb{E}_t < \mathbb{E} < \mathbb{E}_c) = N_t$ is a constant, then

$$n_t = N_t k_B T_e \exp\left\{\frac{-\mathbb{E}_h}{k_B T_e}\right\}\left(\exp\left\{\frac{\Delta\mathbb{E}}{k_B T_e}\right\} - 1\right)$$

or

$$n_t \simeq C_2(T_e) \exp\left\{\frac{-(\mathbb{E}_h - \Delta\mathbb{E})}{k_B T_e}\right\} \qquad (12.13a)$$

for $\Delta\mathbb{E} \gg k_B T_e$ with

$$C_2 = N_t k_B T_e \qquad (12.13b)$$

which again is a relatively slowly varying function of temperature.

For this breakdown model to be appropriate, electron–phonon interactions within the band of traps must dominate so that n_t must be much greater than n_c; a condition which can be seen to be met by the following argument. From eqns. 12.10a and 12.13a we have:

$$\frac{n_c}{n_t} = \frac{C_1}{C_2} \exp\left\{\frac{-\Delta\mathbb{E}}{k_B T_e}\right\} \qquad (12.14a)$$

which becomes, using the above approximate value for C_1 at room temperature:

$$\frac{n_c}{n_t} \simeq \frac{10^{25}\,\text{m}^{-3}}{N_t k_B T_e} \exp\left\{\frac{-\Delta\mathbb{E}}{k_B T_e}\right\} \qquad (12.14b)$$

In order to estimate N_t consider that each 'lattice imperfection' contributes at least one trap state to the trap band. Since the number density of trap states is:

$$\int_{\mathbb{E}_t}^{\mathbb{E}_c} N(\mathbb{E})\,d\mathbb{E} = \int_{\mathbb{E}_t}^{\mathbb{E}_c} N_t\,d\mathbb{E} = \int_{\mathbb{E}_t}^{\mathbb{E}_t + \Delta\mathbb{E}} N_t\,d\mathbb{E} = N_t\,\Delta\mathbb{E}$$

then $N_t \simeq N_i/\Delta\mathbb{E}$ where N_i is the number density of lattice imperfections. Thus from eqn. 12.14b:

$$\frac{n_c}{n_t} \simeq \frac{10^{25}\,\text{m}^{-3}}{N_i} \frac{\Delta\mathbb{E}}{k_B T_e} \exp\left\{\frac{-\Delta\mathbb{E}}{k_B T_e}\right\}$$

For a polymer almost every atom is likely to contribute a trap state. However even if we assume a 'worst case' that only one in 1000 atoms contribute, i.e. $N_i \sim 5 \times 10^{25}\,\text{m}^{-3}$, then for all reasonable values of $\Delta\mathbb{E}$, $n_c \ll n_t$. (For example if $\Delta\mathbb{E} = 0.5$ eV, $n_c/n_t \sim 10^{-8}$ and as $\Delta\mathbb{E} \to 0$, $n_c/n_t \to 1/5$.) Considering that the number density of traps is likely to be much higher than 1 in 1000 atoms, and that $\Delta\mathbb{E}$ is likely to be greater than 0.1 eV, then the condition $n_c \ll n_t$ is always easily satisfied. The rate of energy loss at $T_e = T_0$ is therefore mainly due to electron–phonon interactions with the trap levels, and since electron–phonon interactions may either cause the emission or absorption of phonons it is the net loss that must be estimated.

The rate of phonon absorption W_a per unit volume (the rate at which the electron system gains energy in units of $h\nu$) is given by:

$$W_a = P_{tran}(\mathbb{E})P(\mathbb{E})n_\nu(T_0) \qquad (12.15)$$

at the lattice temperature T_0. Here $P_{tran}(\mathbb{E})$ is the probability per unit time that an electron will make the transition from energy $\mathbb{E}$ to $\mathbb{E}+h\nu$, $P(\mathbb{E})$ is the Fermi–Dirac distribution (eqn. 2.5) and $n_\nu(T_0)$ is the concentration of lattice quanta of frequency ν at temperature T_0 (eqn. 12.4). The reverse process of phonon emission (i.e. the rate at which the electron system loses energy in units of $h\nu$) is given by:

$$W_e = P_{tran}(\mathbb{E})P(\mathbb{E}+h\nu)(1+n_\nu(T_0)) \qquad (12.16a)$$

and noting from eqn. 12.4 that $1+n_\nu(T_0) = n_\nu(T_0)\exp\{h\nu/(k_B T_0)\}$ then

$$W_e = P_{tran}(\mathbb{E})P(\mathbb{E})n_\nu(T_0)\exp\left\{\frac{h\nu}{k_B}\left(\frac{1}{T_0}-\frac{1}{T_e}\right)\right\} \qquad (12.16b)$$

so

$$B(T_e, T_0) = h\nu\sum_{\mathbb{E}_t}^{\mathbb{E}_c}(W_e - W_a)$$

$$= h\nu n_\nu(T_0)\left(\exp\left\{\frac{h\nu}{k_B}\left(\frac{1}{T_0}-\frac{1}{T_e}\right)\right\}-1\right)\sum_{\mathbb{E}_t}^{\mathbb{E}_c}P_{tran}(\mathbb{E})P(\mathbb{E}) \qquad (12.17)$$

Assuming $P_{tran}(\mathbb{E}) \approx 1/\tau_c$ (i.e. independent of $\mathbb{E}$) and since $\sum_{\mathbb{E}_t}^{\mathbb{E}_c} P(\mathbb{E}) = n_t$ then substituting from eqn. 12.13 gives:

$$B(T_e, T_0) = \frac{h\nu}{\tau_c}C_2 n_\nu(T_0)\exp\left\{\frac{-(\mathbb{E}_h-\Delta\mathbb{E})}{k_B T_e}\right\}$$

$$\times\left(\exp\left\{\frac{h\nu}{k_B}\left(\frac{1}{T_0}-\frac{1}{T_e}\right)\right\}-1\right) \qquad (12.18)$$

If the function $A(T_e, E, T_0)$ (eqn. 12.11) and $B(T_e, T_0)$ are plotted as a function of T_e (i.e. electron temperature or mean kinetic energy) then a very similar plot to Fig. 12.1 results with a maximum value of B as a function of T_e and a critical value of field at which the values of A and B are equal as are their derivatives with respect to T_e. Using this criterion the 'collective' breakdown field, E_{coll}, can be established. Fröhlich[622] showed that the critical electron temperature was given by:

$$\frac{h\nu}{k_B T_0} - \frac{h\nu}{k_B T_e(E = E_{coll})} \simeq \frac{h\nu}{\Delta\mathbb{E}} \qquad (12.19)$$

This can be substituted back to give:

$$E_{coll} = C\exp\left\{\frac{\Delta\mathbb{E}}{2k_B T_0}\right\} \qquad (12.20a)$$

with

$$C = \left(\frac{m^* n_\nu(T_0)}{\exp(1)\,\Delta\mathbb{E}}\frac{C_2}{C_1}\right)^{1/2}\frac{h\nu}{e\tau_c} \qquad (12.20b)$$

For this mechanism then the breakdown strength drops rapidly with increasing temperature such as has been observed in various amorphous materials (e.g. Fröhlich[622]). A direct consequence of this model is that for fields less than the breakdown strength, the conductivity is expected to rise:

$$\sigma = \sigma_0 \exp \left\{ \left(\frac{E}{E_{coll}} \right)^2 \frac{\mathbb{E}_h}{\Delta \mathbb{E}} \frac{1}{\exp(1)} \right\}$$ (12.21)

where σ_0 is the ambient temperature (T_0) low-field conductivity.

Fröhlich[622] cited as evidence for his theory work carried out by Austen and Pelzer[624,625] on PE and PVC. Since the long chains of PVC contain strongly-dipolar $H-\overset{|}{\underset{|}{C}}-Cl$ side groups which would act as scattering centres, the mean free path for electrons in PVC should be much less than for PE and so at low temperatures the Fröhlich theory predicts a higher breakdown strength which is found. Furthermore because PVC is amorphous and polyethylene is semicrystalline the transition to a negative temperature coefficient of breakdown strength should be at a lower temperature; this was also found by Austen and Pelzer. This qualitative data however could also be interpreted as evidence for other theories (e.g. electromechanical breakdown). Quantitative agreement with Fröhlich's ideas however have certainly been found in other materials such as soda-lime glass[625].

12.2 Avalanche breakdown

In this section we shall give a simple description of avalanche breakdown in which a high-energy electron collides with a bound electron thereby producing a pair of free electrons. In the presence of a high electric field this pair of electrons acquire sufficient energy to produce two more pairs of free electrons. Repetition of the process increases the number density of free electrons, and since it is only free electrons which can acquire energy from the field, the avalanche can lead to a very high local energy dissipation causing local lattice disruption after a sufficient number of generations. The description of avalanche breakdown is in two parts; first the number of generations of ionising collisions required to cause damage must be estimated, and secondly the corresponding field must be evaluated. After the description there follows some observations on avalanche breakdown and, in particular, some of the simplifications which have been made. These simplifications however do not appear to alter the conclusions sufficiently to justify the massive increase in model complexity that a more detailed treatment would require.

12.2.1 The critical number of ionising generations
The critical number of generations, i, was calculated as ~40 by Seitz[152] and ~38 by Stratton[615] using similar insulator dimensions and properties. The actual value of i does not influence the breakdown field strongly and we will follow a derivation which is similar to Seitz[152].

Let us consider an electron emitted from the cathode and which drifts towards the anode causing ionising collisions *en route*. As it drifts from anode to cathode it will also tend to move by diffusion processes in a plane orthogonal to its drift direction. Similarly the secondary electrons will also tend to diffuse sideways so that the avalanche tends to spread out in a cone shape as it approaches the anode. Let us assume that the average extent of this sideways diffusion, l_{diff}, of the primary electron is also approximately the radius of the avalanche in the plane orthogonal to the drift as it approaches the anode.

If we consider a critical avalanche, i.e. one which is just big enough to cause breakdown, which has i generations of ionising collisions, then it will have 2^i electrons as it approaches the anode. It is considered that it is this ith generation of electrons which are sufficiently concentrated to cause catastrophic lattice disruption. If the insulator thickness is d then this must completely contain the i generations. Thus a new generation of electrons is produced each time the previous generation of electrons are accelerated through a distance $d/(i+1)$. (This assumes that the ith generation of electrons are also accelerated through this distance before reaching the anode in order for them to gain sufficient energy from the field to cause the damage required for breakdown.) The last generation of electrons are therefore contained in a cylinder of insulation of area πl_{diff}^2 and length $d/(i+1)$. (An alternative approach[615] which gives a similar solution is to consider that the avalanche is shaped like a cone of length $d/(i+1)$ and radius l_{diff}.)

Each electron gains, as it moves a distance, $d/(i+1)$ through the field, E, an energy $eEd/(i+1)$, so the energy acquired by all the electrons in this volume is $2^i eEd/(i+1)$. As the avalanche is generated very quickly it is reasonable to assume that thermal impulse conditions apply, i.e. there is no significant heat flow out of this volume and that all this energy may cause lattice disruption. An estimate of the energy required may be made by assuming that each atom in this volume requires a given energy, $\mathbb{E}_{atom}$, to remove it from the lattice structure; where $\mathbb{E}_{atom}$ is of the order 1—10 eV. If there are N_a atoms per unit volume, then the total energy required for breakdown in the volume occupied by the final generation of avalanche electrons is $\mathbb{E}_{atom} N_a \pi l_{diff}^2 d/(i+1)$ so that the critical condition for breakdown ($E = E_{aval}$) in terms of i is:

$$2^i eE \frac{d}{i+1} \geq \mathbb{E}_{atom} N_a \pi l_{diff}^2 \frac{d}{i+1}$$

or

$$i = \log_2 \left\{ \frac{\pi l_{diff}^2 N_a \mathbb{E}_{atom}}{eE_{aval}} \right\} \tag{12.22}$$

The value of $l_{diff} \simeq (D \cdot t_f)^{1/2}$ where D is the diffusion coefficient appropriate to electrons of *mean* energy (e.g. energy $\mathbb{E}_1$ in Fig. 12.1 for field E_1, i.e. the lower, stable, energy solution of A, eqn. 12.3, equal to B, eqn. 12.5). The value of D may be estimated in various ways. Seitz[152] and Franz[2] assume

$D = \frac{1}{3} s_{th} l_c$ where s_{th} is the mean thermal speed of the electron system (not only those involved in the avalanche) and l_c is the constant *mean* free path between collisions (which is much less than that for the electrons involved in the avalanche process). The mean free path, l_c, is difficult to estimate but it is likely to be several, perhaps 1—100, interatomic spacings ($= N_a^{-1/3}$). The time to form the avalanche, t_f, is simply the time for the primary electron to cross the insulation thickness, $t_f = d/v_d$ where the drift velocity, $v_d = \mu \cdot E_{aval}$ and μ is a mobility appropriate to hot electrons in a conduction band. Typically i is found to be in the range 30—40, values of 38 or 40 are quoted for i as if they were definitive whereas i is not, in fact, well defined. Fig. 12.3 shows examples of values of i based on the above calculations for thicknesses from 1 μm to 0·1 m and for $l_c = 1$, 10, and 100 interatomic distances for $E_{aval} = 10^8$ V · m^{-1}, $\mu = 10^{-4}$ m^2 V^{-1} s^{-1}, $N_a = 5 \times 10^{28}$ m^{-3}, $m^* = 0·1\, m_e$, and $T_0 = 293$ K which may be typical for hot electrons.

12.2.2 The avalanche breakdown strength

We will follow the simpler of the two approaches given by Stratton[615] here. A free electron acquires kinetic energy from the field at a constant rate. Since on average the electron system has no net velocity in the absence of an electric field, there will be an average time, t_i, taken by a free electron to acquire sufficient energy to produce an ionising collision. Assuming that ionisation occurs immediately the electron acquires sufficient energy, the mean rate of ionisation (per electron) is $1/t_i$. We could also define an average distance the electron must 'fall' through the field to acquire sufficient energy

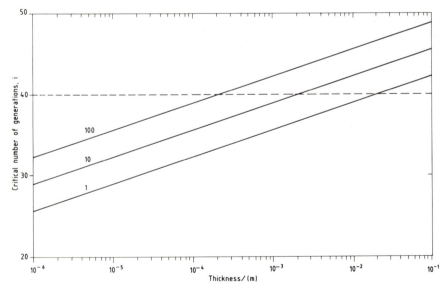

Fig. 12.3   The critical number of avalanche collisions for breakdown to occur as a function of mean free path, l_c, expressed as a number of interatomic distances, based on the typical values given in the text

for ionisation. Using the values from the previous section this distance, l_i, is of the order $10\,\mathrm{eV}/10^8\,\mathrm{V}\cdot\mathrm{m}^{-1}$, i.e. $\sim100\,\mathrm{nm}$ which is likely to be much further than the mean free path between collisions, $l_c\sim1\,\mathrm{nm}$. Since $l_i\gg l_c$ and therefore $t_i\gg\tau_c$, the mean time between collisions, the probability, $P(t_i)=\tau_c/t_i$, that it can travel for a sufficiently long time, t_i, for ionisation to occur is small. The mean rate of ionising collisions per electron is therefore:

$$\frac{1}{t_i}=\frac{P(t_i)}{\tau_c(\mathbb{E}_1)}$$

where τ_c depends on the mean energy of the electron system corresponding to $\mathbb{E}_1$ for field E_1 in Fig. 12.1. Since the drift velocity is given by $\mu(\mathbb{E}_1)\cdot E$ the mean number of ionising collisions it makes *per unit length* is given by:

$$\alpha=\frac{P(t_i)}{\tau_c(\mathbb{E}_1)\mu(\mathbb{E}_1)E} \tag{12.23}$$

Since avalanche breakdown requires a critical number, i, of generations over the insulation thickness, d, at breakdown we have:

$$\alpha(E=E_{aval})=\frac{i}{d}=\frac{P(t_i)}{\tau_c(\mathbb{E}_1)\mu(\mathbb{E}_1)E_{aval}} \tag{12.24}$$

Note that this places a lower limit (perhaps of the order of microns or more) on the thickness for which this mechanism can be operative.

$P(t_i)$ may be evaluated by noting the general form of $P(t)$. The probability of a collision in a short time δt may defined in terms of the collision rate, $1/\tau(t)$, as $\delta t/\tau(t)$. The probability of not having a collision in this time is therefore $P(\delta t)=1-\delta t/\tau(t)$. Since $P(t+\delta t)$ must be equal to $P(t)\cdot P(\delta t)$, we have $P(t+\delta t)=P(t)[1-\delta t/\tau(t)]$ and:

$$\frac{dP(t)}{dt}=\frac{P(t+\delta t)-P(t)}{(t+\delta t)-t}=-\frac{P(t)}{\tau(t)}$$

so that

$$P(t_i)=\exp\left\{-\int_0^{t_i}\frac{dt}{\tau(t)}\right\} \tag{12.25}$$

$\tau(t)\neq\tau_c$ in this case as the electron is being accelerated by the field, $dv/dt=eE/m^*$, and $\tau(t)$ will therefore decrease with time and $P(t_i)$ is field dependent. Assuming that $v=0$ at $t=0$ and defining v_i such that the ionisation energy $\mathbb{E}_i=\frac{1}{2}m^*v_i^2$ the integral can be transformed to be field independent and:

$$P(t_i)=\exp\left\{-\frac{m^*}{eE}\int_0^{v_i}\frac{dv}{\tau(v)}\right\} \tag{12.26}$$

Thus:

$$P(t_i)=\exp\left\{-\frac{H}{E}\right\} \tag{12.27a}$$

with

$$H = \frac{m^*}{e} \int_0^{v_i} \frac{dv}{\tau(v)} \tag{12.27b}$$

Substituting for $P(t_i)$ into eqn. 12.24 for $E = E_{aval}$ and rearranging gives:

$$E_{aval} = \frac{H}{\log_e \left\{ \dfrac{d}{E_{aval}\mu(\mathbb{E}_1)\tau_c(\mathbb{E}_1)i} \right\}} \tag{12.28}$$

Various forms of H can be envisaged and some have been discussed by O'Dwyer[82] and Stratton[615]. Baraff[626] has found more exact numerical solutions using the Boltzmann equation and, in the light of these, O'Dwyer recommends that H should be treated as an empirical constant rather than the result of a simple physical model. The characteristic feature of this type of electronic breakdown is that the breakdown field is thickness dependent (cf. 'intrinsic' breakdown, Section 12.1), although, because of the logarithmic function, specimens with a wide range of thicknesses need to be tested to prove that this mechanism is operative. For example Mizutani *et al.*[627] have attributed breakdown to this avalanche mechanism in polyphenylene sulphide films after plotting reciprocal breakdown field versus log (thickness). Although reasonable physical parameters were evaluated the three film thicknesses of 6 μm, 13 μm, and 25 μm used cannot be considered by themselves to be sufficient to verify eqn. 12.28.

12.2.3 The statistical time lag

The initiation of an avalanche depends on the chance existence of an electron with a sufficiently high energy, usually near or from the cathode. Its continuation and growth depends on the subsequent probabilities that electrons will acquire sufficient energy in between collisions. Wijsman[628] has shown that the statistical time lag, t_s, such that the probability of a critical avalanche in a period t is t/t_s, is given by:

$$t_s = \frac{1}{\nu_0} \exp \left\{ \frac{N_i}{\exp(\alpha d)} \right\} \tag{12.29}$$

where ν_0 is the rate of electron emission from the cathode (which is highly field dependent, Section 9.2), and N_i is the number of electrons required in the avalanche for breakdown (i.e. $N_i \sim 2^{40}$). Note that because of the second exponential, t_s is highly thickness dependent. A formative time lag, t_f, can be defined as the time taken from the initiation of the avalanche to breakdown. The simplest formulation of this is simply the time taken for the initiating electron to cross the insulation thickness:

$$t_f \simeq \frac{d}{E\mu} \tag{12.30}$$

Generally speaking $t_f \ll t_s$ so that once a critical avalanche is initiated, breakdown follows in a few nanoseconds. The field for avalanche breakdown is extremely well defined experimentally in crystalline systems. For example in Zener diodes the current may rise by an order of magnitude for about 0·1% rise in the field. In polymer systems, due to their inhomogeneities, this may not be the case and holding the voltage just below that for avalanche breakdown may result in a very noisy current as sub-critical avalanches are produced and subsequently extinguished. From a comparison of the spectrum of the current noise with the statistical time lag it may be possible to establish whether the avalanche mechanism is operative or not.

Kitani and Arii[629,630] have extensively studied the time lag to breakdown in mica, PE and other polymer films using nanosecond pulse techniques and consider that initiating electrons may be provided from the electrode or the bulk depending on the material and temperature.

12.3 Critique of intrinsic and avalanche breakdown

Klein[161] has suggested that E_{vH}, the high-energy criteria breakdown field is generally about 2 or 3 times that of E_{Fr} the Fröhlich low-energy field with, as would be expected, E_{coll}, that for the collective model in between these two values. Seitz[152] has suggested that typically the avalanche breakdown field, E_{aval}, may be perhaps only a fifth of E_{vH}. Whilst it seems unlikely that E_{aval} can be much less than E_{Fr} since E_{Fr} defines the minimum field at which significant ionisation can occur, it is possible however, that, due to stochastic effects, an avalanche may be precipitated for fields less than E_{Fr}.

The main simplifying assumption made in the above derivation for the avalanche breakdown field has been the ignoring of the space charge built up by holes or positive ions formed from ionising collisions. O'Dwyer[631] has shown that in a typical system, on the basis of the above model, the resulting space charge may lead to fields of $>10^{11}$ V·m^{-1} which is obviously unreasonable. In this case massive injection of charge from the cathode would result and cause breakdown. This has been observed through light emission in alkali halides by Paracchini[632]. O'Dwyer[82] has reviewed this subject in detail and, by using current continuity instead of field continuity as the basis for his calculations arrives at expressions which can only be solved numerically. Both O'Dwyer[82] and DiStefano and Shatzkes[633] have shown that this criterion predicts negative charge injection from the cathode leading to breakdown rather than massive avalanching. Instead of the linear relationship between $(1/E_{aval})$ and log (thickness) expected from eqn. 12.28 a curve results. Although O'Dwyer's analysis has been shown to fit data well for NaCl it is unlikely to be so clear in polymers due to both the inherent scatter of results and the problem of measuring a breakdown strength attributable to the same mechanism over a sufficiently wide range of thicknesses. The tests of Bradwell *et al.*[634] on PE involving both DC and impulse measurements with and without prestressing support the collision-ionisation induced field enhancement near the cathode with an electron mobility of approximately 10^{-11} m^2 V^{-1} s^{-1}. The effect of space charge on breakdown in various materials including polymers has been reviewed by Inuishi[524].

The analysis has also been simplified by using a 'one electron' type model in which the behaviour of an average electron is considered both representative of others and of those causing the breakdown. For example in calculating the probability that an electron acquires an energy $\mathbb{E}_i$ the spread of initial energies about $\mathbb{E}_1$ has been ignored and the averaging process rather simplified. Taking these factors into account will tend to enhance the value of $P(t_i)$, resulting in a somewhat lower value for E_{aval}. The interested reader is referred to Stratton[615] for a more thorough analysis, however again the theoretical results are not likely to show a great improvement in fitting the data from polymeric breakdown.

Measurements on PE at low temperatures indicate that breakdown strength increases with oxidation[635], with increased doping by materials containing π electrons (i.e. aromatics) and with decreasing crystallinity[636] which, together with time-lag experiments[629,630] which suggested an initial electron was supplied from the electrode, lent support to an avalanche model. At temperatures well above the glass-transition temperature other mechanisms come into play and the breakdown strength starts to drop rapidly with increasing temperature. This appears to be related to weaknesses in the amorphous regions since single crystals of hexatriacontane (n-$C_{36}H_{74}$), often used as a simple model for the crystalline part of PE, continue to show an independence of breakdown strength on temperature typical of avalanche breakdown right up to the melting point[637]. Since the formative time lag in polyethylene is typically[428,588–594,598,638] $<10^{-7}$ seconds it seems likely that the mechanism may be electronic. A striking example of electronic breakdown at low temperatures giving way to thermal breakdown at higher temperatures has been given by Nagao *et al.*[597] for polypropylene films. The transition is clearly delineated but in the high temperature region there is some evidence that electronic breakdown may be coupled with thermal breakdown through the dielectric transient current.

Despite the large number of breakdown strength measurements on polymers suggesting electronic breakdown at low temperatures the exact mechanism in any insulating system tends to be controversial[159]. For example Ieda[620] quotes six different possible breakdown mechanisms for polyethylene depending upon the temperature. There appear to be three reasons for this: (i) sample history and preparation is critical (for example space charge build up is likely to have significant effects[634,639]); (ii) the variation of breakdown strengths observed from apparently identical samples (see Part 4 of the book); and (iii) more than one breakdown mechanism may be operating at any one time.

Chapter 13
Partial discharge and free volume breakdown

In these breakdown mechanisms charge carriers are accelerated by the electric field through spaces in the dielectric. In partial discharge breakdown sparks occur within voids in the insulation causing degradation of the void walls and progressive deterioration of the dielectric. In free volume breakdown carriers are accelerated through spaces within low-density amorphous regions; the energy thereby gained is lost through collisions and various mechanisms have been proposed as to how this may lead to breakdown. The voids necessary for partial discharges may be thought of as extrinsic since they are artifacts of the process used for manufacturing the insulating system. On the other hand free volume is an intrinsic feature of polymeric insulation in which there are always variations of density between crystalline and uncrystallised regions. Free path lengths of up to a few tens of nanometers may be possible through free volume at room temperature[356]. Voids may range almost up to a millimeter in poorly made material and are generally not recognised below a few tens of nanometers. The latter dimension may therefore be used as a 'rule of thumb' dividing line between voids and free volume. In this chapter we will describe these two breakdown mechanisms.

13.1 The nature of partial discharges

Voids are difficult to completely eliminate in polymeric materials (e.g. Stevens *et al.*[105]). They may result simply from non-uniform contraction but particularly difficult to overcome are those produced in the slow chemical reactions of thermosetting occurring after the main manufacturing process. For example steam-curing of polyethylene cables results in small voids ($\sim \mu$m) in the water halo (Section 4.4) and refinements such as nitrogen curing may still result in voids from the crosslinking reaction by-products. In thermosetting resins the degassing procedure is critical for the successful elimination of voids. Voids may also occur next to the electrode, though the use of triple extrusion has largely overcome this in extruded medium and high voltage power cables.

Gas-filled cavities in an insulator generally have a lower permittivity and a lower breakdown strength than the insulation material. Since the lower permittivity will give rise to an enhanced electric field in the void (by a factor of up to the relative permittivity of the dielectric depending on the void shape[640,641,642]), the gas (e.g. air) in the void will generally breakdown before the insulation material as the applied voltage is raised. This was recognised by Petersen[643] in 1912 as potentially damaging to the insulation. The development of the Schering bridge in the 1920s enabled power loss to be measured in cables as a function of voltage[644]. Using this technique it was shown that above a certain voltage, which corresponded to the partial discharge inception voltage, the loss increased above the expected value in proportion to voltage squared[645–648]. For a brief historical review of subsequent instrumentation see Bartnikas[649].

Partial discharge will be considered under two headings: the formation of the electrical discharge, and the degradation caused by the discharge. Techniques for evaluating partial discharges will be briefly outlined in Section 19.2. Partial discharges may evolve into electrical trees; this has already been discussed in Chapter 5.

First we outline the major common features of the nature of the electrical discharge. In practice discharges depend upon the shape of the voids and these are distributed in shape and size. For example Sedding *et al.*[650] have observed that, in hydrogen-cooled turbines with cast-epoxy insulation, the partial-discharge activity immediately increases if the pressure drops. This is attributed to a change in void size since changing the gas (e.g. to air) takes five to six hours to affect partial discharge activity; this is in agreement with diffusion theory. Furthermore the discharge itself will change the void. The gases, which can only diffuse slowly into or out of the void, will be changed by chemical reaction, for example oxygen may be consumed and new species generated such as ozone, atomic oxygen and nitrogen oxide. The pressure may also change, particularly as the void temperature may increase. The walls of the void are likely to become damaged as the high energy density of a discharge may be localised, and they are likely to become more conducting. Charge dumped on the surfaces may form space charge clouds in the dielectric which are slow to dissipate. The discharges are therefore likely to be variable and may start and stop for reasons which are not entirely clear.

It is found that there is a voltage below which breakdown will not occur across a gas-filled void under given conditions. This has been observed for a wide range of polymers[459]. This voltage is known as the inception voltage V_i, and depends on the nature of the insulation and the gas, the shape and size of the void, and the pressure and temperature. If the voltage is raised above the inception voltage then breakdown of the gas may ensue by avalanche ionisation of the gas molecules in a not-dissimilar way to that occurring in solids. For the avalanche to occur a free electron must exist near the cathode (i.e. the cathodic void–insulator interface). There will not be any free electrons already in the gas as the electric field will have swept them all to the anodic end of the void; thus an avalanche initiating electron can only exist by being emitted from the cathode. This is a stochastic process and results in a time lag occurring after the inception voltage has been reached before an avalanche starts.

13.1.1. The statistical time lag
To be emitted from the cathode, electrons must attain sufficient energy to overcome the barrier of the work function (see Section 2.2.2). This is

unlikely to occur as a result of spontaneous thermionic emission at temperatures which are below the melting point of most polymers and at the relatively low breakdown field of the gas it is unlikely that the field would significantly enhance the probability due to the Fowler–Nordheim or Schottky effects (Section 9.2). The production of such electrons will therefore be from cosmic radiation, background radioactivity, and UV phonons incident upon the cathode surface. This will be a random process and so the probability of electron emission in an incremental time δt will be given by:

$$P(\delta t) = \frac{\delta t}{\tau_s} \tag{13.1}$$

which defines τ_s, the mean statistical time lag. By noting that the probability $\bar{P}(t+\delta t)$ of an electron *not* being emitted in time $(t+\delta t)$ is equal to $\bar{P}(t) \cdot \bar{P}(\delta t)$ and $\bar{P}(\delta t) = 1 - P(\delta t)$ and by following a similar argument to that in Section 12.2.2 in which $P(t_i)$ was determined, it is easily shown that the probability of an avalanche starting in time t is given by:

$$P(t) = 1 - \exp\left\{-\frac{t}{\tau_s}\right\} \tag{13.2}$$

which has been verified by Devins[381]. Typical statistical time lags are of the order of milliseconds and so are by no means negligible in the discharge process; indeed they are generally orders of magnitude greater than the formation time of the avalanche.

13.1.2 Equivalent circuit of void

The existence of a time lag between the inception voltage being reached and the breakdown being precipitated implies that it is possible for a voltage greater than V_i to exist across the void, at least for short periods of time. In order to consider the relationship between the void voltage and the voltage applied to the insulation it is useful to use an equivalent circuit such as the one shown in Fig. 13.1. The void is represented as a capacitor, C_v, in parallel with a spark gap, which discharges, at least partially, when the voltage across it exceeds the inception voltage. The void must be charged through the insulation in series with it which is represented by a parallel capacitor, C_s, and resistor, R_s. The remainder of the insulation is represented as a parallel capacitor, C_p. Typically $C_s \ll C_v \ll C_p$ and the charging time constant, $R_s \cdot C_v$ is of the order of seconds as the insulation leakage resistance R_s is very high. A resistor could similarly be included in parallel with C_p but, as this plays no part in the discharge characteristics, this has been omitted for clarity. The equivalent circuit is only an approximation since in real insulating systems the capacitive impedances in series and in parallel with the void would be distributed. (Pederson[651] has considered this in more detail.) If an AC voltage of frequency f and magnitude V_a is applied across the insulation, then, assuming that $R_s C_v \gg 1/(2\pi f)$, the voltage will be capacitively divided across the void and the insulation in series with it such that

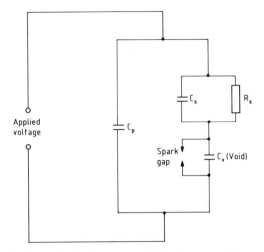

Fig. 13.1 An equivalent circuit for a partial discharge void in a polymer

the void voltage is:

$$V_v = V_a \frac{C_s}{C_v + C_s} \tag{13.3}$$

If the void voltage exceeds the inception voltage then it is likely to discharge until its voltage drops to below the extinction voltage, V_e, which is less than the inception voltage. This will happen in times of the order of a microsecond. The situation is shown for both DC and AC applied voltages in Fig. 13.2a and 13.2b in which the *overvoltage*, ΔV, is also defined. In the AC case it can be seen that partial discharge activity is greatest on the fastest changing parts of the voltage waveform. This has also been observed in impulse breakdown in epoxy resins[652]. The variation of void voltage due to discharging has been substantially exaggerated for the sake of clarity. Notice that both the overvoltage and the extinction voltage are different from discharge to discharge. The void voltage waveforms in the two cases are substantially different since in the AC case where $R_s C_v \gg 1/(2\pi f)$ the rate of voltage rise depends on the applied voltage waveform whereas in the steady-state DC case it primarily depends on the charging time constant $R_s C_v$. Since it is likely that the statistical time delay is not highly dependent upon the overvoltage[381] the overvoltage will depend on the rate of rise of voltage and the statistical time lag. In the cases of power AC frequencies and steady-state DC this is likely to result in relatively small overvoltages (of the order of 1% or less of the inception voltage).

We will now examine the nature of the discharge and the criterion for a self-sustaining discharge using the classical theories developed in the early part of this century by Townsend[653]. These theories will be related to the semi-empirical Paschen's law (established over a hundred years ago[654]) from which the breakdown voltage across a void may be determined. The Townsend discharge results in a space-charge distortion of the field in the void

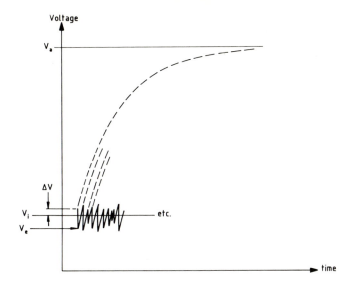

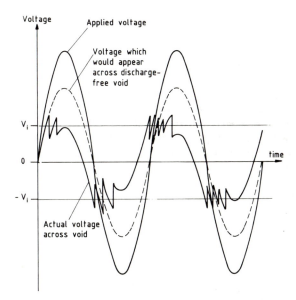

Fig. 13.2   The voltage across a discharging void under DC and AC conditions
(*a*)   DC conditions. The voltage across the void increases approximately exponentially according to the equivalent circuit given in Fig. 13.1. The equation of the void voltage is $V_a - (V_a - V_i) \exp\{-t/(R_s C_v)\}$ with t measured from the previous time that $V = V_i$
(*b*)   Voltage appearing across the void under AC conditions

which can be observed in the shape of the current waveform and which serves either to extinguish the discharge or to enhance it to a more powerful streamer-type discharge. The onset of such streamers can be clearly observed as the current magnitude is greatly increased. Such streamers have very high energy densities and are likely to severely damage the polymer surface where they strike. Streamer production in large gaps has been modelled since Townsend by various workers including Raether, Loeb, and Meek (see for example Rees[655]) but it is likely that the streamer mechanism is slightly different in the small voids producing partial discharges[381].

During the course of a discharge, charge is transferred across the gap and, in the case of a void in the bulk of an insulator, the charge is deposited on the void surface where it can only dissipate slowly through the insulation. This charge therefore causes a potential across the void which opposes that applied. The net voltage therefore decreases until the discharge can no longer be maintained. The charge then dissipates slowly through the insulation until the net voltage builds up to the inception voltage and the process repeats (in the AC case the applied voltage may also be increasing).

We will follow the field distribution in Fig. 13.3 in which the void is shown next to the cathode and in series with the insulation (cf. Fig. 13.1). Although in practice the void is likely to be within the insulation, the representation used in the diagram allows easier visualisation of the division of the voltage between the void and the dielectric. At this stage of the argument, the void is about to breakdown, with the application of a small overvoltage, Fig. 13.3*a*. We assume that there is no space charge anywhere in the system and a uniform flux density, D, so that, since $D = \varepsilon_0 \varepsilon_r E$, the field is inversely proportional to the relative permittivity.

13.1.3 The Townsend discharge

The Townsend model[653] of a discharge is that of a self-sustaining avalanche. An initial electron near the cathode causes an initial avalanche. This avalanche alone would cause a discharge which was not self-sustaining, i.e. the spark would extinguish very quickly. For various reasons discussed below the initial avalanche may, indirectly, produce other so-called 'secondary' electrons near or from the cathode (i.e. other than by the normal ionisation processes in the avalanche itself) which may lead to the production of secondary avalanches. This process may then become self-sustaining, i.e. a current will continue to flow even after the carriers produced by the first avalanche have drifted across the gap. The inception voltage for Townsend discharges is a function of the size and shape of the void and the gas type and number density of gas molecules (i.e. pressure). For a particular gas and an ideal parallel plate gap the breakdown voltage is given by Paschen's law (see next subsection, 13.1.4) which may be derived from the Townsend model.

In the Townsend theory an electron injected from the cathode gains kinetic energy from the field. If the mean free path between collisions in the direction of the field is sufficiently long and the field sufficiently high, then the energy so gained may be enough to cause ionisation upon impact with a molecule, thereby liberating another electron. These two electrons

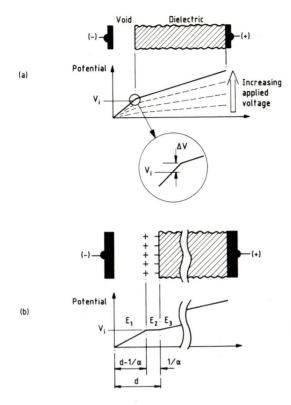

Fig. 13.3 The potential through an ideal void in series with a dielectric (*a*) before discharging; (*b*) immediately after a discharge has extinguished

will therefore continue in like manner and an avalanche may result. Townsend defined a primary ionisation coefficient, α, as the number of electrons liberated by an initial electron in passing through unit distance of gas in the direction of the field. In passing through a distance dx the increase in the number of free electrons, dn, is therefore given as $n\alpha dx$ where n is the number of free electrons at x. Thus:

$$\frac{dn}{dx} = \alpha n \tag{13.4}$$

is a formal definition of α. Integrating $dn = \alpha n dx$ and rearranging gives the number of electrons at x as:

$$n(x) = n_0 \exp\{\alpha x\} \tag{13.5}$$

where n_0 is the number of original electrons at $x = 0$, the cathode. The current due to the electron avalanche therefore increases as the anode is approached. The form of the temporal build-up of current due to electrons (being defined here as the integral sum of the currents from cathode to anode) must also be approximately exponential with the maximum current occurring as the electrons are swept into the anode. This is shown schematically in Fig. 13.4i; the curve of the current is more smoothed in practice due to the randomising effects of diffusion.

Up to this point we have ignored the effect of the positive ions created during the avalanche. These will eventually be swept towards the cathode and thereby also constitute a current. Since the highest concentration of positive ions will occur at the end of the electron avalanche, the ion current will not reach a maximum until the electrons are being swept into the anode i.e. as the electron current is terminating (Fig. 13.4ii). The mobility of the ions is lower than that of the electrons and so their current will be lower than that of the electrons. The ion current however lasts much longer than the electron current not just because of their lower mobility but also because they have much further to travel. The electrons only have to travel to the anode a short distance from where most of them are created at the end of the avalanche whilst the ions must travel to the cathode which is a much longer distance since most of them are also created near the anode. Although the absolute charge due to electrons and ions must be equal and opposite, the apparent charge, i.e. the area under the current–time curve, of the ions is therefore considerably greater. This is substantiated by Devins[381] and other references quoted therein who found a linear relation between discharge pulse-width time and cathode to anode separation. Devins also noted a linear relation with reciprocal field up to a value of field at which the ion drift velocity saturated.

So far we have described the evolution of one avalanche. This is clearly not self-sustaining unless there is a plentiful supply of cathode electrons which is not generally the case. Typical times for the completion of such an avalanche are of the order of nanoseconds and clearly this cannot describe the complete discharge which is typically of the order of microseconds (and then only extinguished by the counterfield of the charge transferred across the gap). Townsend proposed that the avalanche would somehow cause further 'secondary' electrons to be supplied from the cathode so that, provided at least one new secondary electron is emitted from the cathode for each avalanche the process will continue indefinitely. The three most likely causes of secondary electron emission are:

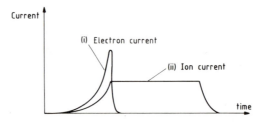

Fig. 13.4   The current observed due to a Townsend-type discharge separated into electronic and ionic components

- *Positive Ion Impact Emission.* Positive ions impacting on the surface of the cathode may cause electron emission. In practice this is unlikely since the kinetic energy of an ion is likely to be low and particularly to be less than that required for the ionisation of other molecules. Furthermore the energy it requires for release of an electron from the cathode must be twice that normally required (i.e. twice the work function) since not only must it cause an emission it also requires an electron for its own neutralisation. Multiple impact collisions may accumulate sufficient energy to release an electron. In the case of a conventional metal cathode the probability of this happening is greatly reduced but, since electrons in a polymer cathode could be temporarily raised to higher energy levels in metastable or trap states, this is not such an unlikely mechanism. A similar mechanism is discussed in more detail in Section 13.3.
- *Photoelectric Emission.* Excited molecules or ions in the avalanche may emit photons in returning to the ground state. Upon impact with the cathode, these photons may give rise to photoelectric emission provided $h\nu \geq \phi$, the work function.
- *Excited and Metastable Atom Impact Emission.* Excited atoms, produced in the avalanche from collisions with electrons $(M + e^- + KE \rightarrow M^* + e^-)$ or metastable atoms (which are simply atoms excited into a metastable state) may diffuse back to the cathode with excess energy. If their total energy exceeds the work function they may liberate an electron.

Other ionisation and electrode processes are possible (e.g. photo-ionisation of gas molecules); some of these contribute to streamer development and will be discussed later. The probability that a secondary electron is produced at the cathode as an indirect result of a primary collisional process in the gap is termed Townsend's secondary ionisation coefficient, γ. The critical (minimum) value of γ for which a self-sustained avalanche occurs corresponds to one secondary electron emission for all the primary collisional processes taking place in one avalanche. That is, each avalanche will indirectly produce a secondary electron at the cathode which will start the next avalanche *ad infinitum.* Since the number of primary ionisation collisions is usually high (in Devins[381] artificial discharge experiments this would have been $\sim\exp(9) \approx 10^4$) a small value of γ can still result in a self-sustaining discharge.

For each electron leaving the cathode there will be $\exp(\alpha d) - 1$ ionising collisions before the electrons arrive at the anode, a distance d from the cathode (from eqn. 13.5). The number of secondary electrons produced will therefore be $\gamma(\exp(\alpha d) - 1)$. Provided this is at least unity, another avalanche will be started and the process self-sustained. The Townsend criterion for a self-sustained discharge is therefore:

$$\gamma(\exp(\alpha d) - 1) \geq 1 \qquad\qquad (13.6a)$$

or

$$\gamma \exp(\alpha d) \geq 1 \qquad\qquad (13.6b)$$

since $\exp(\alpha d) \gg 1$ usually.

Depending on the electron energies available and the type of gas molecules it should be realised that processes other than ionisation can take place upon collision. As well as producing atoms in excited states, attachment occurs in some gases ($M + e^- + KE \rightarrow M^{*-} + $ energy) which may subsequently lead to dissociation of the gas molecule. Attachment and ionisation are competing processes[656]. In attaching gases the primary ionisation coefficient is reduced to an effective value $\bar{\alpha} = \alpha - \eta$ where η is an attachment coefficient defined as the mean number of attachment collisions per unit length of travel of an electron in the field direction. The number of electronic charges reaching the anode in an attaching gas ($M^- + e^-$) is found to be[657]:

$$n_0 \left(\frac{\alpha}{\bar{\alpha}} \right) \exp\{\bar{\alpha}d\} - \left(\frac{\eta}{\bar{\alpha}} \right)$$

which implies that a greater voltage is required to produce a self-sustaining discharge.

13.1.4 Paschen's law

At the end of the last century, Paschen[654], from measurements on air, carbon dioxide, and hydrogen established that the breakdown voltage across a uniformly-stressed gap with metal electrodes was a function of the product of the gas pressure, p, and the electrode spacing, d:

$$V_b = f(pd) \tag{13.7}$$

The initial breakdown behaviour of gas in polymeric voids (before degradation processes have a significant effect) has been shown both theoretically[658,659] and experimentally[660,661,662] to follow this law closely when account has been taken of the field enhancement. Eqn. 13.7 is found to have a minimum voltage at a certain $p \times d$ product. A typical Paschen curve (for air[381]) is shown in Fig. 13.5i. The minimum breakdown voltage for air is 327 V which occurs at $p \times d = 0.756$ Pa · m. As well as the conventional Paschen curve of breakdown *voltage* as a function of $p \times d$, a curve of breakdown *field*/pressure is shown in Fig. 13.5ii. This is a monotonically decreasing function which at atmospheric pressure in dry air is asymptotic to 2.475×10^6 V · m^{-1}. (Atmospheric pressure is approximately 10^5 Pa so the right-hand scale can be converted to breakdown field in V · m^{-1} by multiplying by 10^5. Note that the 'standard' 30 kV/cm breakdown strength for air is actually for a 1 cm gap.)

It is reasonable that the breakdown voltage would be a function of $p \times d$. For an electron to produce ionisation upon impact its kinetic energy must be greater than the ionisation energy of the molecule, eV_{ION}, and the electron's kinetic energy, eEl_c, is determined by the electric field and the mean free path between collisions, l_c. The mean free path length is inversely proportional to the gas molecule number density, i.e. it is inversely proportional to pressure in an ideal gas, so that if the pressure is increased from say p to zp, the mean free path is reduced from l_c to l_c/z. This would reduce the kinetic energy of the electron from KE to KE/z. However if the field is increased to zE, the probability of producing ionising collisions is restored

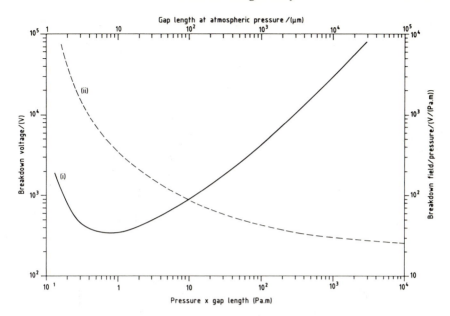

Fig. 13.5   A Paschen curve for dry air showing both (*i*) the breakdown voltage and (*ii*) the breakdown field/pressure as a function of both the pressure–gap length product and the gap length at standard atmospheric pressure

to its original value. Since the total number of collisions per unit length is proportional to pressure, α will be increased to $z\alpha$ when E and p become zE and zp respectively. Hence

$$\frac{\alpha}{p} = f_1\left(\frac{E}{p}\right) \tag{13.8}$$

Similarly γ is proportional to the kinetic energy upon impact of ions on the cathode so that

$$\gamma = f_2\left(\frac{E}{p}\right) \tag{13.9}$$

Using $E = V/d$ (uniform field) and substituting into the Townsend breakdown criteria (eqn. 13.6*b*), gives:

$$f_2\left(\frac{V}{pd}\right)\exp\left\{pdf_1\left(\frac{V}{pd}\right)\right\} = 1 \tag{13.10}$$

which is in agreement with eqn. 13.7. For values of $p \times d$ less than the value for minimum breakdown voltage the low pressure gives a low probability of collision while above this value the mean free path and hence the energy at collision reduces. In both cases, in order to produce a similar size avalanche the voltage needs to be increased. This can be shown quantitatively as follows[663].

For an electron to cause ionisation it must travel a distance l_{ION} through a field E such that $l_{ION} = V_{ION}/E$. If the mean free path lengths are randomly distributed then the probability that the path length l is greater than the minimum for ionisation is:

$$P(l > l_{ION}) = \exp\left\{\frac{-l_{ION}}{l_c}\right\}$$

$$= \exp\left\{\frac{-V_{ION}}{El_c}\right\} \tag{13.11}$$

The number of paths per unit length are l_c^{-1} and the proportion of these sufficiently long to cause ionisation is given from eqn. 13.11 as:

$$\alpha = \frac{1}{l_c}\exp\left\{\frac{-V_{ION}}{El_c}\right\}$$

and since $l_c \propto p^{-1}$ or $l_c^{-1} = Ap$ with A a constant of proportionality, so:

$$\alpha = Ap \exp\left\{-\frac{ApV_{ION}}{E}\right\} \tag{13.12}$$

Using the results of Harrison and Geballe[656] for dry air values of $A = 3\cdot943\ \mathrm{m}^{-1}\,\mathrm{Pa}^{-1}$ and $V_{ION} = 41\cdot3\ \mathrm{eV}$ are found. The latter seems extremely high but is in fact due to the attaching nature of air so that α is really an effective value only in this case. Substituting this into eqn. 13.6b, the Townsend criterion for breakdown, gives:

$$V_b = \frac{A V_{ION} pd}{\log_e\left\{\dfrac{Apd}{\log_e\{1/\gamma(E/p)\}}\right\}} \tag{13.13}$$

which is Paschen's law.

13.1.5 Discharge magnitude

By assuming a uniform field in the void (a more sophisticated model is given in Section 13.1.6) then we can derive a simple expression for the discharge magnitude. Consider that a voltage $V_i + \Delta V$ applied across the gap has resulted in a discharge taking place. Electrons are therefore being swept to the void's anodic dielectric surface where they are building up as they cannot travel quickly through the dielectric. This is equivalent to charge building up on the series capacitor, C_s, in the equivalent circuit of Fig. 13.1. (R_s is effectively infinite on these time scales.) As the electron (negative) charge, Q_n, builds up on C_s then so does the voltage ($V = Q \cdot C$). If we assume that the inception voltage corresponds to an appropriate Paschen voltage across the void then we would expect the discharge to be unsustainable below V_i. Assuming that the applied voltage is constant over the period of the discharge, then if the voltage across C_s increases by ΔV, the voltage across the void must decrease by ΔV to V_i, and the avalanching will terminate. Avalanche formation should thus cease when:

$$Q_n = \Delta V C_s \tag{13.14a}$$

Notice the value of C is that of the equivalent series capacitor of the dielectric, *not* of the void itself.

This situation cannot define the end of the discharge though since it would be in *disagreement* with experimental data in which it is observed that the void voltage drops *below* the inception voltage to an extinction voltage, V_e, before stopping. The space charge distribution at this instant is shown in Fig. 13.3*b*. In this we have assumed: (i) that the mobility of the electrons is much greater than that of the ions so that they have been swept to the anode but the ions remain in the gap more-or-less in the locations where they were created; and (ii) that, since the number density of avalanching collisions doubles every distance α^{-1} as the anode is approached, the average position of the ions is approximately one ionising collision length, α^{-1}, from the anode. Although the avalanche formation has ceased at this point, the discharge will continue, or at least the manifestation of the discharge as a flow of current through the void will continue, as the positive ions drift to the cathode. Since they must drift further than the electrons the *apparent* charge contributed by the positive ions is greater even though their *actual* charge is equal to that of the electrons. This can be seen as follows.

The average distance moved by the electrons is $1/\alpha$ with a drift velocity, $v_{dn} = E\mu_n$ (μ_n is the mobility of the electrons) so that they take an average time $\tau_n = (1/\alpha)/(E\mu_n)$ assuming a reasonably constant field, E. Since the magnitude of the electron current, I_n is proportional to $v_{dn} \propto \mu_n E$, the observed electronic charge, $Q_n = I_n\tau_n \propto 1/\alpha$. Similarly the average distance moved by the positive ions is $d - 1/\alpha$ with a drift velocity, $v_{dp} = E\mu_p$ so that they take an average time $\tau_p = (d - 1/\alpha)/(E\mu_p)$. Since the magnitude of the ion current, $I_p \propto v_{dp} \propto \mu_p E$, the observed charge due to the movement of positive ions is $Q_p = I_p\tau_p \propto (d - 1/\alpha)$ so that for equal absolute numbers of positive and negative charges:

$$Q_p = (\alpha d - 1)Q_n \qquad (13.14b)$$

The total charge observed is therefore:

$$Q = Q_n + Q_p = \alpha d \, \Delta V \, C_s \qquad (13.15)$$

Although the above physical picture is obviously only an approximation Devins[381] has given a formal proof for this expression taking into account the spatial distributions of the charges. He has also verified it for a range of overvoltages in which he found for his rather artificial experimental arrangement that αd was approximately 9, and independent of d. Since most overvoltages are small in comparison with V_i, the primary ionisation coefficient α is reasonably independent of ΔV although for very small values of ΔV ($< \sim 0 \cdot 1\% \, V_i$) the avalanches may be very slow to build up and the value of α may drop. This will give a straight line with an apparent intercept on the ΔV axis if Q is plotted as a function of ΔV. For large overvoltages it may not be possible to ignore the effect of the voltage and Tanaka[380] has shown experimentally that the relation between the apparent charge and the overvoltage is of the form:

$$Q \propto \frac{C_s}{pd} \Delta V^n \qquad (13.16)$$

where $1 \cdot 5 < n < 3$ depending on the gas and humidity. Generally the square law ($n = 2$) appears to be followed by dry gases but n approaches 3 for humid gases.

The above analysis reveals an interesting relation between the overvoltage and the extinction voltage. Let us call the difference between the inception voltage and extinction voltage the undervoltage (this is a convenient term but not generally used). Since the electronic charge causes the voltage to drop by the overvoltage, the ionic charge must cause the further drop in the voltage by the undervoltage. We would therefore expect, for Townsend discharges:

$$\frac{\text{undervoltage}}{\text{overvoltage}} = \frac{Q_p}{Q_n} = \alpha d - 1$$

or

$$V_i = V_e + (\alpha d - 1) \Delta V \tag{3.17}$$

Eqn. 13.15 also indicates that the discharge magnitude will be dependent upon the rate of voltage rise. If we assume that the void voltage increases reasonably linearly through the inception voltage at a rate $\dot{V}$ then the overvoltage will, on average, be reached after the statistical time lag, τ_s, when the overvoltage is $\tau_s \dot{V}$. Thus:

$$Q = \alpha d |\dot{V}| \tau_s C_s \tag{13.18}$$

In the DC case the voltage is increasing at a rate of $(V_a - V_i)/(R_s C_v)$ as it passes through the inception voltage so that

$$Q(\text{DC}) = \alpha d (V_a - V_i) C_s \frac{\tau_s}{R_s C_v} \tag{13.19}$$

In the AC case $\dot{V}$ is a maximum as the applied voltage passes through zero and it can be seen from Fig. 13.2b that the region of maximum discharge activity occurs close to this point.

13.1.6 A criterion for Townsend discharge extinction and the transition to streamer formation

As the discharge progresses an increasing amount of negative charge is deposited on the anodic surface and an increasing amount of positive space charge is built up in front of this surface (Fig. 13.3b). This space charge 'double layer' opposes the applied field thereby reducing the field (E_2 in Fig. 13.3b) within it. As the avalanche discharging progresses E_2 may tend to zero so that all the void voltage appears across a smaller 'effective' width, $d - 1/\alpha$ and avalanching activity is greatly reduced in the region near the anode. This implies a movement of the 'operating point' on the Paschen voltage curve (Fig. 13.5i) to the left, since the voltage is reasonably constant but the effective gap is reduced. If this results in the operating point dropping below the Paschen curve then discharging may cease, whereas if the operating point remains above it, discharging is likely to continue. If the operating point is to the left of the minimum in this curve, (which is at $d \sim 7 \ \mu$m at atmospheric pressure), we would expect to a first approximation the conditions to be unfavourable for discharging to continue whereas if the $p \times d$ product is greater than this the discharge is likely to intensify. However, in the latter case, there is also an 'opposing force' which will tend to extinguish the discharge. This is because, as the discharge intensifies and current starts to flow through the gap, charge is deposited on the anode and the voltage across the gap as a whole is reduced. It is therefore possible to work to the right of the Paschen voltage curve minimum without the Townsend-like discharge developing further. This latter case does not therefore necessarily imply that the Townsend discharge will change to a streamer, however, the continuation of the discharge is obviously a prerequisite for the streamer to develop.

The physical situation is in fact much more complicated than this and the discharge is difficult to model accurately. The reasons for this include: the space charge density is not discontinuous so the field does not change abruptly, negative space charge around the cathode exists in attaching gases such as air, the values of α and γ depend on field and γ may also depend on the size and shape of the gap whose dimensions are effectively changing, and the discharge tends to become filamentary due to space-charge induced field distortion in the plane orthogonal to current flow which also gives rise to local potential changes on the cathode and anode surfaces. In spite of these simplifications it is found that below a certain gap size the discharge is unlikely to develop from the Townsend discharge into a streamer. Devins[381] for example, in his artificial experimental arrangement, found that even with considerable overvoltages, streamers were not formed below a gap size of 100 μm in air. Since much more damage is likely to occur from the more energetic streamers than from the Townsend discharges, these considerations suggest that voids which are sufficiently large, perhaps more than a few microns (corresponding to the Paschen minimum), are likely to be particularly dangerous.

Devins[381] has shown, by considering the space-charge induced field distortion in the direction of the current, that the size of consecutive Townsend avalanches may increase or decrease depending on prevailing conditions (particularly the overvoltage). By deriving an expression for $dS(n)/dn$, where $S(n)$ is the number of generations in the nth avalanche, he uses $dS/dn \geqq 0$ as the criteria for streamer development. His analysis, which is too long to reproduce here, shows that the minimum overvoltage required for streamer development, $\Delta V_{\min}$, is a function of the void inception voltage, V_i, the void size and the ratio of the effective series capacitance (C_s in Fig. 13.1) and the void capacitance. Results from his experiments using an air gap between a metal cathode and glass anode are shown in Fig. 13.6.

13.1.7 The nature of the streamer discharge

It is found that if the overvoltage is increased sufficiently and the void is not too small the discharge current, although starting out like a Townsend discharge, suddenly increases by at least an order of magnitude as streamer discharge is initiated[381,289]. This occurs at about the stage of the discharge that would correspond to the decreases in electronic current in a pure Townsend discharge (see Fig. 13.4i). Having reached a maximum the current quickly decays so that the overall discharge time is comparable to

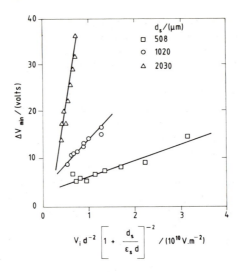

Fig. 13.6 Variation of the critical transition overvoltage, $\Delta V_{\min}$, for streamer formation as a function of air gap inception voltage and size (V_i and d) and series dielectric thickness, d_s, and relative permittivity, ε_s. After Devins[381] Copyright © 1984 The IEEE

the Townsend discharge. It would appear that the size of consecutive avalanches suddenly starts to increase rapidly because of the field distortion produced by the build up of positive space charge throughout the void. As the avalanching accelerates, higher energy photons may cause ionisations in the gas which precipitate avalances from within the gap, in addition to those from the cathode. The surface of the anode may become more conductive in areas of particularly intense activity due to a combination of physical degradation and injection of hot electrons. This, coupled with the positive space charge, tends to further intensify the field in local areas. Avalanches, both from the cathode and those produced from within the gas, will tend to favour these localities and the discharge will break up into filaments. This has been observed using light pattern recordings (Lichtenberg figures) by various workers including Devins[381], Bezborod *et al.*[664] and Morshuis and Kreuger[665]. The positive space charge will drift towards the cathode causing further field intensification as it does so. This is the return stroke of the streamer and upon reaching the cathode a plasma of electrons and positive ions exists in filaments bridging the void. Once such a streamer has been established the effective resistivity of the void gas is low. Charge rapidly builds up on the dielectric surface increasing the voltage across the series capacitance and hence reducing the voltage across the void. This reducing void resistance and decreasing void voltage coupled with the increase in surface conductivity quickly extinguishes the discharge.

The dependencies of the streamer discharge magnitude (Q) on the overvoltage and the void size are markedly different to the Townsend discharge[381]. The Townsend discharge magnitude is proportional to the overvoltage (eqn. 13.15) whilst this dependence is much less strong in the case of streamers. This is because Townsend avalanching stops when sufficient charge has been dumped on the anode surface to overcome the overvoltage, this is independent of the void size. In the streamer case however avalanching continues because of the very high field enhancements at the ends of the positive streamers. Whilst the Townsend discharge magnitude was independent of void size, that for streamers increases with void 'gap' since the charge is now contained in the streamers whose size is dependent on the void size. The much larger discharge magnitude of streamers implies that they are much more damaging and Robinson[666] has given some evidence for a relation between streamer discharge magnitude and lifetime:

$$\text{life} \propto (\text{maximum streamer discharge magnitude})^{-1.4} \qquad (13.20a)$$

Various results[667–671] on time-to-failure as a function of size of stress (well above that for inception) suggest a relation of the form:

$$\text{life} \propto E^{-n} \qquad (13.20b)$$

with $n \simeq 6\text{–}8$.

The discharge magnitudes of Townsend streamers are distributed exponentially because $Q \propto \Delta V \propto \text{time lag}$ (eqn. 13.2) which is randomly distributed. Robinson[666] has shown that this is not the case for streamers under AC conditions which appear to be Weibully distributed (see Chapter 14) with a shape parameter of about 7, i.e. a peaked distribution. This may be because streamers tend to discharge the void to well below V_e, a complete discharge is sometimes observed[672], and so the time to discharge is determined more by the time for the voltage to be restored to V_i than the statistical time lag. Since the former process is deterministic the discharge magnitudes will be centred round a maximum.

13.1.8 Swarming micro-partial discharge

After aging with partial discharges the discharge activity is often noted to almost disappear prior to breakdown (see Section 19.2). The Japan National CIGRE Task Force 15-06-01 have investigated[380] this and found by careful sensing of the partial discharge by electrical and optical means that the discharge activity consists of a large number of tiny discharges (<1 pC) which cannot be easily detected with conventional techniques. It is thought[380] that changes in the void gas chemistry, perhaps due to the increase of moisture, and perhaps the more conductive nature of the end walls, give rise to a plentiful supply of electrons for initiating small local avalanches while at the same time inhibiting streamer formation. Small overvoltages and undervoltages are observed which would be in accord with this hypothesis.

13.2. Partial discharge degradation

Serious degradation due to partial discharges appears to occur upon the formation of streamers, Townsend discharges do not appear to cause severe damage in polymers. Various routes to degradation from partial discharges can be identified: chemical reactions between excited molecules (particularly oxygen) and the void surface; chemical reactions between metastable molecules (e.g. ozone) and both the surface of the void and the volume of the dielectric into which they may diffuse; bombardment of the surface by high energy ions and photons; and high energy gas production leading to localised heating, melting and degradation. Fracture seems unlikely to occur[382] as the mechanical shock wave is only likely to have energies of the order of 10^{-12} J although Mason has suggested that ozone may be capable of initiating cracks if the material is mechanically strained[383]. The surfaces of the void will eventually become pitted with craters which, due to the field enhancement at their ends, may form the stems of electrical trees which propagate across the dielectric and cause breakdown.

Thus partial discharge degradation processes fall broadly into two categories: those initiated by ion bombardment or by chemical reactions. In air, nitrogen is inert and causes damage by bombardment whereas oxygen is active and causes damage by both mechanisms. A brief summary of these mechanisms are given here.

13.2.1 Ion bombardment

Discharges in nitrogen causes craters or pits[380,407] which appear to be the result of ion bombardment. Although individual ions may not necessarily have energies sufficient to break bonds, the concentrated bombardment from a streamer on a small area, coupled with a local temperature rise[383] which may be several 100 K, is sufficient to cause this effect. Mason[383] estimated, in his work, that erosion proceeded at a rate of approximately 10^{-21} m^3 per discharge, i.e. the volume of a cube of side $\sim 0 \cdot 1$ μm. Mayoux[123] has shown that, at least in polyethylene, degradation only progresses reasonably fast if the charge density of ions exceeds about $1 \cdot 5 \times 10^2$ C $\cdot$ m^{-2}. He also showed that degradation due to electron bombardment was usually insignificant, as electrons required an energy greater than about 500 eV to cause substantial damage. Mayoux further suggested that the results of ionic impact on polyethylene could only account for a part of the partial discharge degradation and that chemical degradation is also necessary. Epoxy resin appears to be more resistant to crater formation than polyethylene[383].

Whilst ion bombardment is unlikely to initiate significant damage, it may play a significant role in worsening existing damage. For example cracks may have been initiated by chemical attack or mechanical strain in which case energies of only a few electron volts are required to deepen the crack[145]. This may lead to a crack runaway and mechanical failure[144] or the formation of electrical trees. The presence of cracks at the sharp corners of the artificial disk-shaped voids used in many laboratory experiments has led to the common observation of trees initiating from these edges[125,459].

13.2.2 Chemical attack

The high energies available within the discharge allow the creation of active oxygen species (O, O_3, O_2^-) and the catalysation of endothermic reactions.

These tend to give a reasonably uniform degradation over the surface of the void since the reactive molecules rapidly diffuse throughout the void in contrast to the localised pitting from ion bombardment. The exact form of degradation depends on the dielectric material. For example polyethylene gives reasonably uniform erosion[380,407] due to oxidation from activated molecules, PVC carbonises, PMMA gives gaseous decomposition products producing carbon deposits[673], and mica cracks since metal ions are removed from the surface which results in the collapse of the crystallographic structure. The effect of the different constituents of air in polyethylene has been the subject of particular study because of its importance in power cable manufacture[123,124]. The most important factor is the presence of oxygen since without this constituent the only possible reactions involve the breaking of C—H or C—C bonds to produce free radicals, double bonds C=C, crosslinks or cyclisation. With oxygen however, provided there is sufficient energy for the first of the following reactions, OH groups may be formed through the following reactions:

$$RH \rightarrow R^{\cdot} + H^{\cdot}$$

$$R^{\cdot} + O_2 \rightarrow ROO^{\cdot}$$

$$ROO^{\cdot} + RH \rightarrow ROOH + R^{\cdot}$$

$$ROOH \rightarrow RO^{\cdot} + {}^{\cdot}OH$$

This chain reaction will stop if all the energy has been consumed or an antioxidant reaction occurs terminating the active chain radical $R^{\cdot}$, e.g.:

$$ROO^{\cdot} + HA \rightarrow ROOH + A^{\cdot}$$

$$R^{\cdot} + A^{\cdot} \rightarrow RA$$

In air discharges in crosslinked polyethylene the experimental work of Gamez-Garcia *et al.*[407] has shown that oxalic acid may be formed due to the moisture, carbon monoxide/dioxide or oxygen and acetophenone content. Since this forms crystals on the void surface which are semiconducting, the void voltage is reduced and the discharge extinguished. This may be why Wojtas[674] has found that cable material resistivity decreased with aging above a threshold voltage. Increase of wall conductivity is in agreement with the theory of Mason[658] and Densley and Savage[675] and experimental results of dielectric measurements of Rogers[676] and Nissen and Röhl[677]. Oxidisation of the surfaces may play a similar role[125] although diffusion of charge through the dielectric may once again increase the void voltage to above the inception value[678]. Shahin[679] has shown that discharges in air may also produce oxides of nitrogen and carbonates which may lead to long-term degradation. Lu Zibin *et al.*[680] have investigated the rate of pit propagation by streamers in various polymers (polyethylene, polypropylene, and polyethylene terephthalate). The detailed chemistry of plasma reactions is beyond the scope of this book and readers are referred to other books on the subject e.g. References 681–683.

As air discharges continue the walls become oxidised and pitted. At first the oxygen-related damage is prevalent causing the walls to become oxidised and the void to enlarge. Since the streamers inject charge locally and produce locally oxidised and therefore semiconductive patches, the streamers too become more localised since they subsequently avoid these damaged areas. Local degradation may therefore be quite intense. As the oxygen is consumed, the damage becomes mainly due to nitrogen ion bombardment which causes the damaged surfaces to become pitted[126,380,684] and swarming micro-partial discharge activity to become prevalent.

13.2.3 Electrical tree initiation by partial discharges

Bahder *et al.*[120] have proposed a physical model for the development of electrical trees in polymeric insulation from discharging voids. We present and develop this model in this section.

The model of Bahder *et al.*[120] is based on the injection of charge from a void into the polymer. Since fluid permeation is possible through polymers, Bahder *et al.*[120] assume that there is a network of minute channels throughout the polymer connecting voids and microvoids. This is consistent with the imperfect nature of polymer chain packing leading to the existence of free volume in such materials (see Chapter 1). The model assumes that space charge, in the form of ionised gas molecules, may be injected from discharging streamers into such microchannels and that this charge will propagate through the branched microchannel network into a tree-like (fractal) structure. At the tips of the 'charge branches' the field is greatly enhanced and this field enhancement causes the charge to continue to move. As the charge continues to split into different branches its density reduces and the field at the branch tips becomes less intense. It is assumed, as a criterion for breakdown, that, for applied fields above the partial discharge inception threshold but below the breakdown strength, the local field at the branch tips eventually drops to below that required for further propagation. For densely-branching structures (this will depend on the polymer material) the charge will diminish more rapidly with the length of the tree-structure and branches which are physically close will tend to provide mutual shielding from field intensification. At this stage the structure is not an active electrical tree since discharging is not likely to occur within the new passivated structure; it is simply a tree shaped, fractal structure containing relatively mobile ions and electrons.

The greatest charge movement is obviously through the 'trunk' of the tree structure and it is reasonable to assume that most damage will occur in this region. The model considers that the diameter of this region is enlarged by such damage until discharging is possible. This presumably corresponds to the 'swarming micro-partial discharging' described by Tanaka[380] in which discharging takes place in the many pits or craters formed. The model of Bahder *et al.*[120] develops this by considering the formation of a tube-shaped crater (i.e. a one-dimensional filament) and derives an equation for time-to-breakdown based on the criterion mentioned above. We will first describe this model and then develop it for the more general case of a tree-like structure rather than a simple filamentary structure.

Both models show some interesting and realistic features. For fast rates of voltage increase (e.g. impulse conditions) one would expect a non-steady-state condition to occur which gives rise to a lower breakdown strength. In this case charge moving via 'easier' routes through the free volume network (such routes might have a larger diameter or a more favourable shape or orientation with respect to the applied field) will move much quicker than charge through 'harder' routes. The field intensification at the charge tip will be greater both because there will be less splitting of charge into side branches and because the mutual shielding effect will be reduced. Thus the charge will continue to move even for lower applied fields. This correlates with the observation of lower breakdown strengths under impulse than slowly-increasing applied stresses. The models will be seen to indicate the existence of a threshold field below which aging does not occur (or at least below which the rate of aging is significantly reduced). Such a threshold voltage is generally thought to exist (e.g. References 481, 685–688), although little uncontroversial data exists in support of this (see Chapter 14 for a fuller discussion). The models predict that above the threshold voltage there is an inverse power-law relationship between the time-to-breakdown and applied voltage which is commonly observed (see Chapter 14). (Bahder *et al.*[120] actually subdivide this into two further regions but for many practical purposes these will approximate to a straight line on a bi-logarithmic plot of voltage versus time-to-breakdown.) The predicted correlation between AC and DC behaviour is also observed.

Bahder *et al.*[120] assume that an avalanche will travel a length proportional to $E_a - E_{th}$, where E_a is the applied field and E_{th} a threshold field, before being extinguished by the counter field, energy losses, charge dissipation etc. This is reasonable in terms of the Townsend theory[381] and is in agreement with the expression derived for the length of the first channel of an electrical tree derived in Section 5.3.1:

$$L_m = \frac{E_a - E_{th}}{b} \tag{5.6}$$

From eqn. 13.5, the positive charge generated in an avalanche over a length, x, is given by:

$$Q_p \propto \exp(\alpha x) - 1$$

so that with $x \propto E_a - E_{th}$ we have:

$$Q_p \propto \exp\{k_1(E_a - E_{th})\} - 1 \tag{13.21}$$

In the Bahder *et al.*[120] model it is assumed that the movement of this charge causes chain scission such that the length of the tubular crater created is proportional to the charge; i.e. $L \propto Q_p$. Although the formation of such a one-dimensional crater is not very realistic, we will pursue their model and later refine it in terms of a fractal tree shape. Under an AC stress, the number of current pulses is proportional to frequency×time, $f \cdot t$, so that

$$L(t) \propto Q(t) \propto ft[\exp\{k_1(E_a - E_{th})\} - 1] \tag{13.22}$$

Bahder *et al.*[120] then deduce an expression for the critical length L_c in terms of the applied field at breakdown, E_b, for which the applied field is enhanced such that it exceeds the local breakdown strength of the material. They then take breakdown to occur at the time for which $L(t)$ in eqn. 13.22 equals L_c under a given field, E_b. The equation they derive for the critical length is of the form:

$$L_c = \frac{1}{k_2 + k_3 \exp\{k_4 E_b\}} \tag{13.23}$$

Although not explicitly stated, Bahder *et al.*[120] appear to have assumed (i) that the channel is effectively conducting when a discharge takes place; and (ii) that the field enhancement is similar to that obtained for a conducting hyperboloid of revolution for which the maximum field is given by

$$E_{max} = \frac{2V}{r \log_e\left\{1 + \dfrac{4d}{r}\right\}}$$

$$= \frac{2V}{D}\left(\frac{D}{r}\right) \frac{1}{\log_e\left\{1 + \dfrac{4(D-L)}{r}\right\}}$$

where V is the applied voltage, r is the radius of the hyperboloid, d is the distance between the hyperboloid tip and the counter electrode and D is the length between the void surface and the counter electrode (i.e $L + d = D$). Breakdown occurs at $E_{max} = E_c$ (a system constant) and for applied field $E_b = V/D$. Assuming r does not change as the channel length increases then:

$$\log_e\left\{1 + \frac{4(D-L_c)}{r}\right\} = \frac{2E_b}{E_c} \frac{D}{r} = k_4 E_b$$

where k_4 is defined as $2D/(rE_c)$. Taking exponentials of both sides and rearranging gives:

$$L_c = \frac{r}{4} + D - \frac{r}{4} \exp\{k_4 E_b\}$$

which defines L_c as the length at which an applied field E_b is enhanced sufficiently to cause breakdown. Making the substitution $k_2 = 1/[D + (r/4)]$ gives:

$$L_c = k_2^{-1} - \frac{r}{4} \exp\{k_4 E_b\}$$

$$= k_2^{-1}\left[1 - \frac{rk_2}{4} \exp\{k_4 E_b\}\right]$$

Since $k_2 \simeq D^{-1}$; $rk_2 \exp\{k_4 E_b\} < 1$; and $L_c > 1$:

$$L_c \simeq \frac{k_2^{-1}}{1 + \dfrac{rk_2}{4} \exp\{k_4 E_b\}}$$

which, upon substituting $k_3 = rk_2^2/4 \simeq r/(4D^2)$, gives Bahder *et al.*'s expression for the critical length, eqn. 13.23.

By equating the length from eqn. 13.22 to L_c and rearranging to give time to breakdown $t = t_b$, Bahder et al.[120] obtained:

$$t_b(E_b) \propto \frac{1}{f[\exp\{k_1(E_b - E_{th})\} - 1][k_2 + k_3 \exp\{k_4 E_b\}]} \quad (13.24)$$

for a one-dimensional channel.

If we now consider a fractal model of the channel system (see Section 5.4) then we would expect the charge to be proportional to the volume of the tree which itself is proportional to the sum of the arc-lengths, $S(t)$, of each branch (S_i) (eqn. 5.8):

$$Q(t) \propto ft[\exp\{k_1(E_a - E_{th})\} - 1]$$

$$\propto S(t) = \sum_i S_i \propto L(t)^{d_f} \quad (13.25)$$

This gives the observed frequency dependence of electrical trees ($L \propto (ft)^{1/d_f}$, Section 5.4.2) and mirrors their stochastic development via repetitive discharges which can be initiated at different places in the trees, not only in the main trunk.

The voltage dependence may also be considered. For a truly vented electrical tree (i.e. one in which there are no complications due to gas pressure rises) it is found experimentally that

$$L(t) \propto V^4 (ft)^{1/d_f} \quad (13.26)$$

and from the above model:

$$L(t) \propto (ft)^{1/d_f}[\exp\{k_1(E_a - E_{th})\} - 1]^{1/d_f} \quad (13.27)$$

Using typical values of $k_1^{-1} = 10^6\ \text{V} \cdot \text{m}^{-1}$ and $E_{th} = 3 \cdot 5 \times 10^6\ \text{V} \cdot \text{m}^{-1}$ with $d_f = 1 \cdot 7$ (cf. Fig. 5.9b) and plotting $[\exp\{k_1(E_a - E_{th})\} - 1]^{1/d_f}$ versus E_a over the range $4\text{—}12 \times 10^6\ \text{V m}^{-1}$ on a bi-logarithmic plot, Fig. 13.7, gives a reasonable straight line with a slope of $4 \cdot 5$ as expected from a power law with an exponent of $4 \cdot 5$. This suggests that the V^4 in eqn. 13.26 may be an approximation to the more general expression in 13.27. This latter expression suggests that a plot of $\log\{L(t)\}$ versus E should be a straight line with a slope of $k_1 \log_{10} e / d_f$. In order to test this the data of length versus voltage given in Fig. 5.4a (in which electrical trees were grown from needles with recess voids; this is a reasonable analogue to a partial discharge induced tree and avoids the complications of pressure changes) has been re-plotted on log-linear scales in Fig. 13.8. This graph shows good straight lines which are all parallel, i.e. the gradient is independent of frequency. The gradients are approximately $1/(6\ \text{kV})$ which, assuming $d_f = 1 \cdot 7$, suggests a value of $k_1 \approx 0 \cdot 65\ \text{kV}^{-1}$ (the plot is for voltage not field). It is not possible to determine a threshold voltage because of the number of proportionality constants and limited range if voltages.

This model can also be adapted to give a deterministic breakdown time if a critical length is used, by again assuming that the 'tree' is a conductor

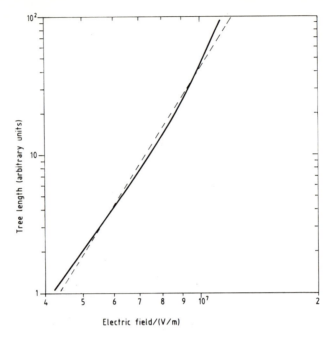

Fig. 13.7   Experimental observations of tree length as a function of electric field are a power law with an exponent of approximately four (eqn. 13.26). The model proposed predicts an exponential dependence (eqn. 13.27) which is plotted here as the continuous line. The dashed line shows that over the field ranges generally under consideration eqn. 13.27 closely approximates to the empirically observed power law

during the discharge. For fields well above the threshold value eqn. 13.26 can be rearranged to give t_b for $L = L_c$, i.e.:

$$t_b \propto \frac{[L_c(E)]^{d_f} V^{-4d_f}}{f} \qquad (13.28)$$

so that for typical values of d_f:

$$t_b \propto V^{-n} \qquad \sim 7 < n < \sim 10 \qquad (13.29)$$

which is reasonable (see Chapter 14). At low fields there will be a cut-off at E_{th} so that:

$$t_b \propto \frac{[L_c(E)]^{d_f}}{f[\exp\{k_1(E - E_{th})\} - 1]} \qquad E \geq E_{th} \qquad (13.30)$$

and at $E = E_{th}$; $t_b \to \infty$. In partial-discharge produced electrical trees from voids, E_{th} is the threshold for the void to discharge. In metal-needle produced electrical trees there is a threshold for initiation ($\sim 4 \times 10^8$ V $\cdot$ m^{-1} r.m.s.) but in this case it is a propagation rather than an inception criterion that is required. Thus here E_{th} may be that of a discharge in the gas along a typical length of a channel (perhaps about 10 μm).

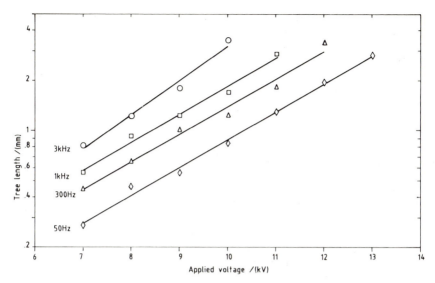

Fig. 13.8 The data of Fig. 5.4*a* for electrical trees grown from needle electrodes but plotted on log-linear axes. The straight parallel lines indicate support for the proposed exponential model with little dependence of the exponent upon frequency

13.3 Free volume breakdown

In 1980 Nelson and Sabuni[477], noted that the breakdown strength of many polymers dropped rapidly above a critical temperature which tended to follow the glass transition temperature. This was not a new observation in itself, e.g. Reference 689 but they discussed breakdown in polymers in terms of their free volume and their cohesive energy density. In the free volume approach they related the electric strength, E_b, to a threshold kinetic energy, W_{th}, of the conduction electrons after their being accelerated through a free volume of average length in the field direction, l_E. They use theory developed by Artbauer[690,691] who had attempted to relate electronic breakdown strength to molecular relaxation processes which could be used to describe the temporally and spatially fluctuating nature of the free volume by:

$$E_b = \frac{W_{th}}{el_E} \tag{13.31}$$

Using this simple (one-electron) approach and by relating the length, l_E, to the specific free volume using WLF theory[692] they predicted a form for the decrease in breakdown strength with increasing temperature above the glass transition temperature which correlated with their experiments on atactic polystyrene. Park *et al.*[693] attributed the increase in breakdown strength with pressure in PET as due to the decrease in free volume. At higher stresses the breakdown strength decreased again and this was attributed to the formation of cracks of the size of the free volume in the amorphous regions. Nelson and Sabuni[477] also noted a correlation between the breakdown strength of many polymers and their cohesive energy density, CED; this is shown in Fig. 13.9. The cohesive energy density is defined by:

$$\mathrm{CED} = \frac{\Delta H_{vap} - RT}{\bar{V}} \qquad (13.32)$$

where ΔH_{vap} is the heat of vapourisation, R the gas constant, and $\bar{V}$ the molar volume. Thus the CED is mainly a measure of the binding forces between different molecular chains although in polymeric materials ΔH_{vap} may also contain a contribution from molecular chain degradation. Note that in Fig. 13.9 the polar polymers' chains are held together by dipolar interactions and also van der Waals' attraction and therefore have higher CEDs and breakdown strengths. Since the temperature characteristics of the CEDs are also different for polar and non-polar polymers one would expect the breakdown strength dependencies on temperature to be different in these two cases; this can be seen in Fig. III.3. More recently Crine and Vijh[694] have demonstrated the correlation between CED and breakdown strength for LDPE by varying the CED by means of changing the temperature, Fig. 13.10.

The dependencies of breakdown strength on both the free volume and the cohesive energy density suggest that the breakdown mechanism involves structural deformation and possibly some chain scission. The Artbauer[690,691] theory used by Nelson and Sabuni[477], eqn. 13.31, suggests that electrons are accelerated to kinetic energies at which impact with molecules is likely to cause chain scission. This would appear to require rather long distances for the acceleration since typical energies for scission are 4–5 eV (Section

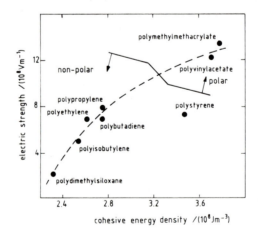

Fig. 13.9 Nelson and Sabuni[477] noted the indicated correlation between breakdown strength and CED for a variety of polymers
Copyright © 1980 The IEEE

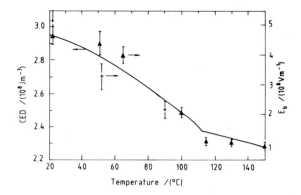

Fig. 13.10 Crine and Vijh[356] have observed a reasonable correlation between CED and breakdown strength of LDPE by varying the temperature. The continuous line defines the variation of CED and the symbols that of the breakdown strength Copyright © 1988 The IEEE

2.2.1) and even at fields of 10^8 V · m^{-1} this implies distances of 40 nm. Although microvoids may be this large, typical free-volume lengths between amorphous chains are about a tenth of this value (e.g. Takahashi *et al.*[695]). Nelson and Sabuni[477] measured the average molecular inter-chain distance as only ~1 nm in atactic (i.e. amorphous) polystyrene although it must be admitted that greater distances than this should be expected as this value is only the mean of the distribution. However it does seem unlikely that electrons could gain sufficient energy within the free volume regions to cause bond scission on exit, though it should be noted that they may travel much further before making a collision with a polymer molecule.

Crine and Vijh[93,356] have noted a linear relation between the dark current activation energy of various polymers and their CEDs; by relating this activation energy to W_{th} (eqn. 13.31) as a breakdown criterion they have evaluated the corresponding length, l_E ($= W_{th}/(eE_b)$), as a function of temperature. Below the glass transition temperature, T_g, they found an almost constant $l_E \approx 0.6 - 0.8$ nm for all polymers[694] whilst above T_g larger values of about 3 nm were obtained. For example in LDPE $l_E \approx 5$ nm between T_g and the melting point; above the melting point l_E increased markedly to about 20 nm and a significant drop in the breakdown strength was noted. These values for l_E suggest that the accelerated charges deform the structure locally, thereby extending the free volume until bond scission can occur.

Crine and Vijh[356] have also pointed out that the amount of charge which can be injected under identical conditions is in inverse relation to the polymer's CED as is the electroluminescence brightness. The results of Sone *et al.*[696] indirectly support this. Using polyethylene with different free volume contents they measured the impulse strength of a point-plane system under both polarities for 'slow' and 'fast' voltage rise rates. (A 'slow' voltage rise rate in the context of impulse testing means that many impulse tests are carried out at the same voltage level (or slowly increasing voltage levels) before proceeding to the next, slightly higher, voltage level.) They claim their samples had very high free volume contents (5—10%). One would expect homo-space charge to be injected from the needle under a negative polarity and for this to eventually form a field-moderating cloud around the needle. They found that the breakdown voltage increased with increasing free volume under slowly-rising negative-polarity needle conditions, was reasonably constant under fast-rising negative-polarity needle conditions, and had a slight tendency to decrease under both fast and slowly-rising positive-polarity needle conditions. In the latter case (positive polarities) they observed electrical tree formation which was assumed to be caused by partial discharging in voids associated with the free volume — a high free volume content presumably leading to a higher rate of degradation this corresponding to the high electroluminescence referred to by Crine and Vijh[356].

At negative polarities electron space charge is likely to be injected from the needle. At low rates of voltage increase this has a chance to form the field-moderating space-charge cloud whereas at high rates of voltage increase the space charge would remain close to the needle. This may explain the increase of breakdown strength at slowly-increasing voltages whilst at fast rates of increase the space-charge field-moderating effect is in competition with partial discharge formation.

The exact mechanism of breakdown is not clear in any of the cases cited since the free volume size would generally be too small for electron energies to be of sufficient magnitude to cause avalanching and partial discharge formation (this corresponds to the extreme left of the Paschen curve where the breakdown voltage is very high). Kao[697], however, has proposed a theory in which bond dissociation takes place near the electrodes even in low-mobility materials in which the mean free path length is small. This leads to the formation of low density (and thus low CED) regions in which impact ionisation is more likely, leading to avalances, partial discharges, electrical treeing and eventual breakdown.

In the Kao[697] model the situations of both positive and negative needle electrodes are considered. In the negative needle case it is considered that electrons are injected by Fowler–Nordheim tunnelling (Section 9.2.2) into the conduction band but quickly become trapped in states in the mobility gap after only a few scattering collisions, because of the small mean free path length and the large localised gap state concentration. However, in making the transition to lower energy states during trapping or recombination, energy typically of the order of a few electron-volts must be released. Kao[697] suggested that this is unlikely to be emitted as photons in amorphous materials and yet phonon energies are typically only a few hundredths of an electron-volt and thus simultaneous multiple phonon emission is also unlikely. It was thus proposed that this energy might be used to promote an electron with an energy close to the conduction band edge to a higher energy level within the conduction band via an Auger-type process; this would give it kinetic energy thereby making it 'hot'. However luminescence in the visible region ($1 \cdot 75$—$3 \cdot 0$ eV) has been observed to accompany charge injection into polyethylene[336,343] and epoxy resins[510], and therefore it seems probable that only a small proportion (<1%) of the electrons follow the above pathway. Nonetheless these hot electrons may then have sufficient energy to collide with a molecule and dissociate it into free radicals:

$$AB + e^- \text{ (hot)} \rightarrow A^{\bullet} + B^{\bullet} + e^- \text{ (cold)}$$

or

$$\rightarrow A^{\bullet} + B^{\bullet} + e^- \text{ (trapped)} + \text{energy release} \qquad (13.33)$$

This process is continuous and implies that a low-density region is formed around the needle. However with a negative needle space charge is quickly injected moderating the local electric field and reducing the probability of electron injection[652]. Such damage has been observed by Cartier and Pfluger[70,68] in thin films of n-$C_{36}H_{74}$ irradiated with hot electrons above a threshold energy of 3·5 to 4·0 eV. Theoretical studies of hot-electron transport in amorphous materials carried out by Sano[355,698] have shown that injected electrons may reach such energies temporarily.

In the case of a positive needle, holes may be emitted into traps in the gap thereby vacating these traps of electrons. An electron from a higher energy level within the gap may drop down to such a vacated level liberating energy to promote an electron from the valence band to an unoccupied gap state and creating a hole in the valence band. A recombination with this hole would release energy which may be used to raise an electron in the gap state to the conduction band. If this now drops back to a gap state it may, in a similar way as before, promote an electron via an Auger-type process to a higher energy level within the conduction band again making it 'hot'. These electrons are accelerated towards the point electrode and cause impact ionisation in the region. This process is likely to be more efficient since there will be less homo-space charge to moderate the field and electron impact ionisation is more efficient under a convergent than a divergent field.

An AC field is likely to be even more efficient in producing low-density (low CED) regions since trapped holes produced in gap states during the positive half cycle will be in collision with the injected electrons during the negative half cycle[697,353]. If the gap states are due to impurities then the energy from electron–ion recombination may be used to dissociate the impurity molecule into radicals:

$$AB^+ + e^- \rightarrow A^{\bullet} + B^{\bullet} \qquad (13.24)$$

The theories associated with free volume breakdown and the relation of such breakdown to the cohesive energy density of a polymer are still very under-developed. General trends have been observed and reasonable descriptive physical theories proposed but very little quantitative work has been carried out either in the development of theory or in comparison of such theory with experiment. The theory of thermal breakdown is well established and recognized to occur under specific conditions. Electro-mechanical breakdown has also been shown to occur albeit under restricted conditions. Theories of electronic breakdown in polymers seem reasonable but little direct uncontroversial evidence exists to support them. The models they are based on are also rather idealised particularly where they use the concept of an energy band gap and it is difficult to anticipate the effect of the very non-ideal band or mobility gaps of polymers. These 'band transport' theories really compete with 'free volume' theories but accurate comparison is difficult for the same reasons that have made proof of the electronic theories difficult, namely the poor understanding of the physical electronics at the molecular level and the effect of the distribution of the structural parameters.

PART 4

THE STOCHASTIC NATURE
OF BREAKDOWN

From the point of view of the individual life is a sequence of choices made for the best of reasons. To an external observer of the whole population the basis for each choice may be impossible to ascertain, and the individual's progress appears to be stochastic; i.e. it can only be described by a probability function.

Introduction

Many theoretical treatments of electrical breakdown in polymeric insulating systems are based on the deterministic models described in the previous part of the book. As well as being applied to 'ideal' insulating systems, they are also applied to systems with defects (such as voids in polymers) and to insulating systems which have been aged or degraded in ways such as those described in Part 2. Although these applications of deterministic models to non-ideal insulating systems give values of breakdown strength which are lower than the 'intrinsic' value otherwise obtained, they generally predict a *specific* breakdown strength (e.g. O'Dwyer[82]) which is not usually found. Such models suggest then that the insulating system will always breakdown when, and only when, a critical breakdown voltage is exceeded; if this critical value is not exceeded the system will last forever. When coupled with a deterministic degradation mechanism a *specific* time to breakdown may, at least in principle, be predicted given known voltage, environmental factors, and insulator history (e.g. Klein[146]).

Whilst such specific values of breakdown voltage may be obtained in some completely crystalline insulating systems and in some very well made amorphous systems, such as the capacitors in some MOS devices[699], this is not found to be the case in polymeric insulating systems. If apparently identical polymeric insulating systems are exposed to identical tests in which the voltage is linearly increased with time, a different breakdown voltage will be observed for each insulating specimen. The breakdown voltages so observed may be characterised by a statistical distribution which describes the *probability of breakdown* as a function of voltage; the parameters of this distribution are dependent upon the test conditions and the specimen construction. If a set of similar tests are carried out at a constant voltage (AC or DC) then the times from the start of the test to breakdown may be observed. In all types of insulating system no specific time to breakdown would be observed; rather each specimen would be found to breakdown at a different time. These times would be found to fit a similar statistical distribution giving the probability of failure as a function of time; the parameters of the distribution would be found to depend upon the value of voltage used and other test conditions including the specimen construction. For this reason the presentation of such data in most publications is usually in the graphical form 'cumulative probability of failure' or 'fraction of samples failed' as a function of 'time to breakdown' or 'breakdown voltage'.

This part introduces the use of the 'Weibull function'[700] in characterising the distribution of times and voltages at breakdown. The physical origins of this distribution are then discussed in Chapter 15 together with other so-called 'extreme value' distributions. In the final part of the book engineering applications of these distributions are considered and lifetime prediction in particular is investigated.

Chapter 14

Statistical features of breakdown

14.1 Statistical description of breakdown

14.1.1 The nomenclature of lifetime distributions

If a constant, AC or DC voltage were applied to an insulating system it would not be possible, even in principle, to predict how long it would be before electrical breakdown would take place; it may however be possible to predict the probability of failure as a function of time. ('Failure', the general term used in the statistical analysis of lifetime, implies electrical breakdown in this context.) An alternative equivalent statement of this principle is that, in testing a large set of identical insulating systems under constant voltage conditions, each specimen will breakdown at an unpredictable and different time; however it may be possible to estimate the fraction of failed samples at a given time within a known degree of confidence. The results of such a test (discussed in more detail later) are shown in Fig. 14.1a as a histogram. The height of each vertical bar represents the number of specimens (left-hand axis) which broke down *within* the time interval defined by the width of the bar. An alternative histogram-type representation of this data is given in Fig. 14.1b. Here the height of each bar represents the *total* number of specimens which have broken down from the time the test started to the end of the time interval defined by the horizontal position of the bar. This latter histogram (Fig. 14.1b) represents the *cumulative* number of specimens broken down as a function of time whereas the former (Fig. 14.1a) shows the number of specimens which broke down within successive time intervals. In either case, by knowing the total number of samples tested, it is possible to display the fraction of specimens failed (right-hand axis) instead of the absolute number.

Given an infinitely large set of specimens the fraction of samples failed corresponds to the probability of failure and the tops of the histogram bars become a smooth curve if the time intervals becomes infinitesimally small; these are shown in Figs. 14.1c and 14.1d corresponding respectively to Figs. 14.1a and 14.1b. These two curves represent the statistical functions known as the *probability density function* ($g(t)$) and the *cumulative probability of failure* ($P_F(t)$). (The equation describing the cumulative probability of failure is sometimes referred to as the *distribution function*.) The total area under the probability density function is unity indicating a probability of unity, i.e. a certainty, that breakdown will occur between times $t = 0$ and $t = \infty$. The area under this curve between two specific times represents the probability that an as yet unaged specimen will breakdown between those two times when tested. The cumulative probability of failure represents the probability

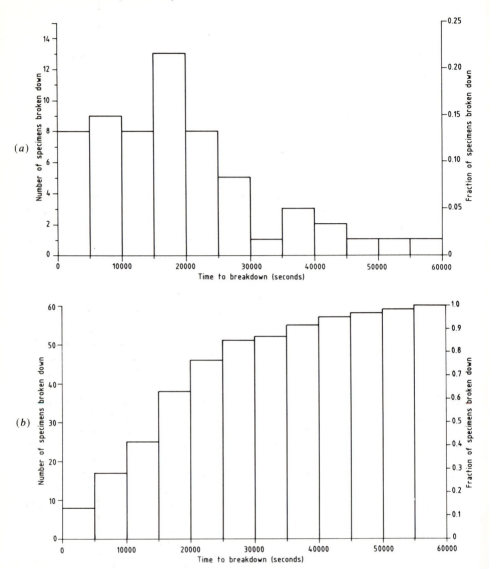

Fig. 14.1   Results of constant-stress tests carried out at 2 kV, 50 Hz by one of us (JCF) on sixty 1 metre wire samples coated with a blend of poly-phenylene oxide and polystyrene

(a)   Histogram representation. The height of each vertical bar represents the number of specimens (left-hand axis) or the proportion of the total number of specimens (right-hand axis) which broke down *within* the time interval defined by the width of the bar

(b)   Alternative histogram representation. The height of each bar represents the *total* number of specimens (or the total fraction of samples) which have broken down from the time the test started to the end of the time interval defined by the horizontal position of the bar, i.e. the *cumulative* number of specimens broken down as a function of time

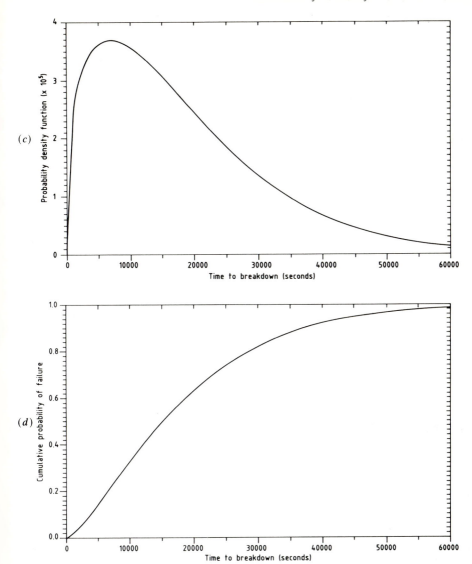

Fig. 14.1
(c)   The *probability density function* $(g(t))$
(d)   The *cumulative probability of failure* $(P_F(t))$

of failure of a new sample as a function of time from the beginning of the test, i.e. the probability that its time to breakdown, t_b, is less than or equal to time, t:

$$P_F(t) = \text{Prob}\,(t_b \leqq t) \tag{14.1}$$

The cumulative probability of failure is therefore the time integral from

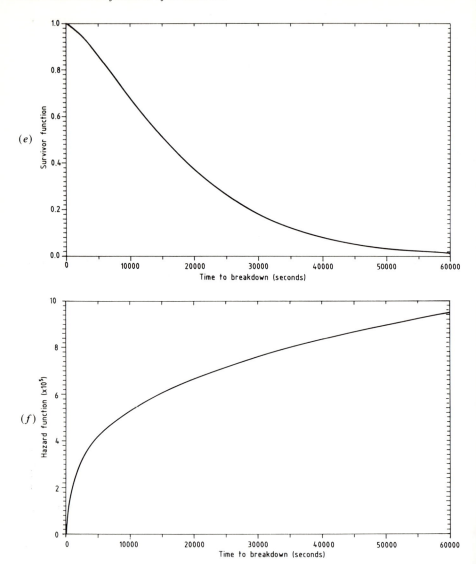

Fig. 14.1
(e) The survivor function (or reliability function)
(f) The continuous cumulative hazard function

$t = 0$ of the probability density function:

$$P_F(t) = \int_0^t g(t)\, dt \qquad (14.2)$$

i.e. the area under the probability density function between times $t = 0$ and t.

Various other statistical representations of lifetime data are also commonly used. The probability of survival of a new sample to time t, $P_S(t)$, is known as the *survivor function* (or sometimes the *reliability function*) and is simply the probability of not failing by time t. Since it is certain that the specimen has either failed or survived, $P_F(t) + P_S(t) = 1$, or:

$$P_S(t) = 1 - P_F(t) = \text{Prob}\,(t_b > t) = \int_t^\infty g(t)\, dt \qquad (14.3)$$

Another commonly used lifetime function is the *hazard function*, $h(t)$ (strictly speaking we are referring to the '*continuous cumulative hazard function*' here). The hazard function specifies the instantaneous rate of failure at time t of a specimen given that it survives to time t:

$$h(t) = \lim\big|_{\delta t \to 0} \frac{\text{Prob}\,(t \leq t_b < [t + \delta t]\,|\,t_b \geq t)}{\delta t} = \frac{g(t)}{P_S(t)} \qquad (14.4)$$

The physical meaning of the hazard function is that in a large set of specimens it equals the failure rate expressed as a fraction of the total number of specimens.

It is sometimes useful to describe the cumulative probability of failure, $P_F(t)$, in terms of the hazard function, $h(t)$. Since, from eqn. 14.2, $g(t) = dP_F(t)/dt$ and from eqn. 14.3, $P_F(t) = 1 - P_S(t)$, then $g(t) = -dP_S(t)/dt$. Substituting into eqn. 14.4 yields the differential equation:

$$\frac{dP_S(t)}{dt} = -h(t)P_S(t)$$

so that:

$$P_F(t) = 1 - P_S(t) = 1 - \exp\left\{ -\int_0^t h(t)\, dt \right\} \qquad (14.5)$$

The survivor and hazard functions obtained from the data given in Fig. 14.1*a* are shown in Figs. 14.1*e* and 14.1*f* respectively. All these functions find general applicability in the statistical analysis of lifetime data. In the next section we will describe the Weibull distribution, a very flexible distribution of wide applicability in the analysis of lifetime data.

14.1.2 The Weibull distribution

(a) The two-parameter Weibull distribution

Firstly we describe the *two-parameter Weibull distribution*[700] which is most commonly used for characterising times to failure for solid insulation. The use of this distribution is not wholly accepted by all workers as being generally applicable and further comments on its use are made in part (b) once an understanding of the function has been established. In the next chapter it will be seen that most physical models do predict this type of distribution for failure as a function of time (but not necessarily of voltage stress).

(i) Definition of the distribution

If, when under constant DC or AC voltage conditions, the probability of failure of an insulating system is described by the two-parameter Weibull distribution, then its cumulative probability of failure is given by:

$$P_F(t) = 1 - \exp\left\{-\left(\frac{t}{\tau_c}\right)^a\right\} \qquad t \geq 0 \tag{14.6}$$

Note that if the exponent a is unity then the distribution simply describes an exponential increase with time constant τ_c to an asymptotic limit of $P_F(t) = 1$. The factor τ_c in the general case is known as the *characteristic time to breakdown* and represents the time at which $P_F(\tau_c) = 1 - (1/e) = 0.6321$ (four significant figures). The characteristic time is a scaling factor which serves to elongate the distribution along the time axis. It is analogous to the mean of a Normal (Gaussian) distribution. The mean, median and standard deviation of the Weibull distribution can be determined (mean $= \tau_c \Gamma(1 + 1/a)$; median $= \tau_c (\log_e 2)^{1/a}$; standard deviation $= \tau_c [\Gamma(1 + 2/a) - \Gamma(1 + 1/a)^2]^{1/2}$) but are seldom used. The parameter a, which is positive, is known as the *shape parameter* (we have just noted that when $a = 1$ the *shape* is that of an exponential). The higher the value of a, the narrower the spread of times to failure. In characterising the distribution of times to breakdown in constant voltage tests the shape parameter is usually found to be in the range $0.5 < a < 3$.

From eqn. 14.6 it can be seen that the probability density function and hazard function of the two-parameter Weibull distribution are:

$$g(t) = \frac{dP_F(t)}{dt} = a\tau_c^{-a} t^{a-1} \exp\left\{-\left(\frac{t}{\tau_c}\right)^a\right\} \tag{14.7}$$

$$h(t) = \frac{g(t)}{P_S(t)} = a\tau_c^{-a} t^{a-1} \tag{14.8}$$

These three functions, the cumulative probability of failure, the probability density function, and the hazard function are plotted in Figs. 14.2a, 14.2b and 14.2c respectively for the five cases of $a = 0.5$, 1.0, 1.5, 2.0 and $a = 3.0$. The time scale has been normalised to time/(characteristic time) so that a normalised value of one indicates a time equal to the characteristic time to breakdown, τ_c. Of these parameters, the characteristic breakdown time, τ_c, is simply determined by the deterministic breakdown mechanism or the aging mechanism. The parameter a however does contain information which can help in the understanding of the fundamental mechanisms involved in the progress of aging and breakdown.

(ii) The effect of the shape parameter 'a' on the distribution

The cumulative probability of failure ($P_F(t)$) graph (Fig. 14.2a) shows the probability that a sample will have failed by time, t, since the start of the test. For a large set of specimens this can be thought of as the fraction of the total number of samples on test which have failed by that time. As would be expected the cumulative probability of failure is zero at zero time (i.e.

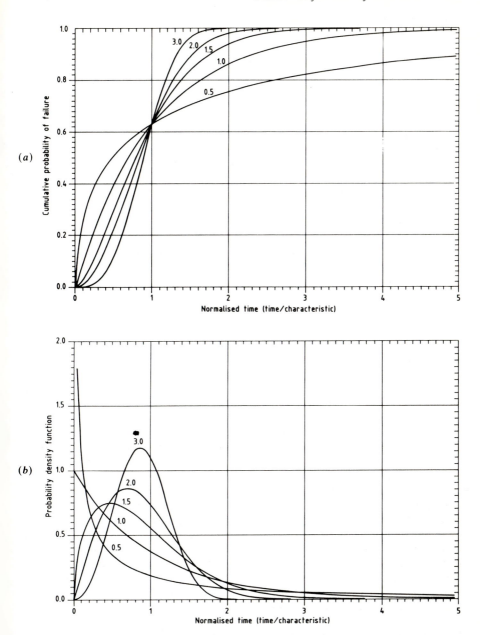

Fig. 14.2 (*a*) The cumulative probability of failure, (*b*) the probability density function, and (*c*) the hazard function of the Weibull distribution for the five cases of $a = 0.5$, 1.0, 1.5, 2.0 and $a = 3.0$. The time scale has been normalised to time/(characteristic time) so that a normalised time equal to one indicates a time equal to the characteristic time to breakdown, τ_c

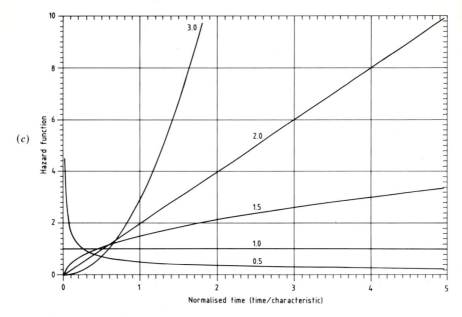

(c)

Fig. 14.2   Continued

no possibility of failure) and tends to unity (i.e. certain failure) at long times. Distributions with high values of the shape parameter, a, approach certain failure faster than those with low values. For example 95% of specimens with a shape parameter $a = 3\cdot0$ will have failed after a time of $1\cdot44\,\tau_c$; in the exponential case of $a = 1$, this takes $3\tau_c$; whilst for $a = 0\cdot5$ it would be $9\tau_c$ before 95% of the specimens had failed. At the characteristic time to failure, $t = \tau_c$, (i.e. at normalised time equal to one), the cumulative probability of failure can be seen to be $P_F(\tau_c) = 0\cdot6321$ irrespective of the value of the shape parameter, a; this is to be expected from the definition of the characteristic time as a scaling factor.

Examining now the Weibull probability density function, $g(t)$ in Fig. 14.2b, it becomes obvious why a is known as the *shape* parameter. The shape of the distribution is peaked for values of $a > 1$ and closely approaches a Gaussian distribution when $a \approx 3\cdot2$. As the value of a is increased the distribution becomes narrower (and therefore taller since the area under this curve is always unity) and its peak (= mode) approaches $t = \tau_c$. For $a = 1$, the distribution is exponential, and as a is decreased from this value the concave curvature becomes more exaggerated with a steeper initial decrease and a gentler decrease at longer times. Thus at time $t < \tau_c$ a greater proportion of the samples will have failed when $a < 1$ than will be the case when $a > 1$; see Fig. 14.2a.

The hazard function ($g(t)$), shown in Fig. 14.2c, has three distinct regimes defined by $a < 1$, $1 < a < 2$, and $a > 2$ (delineated by the two straight-line cases when $a = 1$ and $a = 2$). When $a < 1$ the initial hazard function is very high (infinite at $t = 0$) but rapidly decreases indicating a decreasing failure rate. Many electronic devices and instruments have this type of hazard function and their apparent reliability may be increased by having an initial 'burn-in' period before they are passed to the customer so that defective devices are removed. After the burn-in period the failure rate is considerably reduced. The decreasing hazard function indicates a conditioning process in which the system is less at risk in each successive time interval and suggests that pre-breakdown events, each of which may separately lead to breakdown, retard one anothers' subsequent generation and development. An example of such a process in a solid insulator may be the growth of a tree or the injection of space charge from a field-enhancing protrusion on an electrode. As time advances and the defect grows the local field at the defect tip is moderated reducing the hazard rate. This decreasing hazard function implies that the failure rate monotonically decreases with time and it seems unlikely that this state of affairs could go on *ad infinitum*. Whilst it is observed in some cases for long periods of initial use, an alternative aging mechanism is likely to dominate eventually. The hazard function will then increase at long times giving the so-called 'bathtub' shape, well-known to engineers, and it will not be possible to describe the lifetime statistics in terms of a single two-parameter Weibull distribution.

In the exponential case of $a = 1$ the hazard function is constant, $g(t) = 1$, indicating that attempts at breakdown are statistically independent of each other — i.e. the system has no memory. (Compare this with the exponential behaviour of radioactive decay in which the atomic decay events, which produce the radioactive particles, are independent of each other.) Thus when breakdown testing a set of samples which have a value of a equal to unity, one would not be able to determine whether they had been electrically stressed prior to the experiment.

The monotonically decreasing hazard function corresponding to the case of $a < 1$ is less common than that of the monotonically increasing hazard function which can be seen to be the case when $a > 1$ and in which the system progressively deteriorates. In the limit of large a values, many different pre-breakdown processes are involved in producing the necessary and sufficient conditions for breakdown, leading to a situation which may be thought of as a true catastrophe. When $a = 2$ the hazard function linearly increases since $h(t) = 2(t/\tau_c)$ for this value of shape parameter. For values of a between one and two the *rate* of increase of the hazard function decreases with time so that, whilst the system becomes more likely to fail, the failure rate appears to stabilise. However for shape parameters greater than two the hazard function increases at an ever increasing rate and the system becomes continuously more prone to failure. This suggests that pre-breakdown events produce favourable conditions for further pre-breakdown events to occur so that the rate of pre-breakdown events increases. An example of such a scenario is in avalanche breakdown. In practice most systems have shape parameters between one and two as systems with higher shape parameters are inherently unstable and 'bad designs'. However even in 'good designs' the final processes before failure may change the statistics towards higher shape parameters resulting in a distribution at long times which deviates from the single two-parameter Weibull distribution.

(iii) Obtaining the Weibull parameters

In general scientific investigators tend to like to present their results as straight-line graphs; this makes parameterisation easy and appears to make them content with their theoretical interpretations of the data. Data of cumulative probability of breakdown versus time to breakdown which fit the two-parameter Weibull distribution can be re-plotted in the straight-line form $y = m \cdot x + c$ as can be seen by re-arranging eqn. 14.6 to give:

$$\log_{10}\{-\log_e[1 - P_F(t)]\} = a \cdot \log_{10}(t) - a \log_{10}(\tau_c) \qquad (14.9)$$

so that plotting $\log_{10}\{-\log_e[1 - P_F(t)]\}$ versus $\log_{10}(t)$ results in a straight line of slope a. The characteristic time to failure can also be estimated from the graph since this is the time at which the cumulative probability of failure is equal to 0.6321 (or $\log_{10}\{-\log_e[1 - P_F(t)]\} = 0$). In order to calculate the cumulative probability of failure the data points are first ranked according to the order in which breakdown took place from $i = 1$ to $i = n$ for n samples. Each data point is then assigned a value of P_F according to an equation such as[701,159] $P_F(i, n) = (i - 0.3)/(n + 0.4)$. This has been carried out in Fig. 14.3 for the results given in Fig. 14.1b and the straight line plotted corresponding to the curve of cumulative probability of failure given in Fig. 14.1d. It can be seen that the characteristic lifetime $\tau_c = 1.99 \times 10^4$ s and the shape parameter $a = 1.33$. Whilst this technique is an effective method of presentation and a reasonably good fit to a straight line enables one to have some confidence in both theory and practice, more accurate techniques for

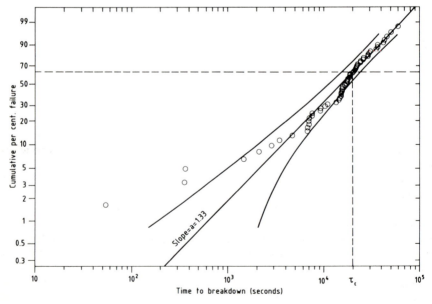

Fig. 14.3   The lifetime data given in Fig. 14.1 plotted on Weibull probability paper. The horizontal dashed line is the 0.6321 quantile and its intercept with the best straight line indicates the characteristic time, τ_c. The slope of the graph is equal to the exponent a. The curved lines bracketing the data represent 90% confidence limits

estimating the Weibull parameters and also for estimating the confidence limits are available; these are discussed in Section 16.2.

(b) Comments on the applicability of the two-parameter Weibull distribution
The times to breakdown observed in constant-stress tests on solid non-crystalline insulation (e.g. References 160, 173, 174, 176, 177, 437, 440, 472, 685, 702–720) are usually found to be distributed according to the two-parameter Weibull distribution. Some of these experiments seem to indicate the existence of more than one statistical distribution of breakdown events which suggests that the deterministic mechanism of breakdown alters as a function of time due to an aging mechanism. There has been some discussion as to whether the Weibull distribution is the most appropriate form (e.g. Dakin *et al.*[721]), particularly as it has very general applicability (i.e. the parameters can be chosen so that it can describe almost anything!), and up until recently there was no physical model to support its use — i.e. it had been used simply because it fitted the data. Recent critical evaluations have generally concluded that the Weibull distribution is the most appropriate for solid insulating systems[173,703,714,722] a view which coincides with the consensus of opinion of the IEEE Electrical Insulation Society, Statistics Technical Committee[723] and physical models are now available which support it[174,173,703] (see also Chapter 15). Various other distributions have been used and these are described briefly below.

(c) The three-parameter Weibull distribution
The two-parameter Weibull distribution described above is a particular case of the general, three-parameter distribution originally described by Weibull[700] which can be expressed in terms of the cumulative probability of failure as:

$$P_F(t) = 1 - \exp\left\{-\left[\frac{t - t_I}{\tau_c}\right]^a\right\} \qquad t \geq t_I$$

$$= 0 \qquad t < t_I \qquad\qquad (14.10)$$

where t_I is an inception time before which breakdown cannot occur. The general term for t_I, is the *location parameter* since it serves to shift the whole distribution along the (linear) time axis thereby re-locating it by a time, t_I.

The three-parameter Weibull distribution often gives a better fit to data points than its two-parameter counterpart. However this may result from its increased mathematical flexibility; it does not necessarily imply that it is the more appropriate distribution in terms of the physical origins of the stochastic nature of the breakdown process. Often breakdown data which has been fitted to the three-parameter distribution can be re-plotted using $t_I = 0$ (i.e. a two-parameter fit) and is still found to lie within 90% confidence limits. Plotting raw, three-parameter data on Weibull probability paper results in a gradually curving, convex (looking from the 'top' or 'left') plot as shown in Fig. 14.4.

There are some laboratory test results which do suggest that a three-parameter distribution is appropriate (see Section 14.3) but these tend to be in cases where aging or breakdown mechanisms with inception times

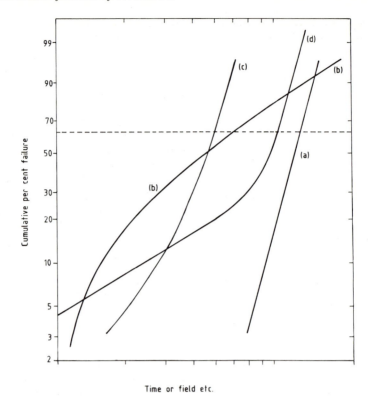

Fig. 14.4   Various functions plotted on 2-parameter Weibull paper:
(a)   2-parameter Weibull distribution
(b)   3-parameter Weibull distribution
(c)   1st asymptotic extreme-value (1AEV) distribution
(d)   multiplicatively mixed Weibull distributions

(such as electrical treeing) have been especially induced. The value of t_I is often assumed to be zero in practical applications.

14.1.3 Other statistical distributions

(a) The first asymptotic extreme-value distribution
The first asymptotic extreme-value (1AEV) distribution is one of three stable solutions of the extreme-value distribution function first defined by Gumbel[248] and described in detail in Section 15.1.2. It is consequently referred to, frequently and perhaps rather misleadingly, as the *Gumbel distribution* or the *smallest extreme-value distribution* or simply *the extreme value distribution*. The cumulative probability of failure given by the 1AEV distribution in terms of time, x, is:

$$P_F(x) = 1 - \exp\left\{ - \exp\left[\frac{x - x_I}{x_c} \right] \right\} \qquad -\infty \leq x \leq \infty \qquad (14.11)$$

where x_I is a location parameter (equal to the 0·6321th. quantile) and x_c is a scale parameter. The 1AEV distribution is directly related to the Weibull distribution since if t has a Weibull distribution then $x = \log_e(t)$ has a 1AEV distribution with $x_c = 1/a$ and $x_I = \log_e(\tau_c)$. By carrying out this transformation the parameters may be estimated using the same techniques as for the Weibull distribution. Data distributed according to the 1AEV distribution results in a concave plot if presented on Weibull probability paper (Fig. 14.4).

It has been suggested that the 1AEV distribution is appropriate for characterising breakdown in fluids[723] but there is some strong evidence for the use of the Weibull distribution in at least some liquid insulating systems (e.g. Frei[724]); this is also to be expected from the theoretical models presented in Chapter 15. The 1AEV distribution may be appropriate for extremely thin insulators in which the device will fail at some defect which acts as the weakest spot. The concentration of defects must be so large that each device contains a number of defects so that a 'weakest-link' model is appropriate[725].

(b) The log-normal distribution

Times to breakdown, t_b, are log-normally distributed if $\log_e(t_b)$ is normally distributed. (Lawless[726] points out that the name 'log-normal' is rather a misnomer, since the log of a log-normal variate actually has a normal distribution.) The probability density function therefore has the form:

$$g(t) = \frac{1}{(2\pi)^{1/2}\sigma} \exp\left\{-\frac{[\log_e(t)-\mu]^2}{2\sigma^2}\right\} \qquad t > 0 \qquad (14.12)$$

where μ is the logarithmic mean (i.e. $\exp(\mu)$ is a scale parameter) and σ is the logarithmic standard deviation. There is no analytical solution to the integral of this equation and so it is not possible to write down an equation for the cumulative probability of failure. This distribution was sometimes used in preference to the Weibull distribution partly because of the arithmetic convenience of having log-lifetimes normally distributed. This is no longer an advantage with the use of computational techniques.

The hazard function of the log-normal distribution is peaked and is zero at $t = 0$ and $t = \infty$. This decreasing hazard function at long times is unlikely to occur in practice as another aging or breakdown mechanism is likely to dominate eventually (cf. the two-parameter Weibull distribution with $a < 1$, Section 14.1.2a(ii)) however the distribution may be useful when large values of time are not of interest. Whilst the use of log-normal statistics have been advocated[727,728] in the past, no reasonable physical model in which the pre-breakdown events have a multiplicative effect has been proposed without assuming a tailored distribution of defect sites[729].

(c) Mixed distributions

In some cases it may be found that a different aging or breakdown mechanism dominates at long times than that which was responsible for early failures. This is particularly likely to be the case if the early failure mechanism had a monotonically decreasing hazard function as is the case, for example, with the two-parameter Weibull distribution with $a < 1$ (Section 14.1.2a (ii)). This may result in a Weibull plot (or another plot such as a 1AEV plot) with two distinct straight lines as is shown schematically in Fig. 14.4[166]. This is known as a multiplicatively-mixed 2-parameter Weibull distribution[159]. Additively-mixed distributions, which are much less common and usually produced artificially, are also described in Reference 159. These may however occur when the test set includes some substandard samples containing defects or contaminants which are not normally present. Care needs to be used before fitting data in this way because, as was the case for the three-parameter Weibull distribution, one always obtains a better fit to data points than by using a single two-parameter distribution simply because of its increased mathematical flexibility. Statistical techniques for discriminating between different distributions are available and are briefly described in Section 16.2.

14.2 The effect of voltage and time on the failure statistics

14.2.1 The effect of voltage on the characteristic time to failure

Up to now we have considered the distribution of times to failure under constant (AC or DC) voltage conditions. It is reasonable to assume that this distribution is primarily a function of the insulating material/system since a deterministic mechanism would predict a constant time to failure. It is essentially the stochastic features of the insulating system, such as intrinsic structural fluctuations[174,26,226,730] or an extrinsic distribution such as void size, which lead to a distribution of times to failure. How is this distribution affected by the level of applied voltage? One would expect from deterministic considerations that the characteristic time to failure would reduce as the voltage is increased or at least when it is increased above some 'threshold' level. It would also be reasonable to expect that the shape of the distribution would remain the same and simply be scaled up or down with the characteristic time to failure since the shape is caused by material/system properties which are not affected by voltage stress.

(a) The inverse power law
The exact form of relationship between the applied voltage, V and the characteristic time to breakdown, τ_c, must depend upon the deterministic mechanism operative at that voltage and under the specified experimental/test conditions. From as early as 1916[731] however, it has been suggested that there exists a *3-parameter inverse power-law* relationship between applied voltage and time to breakdown:

$$\tau_c = \infty \qquad\qquad V < V_{th}$$

$$\propto (V - V_{th})^{-n} \qquad V \geqq V_{th} \qquad\qquad (14.13)$$

where V_{th} is a threshold voltage below which the breakdown mechanism does not operate. In many circumstances the threshold voltage is assumed to be zero so that the time to failure is simply inversely proportional to a power of applied voltage (the *2-parameter inverse-power law*). The value of n is always greater than 2 and typically in the range 5—20 for polymer systems. In the case of a true deterministic breakdown voltage, n would be infinite and V_{th} would represent the breakdown voltage.

A major reason for the continuing general use of the inverse power law is its mathematical ease of handling and that it can be easily incorporated into the Weibull distribution; however care needs to be taken when using it. In particular the values of the constant of proportionality and of the exponent, n, are likely to be different for different breakdown mechanisms and different mechanisms are likely to be operative at different applied voltages. It may therefore not be valid to estimate these parameters at high applied voltages and then extrapolate to working stresses in order to find the expected lifetime of the system. Furthermore these parameters may change over periods of time as low-level degradation takes place. Provided these factors are considered, the inverse power law is useful and we will continue our discussion on this basis. Initially we will also assume that $V_{th} = 0$ so that $\tau_c \propto V^{-n}$ although we will not carry this assumption throughout. The origins and rôle of the threshold voltage are considered in more detail in Chapter 15.

(b) The exponential law
During the mid-1970s exponential relationships between time to failure and applied voltage became popular (e.g. References 732, 733, 688) although this relationship is still not as popular as the inverse power law. The exponential relationship has the form:

$$\tau_c \propto \exp\{k/V\} \tag{14.14}$$

where k is a constant. More complicated forms of this equation have been proposed which generally include a threshold voltage term. One of the most popular is that proposed by Dakin (e.g. Reference 688) which has the form:

$$\tau_c = \infty \qquad\qquad V \leqq V_{th}$$

$$\propto \frac{\exp\{k/(V - V_{th})\}}{V - V_{th}} \qquad V > V_{th} \tag{14.15}$$

This equation is much more difficult to manipulate mathematically, for example it is not possible to estimate the parameters (k and the constant of proportionality) using graphical methods.

(c) Comparison of the inverse power and exponential laws
Stone[734] has compared these two laws and has shown that, given a typical set of data of times to failure versus applied voltage, it is likely that both the inverse power law and the exponential law will result in good fits when plotted graphically. This arises because a very small range of stresses produces a very large range of times to failure. He demonstrated this by considering the data for epoxy insulation given in Reference 177. This is plotted in Figs. 14.5a and 14.5b for the inverse power law (log–log plot) and for the exponential law (lin–log plot) respectively. When analysed using linear regression both functions yield regression coefficients $r > 99\%$ and do not appear curved. However extrapolating to a hypothetical working

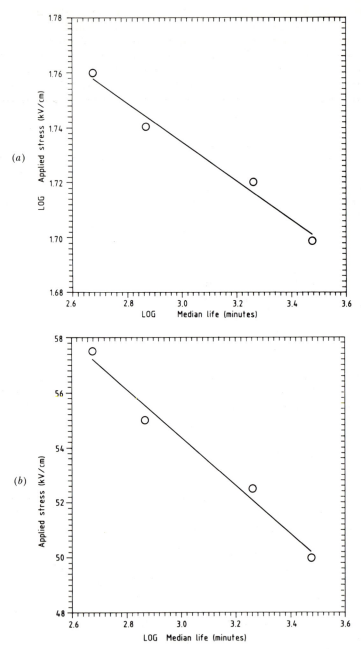

Fig. 14.5 Characteristic time to failure as a function of voltage for epoxy-resin samples[177] plotted as (*a*) a log–log plot (implying an inverse power-law relation) and (*b*) a lin–log plot (implying an exponential relation). Both plots given excellent fits

stress of 2×10^7 V $\cdot$ m^{-1} (only a factor of about 2·5 lower than the test stresses) the simple exponential model yields a median life of 2000 years whilst the inverse power law a median life of only 13 years[734]. We tend to favour the use of the inverse power law in the absence of further data since theoretical physical models of breakdown tend to predict this relationship and not the exponential law. (In Chapter 13 we noted that the Bahder *et al.*[120] model for partial discharges predicted an exponential relationship with a threshold voltage. However their results appeared to show a power-law behaviour; Fig. 14.14.)

(d) The flat-z characteristic
The *flat-z characteristic*[120,685] acquires its name from the sigmoidal shape of the log–log graph of applied stress versus characteristic time-to-failure which, it is supposed in this case, is asymptotic to the three straight lines formed by pulling a letter '*Z*' horizontally so that it becomes flatter.

The three-parameter inverse power law (eqn. 14.13) results in a log–log plot of characteristic time to failure versus applied stress which is asymptotic to two straight lines. At short times (high stresses) the curve is asymptotic to $\tau_c \propto V^{-n}$ (i.e. a line with a slope of $-n$) and at long times the curve tends to $V = V_{th}$ (slope $= -\infty$). In the flat-z characteristic it is proposed that there is a third straight-line asymptote given either by an upper voltage threshold at which failure immediately occurs ($n = \infty$) or by $\tau_c \propto V^{-n'}$ where $n' \gg n$. This is shown schematically in Fig. 14.6. The latter case occurs if the dominant deterministic breakdown mechanism changes at a given stress level.

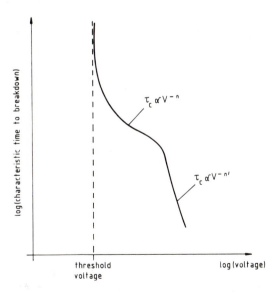

Fig. 14.6 Schematic graph of characteristic time to failure versus applied stress on log–log axes showing the 'flat-z characteristic'

Bahder *et al.*[120] have proposed a relation between characteristic time to breakdown and voltage which can be simplified to:

$$\tau_c = \frac{1}{\omega p[e^{q(V-V_{th})}-1](e^{rV}+s)} \qquad V \geq V_{th}$$

$$= \infty \qquad\qquad\qquad\qquad\qquad V < V_{th} \qquad (14.16)$$

where ω is the frequency and p, q, r, and s are material constants. This has the general shape of a flat-z characteristic with a straight-line asymptote at $V = V_{th}$. However the other two lines forming the 'Z' are no longer straight on a log–log plot of this function. At high voltages $\tau_c \propto \exp(-V/k)$ with k a constant and at intermediate voltages the slope is steeper, as expected in a flat-z characteristic, but the equation is not simple.

14.2.2 The probability of breakdown as a function of time and voltage

(a) Incorporating voltage into the Weibull equation
Using the two-parameter inverse power law relation (eqn. 14.13 with $V_{th} = 0$), voltage or applied field is easily incorporated into the two-parameter Weibull equations describing the probability of failure as a function of time. It is convenient to write this as

$$\tau_c \propto V^{-n}$$

$$= C^{-1/a}V^{-b/a} \qquad (14.17)$$

where a is the shape parameter in the Weibull function of time (eqn. 14.6), $C^{-1/a}$ is the constant of proportionality and b is a new constant which will be seen to be the shape parameter in the Weibull function of voltage. Substituting this into eqn. 14.8 for the hazard function yields:

$$h(t) = at^{a-1}\tau_c^{-a}$$

$$= at^{a-1}CV^b \qquad (14.18a)$$

or

$$h(t) = at^{a-1}C[V(t)]^b \qquad (14.18b)$$

if V is not necessarily constant.

Eqn. 14.5 expresses the cumulative probability of failure in terms of the hazard function so that in general:

$$P_F(V, t) = 1 - \exp\left\{-C\int_0^t at^{a-1}[V(t)]^b \, dt\right\} \qquad (14.19a).$$

and under constant voltage conditions this simplifies to:

$$P_F(V, t) = 1 - \exp\{-Ct^a V^b\} \qquad (14.19b)$$

As one would expect, the shape parameter a is unaffected by the amplitude of the voltage provided it is held constant and the constant b can be seen to be the shape parameter of the Weibull function of voltage only. The shape parameter b cannot be measured directly as one cannot vary voltage whilst holding time constant! If the threshold voltage in eqn. 14.13 is likely to be non-zero then the term V may be replaced with $(V - V_{th})$ throughout eqns. 14.17 to 14.19.

Because the cumulative probability of failure is a function of electric stress and time of applied stress it may be easier to picture it as the surface of a three-dimensional graph. An example of this probability surface is shown in Fig. 14.7 for the case of $a = 0.3$, $b = 5$ and $C = 5 \times 10^{-6}$ which may be typical values for an AC power cable, appropriate to the dimensions of time and electric field shown in the figure. From the surface, which is drawn for the 2-parameter Weibull distribution ($V_{th} = 0$), it can be seen that the probability of failure at low fields is low irrespective of the time, it is for this reason that it is difficult to evaluate a threshold level should it exist, and why such a level may just in fact appear to exist.

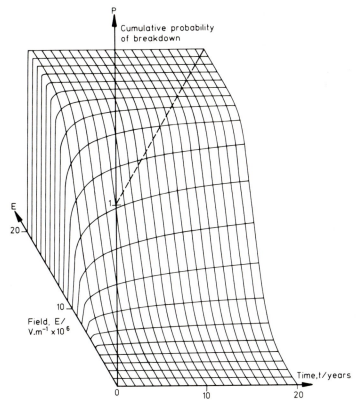

Fig. 14.7 A three dimensional representation of the cumulative probability of breakdown as a function of time and constant applied stress which may be typical of an AC power cable. Notice that below $4 \times 10^6 \, \text{V} \cdot \text{m}^{-1}$ the cumulative probability of breakdown over the time scale shown is very small. Whilst this gives the appearance of a threshold field below which aging does not occur, the graph was in fact evaluated using the 2-parameter Weibull distribution for which it is assumed that there is *no* such field

(b) Estimating Weibull shape parameters

Chapter 16 discusses testing techniques, the estimation of parameters and the interpretation of results in detail. However since the remainder of this chapter presents data reported as evidence for the above probability of breakdown distributions, it is necessary to introduce here the methods whereby the shape parameters are deduced.

Constant stress tests (or *life tests* as they are sometimes known) seem the most obvious method of finding the shape parameter a. These tests generally employ a set of typically ten or more identical specimens which are stressed at a constant voltage. The distribution of times to failure is observed and a value of a is thereby estimated. These results are often plotted on Weibull probability paper in which case, assuming a two-parameter Weibull distribution is observed, the results will fall on a straight line of slope a. If several such constant voltage tests are carried out at different voltages then the exponent n and the constant of proportionality in the inverse power law (eqn. 14.17) may be established. Such results are usually presented on a log–log graph of characteristic time to failure versus voltage which, if the inverse power law holds, shows a straight line of slope $-n$. Since a is known from the Weibull plots, b may be calculated using $b = a \times n$ if required. However there are practical problems with constant-stress tests (see Chapter 16) which make them unpopular, in particular, since a is typically of order unity the times to breakdown are very widely spread (Fig. 14.2).

Progressive stress tests (or *ramp tests* as they are sometimes known) are more popular since they *appear* to measure a breakdown voltage. Typically in such tests the voltage across a specimen is linearly increased from zero until breakdown occurs. This is repeated for a set of identical specimens and the breakdown voltages (or times) are observed. These are generally found to be distributed according to a two-parameter Weibull distribution so that the shape parameter and characteristic breakdown voltage may be estimated by plotting the results on Weibull probability paper and calculating the shape and 0·6312 quantile. If the rate of voltage increase is $\dot{V}$ and the initial voltage is zero then $V(t) = \dot{V} \times t$ in eqn. 14.19 so that:

$$P_F(V, t) = 1 - \exp\left\{ -C \int_0^t a t^{a-1} [V(t)]^b \, dt \right\}$$

$$= 1 - \exp\left\{ -C \int_0^t a t^{a-1} \dot{V}^b t^b \, dt \right\}$$

$$= 1 - \exp\left\{ -C \frac{a}{a+b} \dot{V}^b t^{a+b} \right\} \tag{14.20a}$$

in terms of time-to-failure, or:

$$P_F(V) = 1 - \exp\left\{ -C \frac{a}{a+b} \dot{V}^{-a} V^{a+b} \right\} \tag{14.20b}$$

in terms of breakdown voltage. These can be written in a form analogous to eqn. 14.6:

$$P_F(t) = 1 - \exp\left\{-\left(\frac{V}{V_c}\right)^\beta\right\} \qquad t \geqq 0 \qquad (14.21)$$

i.e. as a two-parameter Weibull distribution with shape parameter $\beta = a + b$. Weibull plots of progressive-stress test results will therefore have a slope equal to $a + b$ whether they are a function of voltage or time.

If several progressive-stress tests are carried out at different rates of voltage increase, $\dot{V}$, a log–log graph of rate of voltage increase as a function of characteristic voltage will result in a graph of slope $(a+b)/a$; i.e. $\dot{V} \propto V_c^{(a+b)/a}$. Such tests can therefore be used to determine a, b and C without recourse to constant-stress tests.

14.3 Laboratory studies

In this section we discuss results of laboratory studies presented in the literature which support the distributions introduced in Section 14.1. We have not attempted to refer to all the data reported exhaustively, rather we have chosen a representative selection. We present these in much the same order as that in which the distributions were presented in Section 14.1.

14.3.1 Constant-stress tests

Conclusive results of constant-stress tests (variously known as 'life', 'voltage endurance' or 'static' tests) on polymeric insulation are not abundant in the literature. Indeed when reviewing the literature it is surprising how little is reported of such tests considering the implications for lifetime prediction.

(a) Results in support of the Weibull distributions
Three sets of constant-stress tests data have been chosen which appear to be represented by the Weibull distribution.

The first set of data comes from experiments carried out by Stone *et al.*[177] and has been referred to by many authors (e.g. References 726, 735). Three constant-stress tests were carried out at different voltage levels (52·5 kV, 55·0 kV and 57·5 kV; 60 Hz), each on twenty epoxy specimens with razor-blade electrodes. The results are shown in Fig. 14.8a. It is important to note that, in order to separate the three sets of test results for the sake of clarity, the results for the 57·5 kV test have been shifted down a decade in time and the results for the 52·5 kV up a decade. The razor blade caused electrical trees to be seen after a voltage-dependent inception time. One would therefore expect to observe a 3-parameter Weibull distribution (eqn. 14.10) and the graph points, especially at the lower voltage, do appear to lie on a convex curve as such a distribution would give. However plotting 90% confidence limits for a 2-parameter (zero inception time) straight line fit reveals that virtually all the data (all except one of the 54 points) are contained within these limits. If the data is plotted after subtracting an optimum, voltage-dependent inception time from each point then better fits are obtained[177]. However this would virtually always be the result of such a

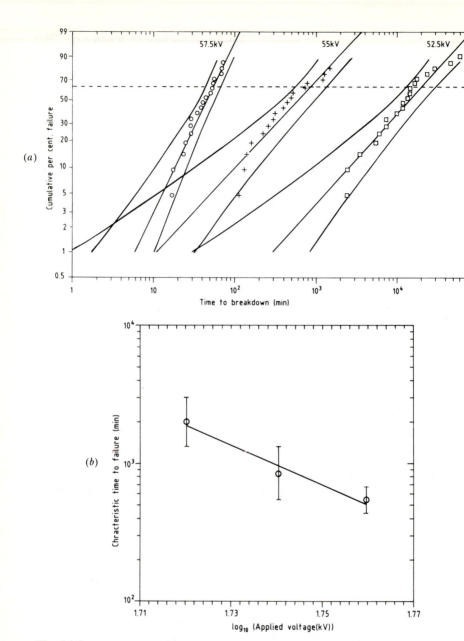

Fig. 14.8

(a) Weibull plots of lifetimes at three stress levels of epoxy specimens with razor-blade electrodes from the data of Stone *et al.*[177]. For the sake of clarity the results of the 57·5 kV test have been shifted down a decade on the logarithmic time axis and the results of the 52·5 kV test up a decade. The best straight line and 90% confidence limits for each set of data are shown

(b) A log–log plot of characteristic time to failure versus applied stress from the data of (a). The 'error bars' represent 90% confidence limits and the slope is −14·4

manipulation. In this case, whilst a reasonable *physical* model would predict a 3-parameter Weibull distribution (i.e. an inception time for tree initiation has been experimentally observed), there is insufficient data to demonstrate *statistically* that the simpler 2-parameter Weibull distribution is too specific. The data illustrates the importance of both a physical understanding of the breakdown mechanism[736] and of plotting confidence limits even though this may be a time-consuming process.

The values of a and τ_c, corresponding to the shape and 63·21 percentile, may be estimated and are shown in Table 14.1.

The time exponents, a, at the two lower voltages are both very similar and close to unity indicating a distribution close to that of an exponential. The a value for the highest voltage is, however, significantly different indicating that the breakdown mechanism is likely to be different in this case. For this highest voltage, $a \simeq 2$ so that the hazard function ($\propto$ failure rate) linearly increases with time. One might expect this change of breakdown mechanism with voltage to lead to a deviation from a straight line when plotting $\log \tau_c$ versus $\log V$, Fig. 14.8*b*, but as there are only 3 points in this case it is impossible to tell. The slope of this log–log plot is $-n = -14.4$, a fairly typical value for this type of insulation.

The second set of data, already presented in Fig. 14.3, is previously unpublished and represents constant-stress tests at 2 kV, 50 Hz which were carried out on 60 wire samples coated with a blend of poly-phenylene oxide and polystyrene. The figure shows the results plotted on Weibull paper with a best straight line estimation and 90% confidence limits (maximum likelihood estimation and confidence limits generated using a Monte-Carlo simulation technique, see Chapter 16). Again a reasonably good fit to a 2-parameter Weibull distribution is obtained with 88% of the points lying within the 90% confidence limits which is approximately what one would expect. However there does appear to be a 'tail' of data, consisting of the 10 or 11 shortest time-to-failure points, which may be a separate distribution. As with the previous sets of data on epoxies, there is really insufficient data to prove this. Test engineers would argue that 60 tests of this nature constitute a large effort and, if these were power cables for example, considerable resources would be required.

This emphasises the difficulty with such tests and the interpretation of results. Drawing confidence limits does not necessarily help in deciding

Table 14.1 Maximum likelihood estimates of τ_c and a assuming a 2-parameter Weibull distribution for the data of Stone *et al.*[177]

Applied voltage (kV)	Estimated value of τ_c (min)	Confidence limits (90%) for τ_c (min)	Estimated value of a	Confidence limits (90%) for a
52·5	2010	1330—2950	1·08	0·73—1·40
55·0	843·5	551—1320	1·06	0·69—1·39
57·5	544·7	438—674	2·05	1·36—2·67

whether a plot represents data from a 3-parameter, 2-parameter, or mixed distribution. Data lying on a curve will cause a biased estimation of τ_c and a so that the estimated confidence limits will tend to move to accommodate the curve. There are no simple statistical techniques for drawing conclusions here and a 'by eye' technique may be a reasonable approach (Chapter 16). Arguing that there is a 'tail' of data implies a mixed distribution which requires 4 parameters to describe it, i.e. it is even more flexible than the 3-parameter Weibull distribution. Again caution must be used before concluding that data is not adequately represented by the 2-parameter distribution since it will always be fitted better by the other two.

In practice a mixed distribution will give a Weibull plot which is asymptotic to two (or perhaps more) straight lines. This appears to be plausible in Fig. 14.3 and taking the lower 10 points and the upper 50 points as separate sets of data, these can be re-plotted as two separate distributions, Fig. 14.9. This results in two much better fits (97% of the points inside the 90% confidence limits) although it may be argued that perhaps the split should have been 11/49 instead of 10/50. The values of the parameters are $a_1 = 0.98$, $\tau_{c1} = 2790$ s for the lower distribution and $a_2 = 1.98$, $\tau_{c2} = 2.45 \times 10^4$ s for the upper distribution. Assuming that there are genuinely two mixed distributions then an explanation of the behaviour might be as follows. The lower distribution, with $a \simeq 1$ is approximately exponential indicating random failures (constant hazard function equal to unity) whilst the upper distribution has a linearly increasing hazard function since $a \simeq 2$. As discussed in

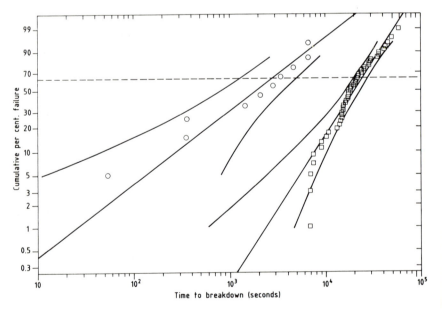

Fig. 14.9 The lifetime data of Figs. 14.1 and 14.3 plotted on Weibull probability paper as two separate distributions assuming that they were additively mixed. Virtually all the data points lie within the 90% confidence limits

14.1.2*a* (ii) the latter implies that different pre-breakdown processes are involved in producing the necessary and sufficient conditions for breakdown. This aging obviously takes time to cause breakdown and in this case another single process dominated at short times in some samples causing occasional random failures.

Whilst both the above examples are typical data and both show integer values of *either a* equal to 1 *or* 2, this not always the case. Generally values of *a* between ~0·5 and ~3 are commonly found.

The next set of data, Fig. 14.10, is from experiments in which a void has been created using a CIGRE cell[737], the material is again an epoxy resin. The construction of the specimens is such that the process of partial discharging tends to cause further partial discharges to appear, i.e. the hazard function increases at an increasing rate. This results in remarkably high values of *a* (i.e. *a* > 3) of 5—6. Such values were found because the cell used encouraged partial discharging which progressively deteriorated the system. Such high values are not observed in commercially-viable systems. Indeed if slopes of this magnitude are observed on Weibull graphs as a function of time it usually indicates that a progressive-stress test has been carried out and the time-to-failure rather than the voltages-at-failure have been recorded (i.e. with a slope of *a* + *b*).

Constant-stress tests on polymers have also been reported by other workers e.g.: Lawless (Reference 726, page 202), Tanaka and Greenwood (Reference

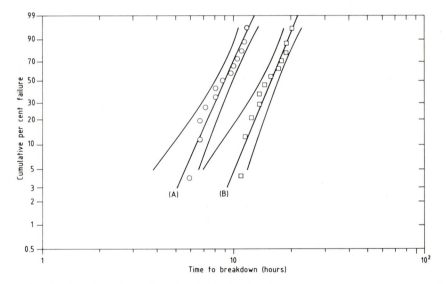

Fig. 14.10 Lifetime data plotted as Weibull plots from constant-stress tests carried out by Schifani[737] on epoxy resin samples using a CIGRE cell in which a void has been deliberately created. The resins were from CIBA-GEIGY Ltd and were Araldite™ CY225 with a hardener of (*a*) CY925 and (*b*) CY227. The resins were tested at temperatures of 80°C = (A) and 45°C = (B). The best straight lines and 90% confidence limits are shown

166, page 95, this data needs to be treated with caution, various sets of data from different experiments and laboratories have been 'transformed' to produce common 'equivalent 50 Hz lives'), Hirose[738] (only a few points but at 7 voltage levels, see also Section 14.3.3), and in the 'IEEE Guide for the Statistical Analysis of Electrical Insulation Voltage Endurance Data'[723] (although surprisingly few examples considering its title). Other examples of mixed distribution have been given by Rowland[719] (large number of data using self-healing breakdown technique) and discussed with examples by Seanor[71] (the best straight line does not appear to have been fitted using appropriate techniques here).

(b) Results in support of other distributions
The 1AEV distribution is most often used to represent breakdown in fluids (e.g. Reference 739) although fluids sometimes display the Weibull distribution[724]. There are some instances, however, particularly in thin-film insulators, in which the 1AEV distribution is appropriate to solids. For example very large scale tests by Wolters and Zegers–Van Duynhoven[725] on metal-oxide-semiconductor (MOS) capacitors clearly demonstrate 1AEV behaviour. However this is not generally observed.

Constant-stress test data is rarely presented now as being represented by the log-normal distribution. Some log-normal data is given as an example in the IEEE guide[723] (their Table 3) but it is not clear where this data comes from and, furthermore, possibly because there are not many points, the data also gives an excellent fit to a 2-parameter Weibull distribution.

14.3.2 Progressive-stress tests

Progressive-stress tests are frequently carried out and the results are almost invariably presented on Weibull probability paper. Many examples for polymeric insulation are to be found by a cursory look through the literature. These progressive-stress tests may be carried out in various ways (see Chapter 16) but provided the voltage is increased in a reasonably linear way, at least close to and through the breakdown region, reasonably straight-line fits, i.e. 2-parameter Weibull fits, are usually observed. It is only recently that workers have started to use confidence limits but often it is self-evident that a reasonably good fit has been obtained. Since many examples are easily found, we only consider two sets of data here, both of these are developed further in the next section.

The first set of data is taken from Bahder *et al.*[706] in which they tested miniature, triple extruded, XLPE cables with 1·25 mm thick insulation. Some of their data is shown in Fig. 14.11 for progressive-stress tests on new and aged cables. The aging involved growing water trees and temperature cycling and the cables were tested either wet or dry. The new unaged cable, as would be expected, exhibited the highest characteristic breakdown strength ($E_c = 5\cdot98 \times 10^7$ V · m^{-1}) and the Weibull plot has the highest slope or shape parameter (=14·1) indicating the narrowest distribution of breakdown stresses. It was shown from eqns. 14.20*b* and 14.21 that the slope of such a plot is equal to the sum of the time and field exponents ($=a+b$) in the combined Weibull equation (eqn. 14.19). Assuming that *a* is of order

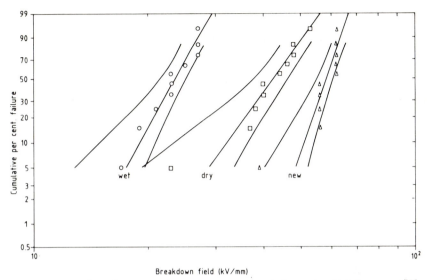

Fig. 14.11 Results of progressive-stress tests carried out by Bahder *et al.*[706] on miniature XLPE cables. The three sets of data correspond to 'as new' cables, and cables which have been subject to water treeing under temperature cycling and tested when both wet and dry. The best straight lines and 90% confidence limits are shown

unity then $b \simeq 12$—13. Values of $5 < b < 15$ are commonly found for unaged polyethylene. The aged cable has a lower characteristic breakdown strength ($24 \cdot 6$ and $44 \cdot 7 \times 10^6 \, \text{V} \cdot \text{m}^{-1}$ for wet and dry cables respectively) and a lower slope ($8 \cdot 7$ and $6 \cdot 8$ respectively). The slopes in this case are quite similar (90% confidence limits are $4 \cdot 5$—$12 \cdot 3$ and $3 \cdot 5$—$9 \cdot 5$ for wet and dry respectively) and indicate reduced values of b. The values of a may also have changed but, assuming a of order unity, this could not solely account for the overall change of $a + b$ of about 6—7. The effect of water trees on the breakdown mechanism has already been discussed in Section 6.1 and a detailed description of the effect on the breakdown statistics is given in Section 15.5.3 and will not therefore be discussed further here.

The second set of data[174] is from four sets of progressive-stress tests carried out on vacuum oven dried and degassed, steam-cured, Rogowski-profiled plaques of power-cable grade XLPE with an insulation thickness of $0 \cdot 5$ mm and an area of $19 \cdot 6 \, \text{cm}^2$. Each set of progressive-stress tests were carried out using small voltage steps at average linear ramp rates of $2 \cdot 0 \times 10^5$, $2 \cdot 22 \times 10^4$, $2 \cdot 22 \times 10^3$, and $1 \cdot 11 \times 10^2 \, \text{V} \cdot \text{m}^{-1} \cdot \text{s}^{-1}$. The results of these tests are shown in Fig. 14.12a and it can be seen that, as predicted by eqn. 14.20b, they all have similar slopes of approximately $7 \cdot 15$ but the characteristic breakdown strength decreases with increasing ramp rate. These results will be discussed further in the next section.

Results of progressive-stress tests are rarely plotted on any axes other than those of Weibull probability paper and so it is difficult to find evidence

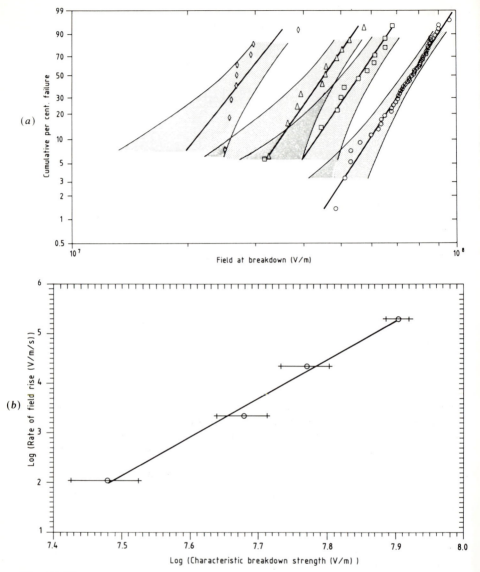

Fig. 14.12

(a) Results of four progressive-stress tests carried out by us[174] on XLPE specimens at different rates of voltage stress increase (V · m^{-1} s^{-1}): ◇ 1·1 × 10^2; △ 2·22 × 10^3; □ 2·22 × 10^4; ○ 2·0 × 10^5. The latter set of points (2·0 × 10^5 V · m^{-1} s^{-1}) was derived from many different tests which were used as regular controls for testing the experimental procedure. The best straight lines and 90% confidence limits are shown. Note that the lines are parallel as predicted by eqn. 14.20*b*

(b) A log–log plot of rate of field rise versus characteristic breakdown strength from (*a*). The error bars represent 90% confidence limits and it can be seen that a good straight line fit is obtained

of distributions other than the Weibull distribution for representing such tests. However mixed-distributions are sometimes observed where a breakdown mechanism at low fields is dominated by another mechanism with a much higher b value at higher fields; Fischer and Nissen[160] have proposed this explanation for many polyolefins. Dissado[740] has also suggested that partial discharge failures in progressive-stress tests are distributed according to the 1AEV distribution, and this is discussed further in Section 15.5.2.

14.3.3 Effect of voltage on lifetime

Graphs, on various scales, of electrical stress versus time are not uncommon in the literature. Care must be taken however as they are often not results showing the variation in characteristic lifetime as a function of stress; instead they often refer to experiments in which an insulating system has been aged in some way (e.g. by subjecting it to water treeing) and its characteristic breakdown strength subsequently measured by a progressive-stress test. In such cases there is no obvious reason to suppose that the stress and aging time should be related by inverse-power or exponential laws (see Section 15.5.3). It is therefore important to check the details of the experiment carefully before drawing conclusions from such results.

There do not appear to be conclusive results for either inverse-power, exponential or any other lifetime versus stress relationship; this has already been discussed briefly with reference to Fig. 14.5. It seems unlikely that this situation will be resolved quickly as careful, long-time experiments would be required with large numbers of specimens and would consequently occupy considerable resources. Furthermore the results may still not be applicable to other insulating systems and, worse still, may still not be conclusive.

The results presented in Fig. 14.12a of breakdowns from sets of progressive-stress tests have already been discussed briefly and it was noted that the slopes were equal in agreement with eqn. 14.20b. This equation, derived from the inverse power law of lifetime as a function of field and the 2-parameter Weibull distribution also predicts a power law relationship between rate of stress increase and characteristic breakdown voltage and, as was shown in Section 14.2.2b, a log–log graph of rate of voltage increase versus characteristic voltage is predicted to be a straight line of slope $(a+b)/a$. Such a graph is presented in Fig. 14.12b and can be seen to be a good straight line (error bars indicate 90% confidence limits) with a slope of 7·76. Knowing from Fig. 14.12a that $a+b=7·15$ it can be seen that values of $a=0·92$ and $b=6·2$ are obtained. Therefore $\tau_c \propto V^{-n}$ with $n=b/a=6·8$, eqn. 14.17. These seem reasonable values for these exponents.

Hirose[738] has presented data from seven sets of constant-stress tests carried out at different stress levels on varnished polyimide films on coils used in rotating machinery and has interpreted the results as supporting a flat-z characteristic (Section 14.2.1d). The times-to-failure of each of the seven constant-stress tests were plotted on Weibull paper and gave essentially seven parallel lines each with a slope $(=a)$ of about two (within their respective 90% confidence limits). This is what one would expect if the breakdown mechanism in each case were the same. Hirose then plotted all these times to failure on a log–log plot of stress versus time to demonstrate the flat-*z* characteristic, a similar plot based on his data is given in Fig. 14.13*a*. (It seems common to plot such results as stress versus time, however, as stress is the variable and time that which is measured, these graphs should be plotted as time versus stress as we do below.) Plotting such graphs using *all* the times to failure causes two problems:

(i)   Since the data points at any given stress are Weibully distributed, they are not symmetrical about the characteristic (or mean) time at that stress and it is tempting to draw an incorrect line through them or to think that a correctly-drawn line is incorrect; and

(ii) Because the data points form horizontal lines, it is tempting to assume from a purely visual inspection that the graph 'flattens out' at the ends to form the flat-*z* characteristic (as shown in Fig. 14.13*a*).

Since it is the characteristic time (i.e. the distribution's scaling factor) which changes with stress, and not the shape of the distribution, it is clear that it is the characteristic time that should be plotted as a function of stress as shown in Fig. 14.13*b*.

On first inspection this graph appears to be a reasonable straight-line fit to the data on log–log axes. However the 'error-bars' on the figure actually represent calculated 90% confidence limits and the best straight line lies outside four or five of these seven confidence limits. Hirose suggests that at the lower fields the 3-parameter inverse power law (eqn. 14.13) is operative, i.e. there is a threshold stress below which τ_c is infinite. He calculates this to be $2 \cdot 77 \times 10^6 \ \text{V} \cdot \text{m}^{-1}$ by assuming all the *a* values are equal (this is a similar technique to that advocated by Lawless[726], pp. 188–190) and using the Newton–Raphson numerical analysis technique. Fig. 14.13*b* has been re-plotted in Fig. 14.13*c* with this threshold field subtracted from the applied field. The solid line indicates the best fit which, if good, would indicate that a 3-parameter inverse power law were applicable. Hirose suggests that this is only applicable at the lower fields and that a different breakdown mechanism dominates at high field with a higher inverse power law exponent 'n' (eqn. 14.13); i.e. the data is represented by a flat-*z* characteristic. The best-fit (solid) line is still not good considering the confidence limits although it is considerably better than without the threshold value. The dashed line represents Hirose's suggestion; the five characteristic times-to-failure of the lower fields do appear to be fitted well by a 3-parameter inverse power law with the two higher-field points deviating from this law to form the flat-*z* characteristic. It is not possible to comment on the form of the high field relationship with lifetime since there are insufficient data.

Hirose has also plotted his data according to an exponential law[738]. The data does not fit a simple exponential relationship such as $\tau_c \propto \exp\{-(E - E_{th})/E_0\}$ but Hirose has shown that it may be fitted to the more complicated expression proposed by Simoni[481]. This Simoni equation uses six parameters so it is not surprising that it can be parameterised to fit the data. The flat-*z* characteristic uses at least four parameters (usually five), and it may be difficult to even justify this over the 3-parameter inverse power law fit. In this respect the Hirose data is particularly interesting; the

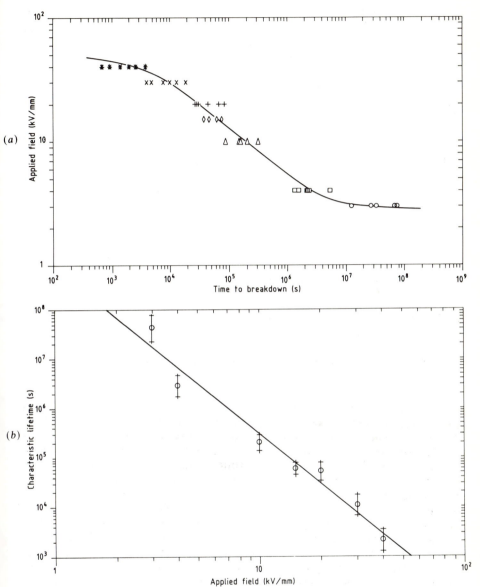

Fig. 14.13
(a)   The results of constant-stress tests on polyimide coated wires at 180°C carried
     out at seven different stresses by Hirose[738]. Plotting *all* the points in this manner
     may be misleading for reasons discussed in the text
(b)   The same data as (a) plotted as a log–log plot of characteristic time to failure
     (s) versus applied field (kV · mm^{-1}), the error bars indicate 90% confidence
     limits. Whilst, at first glance, this appears to be a good straight-line fit, in fact
     such a line is outside most of the confidence limits

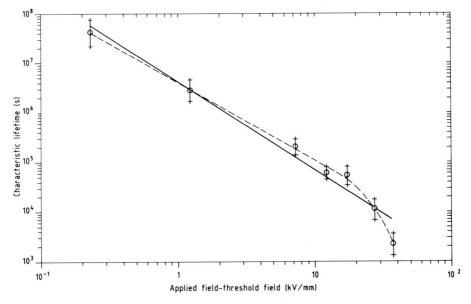

Fig. 14.13
(*c*)   The same data plotted as (*b*) with a threshold field (2·77 kV/mm) subtracted
from each data point. The solid line represents the best fit to all the points,
and whilst being a better fit than in (*b*), it still remains outside the 90% confidence
limits of most of the points. The dashed line represents the flat-z fit suggested
by Hirose[738]

data is drawn from a large set (~40 points) and is optimally chosen to
produce the most conclusive results possible from that quantity of data.
Nevertheless the stress–lifetime relationship is still in contention.

The five parameter, flat-z type equation developed by Bahder *et al.*[120]
(Section 14.2.1*d*, eqn. 14.16) for lifetime versus stress was derived by considering the way space-charge movement affected electrical tree growth
(Section 5.3) and generated a local field (Chapter 13 and Section 15.3). In
a series of papers[706,120,702], this model was developed and experimental
evidence accumulated. Fig. 14.14 is taken from their last paper[702] which
contains the most supportive data for their equation. The solid line shows
their (five parameter) model's prediction (eqn. 10 in Reference 142) and
the dashed line shows the best fit using a 2-parameter inverse power law;
both lines are a good fit. The data points are based on a variety of techniques,
fully described in Reference 702, including progressive-stress and constant-
stress tests at different frequencies, and an impulse test; some are scaled to
'equivalent 60 Hz lives'. It is therefore likely that some errors are introduced
by the assumptions made. It is difficult to assess confidence limits in this
case but likely typical limits are shown on the graph (±5%—±10% for the
breakdown voltage measurements and ±50% for the time-to-failure
measurements). Including these limits makes it difficult, at least on a statistical basis, to justify the use of the 5-parameter equation instead of the much
simpler 2-parameter inverse power law.

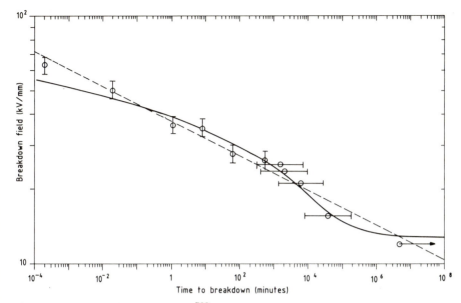

Fig. 14.14 Data of Bahder *et al.*[702] for miniature triple-extruded XLPE cable at 22°C. The data points represent a variety of tests and should not be assumed to be simple constant-stress test results, refer to the original reference for details. The solid line is the workers' fit of their equation (eqn. 14.16) and the dashed line represents a simple 2-parameter inverse power law. The error bars are our approximate estimates of the 90% confidence limits

Before summarising evidence for the different breakdown statistics presented and the form of the relationship between working lifetime and applied stress, the few available published results from service experience will be presented in the next section.

14.4 Service and field studies

Rather disappointingly, very little has been carried out systematically in terms of service and field studies. Thue *et al.*[741] have analysed the failure rate of 5—35 kV PE and XLPE underground residential distribution cables and their data has been reviewed by Kiss *et al.*[192] and Bahder *et al.*[706]. The cumulative probability of failure per 161 km (100 mile) section of PE cable has been derived from this data and is shown in Fig. 14.15. The data indicates support for a Weibull distribution at constant stress (eqn. 14.6) and indicates a time exponent, *a*, of 1·7. The slight deviation from the straight line at longer times may be a statistical artefact due to censoring[719] or may possibly be due to the influence of long-term low-level degradation. Whilst the interpretation of this data appears straightforward, there are two problems associated with it: (i) whilst the number of cables installed each year is known, it is not known how old a particular cable was when it broke down; and (ii) when a cable breaks down it is repaired and so the

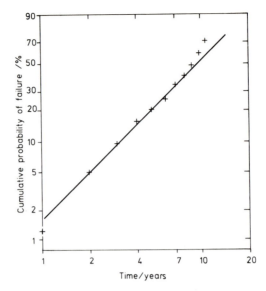

Fig. 14.15 The cumulative probability of failure per 161 km (100 mile) section of polyethylene cable in service. The data is taken from Kiss *et al.*[192]. The slope is 1·7

aging process in that cable continues unlike in a laboratory specimen where it replaced with a new one. Only if the same section of cable breaks down frequently is it likely to be replaced. In order to illustrate the effect of these problems, we have re-plotted some estimates by Bahder *et al.*[706] of the number of failures per 161 km per year as a function of estimated number of years in service on a log–log plot, Fig. 14.16. This develops into a not-unreasonable straight line which is what one would expect if one assumed that the number of failures per 161 km per year were proportional to the hazard function since, from eqn. 14.8, $h(t) \propto t^{a-1}$. However the slope of the graph in Fig. 14.16 is 7·3 suggesting a value for a of 8·3 which is unreasonably high. The reason for this could be due to a number of factors. It is possible that the graph would be better plotted with a linear time scale; this also results in a reasonable fit (see Bahder *et al.*[706]) and would suggest an exponential (not a power law) relationship between the number of failures per 161 km per year and the number of years in service. This may or may not be true but it is certainly true that the number of failures per 161 km per year is not directly proportional to the hazard rate. This is because a cable is normally repaired upon breakdown so that in eqn. 14.4, i.e. $h(t) = g(t)/P_S(t)$, the reliability function $P_S(t)$ is close to one (although the cable will eventually be discarded). The curve may then be more related to $g(t)$, the probability density function, than $h(t)$. In this case we may be seeing the beginning of the increase of a curve similar to that in Fig. 14.2c with a equal to between 1·5 and 2·0. Whilst the data of Thue *et al.*[741] excludes known causes of breakdowns such as dig-ins, there are other complicating factors associated with the cable history. For example the earlier cables were

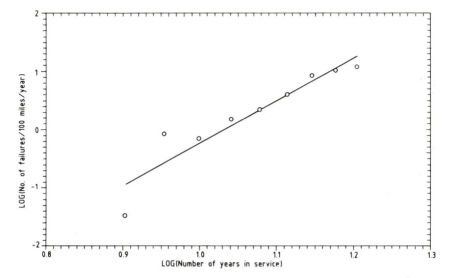

Fig. 14.16 A log–log plot of the number of failures/100 miles/year of polyethylene cables in service as a function of the number of years in service taken from estimates by Bahder *et al.*[706]

operated at higher stresses and were more vulnerable to water treeing, later on, with the introduction of the triple extrusion process, water treeing was considerably inhibited.

Most manufacturers do not publish widely the results of progressive-stress breakdown tests on their 'as-normal' cables. There are however a reasonable number of such reports from laboratories which have studied the effects of various aging techniques on their cables. Generally these are displayed as Weibull plots and exhibit reasonably good fits again supporting the use of the Weibull distribution. Jackson and Swingler[172] have presented results of progressive-stress tests carried out by Williams (reported in a private communication to them) on samples cut from a single length of 15 kV polymeric cable. Some of these have been re-plotted and are shown in Fig. 14.17. The parameters of this graph are $V_c = 34\cdot5\,\text{kV}\cdot\text{mm}^{-1}$, (eqn. 14.21) and slope, $a + b = 5\cdot23$ (these are maximum likelihood estimates and are slightly different from Jackson and Swingler's original estimates). This is reasonable if rather low, assuming a value of a of about unity then a value of $b \sim 4$ is obtained. However Jackson and Swingler argue that the characteristic breakdown strength varies slightly along the length of a cable so that the perhaps rather low slope of Fig. 14.17, representing a slightly wider than expected Weibull distribution, reflects to some extent a genuine variation in breakdown strength and not simply a stochastic variation in the results. They justified this by estimating the field exponent, b, using different lengths of cable specimens. This is possible since, assuming that the probability of failure per unit time (Fig. 14.1a) is proportional to length, then the expression for the *cumulative* probability of failure under constant-stress

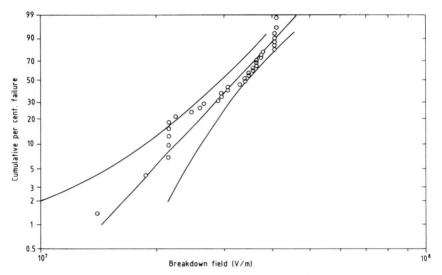

Fig. 14.17   Progressive-stress results carried out on a 15 kV cable by Williams and reported by Jackson and Swingler[172]. The best straight line and 90% confidence limits are shown

conditions (eqn. 14.19*b*) may be modified to:

$$P_F(V, t) = 1 - \exp\{-C' l t^a V^b\}$$ (14.22)

where C' is a modified constant and l the length of the specimen. They carried out two sets of tests on these 15 kV cables, each with only two lengths of cable, and estimated b to be of the order 7—15. This is much higher than the slope estimate and indeed supports their argument that, due to variations in the cable along its length, there is a genuine distribution of characteristic stresses in such systems.

As far as the authors are aware there has not been any systematic study on polymeric cables *in service* to demonstrate the validity (or otherwise) of the inverse power law. However the inverse power law does appear to be assumed widely, although many workers are either skeptical of it or assume there is a threshold voltage below which aging does not occur. Frequently a life exponent, n, of four is used without justification; this seems to have stemmed from the value estimated by Montsinger[324] in 1925 for an oil/board system.

Naybour[134] has studied polymeric power cables, identical to those used in service, subject to water treeing and has identified a power-law characteristic with threshold voltage. The cables were triple extruded with a 3·4 mm XLPE insulation. He carried out constant-stress tests with different fields in the range 4—25 × 10^6 V · m^{-1}. These were fairly exhaustive tests with reasonable size populations (typically 10—15 samples at 5 levels of stress resulting in times up to ~10^4 hours) and no artificial acceleration. He found that above a stress of 8 × 10^6 V · m^{-1} there was obvious power-law behaviour with $n \approx 4·7$. Below this stress the lifetime was much longer than that predicted by this power law and it was estimated that there was either a threshold voltage or the exponent n had increased to about 25. This appears to support the laboratory evidence already discussed for a three-parameter inverse-power law, at least over a limited range of stress, and in the next chapter, we look at physical models which suggest that this is a reasonable form of behaviour.

Chapter 15
Stochastic models of breakdown

15.1 Statistical and physical connections

The deterministic models of breakdown discussed in Part 3 are characterised by a critical field below which no breakdown takes place and a specific time to breakdown for fields in excess of the critical value. In practice however breakdowns are distributed in time at a given field and over a range of fields, and Chapter 14 discusses the type of statistics that have been empirically found to describe the experimental data. It is the aim of this chapter to discuss those models of breakdown which attempt to develop a physical origin for the breakdown statistics. Before moving onto a detailed description of the various models however we will first identify some of the general features of a breakdown process that may give rise to stochastic behaviour. This will be followed by a brief exposition of extreme value statistics which yield the statistical functions appropriate to breakdown, in an attempt to determine the conditions that lead to their adoption.

15.1.1 Physical origins of statistical behaviour
The progress of a breakdown process following the application of a high voltage is generally characterised by three stages, namely initiation, propagation and runaway. In a deterministic mechanism each stage generates the conditions for the occurrence of the succeeding one in a fixed time determined by those conditions (e.g. of voltage, temperature, material). When the mechanism is stochastic however the time required for the different stages is statistically distributed, leading to a range of times to failure (t_b) under the same conditions. There are two possible reasons for this statistical behaviour:

(i)   breakdown is a local process which may occur in different regions of the material each differing in their local conditions;
(ii) breakdown involves alternative sequences of events with each step in the sequence not being automatically generated by the completion of the previous step.

Case (i) above originates in the material inhomogeneity which is a basic feature of polymeric insulation, and its resultant statistics (which we will call class I) will be determined by the spatial distribution of the parameters that control t_b. In case (ii) the distribution of t_b will depend upon the dynamic interplay of the developing breakdown and the variation in the event sequence. Its statistics (which we will call class II) is therefore likely to be determined by the nature of the mechanism itself rather than the material system.

All stages in the breakdown process may contribute independently to the distribution in t_b, and a general description of the breakdown statistics can be expected to involve a complicated convolution. However it is often the case that one of the stages dominates the statistics and hence simplifies matters. The conditions under which this occurs are as follows.

(a) *Initiation* When the time required for propagation and runaway is very short compared to the shortest initiation times, the breakdown statistics will be determined by those of the initiation stage. If this stage depends upon factors such as local field enhancements at sites in the polymer its statistics[398] will be of class (I), whereas when initiation is the result of probabilistic events at similar (or the same) sites such as discharges in voids they will belong[381] to class (II).

(b) *Propagation* Often at high voltages propagation is the rate determining step, as for example in the case of electrical trees (see Chapter 5). Usually propagation is via local filamentary paths with its spatial extent determined by the stochastic growth process[390] (class II), although sometimes it may be governed by material inhomogeneities[434,163] (i.e. weak paths).

(c) *Runaway* This stage will only determine the breakdown statistics on those rare occasions when the preceding stages do not achieve the necessary and sufficient conditions for breakdown to occur without the aid of an ancillary process. Such a process will contribute its own statistics to the breakdown and the most likely candidates are dynamic fluctuations either in the environment of the pre-breakdown entity or within it. For example density fluctuations at the surface of a bush-type tree in liquids[330,331] or charge density fluctuations around the tips of an electrical tree in solids[120] may fit into this category.

As yet research in this area is insufficiently developed to allow any combination of the statistics due to these three categories to be made in any meaningful way, and the stochastic mechanisms reviewed in the succeeding sections essentially concentrate on just one of them. It is essential to note that their results are limited to a given stage and in practice a cross-over to a different stage may occur leading to different statistics even though the fundamental breakdown mechanism may not have changed. Thus different statistical forms may be obtained according to the type of test used. For example in thin laboratory samples low-field life tests and fast ramp tests are likely to probe initiation primarily, whereas high-field life tests are more apt to be dominated by the propagation stage. On the other hand breakdown in service systems with thick insulation will tend to be controlled by the propagation stage except during rapid voltage ramps.

15.1.2 Extreme value statistics

Consider a system which possesses a range of regional properties, or may experience a range of events of different magnitude. The distribution of such entities (events or regions) available to the system is describable by a probability function $(G(z))$, termed the initial distribution, which is a function of a scaled (dimensionless) variable z which characterises the properties of the entity (e.g. local breakdown strength, event magnitude etc.). In any finite-sized specimen only a selection of regions from the initial distribution will be present, or equivalently it will experience only a selection of events over a finite period of time. The maximum and minimum values of the variable z found in such selections will be finite and are termed the extrema of the selection. Different selections will give different extrema which will themselves be distributed. Their distribution function, where it exists, is termed an extreme value distribution function and is definable asymptotically for an infinite number of selections. This branch of statistics has been extensively researched and is thoroughly reviewed by Gumbel[248] to whose book the reader is referred for details. Our aim here is to obtain some insight into the physical nature of the initial distributions that give rise to the different classes of extreme value distribution.

Firstly we note that only three classes of extreme value distribution can exist. This is most conveniently demonstrated by means of the stability postulate[248,742]. Here a set of $M \times N$ selections is subdivided into N sets of M selections. The extrema of the original set of NM selections must also be the extrema of the distribution of N extremes taken from the N sets of sub-selections (size M). Thus the condition for the extreme value distribution function to be stable is

$$\Lambda(Z) = \Lambda^N (a_N Z + b_N) \qquad (15.1)$$

where $\Lambda(Z)$ is the cumulative probability (distribution function) that the largest selection of z is less than Z. The right-hand side of eqn. 15.1 is the joint probability that $z < Z$ in all N of the sub-sets. Stability requires that the extreme value distribution has the same functional form, $\Lambda(\ )$, in each sub-set as in the original NM set. The change in number of selections available (NM to M) on sub-division will however introduce a linear change in the variable from Z to $a_N Z + b_N$. Only three types of solution for $\Lambda(Z)$ exist corresponding to

$$a_N = 1 \qquad b_N \neq 0$$

$$a_N < 1 \qquad b_N = 0;$$

and

$$a_N > 1 \qquad b_N = 0.$$

These are:

$$\Lambda_1(Z) = \exp(-e^{-Z}) \qquad -\infty < Z < \infty \qquad (15.2a)$$

$$\begin{aligned} \Lambda_2(Z) &= \exp(-Z^{-\beta}) \qquad Z > 0 \qquad \beta > 0 \qquad (15.2b)\\ &= 0 \qquad\qquad\quad Z \leq 0 \end{aligned}$$

$$\begin{aligned} \Lambda_3(Z) &= \exp[-(-Z)^\beta] \qquad Z \leq 0 \qquad \beta > 0 \qquad (15.2c)\\ &= 1 \qquad\qquad\qquad\ Z > 0 \end{aligned}$$

where the subscripts 1, 2, and 3 denote the 1st, 2nd, and 3rd asymptote respectively as defined by Gumbel[248]. Unfortunately the literature is not consistent in nomenclature[742], and the first asymptote is also often confusingly called *the* extreme value distribution.

The initial distributions, $G(z)$, that yield a specific extreme value distribution are said to belong to its 'domain of attraction'. Their nature is determined by the restrictions on the constants a_N and b_N in the transformation of the variable z due to the subdivision. In the case of $\Lambda_1(Z)$ the required 'scale' transformation leaves the form of $G(z)$ invariant to a linear translation, and hence $G(z)$ must have an exponential form as can be seen from the exponent of eqn. 15.2a. The sort of system that would exhibit this type of distribution will be one where z refers to an aggregated property built up by the random addition of identical units, as for example the clusters formed in a random defect system[398]. Other distributions belonging to this domain are those which fall off exponentially or faster, such as the normal (or Gaussian) and are typically what might be expected from material inhomogeneities.

The initial distribution for the other two asymptotes (eqns. 15.2b and 15.2c) are characterised by being invariant to scale changes for large values of z. More specifically $G(z)$ belongs to the 'domain of attraction' of $\Lambda_2(Z)$ if

$$\lim_{z \to \infty} \frac{1 - G(z)}{1 - G(hz)} = h^{\beta} \tag{15.3}$$

for every $h > 0$, and to $\Lambda_3(Z)$ if

(a) there exists a z_0 such that:

$$G(z_0) = 1 \quad \text{and} \quad G(z_0 - \varepsilon) < 1 \tag{15.4a}$$

for every $\varepsilon > 0$; and
(b) the following condition is met:

$$\lim_{z \to -0} \frac{1 - G(hz + z_0)}{1 - G(z + z_0)} = h^{\beta} \tag{15.4b}$$

The conditions (15.3) and (15.4) imply that the initial distributions that belong to the 'domains of attraction' of $\Lambda_2(Z)$ and $\Lambda_3(Z)$ are characterised by a power law in their probability density function $(dG(z)/dz)$ as the limit of large z is approached. The two differ in their range of applicability and in the requirement that $G(z)$ have an upper bound if it is to lead to Λ_3. It should be noted that a linear transformation of the variable (i.e. $Z \to Z_0 - x$ in the case of Λ_3) results in an effective variable which is not constrained to be less than zero as in (15.2c) or greater than zero as in (15.2b), nonetheless there must still be an upper bound for Λ_3 to apply. This form of initial distribution is what would be expected of self-similar (fractal) systems, both in their spatial organisation[743,744,745], and in the distribution of events in time (e.g. fractal time processes[746,747] for which the distributions of waiting times has a long-time power-law tail, $t^{-(1+\alpha)}$). Dynamic stochastic processes such as the propagation of electrical trees[390] (Chapter 5) and discharges[289] fall into this category and can be expected to lead to extreme value statistics of the Λ_3 type[748].

It should be noted that the exponential distribution which is a particular case of Λ_3 (i.e. $\beta = 1$) corresponds to a trivial case of self-similarity in the initial distribution, namely one in which the probability density is a constant. In this case the probability of finding z in the interval between z and $z + dz$ is independent of the value of z, i.e. the initial distribution is that of a random white noise process in z.

The different extreme value distributions can also be interchanged by means of a transformation of the variables, in particular a logarithmic transformation will convert Λ_1 to either Λ_2 or Λ_3. For example an exponential (Weibull with $\beta = 1$) extreme value distribution in time, will become a first asymptote (Λ_1) distribution in field if the characteristic time (τ_c) is a field activated process[704] (i.e. $\tau_c \propto e^{E/KT}$). The appropriate statistic may thus depend upon the choice of variable whose extrema is required. In the case of breakdown statistics this will be the property which leads inevitably to failure once it reaches a critical level.

The equivalent distribution of smallest values, i.e. the probability that the smallest value is less than Z is obtained by subtracting $\Lambda(Z)$ from unity, and in the second and third asymptotes reversing the range of validity. It should be noted that whereas the stability postulate ensures that a given extreme value distribution for largest (smallest) is also a valid initial distribution for the same statistic, only the first asymptote for the *largest* (*smallest*) is a valid initial distribution for the alternative extreme (*smallest/largest*) statistic of the same type (Λ_1). If the third asymptote Λ_3 of *largest* (*smallest*) is used as an initial distribution the extreme value distribution of *smallest* (*largest*) is the first asymptote Λ_1. However if the second asymptote of *largest* (*smallest*) is the initial distribution there is no stable distribution of the *smallest* (*largest*) extreme.

The distribution of smallest values equivalent to $\Lambda_3(Z)$ is

$$P_3(Z) = 1 - \exp\left[-Z^\beta\right] \qquad Z \geq 0 \qquad (15.5)$$

where $P_3(Z)$ is the probability that the smallest value of z selected is smaller than Z. This is the distribution developed by Weibull[700] to describe breakdown statistics and has come to be of major importance in the analysis of dielectric failure (see Chapter 14). The application of this particular extreme value statistic to breakdown can be rationalised as follows. The distributed variable is assumed to define regions in terms of their times to failure following the application of a field or alternatively their breakdown strengths. Thus the probability that a breakdown has occurred prior to (or below) the time (field) represented by Z is that of the smallest selection of z available being less than Z. That is, an inevitable breakdown occurs if the weakest region fails as for example mechanical failure in the weakest link of a chain. This line of reasoning leads immediately to $P_3(Z)$ for the cumulative probability of failure as long as the necessary conditions on the initial distribution can be satisfied. Here $[1 - P_3(Z)]$ is the probability that the smallest z is greater than Z, and is thus that of survival to Z. It should be noted that the appropriate $G(z)$ applies only to the initial distribution of locally initiated whole system breakdowns and not to that of isolated local degradation (e.g. from void discharges), sometimes also called local breakdowns.

The analysis presented in this section shows that the basic condition for the applicability of Weibull statistics to breakdown is that of self-similarity in an initial distribution of potential breakdowns bounded at its upper end. In other cases the statistics are likely to be of the first asymptote, although most disconcertingly there may be situations in which no stable distribution function exists. This latter case is an example of chaos in which infinitesimal changes can lead to vastly different results, and predictions cannot be made.

15.2 The fluctuation model

Non-crystalline materials possess a certain amount of free volume, or equivalently configuration entropy (see Chapter 1). As a consequence the ions, atoms, or molecules do not occupy a given 'lattice' site continuously but instead move around from site to site in time whilst maintaining the total free energy at a constant value[730,749]. Such fluctuations in site occupancy will redistribute local mechanical strains or dipole orientations (if polar species are present) about the system, and hence will influence the relaxation currents involved in linear response theory[749]. A general model has been developed for the distribution density of a system variable whose magnitude is controlled by the fluctuations, and the resulting expression applied to the determination of dielectric[26,448,730,27] and mechanical[750] response functions. It has also been pointed out[25] that when the variable concerned was a local electric current the fluctuations would give rise to a noise component with power law, $f^{-2(1-m)}$, $(0 < m < 1)$ spectral characteristics, which would be superimposed on a small DC current driven through the system. The exponent $2(1-m)$, usually denoted by α, depends upon the shape of the fluctuation distribution through m and was found to lie betwen zero (white noise) and two (Gaussian noise) in accordance with experiment[751,752].

In this model the material is divided into local structural units, termed clusters, which are composed of structural elements (atoms, molecules, chain segments etc.) whose displacement motions are connected together[753], Fig. 15.1. Any chosen cluster may aggregate by connecting the motions of its elements to those of its neighbours or fragment by means of internal disconnections. In this way the number of elements and their configuration in a particular cluster will fluctuate in time under the constraint that the free energy is maintained at a constant average value for a given volume of material. The fluctuations may therefore be regarded as local density fluctuations and molecular reconfigurations which occur in liquids[330] and non-crystalline solids because of the presence of free volume. However other cases may be found where the rearrangements do not have such a clear structural interpretation, for example connection switching in a matrix such as a percolation or hydrogen-bond system.

The probability of any given cluster adopting a particular configuration (density etc.) will be governed by a probability density $g(y)$. In the cluster

(i)

(ii)

(iii)

(iv)

(v)

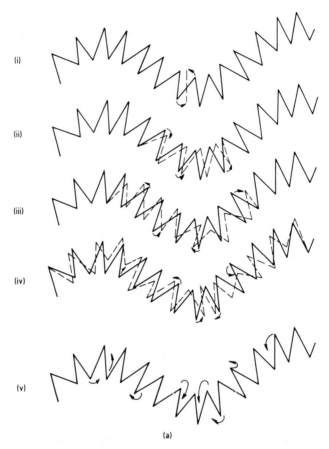

(a)

Fig. 15.1   Examples of the connected motions that give the dynamic fluctuations in cluster configuration[749]. (*a*) A twisting chain, (*b*) A distorting lattice defect (only the motion of one site is shown). The cluster displacements are constructed cumulatively from a hierarchy of sub-group motions (e.g. (i) to (iv)). At each level of the hierarchy the sub-groups share the same fixed energy among their constituent elements

model $g(y)$ was derived by assuming that a given amount of energy[730] or information content[754] was shared equally among all the clusters irrespective of their size, thereby constraining the free energy (or equivalently the strain) to be constant. This assumption is equivalent to taking the potential cluster configurations of the distribution to be self-similarly related[754], that is the clusters have the same geometrical construction in terms of smaller sized groups, as these groups have with respect to the component elements (for example see Fig. 15.8). The resulting distribution density is given by

$$g(y) = \frac{y^{(1-m)/m}}{m(y^{1/m}+1)^2} \qquad 0 \leqq m \leqq 1 \tag{15.6}$$

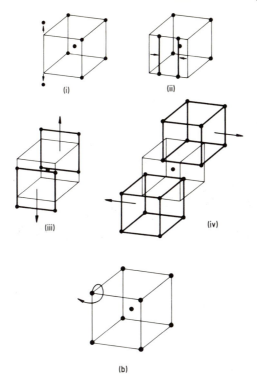

(i) (ii)

(iii) (iv)

(b)

Fig. 15.1 Continued

where y is a cluster variable (e.g. density, polarisation, space-charge density, order parameter) scaled to its value in the characteristic cluster.

The fractional exponent m in eqn. 15.6 is an index measuring the degree of connection between the motions in different clusters[753], that is m measures the extent to which changes in one cluster induce corresponding changes in another and thus redistribute free volume (configuration entropy) and strain about the system. It can vary between zero, in which case individual clusters are isolated from each other with an identical characteristic size and $g(y)$ is a delta function in y; and unity, in which case all clusters are connected with constant probability per connection and $g(y)$ is quasi-exponential. The general form of $g(y)$ varies continuously between these two extremes and some selected examples are shown in Fig. 15.2.

When an electric field is applied to the system those clusters possessing polar groups or space charge will modify its magnitude in their locality through polarisation (or depolarisation) and space charge fields. Local currents arising from the connected motions of space charge in a cluster will also contribute to the local field as long as the system has a finite resistance, even though their average value is zero. The intrinsic fluctuations described above will thus cause the local electric field associated with a given cluster to vary over a range of values with a probability density related to

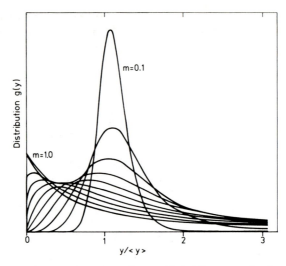

Fig. 15.2   The distribution function of eqn. 15.6. The magnitude of the variable, *y*, has been scaled in terms of its average value for each value of *m*. The parameter *m* has been incremented in 0·1 steps.
After Dissado *et al.*[174]. Copyright © 1984 IEEE

$g(y)$. Over some part of this range the breakdown field may be exceeded, and this led Hill and Dissado[173,703] to suggest that the model would serve as a suitable basis for the derivation of breakdown statistics in a number of cases.

15.2.1 Homogeneous breakdown
This class encompasses those mechanisms in which breakdown occurs, so rapidly as to be effectively instantaneous, once a critical value of the field is exceeded in a local region. Their breakdown statistics are thus governed by the initiation process. Examples of mechanisms which may fall into this class are intrinsic breakdowns[152] (thermal, electrical, or mechanical) initiated in the bulk material, or electrode injection breakdowns controlled by internal space charge fields[434,755,756].

The variable *y* of the fluctuation model can be related to the local field by noting that in the cluster model of dielectric relaxation it is proportional to the scaled magnitude of a cluster relaxation current[749,753,754] and hence to the electric field experienced by the cluster. By equating the energy density (due to space charges, displacement field, and local currents) in the characteristic cluster with the average value arising from the applied field (E_a), the magnitude of the electric field in the characteristic cluster can be found to be $\sim (E_a - E_1)$. Here E_1 is the RMS value of the field corresponding just to currents. Thus

$$y = E/(E_a - E_1) \tag{15.7}$$

where E is the local field.

The probability of survival, P_S, of a given cluster will be the probability that $y < y_0$ where y_0 denotes local enhancement to the critical field E_0 for initiation of breakdown. P_S can be found by integrating $g(y)$ over the range zero to y_0 and gives[173]

$$P_S(E_a) \approx \exp\left[-\left(\frac{E_a - E_1}{E_0}\right)^{1/m}\right]$$ (15.8)

It is interesting to note that $g(y)$ belongs to the class of initial distributions that satisfies eqn. 15.3. Hence the probability that the largest value of y selected will be less than y_0 (i.e. the *survival* probability) can be written directly in the form of eqn. 15.8 by equating $(1/m)$ with β of eqn. 15.2b.

If the number of clusters at risk is N (equal to the system volume divided by the cluster volume, $V/(2\xi)^3$, when every cluster is a potential breakdown site), the survival probability is the joint probability of all N clusters surviving i.e. $[P_S(E_a)]^N$. Allowance must also be made for the number of times a breakdown is attempted during the time, t, over which the stress is applied. Taking this to be νt, where ν is an attempt frequency, the probability of survival to time t becomes $[P_S(E_a)]^{N\nu t}$, giving the failure probability $P_F(E_a, t)$ as

$$P_F(E_a, t) = 1 - [P_S(E_a)]^{N\nu t}$$
$$= 1 - \exp\{-\nu N t(\beta)^{1/m}[(E_a - E_1)/E_0]^{1/m}\}$$ (15.9)

Here β is a geometrical field enhancement factor of order unity. In some cases the frequency ν may be an intrinsic factor, such as the rate of production of electrons for avalanche initiation or the rate at which a breakdown generating cluster configuration is attempted. However ν may also appear because the system is repetitively stressed as in AC fields or a sequence of pulses.

The derived breakdown probability, eqn. 15.9, has the form of a 3-parameter Weibull distribution (i.e. $1 - P_F(x) = \exp\{-[(x - x_0)/x_c]^b\}$, see Chapter 14) with a threshold field, E_1, and values of $a = 1$, $b = 1/m$ and $C = \nu N(\beta/E_0)^{1/m}$. As $a = 1$ this represents the special case of an exponential distribution in time, and though we might expect the model to apply to breakdown in thin films[378], it will not be possible to distinguish it from a process randomly initiated in time with a time to breakdown which is a power of the field.

15.2.2 Aging or tree assisted breakdown

In the previous subsection the system was considered not to have changed in between the attempts at breakdown, i.e. aging was not allowed. Here this restriction is relaxed. The type of aging considered is one in which a progressively developing imperfection in the insulation increases the number of clusters which are at risk at any time. Breakdown however is still a rapid intrinsic process once the field in a cluster at risk exceeds E_0. This type of model would be applicable to situations in which the imperfection supplies a factor essential to the ability of a cluster fluctuation to enhance the local field. For example electrical trees in liquids or solids may supply space charge to the clusters at their surface, and fluctuations could then generate a runaway breakdown as in bush-type trees (Section 6.2). The number of clusters at risk in this case would be proportional to the surface area of the tree, Fig. 15.3. Similarly water trees may generate charges and deterioration on their boundary leading to electrical tree formation (Section 6.1).

Taking

$$L(t) \propto t^{1/d_f} E_a^s \tag{15.10}$$

as the general expression for tree propagation (Chapters 4 and 5), the number of clusters at risk will be proportional to the surface area $L(t)^2$, and eqn. 15.9 becomes

$$
P_F(E_a, t)
$$

$$
= 1 - \exp\left\{-\frac{\gamma}{(2\xi)^2} E_a^{2s} t^{1+2/d_f} \beta^{1/m} \left[\frac{E_a - E_1}{E_0}\right]^{1/m}\right\} \tag{15.11}
$$

where γ is a factor which includes the proportionality constant of eqn. 15.10. Note that this expression differs from those previously published[173,174] by an extra power of t since they did not take into account the failure attempt

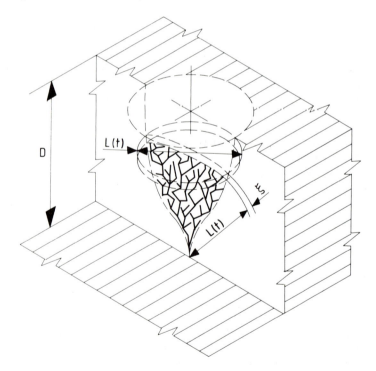

Fig. 15.3   Schematic drawing of electrical tree of length $L(t)$ growing in a material of thickness D. The surface shell of width ξ contains the volume elements that are placed at risk of initiating a runaway breakdown by the presence of the tree

rate. Here again a Weibull distribution is obtained, but now the time exponent a (equal to $1+2/d_f$) is greater than unity as would be expected for a system which ages electrically. The values of a predicted lie between 1 and 3, for example water tree deterioration would give a value of $\sim 1\cdot 7$, whereas a for electrical tree aging would lie either around $1\cdot 8$ or $2\cdot 3$. The field exponent b is now supplemented by the field dependence of the tree propagation (eqn. 15.10), and becomes $(1/m+2/3)$ for water trees, and $(8+1/m)$ for electrical trees.

In the foregoing we have used trees as an example of progressive deterioration, however the model is not necessarily restricted to this situation. For example if the aging involves the formation of defect clusters[132,133] containing dipoles and charges, the breakdown statistics would take the Weibull form when the number of such defects increased as a power of the time and field.

In the case of vented water trees the number of defects that may initiate breakdown at high fields (see Section 6.1) will be proportional to the volume of the tree and a will equal 2. This will also be the value of a if $N \propto t$ such as may be the case for internal void discharges, but its value would approach 4 if N were proportional to the total volume of bow-tie water trees (see Fig. 4.20). Alternatively aging may reduce the magnitude of the critical field E_0 throughout the system, and when the effective critical field is proportional to an inverse power of the time, the breakdown statistics will take a form similar to that of eqn. 15.11. It should however be remembered that in the models of this section, the aging process is not of itself sufficient to cause breakdown. Instead it alters the local conditions in such a manner as to enhance the probability of a fluctuation achieving the critical field for the initiation of a runaway process, the mechanism of which could in principle occur in the unaged material.

15.2.3 Filamentary conduction paths and tree initiated breakdowns

In the previous two sub-sections the breakdown statistics were related to the distribution of conditions required for the initiation of a runaway process. Here the time required to propagate a conducting path to a critical length is the basis for the statistics instead. Firstly it is assumed that the length of the path is a power law function of the local field, E, and the time of growth $(t-t_I-t_1)$, i.e:

$$L(t) = KE^s[(t-t_I+t_1)^{1/d_I} - t_1^{1/d_I}] \qquad (15.12)$$

where t_I is the tree inception period during which the tree does not grow[122,30]. A further parameter, t_1, is included, as in water trees (see Chapter 4), so that an infinite growth rate at $t=t_I$ is precluded. The field power in eqn. 15.12 appears because the process is field driven, and the exponent s will vary with the mechanism, taking a value around 4 for electrical trees. Alternatively a value of 2 may be expected for power driven processes although stronger field dependencies may also[162,588,589] occur. It is unlikely that the linear dependence of ohmic processes would be involved in the formation of such a path during dielectric breakdown.

It is now suggested that breakdown occurs either immediately the path crosses the system (thickness D), or alternatively when a rapid runaway stage is initiated at a critical length D. Inversion of eqn. 15.12 thus yields a time to breakdown as a function of the local field that governs propagation. When this field is subject to fluctuations, application of eqn. 15.8 gives the probability that the longest path at t is less than critical for a given externally applied field, E_a, i.e. the probability of survival to time t. Thus the probability of failure becomes

$$P_F(E_a, t) = 1 - \exp\left\{-N\beta^{1/m}(E_a - E_1)^{1/m}\left[\frac{(t - t_I + t_1)^{1/d_f} - t_1^{1/d_f}}{D/K}\right]^{1/ms}\right\}$$

(15.13)

which has a Weibull form possessing both field and time thresholds. At large fields the strong inverse field dependencies of t_I and t_1 reduce them effectively to zero (see Chapter 5) and the Weibull exponents are $a = (d_f ms)^{-1}$ and $b = 1/m$, i.e.

$$P_F(E_a, t)|_{E_a \to \text{large}} = 1 - \exp\{-N\beta^{1/m}E_a^{1/m}t^{1/(msd_f)}(D/K)^{-1/(ms)}\} \quad (15.14)$$

Here breakdown is governed entirely by the propagation of a conducting path through the system, with the critical factor a length D rather than the field E_0 of the previous sections.

The material properties can be expected to enter expression 15.14 through the proportionality constant K of the propagation eqn. 15.12. In contrast to Section 15.2.2, the growth of a conducting path is itself the breakdown mechanism, rather than a means of enhancing the probability of its initiation. For this class of behaviour the Weibull parameter a will usually be less than unity unless b is very large, with $a \propto (d_f s)^{-1} b$ where the exponents d_f and s will be determined by the type of mechanism and its propagation dynamics. If for example the conducting filament were formed by an electrical tree type process, s could be taken to be 4, d_f either 1·7 or 2·5 (see Chapter 5), and hence a would lie in the range 0·2 to 0·3 for a value of 2 for b ($m = 0.5$). Only when $m < 0.15$ (i.e. $b > 6.5$) will a exceed unity. Thus this type of process may be thought of as 'conditioning' the material rather than aging it, as in Section 15.2.2. A value of $d_f > 1$ (branched propagation) contributes significantly to this result since it ensures that the growth rate of the tree reduces with time and thus introduces the possibility of a reduction in the hazard rate when failure is due to the more slowly propagating trees. However the essential feature is that the time to breakdown decreases more rapidly with increasing field than does the probability of realising the field, and hence the shorter the time to breakdown the more probable its occurrence.

Processes which belong to this category of statistics are those for which the dominating stage is the propagation of a conducting filamentary system. They include filamentary thermal runaway[588,589] and filamentary discharges involving sequential local avalanching[346]. The theory of partial discharge tree-initiated breakdowns proposed by Bahder *et al.*[120], may also lie in this class, since the criterion for breakdown is growth of a current carrying tree to a critical length. The discharge inception field which is an important feature of this theory[685] would be equivalent to the threshold field for the initiation (i.e. $t_I < \infty$) of the path in the present case. In addition the critical length D may also be a function of the applied field[120] (see Chapter 13).

15.2.4 Comparison with experiment

A summary of the Weibull exponents derived on the basis of the various models is given in Table 15.1, in terms of the model parameters.

In all of the above situations the Weibull exponent b approaches infinity when m is zero and the intrinsic fluctuations are non-existent. The initial distribution reduces to a delta function (Fig. 15.2) in this case and the breakdown process is deterministic (see Part 3, Fig. III.1), with a specific field for breakdown in either constant stress or ramp tests where the slope of a Weibull plot $(a + b)$ would be infinite. In practice such behaviour can only be expected for gases or zener diodes, although the high field (intrinsic) breakdowns of many systems including polymers may closely approach it. The mechanisms involved in this latter case are likely to be spatially uniform, such as electro-mechanical breakdown[138], electro-ionisation[152], or thermal runaway[757]. Structural fluctuations will occur in most non-crystalline materials, however, causing local changes in molecular arrangement, and in general we may expect $b < \infty$ (i.e. $m > 0$), giving a distribution of breakdown fields about a most probable value. It should be noted that in this context phonon modes are excluded since they refer to vibrations about fixed sites whereas the fluctuations referred to relate to displacement of the centres of vibration themselves.

(a) Predicted Weibull exponents for different types of fluctuation
One difficulty in relating the experimental values of b to the model predictions is that of ascertaining a suitable means of determining m. This occurs because polymers are morphologically complex systems, containing a hierarchy of structural features within the same material. For example the semi-crystalline polymers possess a chain molecular structure, a random chain entanglement network in amorphous regions, chain folding to form crystal lamellae, close-packed and twisted lamellae and in many cases spherulitic structure. High molecular weight amorphous polymers replace

Table 15.1 Summary of Weibull time a and field b exponents, in terms of the fluctuation exponent m, the fractal dimension of propagation of the pre-breakdown process d_f, and its deterministic field exponent s

Model	a	b	
Homogeneous breakdown	1	$1/m$	
Aging and tree-assisted breakdown	$1 + 2/d_f$	$2s + 1/m$	(2-D)
	$1 + 3/d_f$	$3s + 1/m$	(3-D)
Conduction path breakdowns	$1/(msd_f)$	$1/m$	

the spherulitic crystalline organisation with highly entangled, densely packed, knotted regions, termed 'blobs' by de Gennes[758,743].

Fluctuations of the local magnitude of macroscopic properties such as density, heat content, polarisation etc are intrinsic to the thermodynamic description of any system and do not require externally applied forces for their existence. In non-crystalline materials they can occur at any level of the structural organisation, and it is not obvious which one will have a dominant effect upon the breakdown mechanism. The homogeneous free volume breakdown mechanism[759,476] (see Chapter 13) for example can be expected to be influenced by free volume fluctuations resulting from large-scale segmental motions and intra-segmental configuration changes in the amorphous regions. Alternatively water and electrical tree aging will probably also depend upon fluctuations at the amorphous-crystalline interface[457] where the void formation and trapped charge can be found[164]. Such fluctuations occur on a coarser level than that of segmental rearrangements and involve the reorganisation of the surface chain structure of individual lamellae together with the packing arrangement of interwoven lamellae as well as free volume changes. Additionally density fluctuations may not remain confined to the amorphous (or loose packed) regions, but instead may cause spherulites (or high-density blobs) to fluctuate in size and shape, all of which may affect both aging and the inception and propagation of conducting paths[434].

In principle a value for m can be obtained from the low-frequency dependence of the linear response to an applied field such as is given by the dynamic mechanical compliance[750] or the dielectric susceptibility[26,448,730,27]. The form of the theoretical response function of the cluster model[26,448,730,27,750] is shown in Fig. 15.4a, where it can be seen that the low-frequency loss component (χ'') varies as $(2\pi f)^m$ and the real component (χ') as $\chi'(0) - Af^m$. It might be thought that m ought to be derived from the dielectric susceptibility since this response involves only those fluctuations that affect dipole containing clusters and hence the local polarisation field. However structural fluctuations which do not involve dipoles in the intrinsic material will still give field fluctuations at high fields when space charge is present. Since the breakdown models are mostly concerned with this situation it is to the mechanical response, which embraces a wider range of displacements, that we should turn to determine m. This is particularly the case for non-polar materials such as polyethylene, which exhibit only a weak dielectric response[457].

Fig. 15.4b shows the dynamic compliance of polyethylene plotted in the form of a master curve[730,760] assembled from data at 25 temperatures, together with the fitted response function. The relaxation process (γ) which is responsible for this response is still active at low temperatures (~ 100 K) where it has an activation energy of ~ 0.26 eV, and is thought to originate with main-chain crankshaft motions in the amorphous regions[457], probably combined with dislocation and defect migration within the lamella crystals[457,761]. The observed value of 0.065 for m should thus be be associated with free volume breakdown, giving a value of 10—15 for b ($=1/m$) which is consistent with the range 7—12 observed experimentally[759]. Such fluctu-

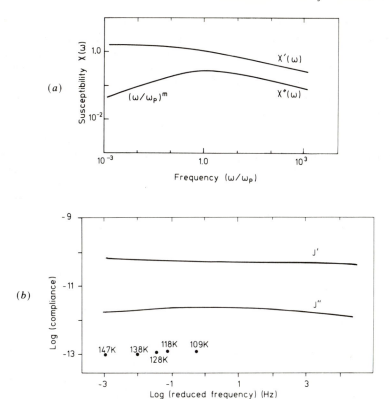

Fig. 15.4
(a) A log-log plot of dielectric susceptibility showing a typical dependence upon rate normalised frequency. The value of the parameter m is[174] 0·3
After Dissado *et al.*[174]. Copyright © 1984 IEEE
(b) A log-log plot of the compliance of crosslinked polyethylene as a function of the temperature reduced frequency. From this plot it can be determined that $m = 0·065$. The points towards the bottom of the figure show the logarithmic (relative) shifts in frequency and amplitude required to bring the response data at the temperatures quoted into coincidence with that for 147°K
After Dissado *et al.*[174]. Copyright © 1984 IEEE

ations may also apply to charge de-trapping in electrical tree initiation[163,164] which also has large values[324] of b (≥ 10). These are processes in which the fluctuations cover a rather small range about a characteristic value. For example the 'full width at half maximum' of the Weibull distribution density, $dP_F(x)/dx$, is approximately[703] $1·5x_p/b$ for $b \geq 6$, where x_p $(= (1 - 1/b)^{1/b}x_c$; x_c = characteristic value) is the value of x at which $dP_F(x)/dx$ has a peak.

By measuring the response due to relaxation on different levels of the hierarchy which will usually appear as separate processes, values of m appropriate to the level may be determined. However many breakdown mechanisms will be affected by fluctuations from a range of structural levels, and the value of m involved in their breakdown statistics may well be a composite of all those observed in response measurements. It is possible that electrical current noise follows this pattern of behaviour. Such measurements have the virtue of investigating directly the fluctuations experienced by a DC current in the material and should therefore be applicable to breakdown via the formation of conducting filaments, and possibly to electrical tree aging. Although the experiments are difficult to perform, polystyrene and polyethylene have been investigated by Pender and Wintle[762], and values of 0·31 and 0·4 respectively obtained[25] for m. These values are more typical of responses originating in segmental motion in the amorphous (α_a) and amorphous-crystalline interface (α_c) regions of polymers[760,763,764]. They imply that b $(=1/m)$ has the small values of 2·5 and 3·3 for polyethylene and polystyrene respectively when the model of Section 15.2.3 is operative. Much higher values ($\sim$10—11) would result in the less likely event that these fluctuations were involved in the aging case.

(b) Effect of changes in morphology
In view of the difficulty of assigning an experimental value of m to a particular breakdown process a better evaluation of the model can be made through the morphological dependence of the Weibull exponents. Fig. 15.5 shows the results of progressive stress tests carried out on crosslinked polyethylene whose lamellae have been thickened with respect to the control[765] by annealing in the mould at 100°C for 63 hours. This procedure also has the effect of increasing the crystallinity substantially (i.e. about 49% to 53%). A change in the slope $(a+b)$ of the Weibull plot from 7·4 to 2·8 was found. Since the value of a was known to be of order unity for the control and other tests indicated $a=0·3$ for the modified system, the major

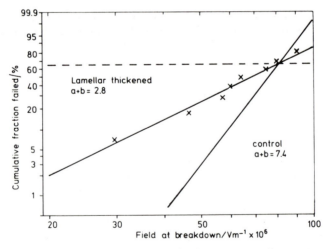

Fig. 15.5 A Weibull plot showing the results from progressive stress tests carried out on polyethylene by Fothergill *et al.*[765]. The lamellae have been artificially thickened with respect to the control without altering the spherulite size. The control data is that shown as ○ in Fig. 14.12

part of the observed difference in $a + b$ must be due to changes in b. If b is taken to be 6·4 and a to be 1 for the control, then the plausible values of $b = 2·4$ and $a = 0·375$ for the thickened system are consistent with the conducting path model. In this case the breakdown mechanism is the same in the two systems, as suggested by their equal characteristic breakdown fields, but differ in the fluctuations that influence them.

In the control relatively weak ($m \approx 0·15$) fluctuations such as those of the γ-relaxation process, Fig. 15.4b, or perhaps the branch[457,761] (β) relaxation, dominate, whereas lamellae thickening appears to have brought all the fluctuations measured in the electrical noise into play. Since the value obtained for sd_f (6·4) is also consistent with a discharge tree (i.e. $s \approx 3·8$, $d_f = 1·68$), this data provides strong circumstantial evidence for a mechanism of the form of Section 15.2.3. However it can be argued that the lamella thickening has led to a greater irregularity of lamellae and inter-lamellae regions, and thus an enhanced distribution of strains and fields due to changes in material inhomogeneity. It should be noted though that the inhomogeneities *must* form a self-similar system if eqn. 15.14 is to be obeyed. This is possible if the dynamic fluctuations implied by $g(y)$ occurred during the formation of the modified material giving an instantaneous range of selections from the cluster possibilities described by $g(y)$, which are then 'frozen-in'. The alternative pictures are thus not necessarily mutually exclusive in their origins and effect upon breakdown.

A somewhat different type of behaviour is shown in Fig. 15.6 where the results of 'self-healing' progressive-stress tests on thin polyethylene films[434,435,611] are plotted. The films were cast from solution and samples were tested either 'as-grown' or 'heat-treated'. The former were shown to be semi-crystalline with a randomly-orientated lamellae-like structure, whereas the heat-treated samples which were cooled very slowly through the melting range exhibited two-dimensional spherulite-like patterns with well-defined boundaries. Here the morphological change has no significant effect upon the slope of the Weibull plot ($\sim$11), but did reduce the characteristic breakdown field by about 50%. In the heat-treated samples the breakdowns were observed to take place almost exclusively through the spherulite boundaries[435] which in these thin samples join the electrodes. It is likely therefore that in both types of samples breakdown occurs in amorphous inter-lamella regions, and that heat-treating has transferred impurities and structurally defective material, from throughout the sample to the spherulitic boundaries[435], where it reduces the resistance to breakdown. Since the fluctuations originate with the same range of chemical and physical inhomogeneities, albeit now concentrated in different locations, the parameter m will be substantially unchanged. The Weibull parameter b will thus remain effectively constant, particularly if the breakdown is the consequence of electrical tree aging (Section 15.2.2), where m would take its electrical noise value of $\sim$0·4, and b would be about 10·5. A conducting discharge is however more likely[434] with the value of m appropriate to that of charge de-trapping via the γ-relaxation. Here it may be noted that the magnitude of this relaxation would increase with such treatment[761] but the shape of its response will be unchanged.

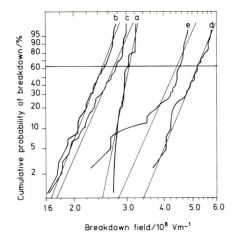

Fig. 15.6   Weibull plots showing the cumulative probability of breakdown of solution grown polyethylene films using DC 'self-healing' progressive stress tests. The results are taken from Kitagawa *et al.*[434,435] and successive experimental data points have been connected by straight lines for the sake of clarity. The curves represent different samples preparations or rates of voltage increase (see text for details). Listed here together with these experimental details are the exact film thicknesses, the characteristic breakdown strengths in units of $10^8 \text{ V} \cdot \text{m}^{-1}$ and the shape parameters, β $(= a + b)$

(a) $10^2 \text{ V} \cdot \text{s}^{-1}$, heat treated, $0.77 \ \mu\text{m}$, $E_b = 3.05$, $\beta = 23$
(b) $10^4 \text{ V} \cdot \text{s}^{-1}$, heat treated, $0.74 \ \mu\text{m}$, $E_b = 2.49$, $\beta = 12$
(c) $10^6 \text{ V} \cdot \text{s}^{-1}$, heat treated, $0.72 \ \mu\text{m}$, $E_b = 2.72$, $\beta = 11$
(d) $10^2 \text{ V} \cdot \text{s}^{-1}$, as grown, $0.69 \ \mu\text{m}$, $E_b = 5.17$, $\beta - 11$
(e) $10^6 \text{ V} \cdot \text{s}^{-1}$, as grown, $0.74 \ \mu\text{m}$, $E_b = 4.40$, $\beta = 11$

(c) In-service failures of power cables
Many premature failures of LDPE power cables in service have been attributed to water trees[192] and the value of 1.7 for a obtained from the Weibull plot, Fig. 14.15, Section 14.4, is consistent with this type of aging mechanism. Although such a value may also result from bush-type electrical-tree aging on the model of Section 15.2.2, detailed inspection of in-service cables[187] seems to rule this option out. A comparison with other data[175,176], such as that given in Fig. 6.2, is however not simple, since these experiments measure the residual life after a period of water-tree formation. Under these circumstances the expressions derived in this section do not apply because they refer to aging produced during the test such as occurs with the service specimens. It does however appear, Fig. 15.7, that in addition to a reduction in breakdown strength the water trees have reduced[176] the value of $a + b$ to ~ 4.5. This reduction corresponds to an increased range of initiation sites for failure as discussed in Section 6.1, which would be consistent with the present model.

Breakdown tests have also yielded values of 1.3—1.8 for the exponent a under conditions in which water trees cannot be formed[685] (see Chapter 14). Here the model suggests that electrical tree or partial discharge[120,685]

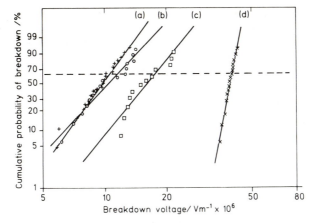

Fig. 15.7 The cumulative probability of breakdown of miniature cable samples as a function of AC voltage measured using progressive-stress tests. The cable samples were 'aged' in electrolyte to form water trees for the following times, the exponents $a + b$ are shown in brackets: (*a*) 1000 hours (5·2); (*b*) 790 hours (4·0); (*c*) 500 hours (4·6); (*d*) not aged (19)
After Bulinski and Densley[175]. Copyright © 1981 IEEE

aging is the relevant breakdown mechanism via the production of a bush-type defect. However other types of cumulative damage (see 15.4.2) cannot be ruled out.

(d) Summary
Values for the Weibull exponents derived on the basis of the analysis of experimental data carried out in this section are quoted in Table 15.2. In the cases of homogeneous and filamentary breakdown the value of *b* is crucially dependent upon the type of fluctuation assumed to affect the

Table 15.2 Summary of Weibull exponents deduced for polyethylene in Section 15.2.1

Mechanism	d_f	a	b	b/a	Comments
Homogeneous *free* volume breakdown		1	10—15	10—15	
Aging Water trees	3	1·7	3—7	2—4	
Electrical trees					
• bush	~2·5	1·8	10—20	5—10	*b* moves to high values
• branched	1·6	2·3			as the fluctuations approach the molecular scale
Filamentary *conduction*	1·6	1	6·5	6·5	Segmental fluctuation
		0·4	~2·5		Current noise

breakdown mechanism, and hence a range is quoted. However the life index b/a in the filamentary mechanism only depends on fractal dimension, d_f, and the field exponent and thus can be predicted reasonably accurately.

The conclusion to be drawn from these experimental results is that they lend qualitative support to the fluctuation model in specific cases. In particular an observed tendency for the exponent a to increase as the exponent b increases is predicted by the conducting path model. However it should be noted that a meaningful comparison with theory often cannot be made because the experimentally determined parameters a and b are too imprecise in consequence of sample sets which contain too few specimens[766]. In general about 40 failures are required to obtain a precision of about 10—20%, and thus it may be difficult to obtain a more quantitative assessment.

15.3 Fractal description of breakdown

In the earliest breakdown theories failure resulted because a bulk process could no longer maintain an energy balance at the applied field. Various critical criteria were proposed. For example Joule heating could supply more energy to the lattice than it could dissipate leading to thermal runaway[82] or the field acceleration (promotion) of electrons in (to) the conduction band may be greater than their loss of energy (number density) by collision with the lattice (recombination), i.e. electrical runaway[618,151,152].

Such processes can be expected to cause breakdown uniformly throughout the system. This is indeed the case with the electro-mechanical failure of polyethylene[138] and in some instances of thermal breakdown[757]. In general though breakdown is a highly localised process, and this fact has prompted a modification of the basic theories so that the runaway release of energy occurs in a restricted region, e.g. filamentary thermal breakdown[757,594], localised avalanches[378], electrofracture[366]. Since runaway is a self-enhancing process it might be expected that these local processes would produce an almost unbranched track bending occasionally to follow 'easy' routes through the material. Many breakdown systems however exhibit a branched tree-like structure (Fig. 3.5) which the deterministic models appeared to be incapable of explaining. An important step has however recently been taken towards understanding how such mechanisms are involved in the evolution of the complete breakdown process with the recognition that tree-like discharges are examples of branched stochastic fractals[289,767].

15.3.1 What is a fractal?

The concept of a fractal has already been introduced in Sections 4.4 and 5.4 but in view of the reader's probable unfamiliarity with the term the most important features will be repeated here.

For our purposes a fractal is a geometrical figure with a fractional dimensionality[768,392] (i.e. the fractal dimension d_f). In tree-like structures d_f can be defined via eqn. 5.8, i.e.:

$$S = \sum_i S_i \propto L^{d_f} \qquad (15.15a)$$

where S is the sum of the arc-lengths of the individual branches within a tree of overall length L.

Fractals are self-similar structures, i.e. if S is measured over half the system $(L/2)$ in terms of a given unit, the same number of units is counted as would be obtained by measuring the whole system (L) in terms of a unit doubled in size, or a doubled system $(2L)$ in terms of a unit quadrupled in size. This scaling property allows eqn. 15.15a to be written in the dimensionless form

$$S \propto (L/a)^{d_f} \tag{15.15b}$$

where S is now the number of length units (magnitude a) constituting a tree of overall length L.

Deterministic fractals are structures which can be constructed by repeatedly amplifying a given generator shape by the *same* magnification factor. A tree-like example is shown in Fig. 15.8. Here it can be seen that the segments cut-off by the arcs through the branch points will reproduce the overall tree structure if suitably magnified. This feature can be used to deduce an expression for the dimension of the figure. Since each branch point travelling outwards splits the remaining structure into two segments

$$S(\text{total } L) \simeq 2S(\text{segment } L/b) \tag{15.16a}$$

for a given overall length L, where $b\ (>1)$ denotes the reduction in length at each branch point. Taking eqn. 15.15b to apply to both the whole figure and each *self-similar* segment of length L/b, then eqn. 15.16a gives:

$$1 = 2/b^{d_f} \tag{15.16b}$$

i.e.

$$d_f = \frac{\log_e 2}{\log_e b} \tag{15.16c}$$

A generalisation of this result to structures with different branching ratios replaces 2 in eqn. 15.16c by the number of branches, n_b, produced at each branch point, i.e.: $d_f = \log_e n_b / \log_e b$.

The exact self-similarity of the deterministic fractal to scale changes which are multiples of a fixed quantity is a consequence of its artificial (man-made) construction. In nature, fractals are formed when bonds (branches) are added at random to the tips of a growing figure such that the instantaneous shape of the figure determines the probability density of each tip being chosen[743,307]. This type of fractal is termed 'stochastic' and examples are the backbone of the incipient infinite percolation cluster[743], and particle aggregation[307]. In the case of the tree-like structure considered here (Fig. 15.8) a stochastic construction would lead to missed or even extra branches at the branch points, and a range of reduction factors b. No longer will any portion of the tree bear an *exact* similarity to the complete structure. Instead an approximate self-similarity occurs which is retained on average throughout the tree, thus allowing a value to be assigned to d_f using eqn. 15.15b. If the generation of the tree is repeated to give a second representation (termed realisation) the value found for d_f will be slightly

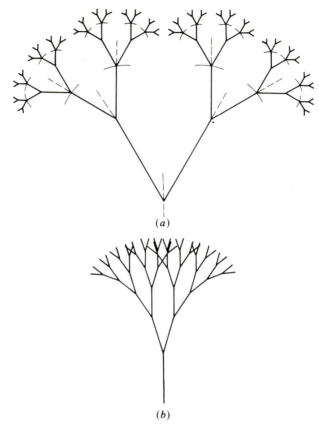

Fig. 15.8   Deterministic tree-like fractals
(*a*)   Self-avoiding planar structure. Cutting through the vertices of the system with arcs (centred on the initiation point) will decompose it into self-similar sub-structures (segments). Splitting the structure along the dashed lines will also generate self-similar segments
(*b*)   Alternative tree-like form with a stem. The branches are not self-avoiding and this construction should be taken as a pictorial representation of a tree growing in 3-dimensions with some branches projecting into the paper. Its fractal dimension will however be $d_f < 2$ as in (*a*)

different[748]. However when d_f is averaged over many realisations its value will still be given by eqn. 15.16*c* as long as n_b and b are also replaced by their average values[307,769].

15.3.2 Branched filamentary breakdowns
The lesson to be learned from the observation of branched stochastic structures in discharges is that the process is filamentary and advances in steps with a random choice being made after each step of a new site from which further progress will occur. In an attempt to answer the question of *why* this happens, Zeller and co-workers[69,338,346,345] have examined the

behaviour of three filamentary breakdown processes, namely electrofracture, filamentary thermal breakdown and local avalanching.

In contrast to electro-mechanical failure, which is a spatially uniform mechanical collapse of the polymer brought about by compressive Maxwell stresses between the electrodes[138], electrofracture is an electrically driven crack propagation[770] which ultimately connects the electrodes via a bridging air channel[366]. It was concluded that this mechanism would only be effective in polymers when local space charge was present in the incipient crack[69,141] such as when corona discharges are incident upon surfaces in tensile stress[460].

In thermal breakdown, Joule heating increases the temperature of the specimen, and since the conductivity of insulators has a positive temperature coefficient, a current increase occurs thereby reinforcing the process[757]. In heterogeneous materials the spatial distribution of currents is likely to be non-uniform, and the associated heating will be concentrated in the regions of maximum current density which are thereby further enhanced. As a result the radial extent of the conducting paths will be progressively reduced as the temperature runs away[594], and a filamentary damage structure, caused by melting or degradation, will ensue.

Avalanches are generated when a charge carrier (electron) gains sufficient kinetic energy from the electric field to produce extra carriers by impact ionisation[152], with a succession of ionisations building up a large number of carriers. However the counter (self) field arising from the relatively immobile ions will usually prevent the current from reaching a disruptive level if the generated carriers are considered to spread out uniformly[346,69]. On the other hand, the self-field will tend to prevent the transverse displacement of the charge carriers and concentrate them instead into a filamentary current which can attain a magnitude sufficient to initiate thermal runaway[346].

Each of the above-mentioned filamentary processes will either require or be enhanced by space charge in the presence of an electric field. Zeller has therefore proposed a unified model[346,69] of their development as a result of space-charge injection. Three regimes of behaviour are identified, Fig. 15.9. At fields just above the threshold, E_{inj}, for injection the space charge is trapped close (~ 10—100 nm) to the injecting site (regime I). At fields above that required to sustain charge mobility (i.e. E_{mc}) the mobility of the injected charge rapidly increases (regime II) and the space-charge region extends uniformly into the polymer. An equilibrium space-charge cloud will be established when the field at the injection surface is reduced to E_{inj} and the field at the space-charge boundary to below E_{mc}. This behaviour is termed field-limited space-charge (FLSC) injection[346]. When a current runaway is possible a negative differential-resistance (NDR) will occur as shown in Fig. 15.9. In this regime (regime III) infinitesimally small spatial fluctuations in charge density will generate local currents which rapidly become enhanced in magnitude and restricted spatially through one of the filamentary runaway processes described above. Thus a uniform space-charge region is unstable and will break up into self-enhancing filamentary currents in this regime. The initiating fluctuation may be either thermal in origin, or, in the case of electro-fracture, due to material

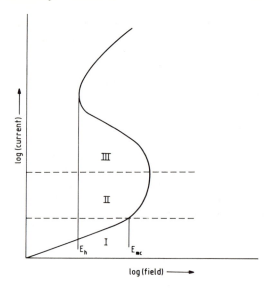

Fig. 15.9 Schematic representation of current-field characteristic in a system exhibiting negative differential resistance (NDR)

inhomogeneity. It may however be very small since its only function is to concentrate the current locally thus allowing the self-enhancing process to start.

Once a filamentary discharge current has been initiated the longitudinal field within a filament will be maintained at E_h ($\lesssim E_{mc}$) the field below which the high conductivity state cannot exist[69,346], see Fig. 15.9. Propagation will continue as long as the field experienced by the polymer at the filament tip is greater than E_{mc}, i.e. sufficient to keep the advancing space charge in the high mobility state[69]. It has been suggested[346] that when injection takes place from a stress-enhancing electrode projection the filament that is formed ceases propagation temporarily when it reaches a point at which the enhancement of the injection site is insufficient to yield a local field above E_{mc}. The subsequent renewal of filament formation will include the possibility of branching.

An estimation of the filament length for a spherical electrode[69] has been made by taking the potential at the tip to be determined by the charge enclosed in the combined filament and electrode system. This construction really refers to an electrode radius (r_0) charged to a level Q by a voltage V (i.e. $Q = V \cdot r_0$), with Q being subsequently shared with a filament of length l giving a tip potential of $Q/[r_0 + l]$ ($= V r_0/[r_0 + l]$), i.e. a system which is conservative of charge. The filament length is determined by equating the potential difference along it to E_h. When the injection point is connected to a power source (i.e. the conductor) this argument is no longer valid, and the positive feedback of the filamentary process will narrow the filament and self-consistently act to produce a field at the tip sufficient to allow propagation to continue.

Alternatively material inhomogeneity has been suggested[390] as an origin for the step-by-step progress of the filament. However computer simulations[346,390] have shown that a spatial distribution of E_{mc} values is insufficient to reproduce a branched fractal structure, when the filamentary discharge is allowed to choose the 'easy' route through the material via the available regions of smallest E_{mc}.

In view of the self-enhancing nature of the processes involved it is unlikely that material inhomogeneities will have much effect except in the case of electrofracture (i.e. most organic polymers have similar properties near their melting point or at ionisation energies). However it is not necessary to invoke features extrinsic to the discharge in order to explain the branching. In a dynamic process, such as a discharge, spatial fluctuations in the current density of a filament are inevitable. Since the process is also self-enhancing the current density of a single filament must reach a magnitude at which it is capable of sustaining a division into two or more branches[311]. Thus the same process that causes a homogeneous space-charge to break up into filaments will also cause the filaments to be unstable with respect to branching[311,312], Fig. 15.10. The formation of branched positive streamers in the breakdown of dielectric liquids has also been described in this way[771].

15.3.3 Computer simulations of breakdown

In the computer simulations of the fractal model[289,390,338], Fig. 5.10, the voltage of the filament tip may be determined from E_h and its length. The voltage drop between the tip and the neighbouring points of the simulation matrix is obtained by a finite difference approximation[289] to Laplace's or Poisson's equation, and taken to represent the tip field along the various directions. For real filaments though the tip field will be determined by their shape and diameter, unless charge is conserved between the electrode and the discharge. However the relative magnitudes of the field strength in different directions from the tip will be reproduced reasonably accurately by this method even though the maximum field is not. In fact the picture presented above suggests that once a filament has grown to a given length it becomes potentially unstable to branching. This length would define the minimum size of bond that could be added to the growth structure before branching possibilities must be considered, and hence provides a natural size scale for the simulation (other than that of insulation thickness). All quantities controlling growth, including the field, would then be related to this size scale. For example the length of the first channel in electrical trees is determined by the distance required for the counter field to suppress an avalanche (eqn. 5.5 and Chapter 13). Under these circumstances the factor governing channel formation (and hence bond addition in *tree* simulations) will be the potential drop along the incipient channel rather than the field at the initiating tip[748,772].

Since the fluctuations in current density are random in space, the choice of a new bond is a random selection from the potential branch sites, weighted by a probability distribution proportional to a power (η) of the voltage drop (local field E_l) along the bond. It has been suggested[69] that this type of field dependence originates with the availability of an electron to initiate an

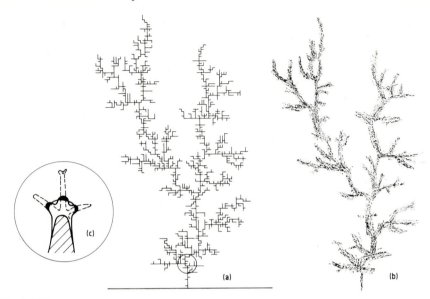

Fig. 15.10

(*a*)   Computer simulation of a discharge grown on a planar grid using a probability weighting of E_l (local field) to determine the choice of growth directions at each branch point ($d_f = 1 \cdot 2$). (Reproduced by courtesy of A. Barclay, National Power TEC, Leatherhead, UK.)

(*b*)   Appearance of the tree under low resolution or during light emission.

(*c*)   Schematic representation of filament and branch formation due to charge density fluctuations shown in black

avalanche. However it has not been firmly established that the relationship between inception time and field implied by this contention applies even in the case of void discharges[381]. A more likely reason for the E_l^{η} weighting is that large values of E_l will allow smaller and hence more probable fluctuations to be effective in forming a filament over the distance of a branch[312]. It should be noted that in systems where the range of noise available is reduced (i.e. the less-probable fluctuations are cut out) the average width of the branches is increased and fewer will be able to attain filamentary runaway over a branch length[312]. The power law form in E_l will follow if the runaway process is driven (at least approximately) by a power of the field, such as the thermal heating rate[594], crack velocity (electrofracture), charge injection, and energy released in an avalanche[69]. The fractal structure of the discharge will now follow from a combination of the stochastic branching possibilities and the shielding of each tip by the others[390].

The fractal structure that is given by these simulations varies both with the conditions inside the filaments (i.e. value of E_h) and the stochastic weighting (i.e. value of η). Thus when η is large and E_h small the runaway of the initial filament is emphasised, only rare fluctuations will succeed in establishing a branch, and the strcture will be nearly linear. On the other hand when $E_h = E_{mc}$ and a NDR regime cannot quite be established the structure is compact and space filling, and thus is not filamentary at all[390]. In between these extremes structures are formed with fractal dimensions between 1 and 2 for planar simulations. When $\eta = 1$ and $E_h = 0$ (i.e. there is no voltage drop in the filament) the dimension is $\sim 1\cdot 7$ if the simulation is performed in a plane[289] but becomes $\sim 2\cdot 6$ for a three-dimensional[748,394] lattice. However a fractal dimension of $\sim 1\cdot 7$ is recovered for the latter case[748,394] if η is increased to about 3, or E_{mc} is increased from zero[390]. These results may not however continue to be true when the discharge sustains a potential drop along its length, and further work is needed to clarify this point. It has also been noted[390] that in planar simulations the overall growth of the discharge can cease when the longitudinal field in the filaments approaches E_{mc}.

The relationship between electrical trees, which are fractal objects, and discharge simulations, has been discussed in Section 5.4. It has been suggested[346,390] that electrical trees are actually the visual aspect of an incomplete filamentary breakdown process. Such an equivalence may be true of DC impulse electrical trees, but the requirement in the simulation that a new branch be added at each step is more appropriate to a discharge in which filaments become unstable to branching fluctuations than to AC field electrical trees which advance by a process of damage accumulation. The simulations do however reproduce some of the typical features of electrical trees, and may thus be held to give an indication of their origin. Thus, for example, they show that the tendency of some trees to be spatially compact (bush-type) and cease growing can be related to the presence of a large potential drop along the filaments. In electrical treeing these local fields result from large concentrations of ionic charge adsorbed on the channel walls. Branch-like trees on the other hand can be expected to form with a fractal dimension between $1\cdot 0$ and $1\cdot 7$ when the channels can support gas discharges controlled mainly by their generated self-fields as in the conducting discharge.

15.3.4 Fractal systems and breakdown statistics

One feature of the fractal model which has not received much attention is its relationship to the failure statistics. Since the filamentary discharge is generated in an inherently stochastic fashion, it is inevitable that structures differing in detail will be formed under identical conditions[773,748]. If the criterion for breakdown is a structure dependent factor such as the time required for the discharge to propagate across the insulation, a sufficient number of repeated simulations will give the form of the breakdown distribution[748]. Because of the approximate self-similarity expected between (and within) each of the different simulations (realisations) of a stochastic fractal, it is likely that the number of times a structure of a given size, L, is formed will be an inverse power of L (i.e. $[L/L_\xi(t)]^{-(\beta+1)}$) as the large L limit is approached[743]. Such a behaviour will obey eqn. 15.3 and hence the probability that the largest tree is smaller than the critical length, L_c, which is the survival probability, is given by eqn. 15.2b as:

$$1 - P_F(t) = P_S(t) = \exp\{-(L_\xi(t)/L_c)^\beta\}$$
$$= \exp\{-(t^{1/d_f}E^s/B)^\beta\} \qquad (15.17)$$

Here the propagation law for the characteristic tree length $L_\xi(t)$ has been introduced from eqn. 15.10, and B is a constant. The Weibull form of eqn. 15.17 derives from the approximate self-similarity of the tree realisations and is identical to eqn. 15.14 if β is identified with $(ms)^{-1}$.

In the fractal model the breakdown statistics arise from the different possibilities inherent in the stochastic growth process itself in contrast to the models of Section 15.2 where it is the intrinsic fluctuations of the material which influence and are influenced by the fractal development. In addition to the distribution of realisations suggested above, fractals can be thought of as a self-similar arrangement of sub-structures. For stochastic fractals the sub-structures will also be distributed with a power-law form[774,745], and Weibull statistics can be expected if breakdown is related to an extreme in the sub-system.

One possibility will be discussed here based on the breakdown model of Bahder *et al.*[120]. These authors consider the formation of an electrical tree by a void discharge to be due to the avalanche generation of space charge in the polymer, similar to filamentary breakdown or electrical tree propagation. They then take the length of the tree to be proportional to the charge Q_t generated over a period of time t, which is equal to the charge produced per avalanche (or discharge), Q_d, multiplied by the number of avalanches or discharges ($\propto ft$). This model treats the tree as a linear (unbranched) entity and must be modified for fractal growth, giving

$$S(t) \propto [L(t)]^{d_f} \propto Q_t \propto ftF(E) \tag{15.18}$$

The field dependent term, $F(E)$, in Bahder *et al.*'s model is derived from an avalanche generation expression $(\exp\{aL_m\} - 1)$ where L_m is the self-field limited length. The form of L_m is that given by eqn. 5.6 and when this is substituted in $F(E)$ a field dependence is obtained for $L(t)$ which has approximately the form observed experimentally (i.e. E^4 for $d_f = 1\cdot7$).

This model and its relationship to electrical treeing is discussed in more detail in Chapter 13. Here we are concerned mainly with the breakdown statistics that result from the fractal form as compared to the deterministic development of the tree. Firstly the critical length required for breakdown must be modified. In Bahder *et al.*'s model[120] this is derived by considering the tree to have essentially the field enhancement of a conducting projection with the shape of a hyperboloid of revolution. Here instead we consider it to be a charged body and use Gauss' theorem to derive the average field at the tree front, E, to be:

$$E \propto (Q_t/A) \propto \frac{[L(t)]^{d_f}}{[L(t)]^{d_f-1}} = L(t) \propto Q_t^{1/d_f} \tag{15.19}$$

where the surface area of the tree front, A, has been given its fractal form, and eqn. 15.18 has been used for Q_t. Note that A is the area of the intersections of a tree with a plane perpendicular to the direction of the field and not its topological surface area. Clearly the average field enhancement is independent of the fractal dimension of the charged structure. If runaway breakdown is assumed to occur when the field at the tree front exceeds a critical value, E_c, which is typical of the material, survival requires the tree length to be less than a critical value L_c (proportional to E_c). Some evidence for a critical length of electrical tree is contained in the data of Wasilenko[325] (Fig. 15.11), where under different conditions runaway propagation is initiated from trees whose length is almost the same fraction of the insulation thickness. A critical length also implies, via eqn. 15.19, a critical charge (Q_c) and hence from eqn. 15.18 an inverse relationship between the time to runaway (i.e. breakdown), t_b, and the charge produced per discharge, that is $t_b \propto Q_d^{-1}$. Such a relationship is very close to that found for void discharge breakdowns[666] where $t_b \propto (Q_{max})^{-1 \cdot 4}$ with Q_{max} the maximum discharge measured for the void.

The surfaces of fractal objects are however not smooth. In general they contain sub-structures with a continuous range of sizes L_s. If the fractal is stochastic it will also exhibit a continuously varying dimension as the unit of scale that it is measured with changes[327,745]. Thus there will be a distribution of field enhancements ($\propto L_s$) or equivalently charge densities ($Q_s L_s^{-3} \propto Q_s^{(d_f-3)/d_f} \propto L_s^{(d_f-3)}$) at different regions on the tree front. In order to deduce the breakdown statistics it is necessary to determine the distribution of field enhancing sub-structures. Because of the average self-similarity of the complete tree it is natural to assume that the sub-structures will be scale related and that their distribution will have an appropriate form[745] such as eqn. 15.3 or eqn. 15.4b. Such a behaviour is indeed found for the sub-structures in viscous fingers and percolation backbones[743,327]. The result of a similar analysis of the alternate-polarity electrical tree of Fig. 5.7c is given in Fig. 15.12. Here the relevant sub-structures are branches whose tips are

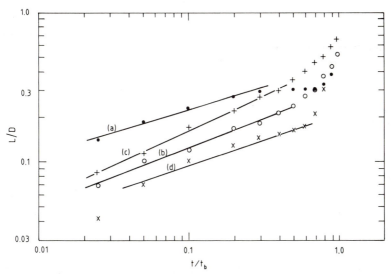

Fig. 15.11 Lengths, L, of electrical trees grown in polyethylene[325]. The plot shows the fraction of insulation thickness D crossed by the tree in a given fraction of the time to breakdown, t_b. (a) 70 kV DC, positive polarity pulses, (b) 50 kV DC, positive polarity pulses, (c) 20 kV AC, (d) 12 kV AC

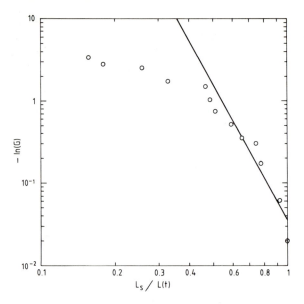

Fig. 15.12   The fraction of sub-structures, G, with a length less than L_S in a tree whose average length is $L(t)$. The data was obtained by analysing the electrical tree structure of Fig. 5.7c.

exposed on the 'rough' tree front. These are characterised by the fraction of the *average* distance from the point to tree front ($\propto L(t)$) that the length (L_s) of the substructure represents. Counting the number of sub-structures of a given $L_s/L(t)$ enables their distribution to be found, and Fig. 15.12 shows that this has the form of $\Lambda_2(L_s/L(t))$, eqn. 15.2b, at large L_s, i.e.:

$$G(L_s/L(t)) = \exp\{-[L_s/L(t)]^{-\beta}\} \qquad (15.20)$$

with $\beta = 5\cdot4$. The distribution density of L_s found by differentiating eqn. 15.20 with respect to $L_s/L(t)$, has a substantial peak close to the average tree length $L(t)$, and indicates that even for this branched tree the surface roughness is quite weak. It can be expected that for the more compact form of a bush-type tree the roughness will be even weaker, and the distribution of L_s more sharply peaked with a higher value of β.

In this stochastic fractal version of Bahder *et al.*'s breakdown model[120], the insulation will survive if the maximum stress enhancing sub-structure has a length less than a critical value L_c (eqn. 15.19). The survival probability, $P_S(t)$, is thus the probability that a selection of L_s from the distribution $G(L_s/L(t))$ has a maximum value less than L_c, and as noted in Section 15.1.2 is given by:

$$1 - P_F(t) = P_S(t) = \exp\{-(L(t)/L_c)^\beta\}$$

$$= \exp\{-[tF(E)]^{\beta/d_f}/B\} \qquad (15.21)$$

Here B is a constant proportional to L_c^β, and the breakdown time is the time required to grow a field-enhancing sub-structure to a length L_c. No allowance is made for the time of the runaway process in contrast to Wasilenko[325] and the model corresponds to case (*b*) of Section 15.1.1. Since the fractal dimension d_f for the chosen tree is 1·4, the appropriate Weibull time exponent, *a*, is 3·8. This value is larger than those usually found for breakdown in power cables ($\sim$0·3—1·7) and it seems that in general the fluctuation model of Section 15.2 gives a better description of the failure statistics in this case. However low-field breakdowns do show some tendency[685] for the field depence to reach a threshold (see Chapter 13) and *a* to move towards a value of around 3 consistent with this model.

Failure statistics specific to electrical tree breakdowns have been obtained by Densley[386] by means of trees artificially induced via the insertion of a stress-enhancing metal needle into polyethylene cable insulation. Both bush-branch and bush trees grow at a common voltage and separating the breakdowns according to the tree shape leads to two well-defined Weibull distributions, Fig. 15.13. The values of the Weibull exponent *a* are $\sim$5·6 and $\sim$4·6 for bush-branch and bush trees respectively. If we assume that the fractal dimension for the bush-branch tree should be taken as $\sim$1·7 and that for the bush tree as $\sim$2·5, the respective values for β can be estimated to be 9·5 and 11·6, which are quite reasonable values for trees of this shape. These results also show that the characteristic breakdown time for bush-branch trees is less than that for bush trees grown at the same field. Such a behaviour follows naturally from eqn. 15.21 and the difference in fractal dimension of the trees, since the characteristic breakdown time (t_b) is given

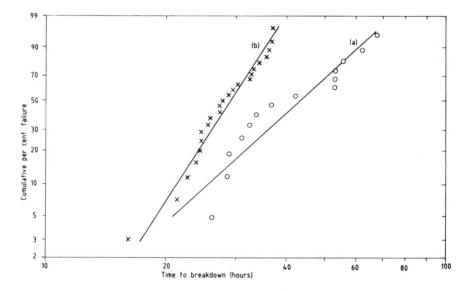

Fig. 15.13 Electric tree breakdowns at 18 kV, 400 Hz
(*a*) Bush-type trees, Weibull time exponent *a* = 4·64
(*b*) Bush-branch trees, Weibull time exponent *a* = 5·64
      After Densley[386]
      Copyright © 1988 American Physical Society

by:

$$t_b = \frac{B^{d_f/\beta}}{F(E)} = \frac{L_c^{d_f}}{F(E)} \qquad (15.22)$$

When grown in the same field, bush-branch trees with d_f in the range 1·7–2·0, will therefore lead to a smaller value of t_b than bush-type trees whose fractal dimension is greater (i.e. $d_f \sim 2·5$).

The field dependence of the characteristic time to breakdown, that is the 'lifeline', predicted by eqn. 15.21 depends upon the form chosen for $F(E)$. If the empirical power law form of eqn. 15.10 is used, a Weibull field exponent b, is obtained with a value of $s\beta$, and the lifeline has a power-law form:

$$t_b E^{sd_f} = \text{constant} \qquad (15.23)$$

with the life index, n, equal to sd_f. In this case n will increase from 4 for one-dimensional (unbranched) trees, through ~6·8 for branch-type trees, to ~10 for bush-type trees. These values are typical of those found in power cables[324].

A similar power-law behaviour is found for the models of Section 15.2, as shown by Tables 15.1 and 15.2. In particular the conduction path breakdowns of Section 15.2.3 give the same value for n, but its time exponent a is in general less than unity in contrast to the large values predicted by this model. This difference should provide a means for identifying which of the two models may be applicable in a given case. However Bahder *et al.*'s theory[120] also implies that the power-law form in field, eqn. 15.10, observed for electrical trees may be just an approximation to a more complex exponential relationship whose special feature is the existence of a threshold level. In this case the tree-related breakdowns of this and the previous section will give a sigmoidal lifeline[120] with a threshold field below which no breakdowns will occur. Some evidence exists for this type of behaviour[685] (see Chapter 13 for a discussion) and it appears that partial-discharge and electrical-tree breakdowns should be regarded as different aspects of the same process as long as failure is dominated by the propagation stage of the pre-breakdown damage pattern.

15.3.5 The utility of the fractal model

It has sometimes appeared that the fractal model is nothing more than a complicated way of getting a computer to draw pretty patterns that look rather like an electrical tree or discharge. Such a comment may be justified in some cases, but if used with forethought, the model can lead to a much clearer understanding as to how the stochastic features of breakdown can influence its overall progress. When a computer simulation uses rules that have a basis in the physics of the process considered, the effect of variation in various parameters can be studied without uncontrolled factors affecting the result. Consequently their ability to produce the features attributed to them can be assessed unambiguously. This is the strength of the fractal description. Its weakness lies in how realistic the simulation rules can be made. Are they perhaps so unrealistic that the results could be completely changed by the introduction of an as yet unconsidered factor? An example is the change of shape as the field in the filaments is increased.

The breakdown statistics are a second area of importance arising from the fractal description. Because of their self-similarity the statistics can be expected to be of the Weibull form (see Section 15.1). As is clear from the example given the Weibull form will however only apply to the critical variable, and a non-power-law relationship between this variable and a controlled parameter such as voltage will alter the form of the statistic. The possibility of using the fractal description to explore the breakdown statistics has only just begun to be exploited. Other theories than the one used as an example can be readily envisaged. For example breakdowns may be the result of discharges within the channels of an electrical tree, with runaway occurring when the discharge is limited to just a portion of the tree, i.e. when it is unbranched. The possibility of such an occurrence would be dependent upon the branching arrangement within the tree. In general different breakdown mechanisms will be related to different sub-structure arrangements and their probability can be derived from an appropriate sub-structure analysis. In this area the fractal description still has considerably more information to yield us.

15.4 Cumulative defect models of breakdown

The basic thesis of the cumulative models of breakdown is that no single deterministic mechanism is responsible for insulation failure, with the exception of short-time breakdown in thin films which may be understood in terms of Fowler–Nordheim electron injection from the electrodes[133] (see Section 9.2.2). Instead several processes combine to generate and extend defects which eventually reach a size and concentration sufficient to sustain runaway damage production.

This form of picture was first proposed by Lloyd and Budenstein[132] who illustrated its concepts by reference to the formation of positively charged defects (V_K centres) in alkali halides by conduction electrons which had acquired enough kinetic energy to participate in an impact ionisation event. Instead of considering this event as the start of an avalanche, it is suggested that the electrons are swept away in the conduction band leaving behind slowly moving defects which eventually accumulate, usually near the cathode. It was then assumed that the local electrostatic field could reach such a magnitude as to cause the solid defect to chemically dissociate into gaseous components, thereby producing a cavity filled with hot conducting gas. The process of damage formation would then become self-enhancing with the high non-uniform field of the conducting cavity attracting further defect centres, both electrostatically (charged defect) and dielectrophoretically (neutral defects).

Although Lloyd and Budenstein[132] only gave the model explicit form for alkali halides, it was suggested that bond breaking in polymers followed by the accumulation of mobile defect centres would have the same effect. These ideas were followed up by Jonscher and Lacoste[133] who pointed out that in uniform fields the breakdown features of materials with very different physical and chemical natures were similar. For example the highest recorded breakdown strength of a wide range of technologically important materials[158-160,162,775-787] (ceramics, polymers, inorganic crystals, etc.) exhibit a very narrow range of values (10^8—10^9 V · m^{-1}), and can be related to charge injection at the electrode interface. As the materials are aged or their thickness increased they all show a similar reduction of two or three orders of magnitude. These results were interpreted by Jonscher and Lacoste as implying that breakdown developed according to a framework which did not rely upon the details of material structure for its operation.

It has been suggested[788] that the unoccupied 'lattice' sites that comprise the free volume of amorphous and semi-crystalline polymers (see Section 3.1, Fig. 3.2) act as 'defects' in these cases. Here breakdown is caused by charge carriers in the low density region of the free volume which attain a kinetic energy sufficient to generate runaway damage[476,788,759,789,477] (see Chapter 13) by the breaking of molecular bonds. The near constant breakdown strength observed for polymers from low temperatures up to the onset of softening[163] would then be the consequence of the weak temperature dependence of the free volume in polymers below their glass transition temperature, T_g (see Fig. 3.3). A similar behaviour could also be expected for inorganic glasses[790,133,778,697] if this type of breakdown mechanism were to be dominant over the appropriate temperature range. Above T_g free volume is generated and hence the breakdown field drops[383,759,476] as the polymer softens. The observed relationship between the breakdown field and the cohesive energy density[477,356] lends further support to a breakdown mechanism in which electrically driven processes locally dissociate[93] the solid polymer.

In the cumulative model[133] it is considered that the severe material rupture associated with breakdown cannot be achieved in a single impact-ionisation avalanche. Electron transport in polymers is regarded as so difficult that it would be impossible for them to reach sufficient kinetic energy in the absence of non-uniform fields. In fact the DC conductivity may be due to ions rather than electrons since ionic transport would be relatively easy in all directions, across strong bonds as well as between them. The observed relationship between cohesive energy density and the activation energy for conduction[93] supports this contention. A consideration of the energies acquired by carriers of either type[133] (when accelerated in the free volume of polymers by an electric field) then leads to the conclusion that the number of carriers produced are insufficient to cause breakdown via runaway damage even at the highest breakdown fields, let alone the much lower fields of technological importance. Jonscher and Lacoste therefore argued[133] that damage would only result if the carriers moved over significantly longer paths than are normally available in the free volume of the material, or alternatively the threshold energy for damage formation is less than that required for material disruption.

Simulations[355] have recently shown that an electron injected into long-chain hydrocarbons can acquire sufficient energy for bond scission ($\sim$4 eV) over a range of 10 nm if the field is large enough ($> \sim 7 \times 10^8 \, \text{V} \cdot \text{m}^{-1}$). As the depth of penetration increases beyond 10 nm the kinetic energy reduces through dissipative phonon scattering processes and the electron is either thermalised or trapped. Thus a field enhancing centre, such as a conducting inclusion or material inhomogeneity, may accelerate free carriers sufficiently to extend existing adjacent free-volume defects[452] by bond breaking although incapable of generating such a defect in an initially perfect lattice. The field threshold for damage producing events is also lowered by the presence of voids in the material. In this case internal discharges occur at fields well below the characteristic breakdown value of the polymer when the void size is greater than that for the minimum in the inception voltage plot[122] (i.e. $\sim 10 \, \mu\text{m}$ at S.T.P.). Some of the carriers generated in the discharge will reach the void surface with energies sufficient to cause bond scission and are therefore capable of extending any free-volume regions in the neighbourhood.

Space charge that is either present initially at a field enhancing centre or accumulated as a result of the defect extending processes will give rise to the possibility of avalanches in alternating fields. Even though the quantity of carriers generated in the avalanche is unlikely to reach the magnitude required for runaway damage formation, their kinetic energy should still be capable of breaking bonds along the avalanche trajectory, rather as seems to occur in the inception stage of electrical trees (Section 5.2).

Again the free volume defect will be extended. Furthermore the resulting space charge will act as a field enhancing defect centre in its own right and thus has the potential of extending other free volume defects in the neighbourhood of the one that first experienced activity. Thus over a period of time a free volume defect may not only be extended but also interact with other closely-spaced defects to form 'clusters' in which the carriers will have a relatively high mobility. The much larger free volume available to the carriers will allow them to acquire higher kinetic energies in the field. In addition it has been suggested[791] that polarisation of the surrounding polymer and its slow relaxation will lead to an autofocussing of the more mobile and hence more destructive carriers leading to an extension of the defect cluster in the direction of the field. At this point the defect clusters are better considered as low-density regions containing field-enhancing space charge of higher mobility than the rest of the polymer. Such a definition can be expected to have widespread applicability to many types of material other than polymers. Eventually a critical size of free volume will be reached such that it can sustain a damage-producing avalanche capable of rupturing the polymer, such as occurs during the first channel formation in electrical trees (Section 5.2). The final stage of breakdown in this model occurs when the material rupture connects up the field-aligned but randomly-distributed free-volume defects along a path of least resistance.

15.4.1 Percolation theory of breakdown

The cumulative model of breakdown is, by its very nature, a stochastic process. A variety of physical phenomena may be involved in extending the

defects by means of repetitive processes which are not likely to be deterministically related (see introduction to Part 3). A statistical element is
therefore introduced through the probability of occurrence of the different
events. In addition however the distribution of *initiating* defects, both
through their spatial location and size, will influence the probability that a
defect will reach a critical size, thereby causing breakdown, in a given time.
These considerations will affect not just the rate at which a single defect
grows but also the possibility of defect amalgamation during 'cluster' generation and the formation of the breakdown path. Because of the range of
possibilities inherent in the combined effect of the various factors, no attempt
was made to deduce the statistics of the cumulative model. However its
essential features are contained, albeit in an idealised form, in the percolation
theory of breakdown[792,398], which is capable of being analysed
quantitatively[793,794,795]. Thus a combination of the two models seems the
most likely way forward.

(a) Construction of the percolation model
This model[793-796] represents the insulator by a grid whose bonds are identical
capacitors which breakdown under the application of a given field (E_0).
That is, each individual capacitor is taken to fail through a deterministic
mechanism (see Part 3). Defects are introduced by randomly replacing
capacitors by conductors. The initial state of the system is thus defined by
the fraction of bonds that are conducting, or equivalently the volume
fraction, p, of the material that is occupied by unit defects. This construction
forms a percolation system of defects. Thus not all unit defects will be
isolated, some will combine in the form of conducting clusters. There will
therefore be an initial range of cluster-defect sizes. However no cluster will
be large enough to cross the insulation when p is below the percolation
limit p_c.

When an electric field below E_0 is applied to the system none of the
remaining capacitor bonds would fail if the field were uniform. However
the conducting clusters will enhance the field in their vicinity, possibly to
such an extent as to cause the failure of a capacitor bond at their perimeter
and thereby extend the cluster. Since the field enhancement of a cluster is
essentially proportional to its longitudinal extent in the direction of the
field[795], extension of the cluster will increase the local field and cause further
bonds to fail. The process is therefore self-enhancing and will runaway to
failure via a path of least resistance linking other clusters on the way.

Computer simulations, Fig. 15.14, have been used extensively to analyse
the percolation model[795]. Here a defect distribution is first generated then
a voltage applied to the system. The capacitive bonds are now searched and
the one experiencing the biggest field in excess of E_0 is failed and incorporated into the defect system. The process is repeated until a system breakdown
occurs. By performing a large number of simulations it was found that the
first failure of a capacitive bond did not always lead to an increased field
enhancement at that particular cluster. Instead some other cluster with a
slightly lower field enhancement sometimes was the one to lead to runaway
failure[398]. However in all cases system breakdown was initiated by a cluster

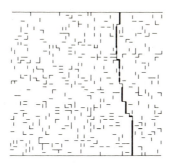

Fig. 15.14 A typical breakdown simulation in the random defect percolation model[795]. Here the volume fraction $p = 0 \cdot 1$, and the system size is 48×48. The breakdown path connecting the original clusters is shown by the thick line
After Beale & Duxbury[795]
Copyright © 1988 Physics Review

whose field enhancement differed by a statistically insignificant amount from that of the most severe cluster, when the limitations of the number of bonds on the simulation grid are taken into account.

Each iteration in the breakdown simulation is effectively equivalent to an equal interval of time. Thus restricting the capacitor bond failures to just the one that is most under stress corresponds to an experiment in which none of the other bonds stressed at field above E_0 has time to fail. In this form the percolation theory represents most precisely an impulse breakdown, although it should also be applicable to ramp tests if they are of sufficiently short duration. In this case the propagation stage of breakdown is rapid and consists of a deterministic sequence of steps initiated by the most severe cluster. Breakdown is therefore dominated by the inception stage (case 'a' of Section 15.1.1) and the survival statistics can be defined in terms of the probability that the most severe cluster is too small to initiate breakdown at the applied field. An expression for the characteristic breakdown strength can also be obtained in terms of the size of the average largest cluster.

(b) Characteristic breakdown strength
(i) Theoretical expression
The characteristic breakdown strength, $E_c(p)$, can be defined as the applied field required to cause E_0 just to be exceeded in a capacitor bond adjoining the average largest cluster for volume fraction p. Duxbury *et al.*[793] consider the clusters to have an ellipsoidal shape with the most severe cluster being the one with the largest length to breadth ratio aligned with its long axis in the field direction. In which case they find[793,795]

$$E_c(p) \propto \frac{E_0}{l(p)[1 + m(p)]} \qquad (15.24)$$

where the *dimensionless* variable $l(p)$ is the average length of the defect clusters measured in units of the capacitor bond length. The other parameter

$m(p)$ relates the dimensionless length of the average most severe cluster $(l_{max}(p))$ to the average cluster length via

$$l_{max}(p) \approx l(p)[1 + m(p)] \tag{15.25}$$

Standard results from percolation theory[793,797] allow $E_c(p)$ to be expressed in terms of the volume fraction p and the dimensionless thickness of the material, D, as

$$E_c(p) \approx \frac{(1 - p/p_c)^{\nu_b} E_0}{1 - \alpha \log_e D/\log_e p} \tag{15.26}$$

where α is a parameter weakly dependent upon p. The numerator of this expression defines the approach of the breakdown field towards zero as p increases towards the percolation limit p_c at which the largest cluster crosses the system. The exponent ν_b is equal to ν the exponent of the percolation correlation length[797] (ξ), when the model is based on a discrete lattice as considered by Duxbury and co-workers[793,794,795] since then

$$l(p) \equiv \xi(p) \propto (1 - p/p_c)^{-\nu} \tag{15.27}$$

however when a continuum percolation system is considered[798] ν_b becomes $\nu + 1$.

In the low loading limit ($p \rightarrow 0$) the *average* cluster size $(l(p))$ reduces towards one lattice unit, that is the *average* cluster is an isolated unit defect. More severe clusters will however continue to occur. The maximum size of these dangerous clusters defines $m(p)$ which can be derived from the largest sequence of consecutive defect bonds that can be formed on average[793] for a volume fraction p of defect sites in material of thickness D. The characteristic breakdown strength now reduces to the form

$$E_c(p = 0)/E_c(p) \simeq 1 - \alpha \log_e D/\log_e p \tag{15.28}$$

where a p and D dependence is retained because of the clusters that continue to exist. Even a single defect in the system will reduce the breakdown strength[793] to $(\pi/4)E_0$.

(ii) Experimental data for metal-loaded materials
By using repeated computer simulations[795,799,800] the form of the analytical expression (eqn. 15.26) for $E_c(p)$ has been verified as a function of D and ξ in the high-loading limit. Such investigations however test only the ability of the theoretical analysis to accurately represent the model and do not show that the model itself is applicable to real systems. Recent work though has verified the low-loading limit of expression 15.28 in a real experimental situation[801]. Here extruded polyethylene was deliberately contaminated with conducting metal particles (aluminium) of a given size (53—75 μm). Rogowski-profiled plaques were then formed and broken down under a ramped (stepped 2 kV every 20 s) AC (50 Hz) stress. The characteristic value $E_c(p)$ was determined as the field for which the *survival* probability was $1/e$, and an uncontaminated super-clean grade of polyethylene used to obtain $E_c(p = 0)$. The relationship obtaining between $E_c(0)/E_c(p)$ and $1/\log_e p$ is shown in Fig. 15.15, where it can be seen that eqn. 15.27 is obeyed.

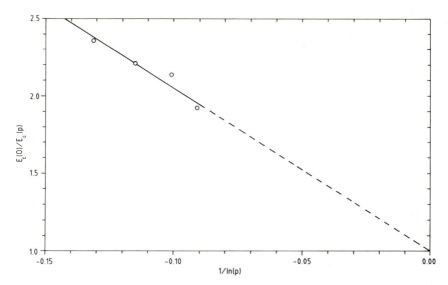

Fig. 15.15 The characteristic breakdown field $E_c(p)$ for volume fraction p of contaminants, as a function of $[\log_e (p)]^{-1}$. After Coppard *et al.*[801]
Copyright © 1989 The Institute of Physics

Dielectric breakdown experiments carried out on solid-fuel rocket propellant[802], which is a random mixture of aluminium (fuel) and aluminium oxide (oxidiser) in a rubbery binder, also show a dramatic reduction in breakdown strength with aluminium content. These results have been associated with the high loading limit and an empirical expression similar to eqn. 15.26 has been used to define the upper limit allowed for the loading (p) in a given field. However a sharp lowering of $E_c(p)$ may not necessarily be a consequence of an approach of p to the percolation limit p_c. At volume fractions as low as 10^{-5} Fig. 15.15 shows that $E_c(p)$ is a half of $E_c(0)$ and thus the breakdown strength reduction is not a negligible effect even in the low loading limit. As yet there has been no experimental verification of the applicability of the high concentration expression (eqn. 15.26) to real systems. However Benguigui[800] has performed 'breakdown' experients on random networks of resistors and diodes in which the characteristic breakdown strength was found to be proportional to the percolation correlation length, ξ, as predicted by the model. Thus the breakdown expression of the percolation theory (eqn. 15.26) can be applied to random defect breakdowns in polymeric insulation with some degree of confidence as regards its validity.

(c) Breakdown statistics
(i) Theoretical behaviour
In the percolation model, system breakdown occurs inevitably and rapidly for an applied field, E, if the most severe (longest) cluster has a length, $l_{\max}$, with

$$l_{\max} = l_c \propto E_0/E \qquad (15.29)$$

from eqns. 15.24 and 15.25. The system will therefore survive if the length of the longest cluster is less than l_c. In percolation systems the cluster sizes are exponentially distributed[398,797], and the distribution density, $g(l)$, of lengths l is

$$g(l) \propto e^{-l/\xi} \qquad (15.30)$$

The probability of survival, $P_S(E)$, at field, E, (i.e. of $l_{\max} < l_c$) thus has the form of the 1st asymptotic extreme value statistic (eqn. 15.2a, Section 15.1.2), i.e.:

$$P_S(E) = 1 - P_F(E) = \exp\left\{-N_c \exp\left[\frac{-l_c}{\xi}\right]\right\}$$

$$= \exp\left\{-N_c \exp\left[\frac{-E_0}{E\xi}\right]\right\} \qquad (15.31)$$

Not all clusters of a given length are of the same severity however and N_c is the number of clusters oriented with their long axis in the direction of the field rather than the total number of clusters. Thus although N_c can be expected to be proportional to the system volume Ω, and will increase with p, i.e.:

$$N_c = c(p)\Omega \qquad (15.32a)$$

$c(p)$ will be an increasing function of p (with $c(p=0)=0$) which is not easily determined theoretically.

(ii) Experimental observations
Dielectric breakdowns are not usually analysed in terms of the 1st asymptotic extreme value (1AEV) distribution which differs from the Weibull distribution in having its mode (distribution density maximum), median and characteristic value coincide. If a set of failures belonging to this statistic are plotted on Weibull paper they can be recognised by a convex curvature in the plot (Fig. 15.16a) which becomes more emphasised as N_c is reduced. An approximate straight line may be drawn which will increase in gradient (Weibull index) as N_c is increased. However the best way to demonstrate the 1AEV distribution is in a plot of $\log\left[-\log_e(1 - P_F)\right]$ as a function of the inverse of the variable (i.e. $1/E$ in eqn. 15.31). Here the y-axis is the same function as used in the Weibull plot, but the x-axis is a *linear* scale of $(1/E)$ rather than a log-scale. In this plot the 1AEV distribution will exhibit a straight line, with a gradient (in the case of eqn. 15.31) given by $(E_0/\xi) \cdot \log_{10} e$.

The statistical data from the experiments[801] on metal-contaminated plaques used to substantiate eqn. 15.28 are plotted against $1/E$ in Fig. 15.16b. Here a series of near parallel[801] straight lines are found for the different volume fractions when the cumulative failure probability is $\geq 20\%$. Since the *average* cluster at these volume fractions ($p \leq 5 \times 10^{-4}$) is an isolated unit defect (i.e. $\xi = 1$), this observation is consistent with eqn. 15.31 for then the gradient of the plot should be a constant ($E_0 \cdot \log_{10} e$). The value of E_0 found in this manner is $\sim 4 \times 10^8$ V $\cdot$ m^{-1}, a figure which is close to the

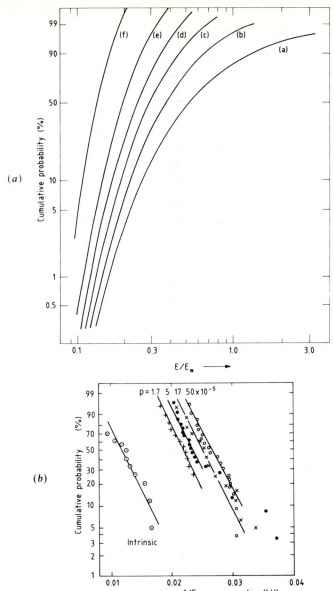

Fig. 15.16

(a) Failure probability, $1 - P_S$, for the 1st asymptotic extreme value distribution, $P_S = \exp\{-A \cdot \exp\{-E_m/E\}\}$, plotted on Weibull paper as a function of E/E_m for a range of values of A: (a) $A = 5$, (b) $A = 10$, (c) $A = 20$, (d) $A = 40$, (e) $A = 100$, (f) $A = 1000$. Note that the curvature is of opposite sense to that of Fig. 14.4c since here the variable x in eqn. 14.11 is proportional to E^{-1} rather than E

(b) Cumulative probability of failure as a function of $1/E$ for the contaminant volume fractions $1 \cdot 7 \times 10^{-5}$, 5×10^{-5}, 17×10^{-5}, 50×10^{-5} as indicated. Ramp tests ($6 \cdot 10^6$ V m^{-1} per min; AC 50 Hz fields) on PE plaques.

After Coppard *et al.*[801]

minimum stress required for the generation of electrical trees in polyethyl-ene under AC fields[129]. It is therefore possible that the failure of a 'capacitor-bond' in this case is related to the growth of an electrical tree channel along the 'bond' with the process being initiated at the surface of the defect cluster. When E_0 is independent of p the parameter N_c $(= c(p)\Omega)$ in eqn. 15.31 is responsible for displacing the cumulative probability plots for different loadings, as shown in Fig. 15.16b. Thus an estimate of the p dependence of $c(p)$ can be obtained from the relative shifts of the straight line plots in Fig. 15.16b with respect to the superclean material. Because of the logarith-mic nature of the cumulative probability scale, the calculation is most conveniently carried out by measuring the separation between the parallel lines in the direction of this axis. The resulting dependence is given in Fig. 15.17 from which it seems that $c(p)$ is proportional to $p^{0.54}$ in this example of the low-loading limit.

An unusual feature of the results shown in Fig. 15.16b is the behaviour found for $P_F \lesssim 30\%$. In this low-field region all the deliberately contaminated sets cross-over to follow the line appertaining to the highest volume fraction used. Here the breakdowns are initiated by rarely occurring regions of high stress enhancement. A comparison with the superclean material shows that these regions are produced by the contaminants with a probability that is essentially independent of concentration. While the possibility exists that cluster alignment may be involved here, results for filtered materials suggest a different explanation[803]. In this case filtering out particles from the large end of the particle size distribution, prior to forming the test samples,

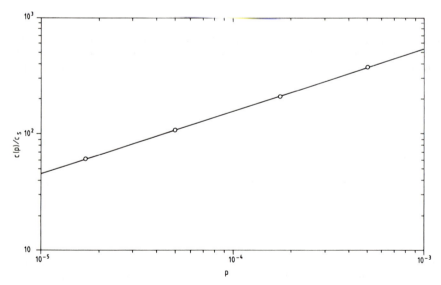

Fig. 15.17   Plot of the parameter $c(p)$ in eqn. 15.32 as a function of volume fraction p. The values of $c(p)$ are taken from the probability plots of Fig. 15.16b, and given relative to the value c_S for the superclean material

removed most of the low-field failures. Thus it seems that large particles (or aggregates) in the tail of the size distribution cause these failures. In practice it can be expected that defects of this type will be rejected, either by inspection, or by proof testing. However such results show that although there is some indication that the 1AEV distribution ought to be used for defect loaded systems, care should be taken over any extrapolation to service stresses.

15.4.2 The relationship between cumulative defect and percolation models

The defects of the percolation model have many similarities to those postulated for the cumulative model. They are regions of the insulations in which charge carriers have a higher mobility than the surrounding insulation, and thus can be regarded as conductors at the high fields near breakdown. Extended defects occur in the form of clusters of adjoining unit defects, and the possibility of a defect path crossing the insulation is allowed for. In this latter case the cluster is called an incipient infinite cluster[797] (IIC) and the breakdown field becomes zero. However the percolation model differs from the cumulative models in that a single additional failure of a capacitor-bond runs away to breakdown rather than leading to a slow extension of the defect-cluster. Thus it is necessary to regard the initial state of the percolation model as having been accumulated slowly over a previous period of aging, such that the most severe cluster will lead to runaway failure at the applied field. The quantitative analysis of the percolation model shows that this process will occur before the extended cluster crosses the system.

(a) Spatially random defect accumulation

If defects can be regarded as being produced in the material by *spatially random* events then a cluster system such as that considered as the initial state in the percolation model will be generated. In this case the volume fraction, p, will be proportional to the time of aging, i.e.:

$$p = t/\tau(E) \tag{15.33}$$

where $\tau(E)$ is the time to produce a new unit defect. In general $\tau(E)$ will decrease as E increases but its functional form is not obvious.

Under conditions of commercial importance (low E and long time aging) a reasonable approximation would be to assume that the characteristic cluster size ξ remains close to a unit defect. In this case the time dependence of the survival probability is determined by N_c in eqn. 15.31, and hence by $c(p)$ in eqn. 15.32a. The experimental results for the metal-loaded polyethylene plaques indicate that $c(p)$ is probably a power law in p, Fig. 15.17, with an exponent close to 0·5. Thus:

$$N_c = c(p)\Omega = \Omega p^\delta \equiv \Omega[t/\tau(E)]^\delta \tag{15.32b}$$

and we should expect a Weibull distribution of times to breakdown with an exponent a (equal to δ) with a value of about 0·5. The model is therefore similar in this respect to the conducting path mechanism of Section 15.2.

The hazard rate decreases with time in this case because the number of clusters increases slower with time than the number of broken bonds. The field dependence however will not have a power-law form. Instead the breakdown time, t_b, will be given by:

$$t_b = \Omega^{-1/\delta} \tau(E) \exp \left\{ \frac{E_0}{E \xi \delta} \right\} \qquad (15.34)$$

which increases strongly at small fields, in a similar way to that predicted for threshold processes[120].

At long aging times the defect concentration, p, will become large and the low-loading approximation breaks down. In these circumstances the term ξ in eqn. 15.31 will start to dominate the statistics. From eqn. 15.33 a maximum possible breakdown time can now be defined

$$t_b(\text{max}) = p_c \tau(E) \qquad (15.35)$$

where p_c is the volume fraction of unit defects at the percolation limit, and $t_b(\text{max})$ is the time required for the defect system to cross the insulation. This result gives a long-time cut-off to the breakdown statistics, which, because of the power-law dependence of ξ and the form of eqn. 15.31, will appear in a Weibull plot as a sharp steepening of the curve at long times (i.e. an apparently large value of a).

(b) Defect extension

As argued by Jonscher and Lacoste[133], it is more likely that the cumulative aging of polymeric systems will take place by the extension of existing defects, than the generation of new defects. In this case the 'new' unit defects generated in the system cannot be modelled by the spatially random addition of defect bonds as in the percolation system. Instead they must be preferentially formed around existing defect-clusters. A model that goes some way towards simulating this situation was proposed by Takayasu[397]. Here again the system is initially represented by a grid of highly resistive (impedance) bonds with a small distribution of resistance magnitudes. At some field, E_0, a bond in this inhomogeneous system will fail to a low resistance state. The fields along each bond are then recalculated and *all* those bonds experiencing a field in excess of E_0 are failed. This model therefore allows sufficient time on each iteration step for all bonds potentially capable of breaking down to proceed to failure. Repetition of the procedure builds up damage clusters of failed bonds in the system which now simulate slow aging in contrast to the impulse simulation of Beale and Duxbury[795].

The initiating defects of the cumulative model correspond to the bonds of slightly lower resistance in the Takayasu simulation[397], and it is from these bonds that the first failures arise. Subsequent failures tend to either extend a failed defect cluster or fail more of the initiating defects. Small defect groups may link-up to form large defect clusters which eventually cross the insulation. In contrast to the path generated during the 'impulse' breakdowns of the Duxbury model (Fig. 15.14) which is nearly linear and unbranched, Takayasu's breakdown structure has the form of an IIC percolation cluster with fractal dimension $d_f \approx 1 \cdot 6$. In the randomly diluted percolation (Duxbury) model similar fractal clusters will only be formed[307] when the initial (or aged) defect system has p approaching p_c ($\xi > 1$), and then their dimension, $d_f = 1·89$, (which determines ν) will be greater[313,397] than that of the Takayasu[397] model. The resulting breakdown structure will be composed of sub-percolation (multiply connected) clusters joined by unbranched links. This form will differ from that of the backbone of an IIC only in that the links are produced deterministically rather than through random addition[743]. On the other hand the Takayasu system[397] fails bonds preferentially around defects thereby extending the clusters. The system thus forms as a set of tree-like defects which amalgamate to give the breakdown structure. Since the extended defects prior to breakdown form a substructure within the final breakdown pattern, they will possess on average its fractal dimension, $d_f \approx 1·6$ (see Section 15.3.1). The state of this system at any time prior to a complete crossing of the insulation can thus be described in terms of the characteristic cluster size (equivalent to ξ of eqn. 15.27), its fractal dimension and the cluster distribution.

(c) Cumulative free-volume breakdown
In the form proposed by Takayasu the model bears a considerable similarity to the fractal discharge theory of the previous section (15.3), with breakdown occurring when the damage structure crosses the insulation. Under these conditions its breakdown statistics will result from fluctuations either in the time to fail a bond or in geometrical arrangement, and may be described through the theories of Sections 15.2 and 15.3. However if the clusters are taken to be extended free-volume defects containing mobile charge carriers, it is likely that a runaway breakdown will be initiated at the applied field prior to the defect system crossing the insulation. In these theories[759,788,477,356,93,476], when the free volume extends to a critical length, $l_c(E)$, charge carriers can be accelerated to a kinetic energy sufficient to initiate runaway damage. The system will therefore only survive as long as the largest extended cluster has a length, l, less than $l_c(E)$, where $l_c(E)$ is inversely proportional to E (see Chapter 13), i.e.:

$$l_c(E) \cdot (eE) = G_0 \qquad (15.36)$$

where G_0 is the damage threshold energy.

As yet no investigation of the cluster size distribution has been made in the computer simulations of this model. However the dimension (1.6) measured by Takayasu for the breakdown defect structure is close to that expected (1.62) for the percolation backbone in a two-dimensional lattice[743]. Since runaway failure connects those clusters on the breakdown path by unbranched linkages, it appears that some idea of the cluster size distribution prior to breakdown can be obtained by considering a percolation system with the connecting bonds removed from the backbone (Fig. 15.18). In this case the cluster distribution density has a power-law form[744] in S, the number of failed bonds in a defect cluster, i.e. $S^{-\tau}$, with $\tau = 1 + d/d_f$ where d is the spatial dimensionality of the matrix in which the defect system is formed ($d = 2$ for percolation in a plane, and $d = 3$ for bulk percolation). By these means the cluster size distribution can be related to the length l through

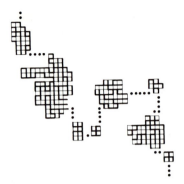

Fig. 15.18   Schematic representation of sub-percolation clusters representative of free-volume defects (after Herrmann and Stanley[743]). Thin lines = simulation grid. Thick lines = failed bonds. Dots = possible route of connecting links in forming a percolation backbone at breakdown

eqn. 15.15*b* as:

$$g(l) \propto \left[\frac{l}{l_\xi}\right]^{-\beta-1} \tag{15.37}$$

where percolation theory gives values of ~ 2 $(d_f = 1\cdot9)$ and ~ 3 $(d_f \approx 2\cdot5)$ for β when the structure is formed in two and three spatial dimensions respectively. Here the dimension of the whole disconnected system ($1\cdot9$ or $2\cdot5$) is used instead of that of the backbone alone $(d_f = 1\cdot6)$. This result has the form of eqn. 15.3 and, since survival requires $l_{max} < l_c$, the breakdown statistic has the Weibull form obtained from eqn. 15.2*b* for the survival probability. Thus:

$$P_S = 1 - P_F \propto \exp\left\{-N_d(t)\left[\frac{l_\xi(t)}{l_c(E)}\right]^\beta\right\} \qquad \beta = 2\text{---}3 \tag{15.38}$$

where $N_d(t)$ is the number of free-volume defect-clusters. If the field is below the level required to initiate new defects as claimed by Jonscher and Lacoste[133], N_d should either remain constant at its initial value or reduce as clusters are amalgamated.

In the present model eqn. 15.10 cannot be used for the time dependence of $l_\xi(t)$ since a number of bonds are all failed in the same time interval. This process is similar to that occurring when successive DC pulses of the same polarity are used to grow electrical trees (see Section 5.4). It is therefore reasonable to assume the same time dependence $(l_\xi(t) = t/\tau(E))$ in eqn. 15.38, which leads to a Weibull time exponent $a = \beta$ (2—3).

This range of values is close to that predicted for electrical-tree aging in Section 15.2 (see Table 15.2) to which the model also bears a physical resemblance. The field dependence of this model will however depend upon that of $l_c(E)$ as well as the time scale for the growth of the defect. As yet no theoretical estimates have been made of this latter factor, however the field dependence of $l_c(E)$ $(\propto E^{-1})$ contributes an E^β component to the exponential exponent of eqn. 15.38. Thus:

$$1 - P_F = P_S(t, E) \propto \exp\left\{-N_d\left[\frac{tEC}{\tau(E)}\right]^\beta\right\}\qquad(15.39)$$

(C is a constant) defines the life statistics applying when aging at a constant field.

If ramp or impulse measurements are made of the residual life following a period, t_a, of aging, the breakdown statistics should take the Weibull form:

$$P_S(E) = \exp\left[-C'[l(t_a)]^\beta E^\beta\right]\qquad(15.40)$$

(C' and t_a are constants) provided that no accelerated aging occurs during the test. The values of field exponent obtained for this type of test[759] on samples that have not been previously electrically aged (i.e. 7—10) are however much greater than the range of β (2—3) deduced from this model. Thus either there has been accelerated aging during the test or the initial distribution of defect sizes is very different from that which exists following aging. In the former case the aging time is proportional to the breakdown field $(E_b/\dot{E})$, and since $\tau(E)$ will reduce with increasing field the effective Weibull exponent $(a+b)$ (see Section 14.2) will be greater than ~ 5. The latter case is however the more likely and the observed exponent (~ 8) would suggest a narrow distribution of initial defect sizes centred on a typical value. Here the defect size and shape is likely to fluctuate as a result of chain (γ) motions and hence the value of the exponent may be best explained by the fluctuation model, see Table 15.2. The constancy of the exponent as the temperature changes[759] from the semi-crystalline state of polyethylene ($\sim 20°$C) to the viscous fluid ($\geqq 100°$C) is consistent with this view.

It has been suggested[476] that in the solid the defects are Matsuoka voids[168] (characteristic size ~ 1 nm) which are zero density regions formed when crystallites cannot occupy the existing space. Neither their size nor distribution should change much until softening starts[476]. At this temperature ($\sim 100°$C) the ramp exponent[759] (~ 19) lies outside the typical range (6—11) observed both above and below it. Here the Matsuoka voids disappear[476] and free volume is created, probably with a narrow range of sizes. At higher temperatures still the free volume sizes thermally equilibrate via coalescence to give a bigger characteristic size[476] and a broader distribution with essentially the same shape as found for the Matsuoka defects in the solid.

If the above argument is correct the distribution in defect sizes (Matsuoka voids or free volume) in *unaged* materials is very different from that suggested by the simulation model as resulting from aging. In this model the initial condition is represented by a distribution of the local (bond) resistivities in both magnitude and space. Such 'resistivity defects' will originate in local density variations (e.g. crystalline and amorphous regions) as well as the void and free volume defects taking part in breakdown. However all the 'resistivity defects' can take part in the aging process thereby generating new free volume centres as well as extending existing ones, and Takayasu's results[397] indicate that the resulting defect-cluster structure is mainly a feature of the aging process and little affected by the initial resistivity distribution. Thus the free volume distribution (in size and space) of an electrically aged material may be very different from its initial state. In consequence the breakdown statistics may change considerably after a period of aging in a way that cannot be replicated in accelerated breakdown tests.

15.5 Some special cases

In the previous sections, stochastic breakdown models have been considered which have been grouped according to an underlying unity in their approach. In nearly all cases the stochastic features of the model result from a propagation stage and thus they can be regarded as examples of category (b) of Section 15.1.1. Even the exception to this approach, namely the percolation model of Duxbury and co-workers[793,794,795], is related to a physical model where a continuous deterioration of the system is the dominant feature. In the following sub-sections some cases of special interest are considered where the stochastic features are either dominated by initiation or inception of defects capable of propagating (category (a) of Section 15.1.1) or the inception of runaway processes capable of leading to a breakdown (category (c) of Section 15.1.1).

15.5.1 Electrical tree inception

Inception may become the dominant stage of electrical tree breakdown when the local field is close to the initiation threshold value (see Section 5.2) where the inception time rapidly approaches infinity, Fig. 5.5. Although the propagation rate will also reduce with applied voltage, observations show that it has a weaker field dependence (eqn. 15.10) than the inception time. Thus in the commercially important low-field regime, propagation may still be much more rapid than inception[177,804]. This will be the case particularly when the stress enhancement of the initiating site is relatively low and the average applied field well in excess of any threshold that may be required for the continuation of propagation. If such a threshold does exist (see Section 13.2.3) it will be more related to the field required for partial discharges in voids than that for the avalanche damage of inception, and hence will be lower in magnitude.

This dominance of the inception stage over that of propagation under certain conditions has been demonstrated using trees initiated from razor blades cast in an epoxy resin[177]. Here the inception time, t_I, lay in the range 100—7000 minutes, while the time required for a tree to cross the insulation was less than 100 minutes. The breakdown statistics were found to be governed by those of tree inception. The distribution of tree inception times was however not simple. When plotted on Weibull paper a sharp rise in cumulative inception probability converted to a very flat curve for which the value of t_I increased rapidly with little change in probability. Similar behaviour has also been found by other investigators[129,342,166], and a selection of their results is shown in Fig. 15.19.

It has been suggested[166] that the form of these observations originate with a Weibull distribution of inception times appropriate to a single needle

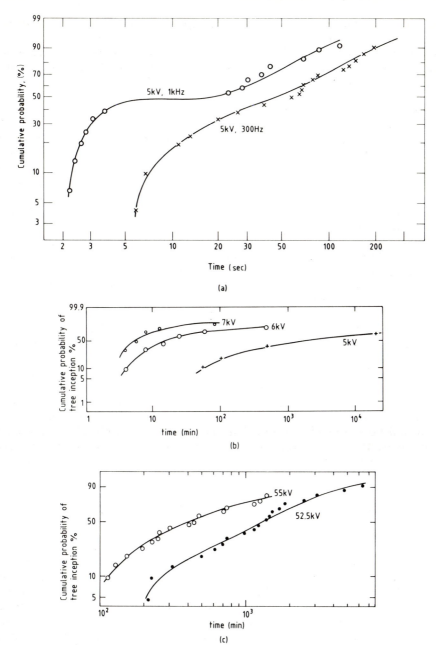

Fig. 15.19 Weibull plots of the cumulative probability of electrical tree inception over a time *t*. Note the log-scale of the time axis. (*a*) From needle electrodes in polyethylene[129] using different field frequencies, (*b*) From steel electrodes in polyethylene with different applied voltages[342], (*c*) From razor electrodes cast into epoxy resin with different voltages applied[177]

electrode. Sample-to-sample variations in the experiments were taken to cause the characteristic value itself to be Gaussianly distributed, and a combination of the two effects was used to fit the experimental values by suitably adjusting the parameters. A similar explanation was given for the distribution of tree inception fields in a ramp experiment[805]. Here the Weibull distribution of breakdown fields in uniform stress was chosen as the starting point. Such explanations are however fraught with danger. Even though there is some justification for a peaked distribution of the sample geometries, the origin of the Weibull distribution assumed to be intrinsic to the process is unknown. In the case of the distribution of inception fields[805] it was taken to be that of the local breakdown field strengths of the dielectric material. However such a distribution may be altered by the stress-enhancing projection which may change the local morphology[937]. Indeed it is possible that the distribution used may have originated with the electrode interface itself which has now been changed, or perhaps with the propagation to breakdown of a bulk process rather than its initiation as with electrical tree inception. It is clear that this interpretation cannot be accepted as more than just a plausible explanation.

In the above convolution of statistics proposed to explain electrical tree inception, it was axiomatically accepted that in the absence of sample variations the distribution would be of the Weibull form. However a simple physical model[806] shows that, on the contrary, a set of needle electrodes with *identical* geometry is sufficient to give the distribution of inception times observed experimentally.

The basis of the model is that electrical trees are initiated by damage-producing injection or extraction currents (see Section 5.2) which start at (or terminate on) a few very small sites ($\leqq 50$ nm) on the electrode surface[256] rather than involving the whole electrode. Such a possibility has already been proposed in the case of electrical trees formed in liquid dielectrics[369] where the injection sites are characterised by low work functions and/or high stress enhancements. In the present case it is assumed that the injection/extraction (active) sites possess a favourable work function with a value typical of a surface imperfection, grain boundary or microscopic protuberance ($\leqq \sim 100$ nm), of the electrode metal concerned.

For any field-enhancing electrode projection, such as a needle, the stress will vary at different points on its surface, and an 'active' site of a given type will experience a field equal (or proportional) to the stress at its location. The electrical tree inception time (which is dependent upon the local field) will therefore also vary with the location of the 'active' site. Since there are very few active sites on a typical needle (their density[369] of 5—250 cm^{-2} gives 1—8 sites on a 30° cone indented by 2 mm) and these are randomly located, the probability of the occurrence of a given inception time will be proportional to the number of site-sized elements of surface area (say 50×50 nm^2) experiencing the same field. This probability can be normalised to unity by recognising that the existence of a threshold field for tree inception (eqn. 5.4) means that there is only a finite region of needle surface over which inception can occur at any given voltage. Thus given a means of calculating the field at a point on the needle surface and the element of surface area at that point, the electrical tree inception time statistics can be derived.

For convenience the needle is taken to be a hyperboloid of revolution about the axis (x) joining the point to the plane (Fig. 15.20a). The field at the electrode surface is thus given by[807]

$$E(x \geq D) = \frac{2 V \zeta_0}{D \log_e \left\{ \frac{1+\zeta_0}{1-\zeta_0} \right\} (1-\zeta_0^2)^{1/2} [(x^2/D^2) - \zeta_0^2]^{1/2}} \tag{15.41}$$

where V is the voltage, D is the point-plane distance and ζ_0 the cosine of the half-angle, θ, subtended by the needle at the ground plane.

Although eqn. 5.4 relates t_I to the applied voltage, details given in the original work[342] allow it to be converted to a function of applied field, i.e.:

$$\log t_I = \frac{\bar{A}}{E} + \bar{B} - \log \left\{ \frac{1}{E_0} - \frac{1}{E} \right\} \tag{15.42}$$

$$\bar{A} = 992 \qquad \bar{B} = -4.04 \qquad E_0 = 4 \times 10^8 \text{ V} \cdot \text{m}^{-1}$$

and hence substitution of expression 15.41 for $E(x)$ gives t_I for an active site located on the needle surface at a given value of x (i.e. $t_I(x)$). The surface area from the needle point to position x is

surface area

$$= \pi \zeta_0 D^2 \left[\frac{1}{\zeta_0^2} - 1 \right]^{1/2} \left[\frac{\sinh 2\phi}{2} - \phi \right]_{\phi = \cosh^{-1}(1/\zeta_0)}^{\phi = \cosh^{-1}(x/D\zeta_0)} \tag{15.43}$$

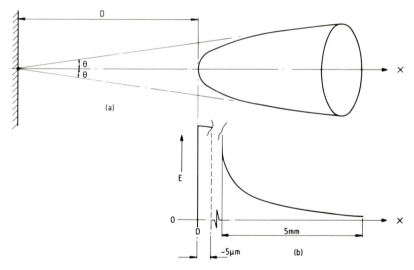

Fig. 15.20
(a) Schematic representation of point-plane geometry with a hyperboloid of revolution for the needle, $\cos\theta = \zeta_0$ in eqn. 15.41
(b) Schematic representation of the electric field at the surface of the needle. Note the change in the length scale between the two sections of the figure

which is proportional to the cumulative probability that an electrical tree has been initiated by a time $t_I(x)$. Division by the surface area up to the value of x for which t_I is infinity gives the cumulative probability as a fraction.

The calculated inception time distribution which is shown in Fig. 15.21 displays all the features of the observed behaviour, namely a sharp threshold at a *minimum* value of t_I followed by a long tail as t_I approaches infinity. This form of behaviour will be found for any form of electrode projection regardless of its geometrical details. A minimum inception time occurs because there is a maximum stress enhancement. The probability rises sharply with very little change in t_I because the field is effectively unchanged in a small region about the needle tip (see Fig. 15.20b). For larger values of x the field drops rapidly towards the threshold value and hence the contribution of the remaining sites that may participate is spread out along the time axis.

The calculations show that 50% of the trees are initiated within a range $(x - D)$ from the tip which is $\sim\frac{1}{3}$ of the maximum distance over which inception can occur. For the half-angles of $< \sim 5°$ chosen to give the same maximum E/V ($= \gamma$) ratio (~ 100) as used by Tanaka and Greenwood[342] it is found that 50% of the trees are initiated within about 5 μm of the tree tip (dependent upon the applied voltage), and within no more than a factor of three times the minimum inception time. However an increase of axial

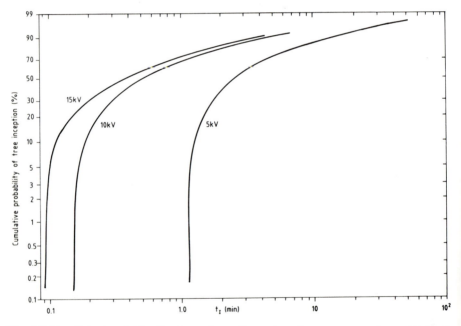

Fig. 15.21   Calculated distribution of tree inception times presented as a Weibull plot. The details are: half-angle $\theta = 3°$; $D = 1$ mm; voltages of 5 kV, 10 kV, and 15 kV (field in kV $\cdot$ mm$^{-1} \approx 10^2 \times$ voltage in kV) 50% of the failures occur for $x \leqq D + 6$ μm at 15 kV, $x \leqq D + 2\cdot 5$ μm at 10 kV, and $x \leqq D + 0\cdot 3$ μm at 5 kV.

displacement from the tip to $\sim 15\ \mu$m will lead to an increase of the inception time by orders of magnitude. These figures apply for a voltage of 15 kV and, although the needle is very sharp (radius of curvature $= D \cdot \tan^2 \theta = 2{\cdot}7\ \mu$m, for $\theta = 3°$, $D = 1$ mm) they show that trees initiated on a short laboratory time scale will grow from almost unresolvable positions close to the tip. Trees initiated only slightly further from the tip require substantially longer times and hence may not occur even in long-time experiments. The possibility of the near simultaneous initiation of a large number of small trees which shield one another is thus unlikely in a needle-like geometry.

Comparison between different experiments should be made using the minimum (threshold) inception time. This parameter is determined by the maximum stress enhancement and will be most responsive to changes in the needle geometry. Thus sample-to-sample variations in needle shape will result in a reduction of the slope of the Weibull plot (i.e. broaden the distribution) in this region but otherwise leave the general features of the probability plot unchanged. In general empirical relationships are determined from the 50% probability value. Fig. 15.22 shows that this parameter has almost the same behaviour as that of the minimum inception time. For ideal needles some errors in estimating $\bar{A}$ and $\bar{B}$ will occur as a result of using the 50% value, but in real systems where a distribution of t_I(min) may exist it is likely that very little improvement can be made. The threshold field though is well defined in either case.

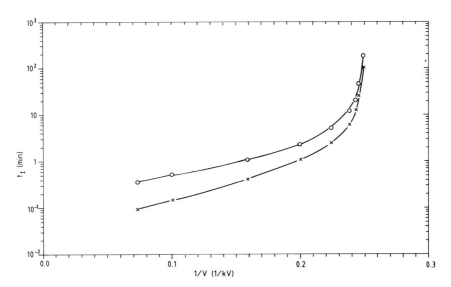

Fig. 15.22   Calculated values of the minimum inception time as a function of voltage (×) from eqn. 15.42 with $V \propto E$, $E = \gamma V$ where γ is the geometry dependent factor that converts an applied voltage to the maximum field E in a given electrode configuration. The time for initiation of 50% of the trees (○) which is the experimental parameter often quoted[129,342] is shown for comparison

The model can also be used to calculate the form of the inception field distribution as determined in a ramp experiment. Here the cumulative probability $P(V_I)$ is determined by the surface area of the needle that has been able to initiate a tree by the time the voltage reaches the quoted value. $P(V_I)$ is converted to a fraction by dividing by the (finite) maximum surface area that is available to initiate trees. Using a value for the injecting surface area arbitrarily chosen to be about 70% greater than that able to generate trees at a voltage of 15 kV gives Fig. 15.23, which shows that the calculated result has the same form as that obtained experimentally[342]. In particular a threshold voltage is found which increases with ramp rate.

The existence of a minimum field for tree inception in eqn. 15.42 is enough to give the voltage threshold, but its displacement to higher values for higher ramp rates occurs because the ramp process itself allows insufficient time for the active sites to respond at lower voltages. This result indicates that the value obtained for the minimum field for tree inception, using rapid ramping to prevent the shielding effects of homo-charge injection[287], may be a substantial overestimate[808].

In the present model space charge effects are neglected. The theoretical development[342] of eqn. 15.42 (see Section 5.2) even neglects image forces. This latter feature would convert eqn. 15.42 into the same function of $E^{1/2}$

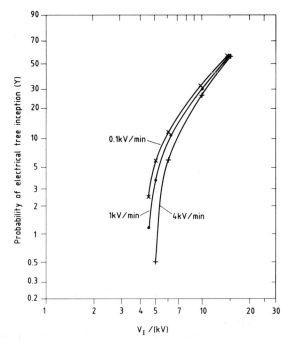

Fig. 15.23 Calculated Weibull plots of the electrical tree inception voltage, V_I, under conditions of a ramped voltage. The electrode configuration has $\theta = \arccos (\zeta_0) = 3°$, and $D = 1$ mm, in eqns. 15.41 and 15.43. Three ramp rates are shown

and otherwise leave the results unaffected. In view of the small range of voltages over which observations can be made without them being dominated by the threshold effects (Fig. 15.22) it is unlikely that the $V^{1/2}$ dependence could be established quantitatively. Thus eqn. 15.42 has been used in the model because of the availability of experimental values for $\bar{A}$ and $\bar{B}$. When trees are generated by an AC field the presence of space charge is not likely to modify the general features of the results. Here the first tree channel is produced by damaging electron extraction avalanche currents flowing to the active site. The previously injected electrons will have formed a homogeneous charge cloud according to the Zeller model[346,338] which now behaves as a heterocharge during extraction (needle positive) and thus enhances the local field according to the spatial distribution of charge around the needle. Since this is proportional to the field distribution around the bare needle during injection[338,347], the conditions of the present model will be recovered.

Under DC conditions a homogeneous space charge cloud will have formed at low field with a boundary at a constant potential and a shape similar to that of the injecting electrode[347]. When charge density fluctuations cause filament formation their location may be influenced by the position of the active sites on the electrode surface and hence lead to a distribution of inception fields under ramp conditions such as calculated here. Further work is however required to justify this contention.

15.5.2 Partial discharge inception

Partial discharges lead to breakdown through the formation of a tree-like system[120,126,127] which eventually either crosses the insulation or initiates a runaway process (see Chapter 13). The stochastic features associated with the propagation stage of such a process will fall into one or other of the models discussed in the previous section, depending upon whether breakdown is caused by a single discharging void (Sections 15.2 and 15.3) or is a cumulative effect of discharges in many voids (Section 15.4). However the relative importance of the propagation and inception stages will depend upon the conditions, such as applied field, insulation thickness, and void size distribution. In general it might be expected that as the field approaches the threshold for discharge inception the initiation stage would dominate. However propagation may be just as slow at low fields and for the large insulation thicknesses often found in cable insulation, damage extension (Section 15.3) and damage link-up (Section 15.4.2c), are likely to dominate the breakdown statistics[324]. At high fields propagation will be fast and discharge-tree inception is probably the rate limiting process[324]. Thus in accelerated ramp tests the inception process could well dominate the statistics and a correlation has been observed[222] between the characteristic breakdown stress and the minimum field required for discharge inception in the voids known to be in the material (polyethylene) tested[405]. It should be noted however that under constant stress a different stage of the *same* breakdown mechanism may be rate limiting giving different statistics.

When partial discharge inception is the dominating process the field (time) for breakdown will be determined by the most severe void in the insulation.

Leaving aside such factors as void shape, which controls the field within the void[641], and non-uniformity of applied field, which alters the local field experienced by the void[809], the most severe void will be the largest void as long as this is bigger than the Paschen curve minimum[122] (~10 μm at 1 atmosphere in *PE*). This void will in general have the lowest value of discharge inception field[122], the largest discharges[381,666], and the largest volume, hence the greatest probability that a charged particle will be available to initiate an avalanche discharge in the gaseous contents[381] and thus the shortest inception time[383].

At a given applied field E a critical void size $L_c(E)$ can be defined such that when the largest void (L_{max}) is less than $L_c(E)$ no breakdowns will occur. The survival probability is thus the probability that $L_{max} < L_c(E)$, and a functional form can be obtained provided the void size can be related to the discharge inception field, and the distribution of void sizes is known.

The relationship between void size and discharge inception field (E_d) is typically given in the form of a Paschen curve[122], but a polynomial fit can be obtained, Fig. 15.24, with the form:

$$E_d = \frac{A}{L^2} + B' + \frac{C'}{L} + \frac{F}{L^{1/2}} \qquad (15.44a)$$

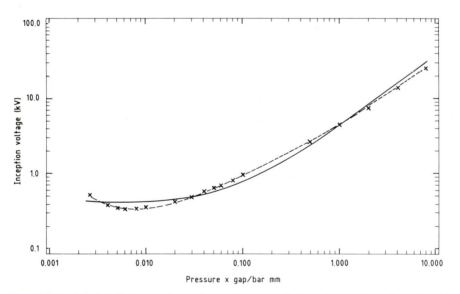

Fig. 15.24 Partial discharge inception voltage (peak) for air as a function of void gap and air pressure. + = experimental data taken from reference[936]. The continuous line is a fit to the Hall–Russek function[659] (expression 15.44b) with $Bp = 3.75$ kV/mm^{-1} and $C = 0.39$ kV. The broken line is a fit to the polynomial expression 15.44a with $A = 0.00101$ (kV · mm), $B' = 2.40$ (kV/mm), $C' = -0.0097$ (kV), $F = 2.244$ (kV · mm$^{-1/2}$). Reproduced by courtesy of A. Barclay, National Power TEC, Leatherhead, UK

which for $L > 10$ μm reduces to the Hall–Russek approximation[659]

$$E_d \approx Bp + \frac{C}{L} \tag{15.44b}$$

where p is the pressure of the gaseous contents. Thus as long as the *largest* void is greater than ~ 10 μm, its size can be related to the critical field for breakdown via:

$$L_c(E) = \frac{C}{(E - E_i)} \tag{15.45}$$

where E_i ($= B \cdot p$) is the asymptotic inception field for an infinite sized void.

Void size distributions in polyethylene power cable insulation are usually taken to be exponential in form[119] as shown by the data given in Table 15.3. Thus the probability that $L_{\max} < L_c(E)$ is given by the 1st asymptotic extreme-value distribution, eqn. 15.2b, which therefore gives the probability of survival at a field E, as

$$P_S(E) = 1 - P_F(E) = \exp\left[-N \exp\left\{-\frac{L_c(E)}{L_\xi}\right\}\right] \tag{15.46a}$$

Here L_ξ is the characteristic void size (see Table 15.3), and N is the number of voids which is proportional to the volume of the system. Using eqn. 15.45 for $L_c(E)$ gives

$$P_S(E) = \exp\left[-N \exp\left\{-\frac{C/L_\xi}{E - E_i}\right\}\right]$$

$$\approx \exp\left[-N \exp\left\{-\frac{E_\xi - E_i}{E - E_i}\right\}\right] \tag{15.46b}$$

where E_ξ is the field at which discharges occur in the average void (cf. eqn. 15.45). This expression (15.46b) shows a breakdown threshold when

Table 15.3 Void distribution in steam-cured (SC) and radiation-cured (RC) crosslinked polyethylene[439]

Size range μm	Measured no.* in 10 mm²		Calculated from $dN/dL \propto L_\xi^{-1} N_0 \exp\{-L/L_\xi\}$	
	SC	RC	SC ($N_0 = 5000$, $L_\xi = 1 \cdot 1$ μm)	RC ($N_0 = 650$, $L_\xi = 0 \cdot 6$ μm)
1—3	>2000	120	2026	118
3—5	~300	3	330	4
5—10	77	0	63	—
>10	4	0	~1	—

* Measured number microscopically counted in a slice of ~ 10 mm thick cable insulation with the quoted surface area. The number of voids, N, in any given size of insulation is assumed to be proportional to the number density quoted in columns 2 and 3.

E is equal to E_i, but at large fields $E_\xi > E \gg E_i$, it takes the form of eqn. 15.31 previously discussed in Section 15.4.1. Thus in this field range a plot of $\log(-\log_e[1 - P_F])$ against $1/E$ should give a straight line.

Failure data taken from ramp experiments performed on cable sections manufactured from the same material whose void content is given in Table 15.3 is shown in Fig. 15.25. These results show a tolerable straight line at high fields with some evidence for a threshold field. The value of E_ξ can be obtained from the gradient of the plot, and the root-mean-square values obtained are $\sim 3 \cdot 80 \times 10^8$ V m^{-1} for the radiation-cured sample and $2 \cdot 23 \times 10^8$ V m^{-1} for the steam-cured material.

Here a stress enhancement factor of $1 \cdot 23$ has been used to convert the average applied field to actual field within the void[439,641]. Although these values are too small to be the inception fields of voids with L_ξ as given in Table 15.3, they are nonetheless of the magnitude appropriate to voids with sizes in the range $1 \cdot 7$—$2 \cdot 5$ μm. These results thus appear to be reasonably

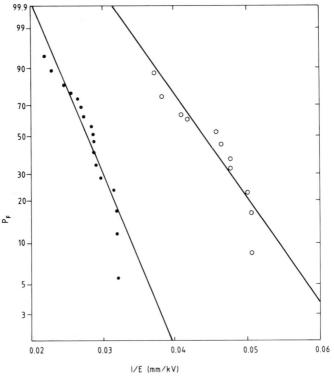

Fig. 15.25 Cumulative probability of failure in cable sections under conditions of ramped voltage plotted as function of E^{-1}. Note that the x-axis is linear in E^{-1}. The data is taken from reference[439] where it was ascribed to a partial discharge breakdown mechanism. Values of the gradients give E_ξ uncorrected for local field enhancement as $3 \cdot 06 \times 10^8$ V $\cdot$ m^{-1} (radiation cured XLPE) and $1 \cdot 81 \times 10^8$ V $\cdot$ m^{-1} (steam cured XLPE). $\bullet$ = Radiation-cured XLPE; $\bigcirc$ = steam-cured XLPE

consistent with the theory, however the relative number density of voids taken from the plots is much higher than that found in Table 15.3 and the threshold fields very high (SC, $\sim 20 \, \text{kV} \cdot \text{mm}^{-1}$; RC, $\sim 30 \, \text{kV} \cdot \text{mm}^{-1}$). One explanation for this latter fact would be that the manufacturing process introduces an absolute upper limit to the void size of around $20 \, \mu\text{m}$ (estimated from the apparent threshold fields, and Paschen's curve) independent of statistical factors. This size is consistent with the data of Table 15.3. It is also possible that this result may be a consequence of the ramp rate (see Section 15.5.1) and it is noticeable that the breakdown field distribution has a similar form to that for the inception of electrical trees, Fig. 15.23. When taken together with the fact that the distribution of breakdown times for a single artificial void has the same form[459,166] as Fig. 15.21, it may indicate that an apropriate version of the electrical tree inception model also applies to partial discharge breakdown.

15.5.3 Influence of vented water trees on breakdown statistics

The effect of vented water trees on dielectric breakdown has already been discussed from a mechanistic viewpoint in Section 6.1. There it was pointed out that the presence of water trees not only reduced the breakdown field but also altered its distribution by increasing the probability of low-field failures (see Figs. 15.7 and 6.2). In Section 15.2.2 a model was described in which water trees aged the insulation by increasing the number of volume elements experiencing breakdown attempts, thereby increasing the probability of failure without otherwise affecting the breakdown mechanism. The resulting breakdown statistics for aging under constant stress are given as a function of time and field by eqn. 15.11. This is the form required for breakdown prediction under operating conditions, however most experimental evaluations[175,176] are based on the residual life determined from ramp tests after a given aging time t_a. In this case eqns. 15.11 and 15.9 show that the characteristic breakdown field for the model is given by:

$$E_b \propto [L(t_a)]^{-2m/(1+m)} \tag{15.47}$$

as long as the length, $L(t_a)$, of the characteristic longest water tree does not have time to increase during the ramp. The distribution of breakdown fields will however remain the same as that of the originating breakdown mechanism (i.e. $a = 1$, $b = 1/m$) and will not be affected by the presence of the water trees.

A relationship between characteristic breakdown strength and the maximum length of vented water trees has often been observed[119,175,810] (see Fig. 6.1) however no agreement has been reached on the form that it takes. Thus Gotoh *et al.*[810] take the characteristic breakdown field to be exponentially related to the maximum expected length $L_{\text{max}}(t_a)$ after aging,

$$E_b \propto \exp\left\{-\frac{L_{\text{max}}(t_a)}{L_\xi}\right\} \tag{15.48}$$

Some workers however regard the change in breakdown strength shown in Fig. 6.1 as evidence of a critical characteristic length[175,939], and yet others find a weak long-time decrease with length, $L_{max}(t_a)$, of a logarithmic[706] or linear[253] form. None of these investigations however consider the effect of water trees upon the breakdown statistics. Their conclusions therefore refer to the characteristic value of the residual life rather than the commercially more important low-field failures.

In Chapter 18 we shall discuss models which can be used to estimate the maximum expected water-tree length from the growth law of the characteristic tree (Section 4.2), the length distribution, and the number density (Section 4.2.1). Since the residual breakdown strength reduces with maximum vented tree length, these estimates can be used to provide a figure of merit for the water-tree degradation of the system. The characteristic or average maximum tree (see eqns. 4.14 and 4.15) can also be regarded in the same way. Here though we are interested in the change in ramp breakdown statistics brought about by water treeing, and in particular the increase in low-field failures. As noted previously (Section 6.1) this change in the shape of the breakdown distribution is retained even when the tree is dried out and the characteristic breakdown strength almost recovers its untreed value[175].

The alteration of the distribution *shape* cannot therefore be attributed to the local stress enhancements of the water-filled microvoids which has been proposed to explain the reduction in characteristic strength. Instead it has been suggested that the formation of the water tree generates 'dielectrically weak' regions between the microvoids which retain their weakness when the stress enhancing water is removed from the microvoids (Section 6.1).

The 'weakness' of the dielectric within the water tree was attributed in Section 6.1 to weakly oxidised regions containing polymer radicals, some broken bonds, some structural distortion and possibly trapped or bound charges and dipoles. Such regions originate as incipient microvoid sites (Section 5.2.3) which have not proceeded to fully form a water-filled microvoid. For a number of reasons, the most prominent being a restricted access to a water reservoir (see Section 6.1), it is likely that these weak regions will be of greatest density and size after long times and in long trees[447,218]. In this incomplete form they would act as extensions of the local free volume, and their influence upon the breakdown statistics may be related to the models of the previous section (15.4). Such a picture would serve to connect the water tree influenced ramp breakdown statistics to those of the untreed material which it can be argued[788,477,356,93] originates with free-volume processes. As noted in Section 15.4.2c the Weibull ramp parameter, $a + b$, of free volume breakdown is high (7—12), prior to cumulative damage, and this is consistent with the values of 12 and 19 shown for the untreed material in Figs. 6.2 and 15.7 respectively.

If the water-tree aging followed the Duxbury model[793,794,795] by randomly generating free volume defects then the breakdown statistics could be expected to be of the 1st asymptotic extreme-value type which appears as a line of large gradient when plotted on Weibull paper (Fig. 15.16a). Instead the gradient observed reduces to about 4·5 (see Figs. 15.7 and 6.2), which suggests that the effect of the tree is to extend existing free volume defects. In some cases it is possible that the extension may even lead to disconnected channels being formed[38,16,39]. Charge acceleration within the largest free-volume region l_{max} in the tree would then cause electrical breakdown under ramp conditions, and not, as has been proposed[258], growth of the water tree itself.

If the extended *free volume defects* form a system of sub-percolation clusters within the tree such as might be expected from the tree conductivity and ability to transport electrolyte (Sections 4.3.2 and 4.3.3) then the Takayasu type model[397] should apply. Here the distribution of defect lengths within a tree, will be given by eqn. 15.37, and its survival probability after an aging time t_a by eqn. 15.40. Thus:

$$1 - P_F = P_S(E) = \exp\left[-N_d C[l_\xi(t_a)E]^\beta\right] \tag{15.49}$$

defines the failure statistics under ramp conditions for a set of vented water trees of a given length produced by aging for a time t_a. Here N_d is the number of defects within a tree and C a constant proportional to the number of trees present. Expression 15.49 is valid for ramp tests as long as the size and number of the free-volume defects remains unchanged during the ramp, and predicts a Weibull form with ramp exponent $(a + b)$ of $\beta = 2$—3. To complete the derivation it is necessary to relate the characteristic free-volume length at t_a (i.e. $l(t_a)$) to the length of the water tree. It is most likely that these two lengths are proportional to one another and hence in a distribution of water tree lengths[234] it will be the longest tree $L_{max}(t_a)$ that contains the most severe free-volume defects. Eqn. 15.49 therefore defines survival as the probability that the largest defect in the longest tree is less than the critical length $l_c(E)$ (eqn. 15.36), with $l(t_a) \propto L_{max}(t_a)$, and hence the characteristic breakdown field in a ramp test will be inversely dependent upon $L_{max}(t_a)$, i.e.:

$$E_b \propto 1/L_{max}(t_a) \tag{15.50}$$

The value of β predicted is near to but below that observed for water tree influenced breakdowns[176] (i.e. ~4·5, see Figs. 15.7 and 6.2), and it may be that the free-volume regions can be extended during the ramp although insufficient time is available for the water tree itself to grow. If it is assumed that extension of the defect is by repeated accumulation of fractal units of damage, as for the DC electrical trees of Section 5.4.2, then:

$$l_\xi(t) \propto l_\xi(t_a)\left[1 + \frac{t}{\tau(E)}\right] \sim \frac{E}{\tau(E)\dot{E}} \tag{15.51}$$

where t is the period under test and $\dot{E}$ is the ramp rate. The resulting breakdown statistic will only have the Weibull form if $\tau(E)$ is a power-law function of the field. This is probably not the case in general, however since $\tau(E)$ will reduce with increasing field an effective Weibull ramp exponent $(\equiv a + b)$ may be obtained, the value of which is estimated to be greater than 4—6. In comparison with the experimental value (~4·5) this variation of the model probably overestimates the exponent as much as eqn. 15.49 underestimates it. We note however that if $\tau(E)$ has an exponential form (e^{-aE}) then the characteristic breakdown field, E_b, is proportional to $A - \log[L_{max}(t_a)]$ ($\equiv A' - \log t_a$) as suggested by Bahder *et al.*[706].

The increase in the effective Weibull exponent found above arises because the size of the free volume defects are assumed to increase with time during the ramp. A contribution to the field exponent thus occurs because of the linear relationship between time and field in a ramp experiment. A similar contribution may be found if breakdown were the result of failure attempts which were random in time, rather than through an automatic initiation of a runaway process once a critical length is reached, as above. In this case eqn. 15.49 takes the form:

$$1 - P_F = P_S(E) = \exp\{-N_d C t [l_\xi(t_a)E]^\beta\} \tag{15.52a}$$

$$= \exp\{-N_d C'[l_\xi(t_a)]^\beta E^{\beta+1}\} \tag{15.52b}$$

where eqn. 15.52b applies during a ramp test in which the defect is not extended. The resulting Weibull exponent lies in the range $a + b \approx 3$—4, and is again close to the experimental values. Here the characteristic ramp breakdown strength is inversely related to a power of the maximum tree length (if $l_\xi(t_a) \propto L_{max}(t_a)$), i.e.:

$$E_b \propto [L_{max}(t_a)]^{-\beta/(\beta+1)} \tag{15.53}$$

All the above versions of the model relate the failure statistics to the distribution of free-volume defects within a water tree and assume that either all trees are identical or that only the most severe (longest) tree is effective. The predictions therefore can only be properly checked by breaking down a set of individually grown trees of the same length. Water tree breakdown studies are however typically carried out on cable sections[492,176,175,811] which have been aged in water for a given period of time. In this case there will be a distribution of water tree lengths. If it is assumed as before that the length $l(t_a)$ of the largest defect in a given tree is proportional to the tree length ($L(t_a)$), then it may be that it is the tree length distribution that governs the breakdown statistics. Should this be so the survival probability (i.e. $l(t_a) < l_c(E)$) is either given by the third asymptotic extreme-value distribution, eqn. 4.15, or the 1AEV, eqn. 4.14. The former option is very difficult to use because of the need to obtain values for the longest possible water tree, but the alternative yields

$$P_S(E, t_a) = 1 - P_F$$

$$= \exp\{-N_T V_T(t_a)B \exp\{-\alpha[l_c(E) - AL_{max}(t_a)]\}\} \tag{15.54}$$

Here A, B, and α are constants, and the product $N_T V_T(t_a)B$ relates the number of defects (N_d) to the number of trees, N_T, and the average tree volume, $V_T(t_a)$. Such a complicated expression is difficult to relate to experimental data particularly when there are a limited number of failures, however some evidence for the onset threshold originating with the ($l_c(E) - AL_{max}(t_a)$) term can be found in the data of References 492 and 811. It should be noted that whenever breakdown studies are carried out on a system containing an ensemble of water trees the number of free volume defects (N_d) will be proportional to the total volume of the water trees which is dependent upon the aging time. Thus N_d introduces an extra time dependence if the failures are measured during water tree aging, but can be regarded as constant during a ramp test.

The models discussed here show that attributing the water tree effect on breakdown statistics to the production of free volume defects within the tree reproduces qualitatively the observed behaviour, in particular the increase of low-field breakdown (reduced Weibull exponent). However quantitative agreement has not been obtained in any one case. The construction of the models vividly illustrates the difficulties arising when a number of independent parameters combine to determine the distribution of the critical variable, and it should be noted that no allowance has yet been made for the local field enhancement of water-filled microvoids[262] identified in Section 6.1 as an important factor in reducing the breakdown strength. Furthermore these models have made use of the water tree length as a parameter quantifying the size of the defects within the tree. However these defects may extend beyond the tree boundary as revealed by staining, as for example does the electrolyte content[216,217]. Alternatively the tree size revealed by fluorescence[225] may be smaller than that observed following staining, while the internal structures that emit may be larger than those of the defects. Therefore until techniques are developed to determine the size and distribution of the damage structures[447] within a water tree it will be necessary to ensure when measuring water tree lengths that all the damage is included.

15.6 Differences and similarities in the model statistics

The models of the preceding sections (15.2–15.5) all adopt the same basic mode of development. First define a critical parameter that will lead to breakdown. Second identify a critical value that gives runaway breakdown at a given field. Thirdly determine the distribution in the critical parameter after a given time of application of the field. Finally define system survival as the probability that the most severe value of the chosen parameter in the system is less than critical.

Each of the models differ considerably in their approach to the problem but can be broadly divided into two classes, namely those for which the parameter at breakdown is part of a self-similar construction and those where it has been produced by random spatial development. In the former case, Sections 15.2, 15.3 and 15.5.3, the statistics have a Weibull form, but in the latter case they are of the 1st asymptotic extreme-value (1AEV) distribution.

The fluctuation and fractal models both concentrate on the propagation stage of breakdown during which a chosen parameter reaches criticality. In both cases the parameter has a fractal geometry but in the fluctuation model its distribution is governed by intrinsic material fluctuations, whereas in the fractal model it is its geometrical arrangement that is important. The Weibull parameters predicted in Table 15.1 and eqn. 15.21 therefore depend upon both the propagation and fluctuation exponents, and the values that can be deduced for them are given in Tables 15.2 and 15.4.

Although the critical parameter in the fluctuation and fractal models is shown to have a Weibull distribution, the dependence of the breakdown

Table 15.4 Summary of the breakdown statistics and their exponents deduced from the models of Sections 15.3 and 15.4. Note that the fractal dimension (1·6) quoted for the Takayasu model refers to the breakdown structure (percolation backbone), whereas the values of 1·9 or 2·5 are used for the pre-breakdown cluster distribution

Model	d_f	Statistics				Comment
		Weibull			1AEV	
		a	b	b/a		
Fractal	1·4	3·8				Possible field
	1·7	~4—6	> ~35	6·8		threshold
	2·5	~4—6	> ~40	10		
Cumulative						
Random percolation		~0·5			1AEV	1AEV in field
Takayasu:	1·6					breakdown structure
Life tests	1·9—2·5	2—3			1AEV	1AEV in field
Ramp tests	1·9—2·5	2—3	2—3			Ramp exponent is $a+b$

statistics upon the observed time and field variables depends upon their relationship to the critical parameter. It was pointed out in Section 15.3 that this need not be a power law in field, and hence breakdown statistics which are Weibull as regards time, may have an exponential (1AEV) form for the field, and even a threshold value may occur. A similar modification would apply in this case to the results of Sections 15.2.2 and 15.2.3.

A blurring of the distinction between the 1AEV and the Weibull distribution also occurs for the percolation models. Here it is essentially the initiation stage that is regarded as dominating the breakdown statistics leading to the 1AEV distribution. However when a preparatory propagation stage is considered, so as to convert the model to one of a cumulative breakdown, the statistics are converted to a Weibull form in time. In the Takayasu[397] version Weibull statistics are predicted for both time and field if the latter is measured in a progressive-stress (ramp) test.

The statistics of electrical-tree initiation form an example which cannot be said to be either of the Weibull or 1AEV form. In general it seems from these models that initiation is much more responsive to material inhomogeneities and hence is likely to have statistics much nearer to the 1AEV form than the Weibull form. However there is no reason why Weibull statistics may not occur in initiation, perhaps for example due to fractal time processes[746,747] in charge de-trapping at a local stress enhancement site.

The values given in Tables 15.2 and 15.4 depend upon a number of factors some of which are known and some of which are estimates. They should therefore be taken as guides to the predictions of the model rather than absolute values. Nevertheless some features can be discerned. There is a tendency for the value of a to increase as b increases, that is the breakdown becomes more deterministic in time when it does so in field. Aging processes which grow a single defect structure (Section 15.2) or extend several defects (Section 15.4) give values of a in the range 1·7—3. Conducting path breakdowns controlled by intrinsic fluctuations, either as the path grows (Section 15.2) or through the *spatially random* generation of conducting segments have values of a less than unity.

Experimental examples of all these types of behaviour can be found depending not just on the conditions but also on the type of test, i.e. constant stress (life) tests or progressive stress (ramp) tests. Many of the models suggest or imply a threshold field for breakdown. In these cases the field (voltage) statistic is only approximately of the Weibull form over a limited range of field. Here the ratio b/a is determined solely by the propagation mechanism and the values of 6·8 and 10 obtained for electric trees, whose growth law is known, correspond quite well to those found for polyethylene insulation in power cables.

Finally it should be noted that self-similarity in any system only extends over a finite range of sizes and times. Thus we should only expect the associated Weibull statistics to apply to a finite range of fields and times. Outside of this range more detailed processes, such as those growing the first channel in an electrical tree, or coarser ones such as the linking of tree-like defects take over. Such processes can be taken to be different mechanisms with their own statistics, and are usually regarded in this way when a composite breakdown statistic is analysed. They may however be different facets of the same mechanism, for example a cross-over from inception to propagation dominated statistics. Two other examples have been mentioned in this chapter, namely the sharp steepening of a Weibull plot as the percolation model becomes dominated by large clusters near the conducting limit, and the threshold in field for partial discharge breakdown and tree inception. However it is clear theoretically that this must occur in *all* cases of Weibull statistics. It is therefore necessary to be extremely wary of extrapolating Weibull parameters over too large a range of time and field.

PART 5

ENGINEERING CONSIDERATIONS FOR BREAKDOWN TESTING AND DEGRADATION ASSESSMENT

Introduction

In earlier parts of this book we have reviewed the scientific understanding of electrical degradation and breakdown in polymers. In this part we wish to review and suggest techniques for assessing polymeric insulation systems. The first chapter in this section reviews current breakdown testing techniques and the procedures which are used for analysing the data. The basis of these tests and procedures are examined and complementary techniques are also suggested. We have noticed in the course of our experience with testing cables under AC and DC conditions that there are significant differences in the statistics describing the results. Chapter 17 analyses these two cases and indicates the differences in the techniques necessary for understanding the data. It also explains commonly observed, empirical relationships between AC and DC breakdown strengths and suggests that the build up of space charge may be the critical factor linking AC and DC breakdown. Since extruded cables are a major user of polymers for insulation purposes, Chapter 18 is devoted specifically to cable assessment procedures. Since these tests are currently being reviewed by many bodies we have not attempted to specify tests in any detail in this chapter but, rather, we attempted to identify the major considerations in carrying out such tests. Finally we have outlined some techniques for the non-destructive diagnosis of electrical polymeric degradation.

Chapter 16

Breakdown testing and analysis

This chapter describes standard tests in which the breakdown stress or time to breakdown is evaluated for a polymeric material or insulating system; the interpretation of these tests is also discussed. Since very few polymers are produced in exactly the same form from one decade to another (for example different additives or different manufacturing methods may be used), such tests may be used by design engineers in an attempt to predict their performance for the next three or more decades. First the tests prescribed by the standards institutions are described. Such tests are generally used for establishing suitability for application and for control of quality. Tests for specific products have also been agreed by manufacturers and customers. Examples of some of the breakdown tests adopted by American and other cable manufacturers are given. In all these tests conditions are carefully specified to ensure reproducibility between test laboratories despite the fact that, in practice, the insulation is unlikely to be operated under such conditions and may therefore display different characteristics.

Most of the tests attempt to measure a 'breakdown strength' which is intended to be representative of the system. In all cases it is recognised that this does not correspond to the ultimate strength or the intrinsic breakdown strength (i.e. one which is independent of thickness) of the polymer. It is also acknowledged that the measured breakdown strength is not reproducible in the sense that there will be a range of breakdown strengths observed.

None of the standards are intended to establish the probability of failure of an insulating system or material as a function of time. The closest approach to this is given by the 'IEEE Guide for the Statistical Analysis of Electrical Insulation Voltage Endurance Data'[723] but even this has the disclaimer 'The mathematical techniques contained in this guide may not directly apply to the estimation of equivalent life'. The guide describes statistical techniques to analyse the data for the times-to-failure from constant-stress 'voltage endurance tests' or breakdown voltages from progressive-stress tests. However the description of experimental procedures used to make measurements was not within the scope of the guide.

Finally we comment on the interpretation of the tests and also discuss ways of predicting the lifetime, or more accurately, the probability of failure of the system as a function of time. This discussion includes other approaches to testing and also methods for analysing the results in order to estimate the parameters describing the breakdown probability of the insulation.

16.1 Breakdown test methods

16.1.1 Standards

The four standards commonly referred to in measuring the breakdown strength of solid insulating systems at power frequencies are given in References 812 and 813.

Various methods exist which vary in the way the voltage is applied and in experimental configuration in order to cater for different specimens such as moulding material, flexible extrusion compound, sheet, tube, rod, and cast and laminated resin systems. Typically the standards specify the following:

(i)   *Experimental configuration.* The specimens are to be 3 mm thick and contacted by metal electrodes with 3 mm rounded edges, the smaller electrode being 25 mm square. Other dimensions exist for tubes and rods (although these are similar) and for tapes. The normal temperature is room temperature. BS2782[812] specifies that the specimens should be in oil but the others generally allow air, other liquids, or the environment in which the polymer is to be used. Whitehead[505] has shown that in order to avoid surface discharges then:

$$E_m \varepsilon'_m \sqrt{(D_m^2 + 1)} > E_s \varepsilon'_s \sqrt{(D_s^2 + 1)} \tag{16.1}$$

where the subscripts s and m refer to the solid dielectric and the immersion medium, E is the breakdown strength, ε' is the real part of the permittivity and D is the dissipation factor.

(ii)   *Number of specimens.* The standards all suggest the use of one trial specimen to establish an approximate breakdown voltage and suitable rate of voltage increase or starting voltage. The subsequent number of specimens required varies from two[812] to five or even ten if there is a significant range of observed results[814,813]. The mean of individual breakdown strengths is to be reported.

(iii)   *Test voltage.* In all cases a sinusoidal voltage is called for with a peak to RMS ratio of $\sqrt{2} \pm 5$—7%.

(iv)   *Procedure.* Various test procedures are advocated which can be broadly divided into those in which the voltage is increased in a step-by-step fashion or as a continuous ramp.

The step-by-step increase was first defined by the British Standard[812] and calls for the voltage to be increased in steps of between 5—10% in the following way. A trial specimen is broken down by increasing the voltage from zero at a reasonably uniform rate such that it breaks down in 10—20 seconds. Subsequent specimens are tested by applying an initial voltage of 40% of the trial specimen's breakdown voltage and then stepping up in 20 second steps in the following way:

Step from (kV RMS)	to (kV RMS)	in steps of (kV RMS)
0·50	0·95	0·05
1·0	1·9	0·1
2·0	4·8	0·2
5·0	9·5	0·5
10	19	1
20	48	2
50	95	5 etc....

The breakdown voltage is defined as the highest voltage for which the specimen withstood the full 20 second step. All the standards have adopted these 'preferred' voltage levels but some give the choice of 60 s or 300 s steps instead of just 20 s.

The continuous ramp tests are either 'short-time' or 'slow rate of rise'. The short time test has voltage rates of rise of 1, 2, or 5×10^n V·s^{-1} with $n = 0, 1, 2, 3, \ldots$ such that all samples breakdown within 10—20 s. The slow rate of rise is the same except that the ramp is started at about half of the estimated breakdown voltage and is linearly increased such that breakdown takes place at times longer than 120 s.

The British Standard[812] is the most prescriptive since it allows virtually no variation from test to test: the only parameters it does not specify are the temperature (room or 90°C) and the thickness in the case of preformed sheets or tubes. The other standards give choice as to the procedure which means that when only the breakdown voltage is reported (this is often the case in tabulated information) it is far less meaningful. The step size, which is between 5% and 10% of the estimated breakdown voltage, means that samples will always breakdown in a reasonable number of steps (typically 14—18) whereas, if a given rate of voltage increase has been specified, the test time would have been very dependent on the breakdown voltage. The step time of 20 s was chosen because it was usually significantly longer than the time taken to change the voltage from one step to the next, because it gave a reasonable total test time (~5 minutes per specimen), and because it was found that the breakdown voltage measured, V_B, could often be interpreted as the voltage which a specimen could withstand for about one minute. The reasons for this rather surprising finding can be explained as follows.

Assume the step size is sufficiently small such that the staircase increase can be approximated to a ramp. In this case the cumulative probability of breakdown is given by eqn. 14.20b:

$$P_F(V, \dot{V}) = 1 - \exp\left\{ -C\frac{a}{a+b} \dot{V}^{-a} V^{a+b} \right\}$$

Typically values for the relatively poor insulation of the 1950s when this standard was first drafted might have been $a \sim 1$ and $b \sim 4$. (This would have given $\tau_c \propto t^{-n}$ with $n = b/a = 4$. Early examples of this were given by Montsinger[324] in 1935 for an oil/board system and by Peek[815] in 1920 for oil-impregnated paper for which a value of ~ 4 was given each time.) The rate of voltage increase $\dot{V}$ in the above equation is approximately that prior to breakdown corresponding to an increase of between 0·05 and 0·1 of V_B over a period of 20 s. That is $\dot{V} = \phi \cdot V_B/20$ where $0·05 < \phi < 0·1$. The breakdown voltage is somewhat arbitrary but may be taken to be that corresponding to failure of half the samples, i.e. when $P_F = 0·5$, although this does not affect the argument. Making these approximations it is possible to calculate how long the same specimen would typically last if a continuous voltage of V_B were to be applied. Using eqn. 14.19b:

$$P_F(V_B, t) = 1 - \exp\{-Ct^a V_B^b\}$$

Putting $P_F = 0·5$ then the exponents of the exponentials in eqns. 14.20b and 14.19b must have the same value:

$$-Ct^a V_B^b = -C\frac{a}{a+b}\dot{V}^{-a}V^{a+b} \tag{16.2}$$

and replacing V by V_B for $P_F = 0·5$ in the ramp expression and $\dot{V}$ by $\phi \cdot V_B/20$ gives:

$$-Ct^a V_B^b = -C\frac{a}{a+b}\frac{\phi^{-a}V_B^{-a}}{20^{-a}}V_B^{a+b} \tag{16.3}$$

Cancelling the $-C \cdot V_B^b$ from both sides leaves:

$$t^a = \frac{a}{a+b}\frac{\phi^{-a}}{20^{-a}} \tag{16.4}$$

Substituting in values of $a = 1$ and $b = 4$ gives $t = 80$ s or 40 s for values of $\phi = 0·05$ or 0·1 respectively so that typical lifetimes at the measured breakdown voltage are likely to be of the order of one minute.

The linear ramps specified as alternative test procedures by the other standards have the advantage that they are likely to give well-defined Weibull distributions (see Section 14.1) according to eqn. 14.20. However in order to predict lifetime the measurement of breakdown voltage alone is not sufficient and it is necessary to either repeat the ramp test at different rates of voltage increase or carry out a set of constant-stress tests at different stresses (Section 14.2.2b). The ASTM standard[816] does state (paragraph 12.1.2.7) 'For research purposes, it may be of value to conduct tests using more than one time interval on a given material.' (i.e. in step-by-step tests) but they do not give any further explanation.

Step tests are frequently assumed to give the same breakdown voltages as ramp tests with equivalent rate of voltage rises. The breakdown voltage measured by a step test may however be significantly biased from the 'true'

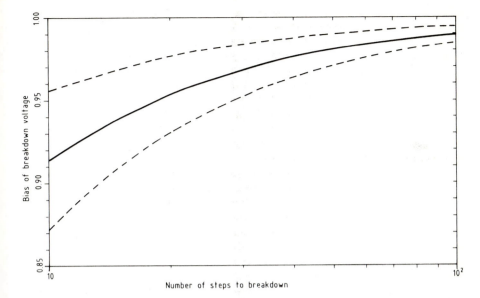

Fig. 16.1 The measured breakdown voltage using step tests is biased with respect to that measured using ramp tests and is dependent upon the Weibull time exponent and the number of steps to breakdown. The continuous line shows the most likely bias however the actual bias may be anywhere between the broken lines depending on the time during the step at which the breakdown occurs

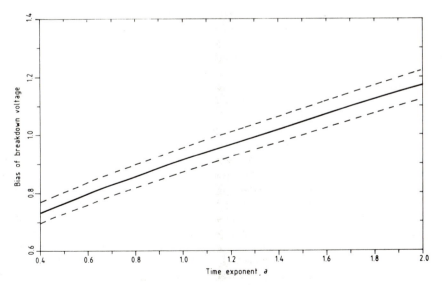

Fig. 16.2 The variation of bias with the Weibull time exponent a for a test in which there were ten steps to breakdown. The broken lines have the same meaning as in Fig. 16.1

value measured by a ramp test. Defining the bias as the ratio of the measured breakdown voltage using a step test to that using a ramp test, Figs. 16.1 and 16.2 show how the bias is dependent upon the number of steps and the value of the time exponent 'a' in eqns. 14.19 and 14.20 (the bias is independent of rate of voltage increase). The figures assume an inverse power law with no threshold voltage and a Weibull distribution (i.e. the probability of failure is given by eqn. 14.19). Fig. 16.1 shows the variation of the bias of the breakdown voltage as a function of the number of steps to breakdown. In this case values of $a = 1$ and $b = 10$ have been used and it is assumed that the steps were of equal size. In a step test if breakdown occurs at *any* time during step $i + 1$ then the breakdown voltage is said to be that of the voltage level of step i. (The first voltage is applied at time = zero, so the voltage does not equal zero during the first step, and the breakdown voltage is defined as the highest voltage for which the specimen survived a complete step.) Hence there is a range of possible values of the bias depending on the time during step $i + 1$ that the breakdown took place. The solid line shows the median value of the breakdown bias and the dashed lines show the possible variation. It can be seen that for a value of $a = 1$ the bias is always less than one. This implies that in such step tests the specimen is more likely to break down by the time V_B has been reached than in the equivalent ramp test. This is reasonable since in order to survive the sample will have to remain at V_B for the full period of one step whereas in the ramp case this value is gradually approached. Since $a = 1$ breakdown is likely to occur near the beginning of the step. This is not always the case however. Fig. 16.2 shows the variation of bias with a for a test in which there were ten steps to breakdown. The solid and dashed lines carry over their meanings from Fig. 16.1. These curves are intended to show that it may be dangerous to assume direct equivalence between step and ramp tests. However for a reasonable number of steps the values of breakdown voltage are unlikely to vary by more than about 10%. Similar inaccuracies are likely to occur in the estimation of a and b from such tests.

Impulse tests are not so well defined. Suitable wave-shapes are discussed in BS923[817] but the procedure is not defined there. A typical procedure is similar to that defined in BS2782[812] but, instead of a step-by-step increase, three impulses are applied at each voltage level. The impulse voltage level is then defined as the highest level at which the specimen withstood all three impulse voltages.

16.1.2 Manufacturers' qualification tests

Various different 'qualification' or 'type' tests are advocated by different manufacturers in which the performance of a specific product and its quality is established. Such tests are usually carried out on a given design which has been produced and they are not normally repeated unless part of the manufacturing procedure is changed. Such a test is the widely-recognised US procedure AEIC No. 5-75[818] for polyethylene and crosslinked polyethylene insulated and shielded power cables rated from 5 to 69 kV inclusive. Two parts of this test involve destructive breakdown measurements: the *high-voltage time test* and the *impulse breakdown test*.

The high-voltage time test consists of the following consecutive stages applied until breakdown:

$7{\cdot}87 \times 10^6$ V $\cdot$ m^{-1} (200 V $\cdot$ mil^{-1}) for 4 hours

$11{\cdot}8 \times 10^6$ V $\cdot$ m^{-1} (300 V $\cdot$ mil^{-1}) for 1 hour

$14{\cdot}2 \times 10^6$ V $\cdot$ m^{-1} (360 V $\cdot$ mil^{-1}) for 1 hour

increases of 30-minute steps of $1{\cdot}57 \times 10^6$ V $\cdot$ m^{-1} (40 V $\cdot$ mil^{-1})

The impulse test is performed at nominal conductor temperatures of 90°C (PE) or 130°C (XLPE) by applying three impulses of both polarities corresponding to the basic impulse level (BIL) of the transformer supplying the cable, i.e.:

Cable rating (kV RMS)	BIL (kV)
5	60
8	95
15	110
25	150
28	150
35	200
46	250
69	350

The voltage is then raised in steps of 25% of the BIL with three negative pulses produced at each step.

Such tests have evolved partly as proof tests (see Section 18.1) and partly out of necessity to place products in an order of merit. However the results of such tests taken alone can be misleading since a high breakdown strength does not necessarily imply a long lifetime. The insulation may be prone to electrical degradation such as electrical or water treeing or non-electrical degradation such as chemical instability. Furthermore the time exponent 'a' in the Weibull eqn. 14.19b may be small even if V_B is large so that the probability of breakdown at times much less than the required lifetime may be significant.

16.1.3 A suggested test for measuring threshold voltage

To definitively establish the existence of a threshold voltage is difficult using standard progressive-stress tests. Although such tests, if carried out with a reasonably linear rate of voltage rise, should result in a convex curve on a Weibull plot (or concave on a 1AEV plot, see below), statistical deviations from the best straight line generally mask such effects unless there are a large number of data points ($\gg 30$). Even given such a plot it is difficult to accurately judge the threshold voltage by graphical or numerical techniques. The technique suggested here may be used to confirm the existence of a threshold voltage having first estimated it using another technique such as progressive-stress testing.

Given that a threshold voltage, V_{th}, exists then the rate of aging of the insulation will be very much reduced, i.e. approximating to zero, when the applied voltage, $V < V_{th}$. The proposal for this test is that, having carried out a standard set of progressive-stress tests, a further set is carried out after the specimens have been subjected to a voltage just below the suspected threshold voltage for a significant period (the *conditioning period*). If the threshold voltage exists this conditioning period should not physically degrade the samples and the subsequent progressive-stress tests should give similar results to those carried out without a conditioning period. If however the threshold voltage is lower than that estimated such that the samples are conditioned at a voltage $V > V_{th}$, they will be degraded and their probability of breakdown in the subsequent progressive-stress tests will be increased. Thus this conditioning period needs to be long enough to give a significant increase in the probability of breakdown if $V > V_{th}$. If we assume a Weibull distribution of breakdown voltages this can be calculated as follows.

From the first set of progressive-stress tests (no conditioning period) estimate the threshold voltage and values for the Weibull parameters, a and b, *assuming no threshold voltage* (techniques for these estimations are described below). From eqn. 14.20b the cumulative probability of failure in a progressive-stress test with no threshold voltage is given by:

$$P_F(V) = 1 - \exp\left\{-C\frac{a}{a+b}\, \dot{V}^{-a}V^{a+b}\right\}$$

At the characteristic breakdown voltage, V_c, $P_F = 1 - 1/e = 0.632$ so that

$$C\frac{a}{a+b}\, \dot{V}^{-a}V_c^{a+b} = 1 \tag{16.5}$$

As this represents a 'significant' probability of breakdown it can be compared with the equation for constant-stress tests (eqn. 14.19b)

$$P_F(V, t) = 1 - \exp\{-Ct^a V^b\}$$

in which, for the same cumulative probability of breakdown:

$$Ct^a V^b = 1 \tag{16.6}$$

Equating eqns. 16.5 and 16.6 gives an estimate for the length of the time of the conditioning period at voltage, V:

$$t \approx \left[\frac{a}{a+b}\, \dot{V}^{-a}V_c^{a+b} V^{-b}\right]^{1/a} \tag{16.7}$$

Equation 16.7 gives the time required for some 63% of the samples under test to be affected by conditioning at voltage V if the apparent threshold V_{th} is a statistical artifact. In practice this value may be unacceptably long. A minimum requirement may be defined through the fraction of samples sufficient to alter the original failure distribution if affected by conditioning. A value of $P_F = 0.33$ can be obtained from eqn. 16.4 and this is likely to be

Table 16.1 Normalised conditioning times for detecting the presence of threshold voltages

	a value		
b	0·5	1·0	2·0
4	30	9·9	4·1
9	$1·1 \times 10^6$	$2·0 \times 10^3$	61
15	$8·1 \times 10^{11}$	$1·7 \times 10^6$	$1·8 \times 10^3$

a minimum acceptable value in most cases. Substituting this value into eqn. 14.20b, instead of $P_F = 0·632$, reduces the expression for the conditioning time to:

$$t \approx (0·4)^{1/a} \left[\frac{a}{a+b} \, \dot{V}^{-a} V_c^{a+b} V^{-b} \right]^{1/a} \tag{16.8}$$

For example Table 16.1 shows the conditioning times required for a selection of values of *a* and *b* if the samples are conditioned at 30% of the characteristic breakdown voltage. The times are calculated using eqn. 16.8 and are normalised since they depend on the rate of voltage increase, $\dot{V}$. In order to calculate the conditioning time required it is necessary to multiply the appropriate normalised time by the time it would take for the ramp voltage used in the progressive-stress test to reach the characteristic voltage when raised from zero. It can be seen that most of the required conditioning times are reasonable although that for $a = 0·5$, $b = 15$ is likely to be longer than the product lifetime!

16.2 Statistical analysis

In Part 4 of the book it was shown how the distribution of breakdown events could be described by the Weibull or 1st asymptotic extreme-value (1AEV) distributions in time and/or field. In this section we will describe statistical procedures for estimating the parameters of these distributions given a set of data from a constant-stress or progressive-stress test. Considering the cumulative probability-of-failure functions of the two distributions in their general 2-parameter forms:

Weibull distribution:

$$F(x) = 1 - \exp \left\{ -\left[\frac{x}{\alpha} \right]^{\beta} \right\} \qquad x > 0$$

$$= 0 \qquad x \leqq 0 \tag{16.9a}$$

1AEV distribution:

$$G(y) = 1 - \exp\left\{-\exp\left[\frac{y-u}{b}\right]\right\} \qquad -\infty < y < \infty \qquad (16.9b)$$

it can be seen that one can be transformed to the other using the relations: $y = \log_e(x)$, $u = \log_e(\alpha)$, and $b = 1/\beta$. This being the case we will mainly describe procedures for estimating the parameters of the Weibull distribution, similar techniques being used for the 1AEV distribution after transformation of the data. The variable x may, for example, represent time-to-failure data measured from constant-stress tests in which case $\alpha = \tau_c$ the characteristic time to failure in the nomenclature of Chapter 14 and $\beta = a$ the shape parameter and time exponent. Similarly if the variable x represents breakdown voltage data from progressive-stress tests then α is the characteristic breakdown voltage and the shape parameter $\beta = a + b$. The Weibull distribution may also be written in a 3-parameter form which includes a threshold value $x = \gamma$ below/before which breakdown is not possible (i.e. $\gamma = 0$ in the 2-parameter form above):

$$F(x) = 1 - \exp\left\{-\left[\frac{x-\gamma}{\alpha}\right]^{\beta}\right\} \qquad \gamma > 0,\; x > \gamma$$

$$= 0 \qquad\qquad\qquad x \leqq \gamma \qquad (16.10)$$

Experimental lifetime and breakdown-voltage data are frequently 'censored'[726]; i.e. not all the specimens under test are broken down. In life (constant-stress) tests this is usually because there is either a maximum time beyond which it is inefficient to run the test or because specimens need to be removed from the life test for other destructive tests such as those to investigate physical or chemical changes which have taken place. In progressive-stress tests the samples may not break down before the maximum voltage of the voltage supply has been reached. *Right censoring* in which specimens beyond a certain time or voltage are not tested is the most common in electrical breakdown tests and *left* or *mixed* censoring will not be discussed here; i.e. by *censoring* we imply *right censoring*. Censoring is also sometimes categorised as *type I* or *type II*. In type I (or *time*) censoring the test is stopped at a predetermined time or voltage whereas in type II censoring the test is stopped after a given proportion of specimens have broken down. The latter is the most common and will be discussed here. However the methods developed may also be applied to type I censored data with little loss of accuracy.

Given a finite set of breakdown data it is possible to estimate the distribution parameters (α, β, and γ in the case of the Weibull distribution) using graphical or numerical techniques. However, even if the data were measured exactly (i.e. with no experimental errors), because the number of data points is necessarily finite, the parameters can only be estimated within determinable confidence limits. These are known as *statistical-confidence* or *s-confidence limits*. For the sake of brevity we will refer to them simply as confidence limits but it should be borne in mind that these do not reflect uncertainties in the experimental data which may lead to further bias of the estimated parameters[172]. For example suppose the parameter α is estimated to be $\hat{\alpha}$ and the upper and lower 90% confidence bounds α_u and α_l are determined such that $\alpha_l < \hat{\alpha} < \alpha_u$. If a large number of experimentally-identical tests were carried out one would expect $\hat{\alpha}$ to lie between α_l and α_u for 90% of the tests.

Graphical and numerical techniques are presented here for evaluating Weibull and 1AEV parameters, confidence limits, and percentiles.

16.2.1 Graphical techniques

(a) Use of probability graph paper

Special graph papers are available for plotting data conforming to the 2-parameter Weibull and 1AEV distributions[819]. Plotting cumulative probability of failure versus breakdown stress on the non-linear scales of such paper enables a straight line to be constructed through the data points. Constructing Weibull probability plots was discussed in Section 14.1.2a (iii). The vertical axis of Weibull probability paper corresponds to $V_W(F(x)) = \log\{-\log_e[1 - F(x)]\}$ and the horizontal to $H_W(x) = \log(x)$. For 1AEV (Gumbel) probability paper $V_G(G(y)) = \log_e\{-\log_e[1 - G(y)]\}$ and $H_G(y) = y$; i.e. a linear horizontal scale.

In order to plot data on these probability papers it is necessary to rank it according to size and assign a cumulative probability of failure, P_F, to each point. An exact calculation of P_F is difficult and unnecessary since an extremely good approximate calculation can be performed[701]. If n specimens have been tested (in the case of a censored test not all n specimens would have broken down) and placed in order of size (i.e. breakdown time or stress) then the cumulative probability of failure of the ith smallest datum is given by:

$$P_F(i, n) \cong \frac{i - 0\cdot3}{n + 0\cdot4} \times 100\% \qquad (16.11)$$

For example the following represent times-to-breakdown (seconds) during a type-II censored life test in which nine of the twelve samples broke down (i.e. $n = 12$): 3310, 33, 138, 283, 1464, 344, 1670, 518, 940. These data have been placed in order and assigned values of P_F in Table 16.2.

(b) Estimation of parameters of distribution

Such lifetime data frequently follows a 2-parameter Weibull distribution. This data has therefore been plotted on Weibull probability paper, Fig. 16.3, and a best straight line fitted 'by eye'. This is a reasonable technique for this data which lies on a good straight line. However fitting data 'by eye' is difficult for data which deviates greatly from a straight line as each graph point carries a different weighting which depends upon the size of its statistical confidence limit as it appears on the Weibull plot. As we will see later this means that data close to $P_F = 50\%$ carries a higher weighting than that further away. From Fig. 16.3, the parameter α is estimated at the $P_F = 63\cdot2$ percentile of cumulative failure probability; the estimated value of α is $\hat{\alpha} = 1950$ s. The Weibull probability paper used here has a scale and 'estimation point' from which β can be estimated. On this paper a straight

Table 16.2 Simulated data from a type-II life test in which the data is distributed according to the 2-parameter Weibull distribution (eqn. 16.9a)

i	x (sec)	$P_F(\%)$
1	33	5·6
2	138	13·7
3	283	21·8
4	344	29·8
5	518	37·9
6	940	46·0
7	1464	54·0
8	1670	62·1
9	3310	70·2
	1—12 did not breakdown	

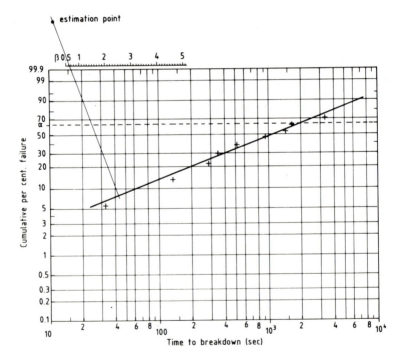

Fig. 16.3 Fitting data by eye to a Weibull distribution using Weibull probability paper and estimation of the characteristic value α, and shape parameter, β

line is constructed through the 'estimator point' at right angles to the best straight line and from this $\hat{\beta}$ is read as 0·63. If such a scale does not exist or the value of β is out of range it may be estimated by evaluating from the graph the values of $\log_{10}(x)$ at the 10 and 90 percentiles and using:

$$\hat{\beta} = \frac{1·34}{\log(x)_{90\%} - \log(x)_{10\%}} \qquad (16.12)$$

(The numerator 1·34 comes from $\log_{10}[-\log_e(1-0·90)] - \log_{10}[-\log_e(1-0·10)]$.) In this example $x_{10\%} = 63$ s and $x_{90\%} = 6900$ s so $\hat{\beta} = 0·66$ which is close to the value of 0·63 previously estimated.

If the data is distributed according to a 3-parameter Weibull distribution, i.e. there is a threshold value of breakdown time or stress ($x = \gamma$ in eqn. 16.10), then there is no quick and simple method for evaluating this from a Weibull plot. Whilst a threshold value will, in theory, give rise to a convex curve on Weibull probability paper, in practice this is unlikely to be noticeable with less than about 100 data points because of their natural scatter. The data given in column (b) of Table 16.3 was obtained from a life test in which nine of the ten samples tested broke down.

The data is plotted as crosses and curve (a) in Fig. 16.4 shows a consistent convex curve. If this data actually corresponds to a Weibull distribution with a threshold value, γ, then it should be possible to subtract γ from each data point and re-plot this as a straight line on a Weibull plot. If this is the case γ may be found using a trial-and-error method. Guessing from curve (a) that $\gamma = 230$ hours then another plot corresponding to $x - 230$ (column

Table 16.3 Data from a life test which shows a threshold value. The cumulative probability of failure is given in column (a) and the raw data in column (b). The other columns are described in the text

Column			
(a) P_F (%)	(b) x (hr)	(c) $x - 230$ (hr)	(d) $x - 200$ (hr)
6·73	240	10	40
16·3	300	70	100
26·0	340	110	140
35·6	390	160	190
45·2	490	260	290
54·8	530	300	330
64·4	590	360	390
74·0	750	520	550
83·7	900	670	700

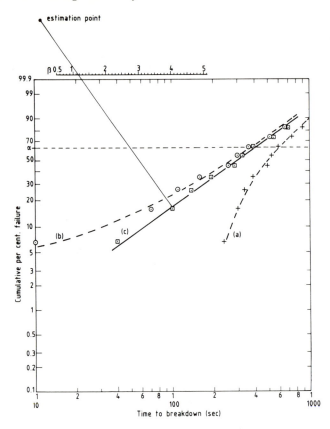

Fig. 16.4 A three-parameter Weibull distribution may result in a noticeable curve on the Weibull plot (*a*). Successive estimations of the threshold value (e.g. (*b*)) until a straight line is produced (*c*), allows the threshold value to be estimated

(c) in Table 16.3) may be plotted. This corresponds to curve (b) in Fig. 16.4 and can be seen to be slightly concave indicating that $\gamma = 230$ is an overestimate. After several tries a value of $\hat{\gamma} = 200$ (column (d) in Table 16.3) was found to give a good straight line fit (curve (c) in Fig. 16.4). From this the characteristic value can be read off as the 63·2 percentile equal to $\hat{\alpha} = 400$ hours, and the slope parameter using the estimation point as $\hat{\beta} = 1·2$.

Having estimated the distribution's parameters it is possible to estimate the lifetime or breakdown stress for any given failure probability from the Weibull or 1 AEV plot. If the cumulative probability of failure is expressed as $z\%$ then the corresponding lifetime or breakdown stress is known as the z percentile. For the Weibull distribution the estimated percentile, $\hat{x}_p$ at a cumulative probability of failure, P_F is given by

$$\hat{x}_p = \hat{\alpha}[-\log_e(1 - P_F)]^{1/\hat{\beta}} + \hat{\gamma} \qquad (16.13a)$$

and for the 1 AEV distribution the estimated percentile, $\hat{y}_p$ is given by:

$$\hat{y}_p = \hat{u} + \hat{b}\log_e\{-\log_e(1 - P_F)\} \qquad (16.13b)$$

For example, for the data in Table 16·2 the 0·1 percentile (i.e. the 0·001 quantile) is

$$x_p = 1950 \times [-\log_e (1 - 0 \cdot 001)]^{1/0 \cdot 63} + 0 = 0 \cdot 0338.$$

This implies that up to a time of 0·0338 seconds the system has a probability of failure of less than 0·1%.

(c) Confidence limits
It is frequently required to estimate the confidence limits of the estimated parameters and percentiles. Simple graphical techniques have been described by Stone and Rosen[820] and adopted in the 'IEEE Guide for the Statistical Analysis of Electrical Insulation Voltage Endurance Data'[723] for calculating confidence limits for tests involving between 6 and 25 samples for the 2-parameters of the Weibull and 1AEV distributions and for the 1, 5, and 10 percentiles. The reader is recommended to consult these if an alternative to the numerical techniques described in the next section is required.

In uncensored tests it is possible to draw approximate confidence limits which are sufficiently accurate for most purposes for percentiles of <1%. This is useful since both manufacturers and customers are frequently interested in confidence limits for lifetimes and stresses at very low probability of failures such as 1% and 0·1%. The method is illustrated for the following uncensored breakdown data[723] of time-to-breakdown in seconds: 25 200, 30 600, 39 600, 43 200, 43 200, 61 200, 64 800, 64 800, 75 600. These data are plotted in the usual way as a Weibull plot in Fig. 16.5. The three solid lines represent the best straight line and the upper and lower 90% confidence limits. The latter were plotted for the sake of comparison using the conventional graphical techniques described above[723] for the 1, 5, 10, and 63·2 percentiles. The dashed lines represent asymptotes for the confidence limits found using a simple technique described below. Note that the difference between these asymptotes and the actual confidence limits are small at the 1 percentile (within 5% in this case) and insignificant in terms of the other errors by the 0·1 percentile. From the best straight line the Weibull parameters can be estimated: $\hat{\alpha} = 55\,600$ and $\hat{\beta} = 3 \cdot 44$. From the best straight line and the confidence limits asymptotes it can be seen that the most likely value of the 0·1 percentile is 7470 s with 90% confidence limits of 1037 s and 13 300 s.

The asymptotes are calculated as follows. Each of the asymptotes goes through the 50 percentile point marked as a dot on the figure. Using eqn. 16.13a or from the best straight line it can be seen that this is 5×10^4 s in this case. It is now necessary to calculate the 90% confidence limits for the slope β. These can be found using the graphical technique referred to above[723] or by using the numerical approximations described below (eqns. 16.23). The limits for β are found to be $1 \cdot 688 < \beta < 4 \cdot 937$. The two asymptotes are constructed though the 50 percentile with these two slopes. In this case the β scale and estimation point on the graph paper were used. If the data is censored, using this technique will give incorrect results for the upper (i.e. right-hand side) confidence limit.

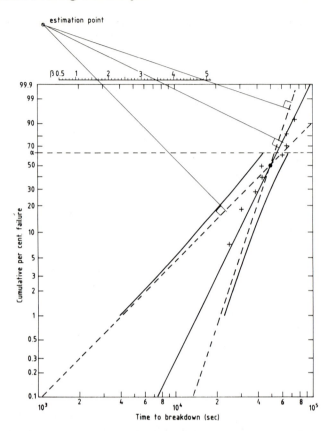

Fig. 16.5 A simple method for estimation of approximate confidence limits for non-censored data. Continuous lines give the actual 90% confidence limits, and the broken lines the approximate limits based on the 90% confidence limits on the Weibull shape parameter.

16.2.2 Numerical techniques

Estimating the parameters of the Weibull or 1AEV distributions numerically is straightforward but requires the use of a computer. The estimation of confidence limits may also be carried out numerically but this is less straight-forward. Although numerical techniques give more accurate answers, the span of the confidence limits is usually much wider than any inaccuracies introduced by graphical techniques.

A numerical technique which it may be tempting to use is simply to fit the points on the appropriate probability paper to a straight line using a least-squares linear-regression procedure. This is not however a correct approach as each point effectively has a different weighting due to the non-linear scales used on the graph. This can be easily envisaged as the confidence limits broaden away from the 50 percentile indicating less weight-ing should be given to extreme points. The relative weighting, as well as being difficult and computationally slow to calculate, depends on the distribution's parameters; a linear regression procedure using weighted data would therefore be slow, difficult and iterative.

(a) Maximum likelihood estimation

The generally accepted technique for estimating the parameters is the maximum likelihood technique[821,822]. Consider a test in which n samples broke down at times or stresses x_i, $i = 1, 2, 3, \ldots, n$ and that these form a random sample from a distribution whose probability density function is $f(x; \boldsymbol{p})$ where $\boldsymbol{p}$ is a vector of the unknown parameters to be estimated. So for the 1AEV distribution $\boldsymbol{p} = u$, b and for the Weibull distribution $\boldsymbol{p} = \alpha, \beta$ and, in the three parameter case, γ. (In general the x_i's may also be vectors which may be useful in the case of, for example, life tests at different stresses.) The likelihood function is defined as:

$$L(\boldsymbol{p}) = f(x_1; \boldsymbol{p}) \times f(x_2; \boldsymbol{p}) \times f(x_3; \boldsymbol{p}) \times \cdots \times f(x_n; \boldsymbol{p}) \qquad (16.14)$$

Because the random sample of data will not be entirely representative of the distribution (i.e. not all the data points will lie exactly on a straight line on the appropriate probability paper) $L(\boldsymbol{p})$ actually depends on the x_i's but this is usually ignored. A value of the parameter vector $\boldsymbol{p} = \hat{\boldsymbol{p}}$ now needs to be found to maximise $L(\boldsymbol{p})$ and the parameters it represents are known as the maximum likelihood estimates.

The maximum likelihood technique cannot generally be used to estimate parameters for a 3-parameter Weibull distribution but it is useful for the 2-parameter Weibull and 1AEV distributions. The technique is not useful for the 3-parameter Weibull distribution because its probability density function tends to infinity as $x \to \gamma$ if $\beta \leq 1$. Setting γ to be equal to the lowest observed value, x_1, and $\beta \leq 1$ will make $L(\boldsymbol{p}) \to \infty$ which is the solution most computational techniques will find. It is therefore necessary to use a separate procedure for estimating γ, subtract γ from all the data values, and then treat these as a 2-parameter Weibull distribution. A numerical technique for estimating γ is briefly described later.

For the Weibull and 1AEV distributions the maximum likelihood value $\hat{\boldsymbol{p}}$ always exists and is unique. It is usually more convenient to maximise $\log [L(\boldsymbol{p})]$ and solve the so-called maximum-likelihood equations:

$$\frac{\partial \log [L(\boldsymbol{p})]}{\partial p_1} = 0$$

$$\frac{\partial \log [L(\boldsymbol{p})]}{\partial p_2} = 0 \qquad (16.15)$$

(etc. for however many parameters need to be found).

The calculation of the expression for the maximum likelihood estimate of the two Weibull parameters is easiest found by making the substitution $A = \alpha^{-\beta}$ so that the Weibull equation:

$$P_F = 1 - \exp\left\{ -\left(\frac{x}{\alpha}\right)^{\beta} \right\}$$

becomes:

$$P_F = 1 - \exp\{-Ax^\beta\} \tag{16.16}$$

Maximum likelihood estimates, $\hat{A}$ and $\hat{\beta}$ can then be evaluated. From eqn. 14.7 the probability density function in terms of A is:

$$F'(x) = \frac{dP_F}{dx} = A\beta x^{\beta-1} \exp\{-Ax^\beta\} \tag{16.17}$$

Taking logarithms of both sides gives:

$$\log_e [F'(x)] = \log_e (A) + \log_e (\beta) + (\beta - 1) \log_e (x) - Ax^\beta \tag{16.18}$$

so the log-likelihood function for n specimens is:

$$\log_e [L(x; A, \beta)] = n \log_e (A) + n \log_e (\beta)$$

$$+ (\beta - 1) \sum_i^n \log_e (x_i) - A \sum_i^n x_i^\beta \tag{16.19}$$

Setting the partial derivatives with respect to A and β to zero yields the maximum likelihood estimates $\hat{A}$ and $\hat{\beta}$:

$$\frac{\partial \log_e (L)}{\partial A} = \frac{n}{\hat{A}} - \sum_i^n x_i^{\hat{\beta}} = 0 \tag{16.20a}$$

$$\frac{\partial \log_e (L)}{\partial \beta} = \frac{n}{\hat{\beta}} + \sum_i^n \log_e (x_i) - \hat{A} \sum_i^n x_i^{\hat{\beta}} \log_e (x_i) \tag{16.20b}$$

Eliminating $\hat{A}$ gives:

$$f(\hat{\beta}) = \frac{\sum_i^n x_i^{\hat{\beta}} \log_e (x_i)}{\sum_i^n x_i^{\hat{\beta}}} - \frac{1}{\hat{\beta}} - \frac{\sum_i^n \log_e (x_i)}{n} = 0 \tag{16.21}$$

By guessing a value of $\hat{\beta} = \hat{\beta}_j$ a better value of $\hat{\beta} = \hat{\beta}_{j+1}$ can be found using the Newton–Raphson technique:

$$\hat{\beta}_{j+1} = \hat{\beta}_j - f(\hat{\beta}_j)/f'(\hat{\beta}_j) \quad \text{where } f'(\hat{\beta}_j) = \left. \frac{df(\hat{\beta})}{d\beta} \right|_{\hat{\beta}_j}$$

$\hat{A}$, and hence, $\hat{\alpha}$ may then be found from eqn. 16.20a to be:

$$\hat{\alpha} = \left(\frac{1}{n} \sum_i^n x_i^{\hat{\beta}} \right)^{1/\hat{\beta}} \tag{16.22}$$

Cohen[823] has shown, rather surprisingly perhaps, that for censored samples in which only r specimens break down out of a total of n tested, expressions 16.21 and 16.22 still hold but the $(n - r)$ surviving specimens are assigned lifetimes or breakdown stresses, x_T, where: (i) in type I censoring x_T is the time the test terminated; (ii) in type II censoring $x_T = x_r$; and (iii) in mixed censoring a censored specimen is assigned a value equivalent to the time/stress of the censoring.

Exact numerical techniques for finding confidence limits are complicated and beyond the scope of this book. They are described in References 324, 726, 824, and 825.

(b) Computer program

Stone and Rosen[820,723] have given a BASIC computer program for estimating the maximum likelihood parameters for a 2-parameter Weibull distribution given an initial estimate for the shape parameter. Appendix 1 gives a listing of a more sophisticated BASIC program which calculates maximum likelihood estimates for both 2-parameter Weibull and 1AEV(Gumbel) parameters without requiring an initial estimate. The program is easily implemented on a personal computer. All the LOG functions in this program represent natural logarithms. Due to the limitations of the range of floating-point arithmetic operations on many personal computers and in many implementations of BASIC it may be necessary to express the values of the breakdown data using small numbers to prevent overflow. For example it may be necessary to express breakdown voltages in kilovolts rather than volts.

The program uses approximations to find the lower and upper 90% confidence limits for alpha and beta. The approximations for the limits of β, β_l and β_u, are simple to implement on a calculator and useful for calculating the confidence limits using the asymptote method described previously. They are accurate to within 1% for values of r (number of samples failed) less than 120. The approximations are as follows

$$\beta_u = \hat{\beta} \times (u_0 + u_1 \cdot r + u_2 \cdot r^2 + u_3 \cdot r^3 + u_4 \cdot r^4) \qquad (16.23a)$$

where

$$u_0 = 1\cdot63617$$

$$u_1 = -0\cdot0272229$$

$$u_2 = 5\cdot78997 \times 10^{-4}$$

$$u_3 = -5\cdot39819 \times 10^{-6}$$

$$u_4 = 1\cdot80269 \times 10^{-8}$$

$$\beta_l = \hat{\beta}/(l_0 + l_1 \cdot r^{-1} + l_2 \cdot r^{-2} + l_3 \cdot r^{-3}) \qquad (16.23b)$$

where

$$l_0 = 1\cdot08509$$

$$l_1 = 6\cdot89715$$

$$l_2 = 13\cdot53525$$

$$l_3 = 15\cdot17619$$

Stone and Rosen[820] give similar approximations which are only accurate for $r < 25$. Computational techniques do exist for estimating exact confidence limits of any percentiles but these are complicated and may not be suitable for use on a personal computer. The program listed in Appendix 1 also calculates the 90% confidence limits for the parameters and the 1, 5, and 10 (and 63·21) percentiles and their confidence limits for 5—120 samples using an approximate technique. The confidence limits obtained using the program are therefore not exact. In practice the confidence limits obtained for the parameter will be within 1% and those for the percentiles within 10%.

(c) Monte Carlo estimation
An alternative technique for finding confidence limits is to use a Monte-Carlo technique. In this technique, which simulates breakdown tests using a computer, the randomness of the breakdown time/stress is accomplished using the computer's random number generator which supplies a cumulative probability of failure for the inverse Weibull or 1AEV functions (eqns. 16.13). The results of such a simulated test may then be processed in the usual way, for example the parameters of the distribution may be estimated using maximum likelihood techniques or a cumulative probability of breakdown plot may be generated. A large number of breakdown tests, typically at least a thousand, are then simulated and the same data processing carried out each time. This process results in a large number of estimates for the parameters and percentiles. These estimates for the parameters and percentiles may then be placed in order, the 'middle' value corresponding to the median of the distribution. The median value corresponds to the most likely value of the estimate provided a statistically large enough number of simulated tests have been carried out. The confidence limits may be found using this technique since, having placed the estimates in order, the 'middle' 90% of the estimates should lie inside the 90% confidence limits. A pseudo-code representation of a program is given below which finds the confidence limits for estimates of parameters and percentiles from tests in which n specimens were tested and r specimens broke down and the data is believed to be distributed according to a 2-parameter Weibull distribution. It was not feasible to write such a program in a standard BASIC language which does not support recursion as the sorting routines would take too long and it was not thought to be useful to give a full listing in other programming languages. Although this may be implemented on a personal computer, memory size is likely to be a problem for a realistic number of simulations. The number of Monte-Carlo simulations is s; 'RND' represents a real random number between zero and one; and 'INT(x)' represents a function to find the *nearest* integer value to x. Beginnings of pseudo-code lines are indicated with a '◇' and end in ';' or ':'.

◇   Set constants α, β, r, n, s;
◇   Declare an array for breakdown data
    $D(s, r)$ of s columns by r rows;
◇   Declare arrays to contain estimated values of α and β
    for each simulated test: $\hat{\alpha}(s)$; $\hat{\beta}(s)$;
◇   FOR sim = 1 TO s (i.e. for each simulation):
    ◇   Simulate a test as follows:

◇ FOR $bd = 1$ TO r (simulate r breakdowns):
  ◇ Generate a breakdown time/stress using $x = \alpha[-\log_e(1 - \text{RND})]^{1/\beta}$;
  ◇ Store x in $\boldsymbol{D}(\text{sim}, bd)$;
◇ NEXT r;
◇ Sort data for this simulation such that $\boldsymbol{D}(\text{sim}, 1)$ contains smallest value and $\boldsymbol{D}(\text{sim}, r)$ the largest;
◇ Estimate α and β for this simulation and store in $\hat{\alpha}(\text{sim})$ and $\hat{\beta}(\text{sim})$;
◇ NEXT sim;
◇ Note: the array $\boldsymbol{D}(s, r)$ is now full. Each row contains data corresponding to percentile $(\text{row} - 0.3)/(n + 0.4)$—eqn. 16.11;
◇ FOR row $= 1$ TO r (i.e. for each row of $\boldsymbol{D}(s, r)$):
  ◇ Sort data such that $\boldsymbol{D}(1, \text{row})$ contain smallest data and $\boldsymbol{D}(s, \text{row})$ the largest;
◇ NEXT row;
◇ Sort data in $\hat{\alpha}(s)$ and $\hat{\beta}(s)$ arrays such that $\hat{\alpha}(1)$ and $\hat{\beta}(1)$ contain smallest values and $\hat{\alpha}(s)$ and $\hat{\beta}(s)$ the largest;
◇ PRINT 'lower 90% confidence limit, maximum likelihood estimate, and upper 90% confidence limit for α are:' $\hat{\alpha}(\text{INT}(0.05 \times s))$, $\hat{\alpha}(\text{INT}(0.5 \times s))$, and $\hat{\alpha}(\text{INT}(0.95 \times s))$;
◇ PRINT 'lower 90% confidence limit, maximum likelihood estimate, and upper 90% confidence limit for β are:' $\hat{\beta}(\text{INT}(0.05 \times s))$, $\hat{\beta}(\text{INT}(0.5 \times s))$, $\hat{\beta}(\text{INT}(0.95 \times s))$;
◇ FOR row $= 1$ to r (i.e. for each percentile):
  ◇ PRINT 'for the percentile' $(\text{row} - 0.3)/(n + 0.4)$ 'the lower 90% confidence limit, maximum likelihood estimate, and upper 90% confidence limit are:' $\boldsymbol{D}(\text{INT}(0.05 \times s), \text{row})$, $\boldsymbol{D}(\text{INT}(0.5 \times s), \text{row})$, $\boldsymbol{D}(\text{INT}(0.95 \times s), \text{row})$;
◇ NEXT row;
◇ END;

(d) Incomplete beta function
Another technique which may be used to find maximum likelihood values and confidence limits is the use of the incomplete beta function. This gives the most likely cumulative probability of failure and its confidence limits for each breakdown datum rather than the maximum likelihood value and confidence limits for each percentile. For the ith of n samples failed ($n = r$, $i = 1$ represents the sample with the smallest breakdown value):

$$B(i, n) = \frac{\Gamma(n+1)}{\Gamma(i)\Gamma(n+1-i)} \int_0^{P_F} x^{i-1}(1-x)^{n-i}\, dx \qquad (16.24)$$

The variable x represents the time to breakdown or the breakdown stress. In order to find the median rank cumulative probability of failure, $B(i, n)$ is set to 0.5 and solved for P_F. Standard computational mathematics packages (e.g. MathCAD[826]) provide simple routes for this solution. Alternatively a Newton–Raphson method may be used within a computer program. This is possible since the gamma functions are of integers and the relation

$\Gamma(j+1)=j!$ may be used. Quantiles other than the median rank may be found by setting $B(i, n)$ to other values. For example to find the upper and lower 90% confidence limits $B(i, n)$ would be set to 0·95 and 0·05 respectively.

(e) Least-squares calculation for threshold value
As explained at the beginning of this section (16.2.2), the least-squares technique is not a valid technique for finding the 'best straight line' on Weibull or 1AEV plots. This is because each point has an effectively different weighting dependent on the slope of the plot. However for data which are found to lie close to a straight line there may not be large differences between the estimated parameter values found using least-squares and maximum likelihood techniques. A technique used to evaluate the threshold value has already been described in the section on graphical techniques (16.2.1) in which successive estimates of the threshold value were subtracted from the data until the convex curve (indicative of a three-parameter Weibull distribution on Weibull paper) 'straightened out' to a straight line. It should be stressed however that a threshold may not give rise to a definite curvature on a Weibull or 1AEV plot (because of the expected statistical deviation from the straight line) unless there are a large number of points. For this reason it is generally difficult to establish conclusively the existence of a threshold value. If the points *obviously* lie on a curve then it is likely that they will fit a straight line well when the threshold value has been subtracted from each of them. It may therefore be possible to use a least-squares type algorithm in this case for finding the threshold value.

The least-squares technique attempts to find the best straight line on suitable probability paper. Consider the case of a Weibull distribution in which there are n samples of which r have broken down. After ordering the breakdown values such that sample $i = 1$ has the smallest value and $i = r$ the greatest then it is possible to assign cumulative probability of failure values to each datum by using the median rank approximation: $P_i = (i-0\cdot3)/(n+0\cdot4)$[701]. However on Weibull probability paper the vertical axis has a non-linear scale. Calling this the 'Weibull' scale then we allocate values on the Weibull scale, $w(i)$ to each breakdown: $w_i = \log_{10}\{-\log_e(1-P_i)\}$. In order to produce a straight-line plot these values must be plotted against $\log_{10}(x_i - \gamma)$ where x_i represents the ith breakdown value and γ is the threshold value (which needs to be estimated along with α and β). This results in a straight line of the form $y = m \cdot x + c$ where y corresponds to values on the Weibull scale, x corresponds to values of $\log_{10}(x - \gamma)$ and so $m = \beta$ and $c = -\beta \cdot \log_{10}(\alpha)$. By following the usual least-squares method it can be shown that the Normal Equations of the problem are:

$$\sum_{i=1}^{i=r} [m \cdot [\log_{10}(x_i - \gamma)]^2 + \log_{10}(x_i - \gamma) \cdot [c - w_i]] = 0$$

$$\sum_{i=1}^{i=r} [c - w_i + m \cdot \log_{10}(x_i - \gamma)] = 0$$

$$\sum_{i=1}^{i=r} \left[\frac{\log_{10}(e)}{x_i - \gamma} \{w_i - m \cdot \log(x_i - \gamma) - c\} \right] = 0 \qquad (16.25)$$

These need to be solved for m, c and γ with the obvious constraints such as $m > 0$ and $0 < \gamma < x_1$. This is quite straightforward using computer-based mathematical packages. An example using the MathCAD package[826] and the data from Table 16.3 is shown in Appendix 2. The values found compare closely with those found using the graphical technique described in Section 16.2.1*b* (graphical values shown in brackets with calculated confidence limits for α and β): $\hat{\alpha} = 413$ $(234 < \hat{\alpha} = 400 < 675)$; $\hat{\beta} = 1 \cdot 16$ $(0 \cdot 65 < \hat{\beta} = 1 \cdot 2 < 1 \cdot 91)$; and $\hat{\gamma} = 199$ (200).

16.3 Temperature and frequency acceleration

Although most breakdown tests use increased voltages to effectively accelerate the aging process other techniques are also used. The other main techniques, which will be described briefly here are: frequency acceleration, temperature and multivariate techniques.

16.3.1 Frequency acceleration

Frequency acceleration usually involves constant-stress tests run at frequencies higher than the power frequency or that at which the insulation is designed to operate. This technique is experimentally difficult as the apparatus required is neither standard nor simple. In particular it is very important to ensure that significant harmonics are not generated which may have more effect than the fundamental. Most elevated-frequency tests assume that aging is accelerated in proportion to the frequency so, if a Weibull distribution is expected with parameters a and b, then the harmonic content must be such that:

$$\sum_{i=2}^{i=\infty} \left[i \left(\frac{V_i}{V_1} \right)^{b/a} \right] \ll 1 \tag{16.26}$$

where i represents the ith harmonic, V_i is the amplitude of the voltage of the ith harmonic and V_1 is the fundamental voltage amplitude.

Results from typical frequency accelerated tests by Bahder *et al.*[702] were described in Section 14.3.3 which showed a good inverse power law fit over twelve orders of magnitude of effective (frequency-accelerated) lifetimes.

16.3.2 Temperature and multivariate tests

Some authors have suggested the use of temperature as the accelerating parameter. In particular Simoni and his group have advocated the use of an Arrhenius-type approach possibly coupled with field acceleration[481,687,827,828,829] in which the inverse power law relationship of eqn. 14.17 becomes:

$$\frac{\tau_B(V_1, T)}{\tau_B(V, T_0)} = \left(\frac{V_1}{V} \right)^{-n} \exp\left\{ A \left[\frac{1}{T_0} - \frac{1}{T} \right] \right\} \tag{16.27}$$

where T is the absolute temperature and T_0 is the absolute ambient or reference temperature. This has mainly been used for non-polymeric solid insulation systems although success has been claimed in an attempt to apply it to low-density polyethylene[828]. However other authors[708] have suggested that the Weibull equation for repeated voltage impulse stressing should be of the form:

$$P(N, V, T) = 1 - \exp\{-C_1 N^a V^b T^c\} \qquad (16.28)$$

where N is the number of voltage impulses before breakdown and T is the absolute temperature. Some (limited) data on crosslinked polyethylene was used to justify this. If eqn. 16.28 can be interpreted as being equivalent to

$$P(t, V, T) = 1 - \exp\{-C_2 t^a V^b T^c\} \qquad (16.29)$$

for a continuous AC stress then this would imply an inverse power-law relationship between time to breakdown and temperature and not an Arrhenius-type relationship. There is some evidence from thin films of an inverse linear relationship between breakdown strength and temperature[162,830] but this is likely to be due to electrode carrier injection related breakdown rather than a bulk-limited behaviour. In summary there does not appear to be a well-accepted accelerated testing procedure using temperature for polymeric insulation. This is not surprising in view of the lack of dependence of breakdown strength on temperature[133,163]. In poly-ethylene insulation in particular the onset of melting at ~95°C strongly reduces the breakdown field[759,831]. This effect essentially leaves too small a temperature range for any expression such as eqns. 16.27–16.29 to be substantiated unless data at cryogenic temperatures is obtained. This may not be the case if highly crosslinked amorphous polymers are studied with glass transition temperatures in excess of 200°C.

16.4 Size scaling

Acceleration factors such as field, temperature, and frequency are not the only features which differ in laboratory development tests from those obtaining in service insulation. There is also the factor of physical size. For example cable systems are usually tested as standard length specimens which may be as small as 10^{-5} of the whole cable length. In other cases scale models are tested for which the dimensions have been proportionately reduced from that of real systems. Therefore if the laboratory tests are to be useful for anything other than order-of-merit comparisons their results must be rescaled to service system sizes. Such a procedure can only be carried out if a number of assumptions are made. These are:

(i)   The laboratory sized specimens must contain a typical selection of breakdown initiating factors. They therefore cannot be too small and should be produced under standard processing or production conditions;

(ii) On converting to a full-scale system all variables that are controlled in the laboratory tests must either scale equivalently or remain constant.

In power cables for example the laboratory specimens should be chosen such that the conductor can achieve service temperatures in life tests;

(iii) The pre-breakdown processes in different laboratory-scale regions of service-scale systems do not interfere with one another;

(iv) The number and type of breakdown initiation sites in the service-scale system are homogeneously distributed.

These assumptions allow service-scale systems to be treated as a set of statistically independent laboratory-scale regions. In this case the probability of survival of the whole system, P_S(syst), is just the joint probability of survival of all of the regions, i.e.

$$P_S(\text{syst}) = [P_S(\text{lab})]^N \qquad (16.30)$$

Here P_S(lab) is the survival probability of a laboratory specimen and N is the number of regions of such size that make up the service system.

The identification of N with the system size is relatively simple when the laboratory specimens are sections of a chosen length (satisfying conditions (i) to (iv)) cut from a cable whose other dimensions are the same as those used in service and where the laboratory specimens effectively sample the distribution of potential breakdown initiation sites. In this case N is given by L_c/L_s with L_c and L_s the service and section cable lengths respectively. When the service system is related to the laboratory specimen by more complicated scale changes it may be necessary to have some knowledge of the manner in which breakdown is initiated. For example when breakdown is initiated at defect sites in the bulk as in the cumulative defect model (Section 15.4) and discharge inception model (Section 15.5.2), N will be given by the volume ratio of service to laboratory specimens. However if breakdown is initiated at defect sites on the electrodes (e.g. electrical trees, Sections 15.3 and 15.5.1, and vented water trees, Section 15.5.3) N becomes the ratio of electrode areas independent of any thickness changes. It may even be possible to find situations in which N is the ratio of lengths along a specific axis, even when scale changes occur in other directions as well. This could occur for example when the system possesses a boundary region in the insulation. The number of defects in such a boundary may well scale with the insulation thickness irrespective of changes in electrode area.

Having determined the form of scale ratio that should be used for N eqn. 16.30 can be used to make lifetime predictions for service systems provided the survival probability of laboratory specimens is known as a function of time and voltage or field. Here it is important that it be ascertained for certain whether the controlling variable is either the voltage or the field since a scaling of the system to service size may alter the field when the voltage is constant, and allowance must be made for this in the calculation. Such a procedure can be applied regardless of the functional form of P_S(lab). However when this has the form of one of the asymptotic extreme-value distributions (Sections 14.1.3 and 15.1) the stability postulate that they all obey (eqn. 15.1) conveniently allows P_S(syst) to be expressed as the same function as P_S(lab) but with a size-dependent characteristic strength[172,739].

When $P_S(\text{lab})$ is the Weibull function:

$$P_S(\text{lab}) = \exp\left\{ -\left(\frac{E}{E_c}\right)^b \left(\frac{t}{t_c}\right)^a \right\}$$

(16.31)

then

$$P_S(\text{syst}) = \exp\left\{ -\left(\frac{E}{E_c(N)}\right)^b \left(\frac{t}{t_c}\right)^a \right\}$$

(16.32)

where the rescaled characteristic field $E_c(N)$ is given by:

$$E_c(N) = \frac{E_c}{N^{1/b}}$$

(16.33)

Where the laboratory investigation concerns cable sections N is given by L_c/L_s and it has been suggested[172] that values of $E_c(N)$ for sections of different lengths be used to determine the exponent b. In principle the values obtained by this route should be the same as those obtained from experiments on constant length sections, however in practice they appear to be higher[172]. Jackson and Swingler[172] assign the difference to a random variation in insulation thickness along the length of production cables which effectively introduces a sample-to-sample field variation which is inadequately allowed for in eqn. 16.32. They therefore propose the size-dependent method as the best way of determining b. However it should be noted that this method of obtaining b depends strongly on the confidence with which E_c can be estimated from the test set, which in turn depends upon the number in the set. For example the range of F_c will be $0 \cdot 87 E_c < E_c < 1 \cdot 12 E_c$ for a set with 15 samples and $b \sim 5 \cdot 5$. This introduces an uncertainty into the estimation of b from eqn. 16.33 which is much greater than that resulting from variations in insulation thickness. Only if the difference in the two estimates of b persist for large sample sets (i.e. 40 samples) would it be justifiable for the variable size method (eqn. 16.33) to be preferred over that applied to constant size sections. It should also be noted that it is necessary to show that a Weibull distribution free from such features as threshold fields does in fact apply to the system in question before the size variation approach can be attempted.

When failures in the laboratory experiments are represented by the Gumbel (or 1AEV) distribution, eqn. 14.11, i.e.

$$P_S(\text{lab}) = \exp\left\{ -\exp\left\{ \frac{E - E_c}{E_0} \right\} \right\}$$

(16.34)

then scaling up to the service system converts E_c to $E_c(N)$ where

$$E_c(N) = E_c - E_0 \log_e N$$

(16.35)

If the rescaling involves a change of electrode area only, then N is proportional to the ratio of system and laboratory specimen areas, and the breakdown strength is linearly dependent upon the logarithm of the electrode area. This was first recognised by Endicott and Weber[739] who showed that the breakdown of transformer oil in uniform fields obeyed this type of statistics.

A modified version of the 1AEV statistic, i.e.

$$P_S = \exp\left\{-N_c \exp\left\{-\frac{E_0}{E}\right\}\right\} \qquad (16.36)$$

has been derived for the percolation defect model (Section 15.4.1) and the discharge inception model (Section 15.5.2). In this case the characteristic breakdown field (i.e. the field for which $P_S = \exp\{-1\}$) is inversely proportional to the logarithm of the number of defects, N_c, and hence the volume, so:

$$E_c = \frac{E_0}{\log_e N_c} \propto (\log_e \text{ volume}))^{-1} \qquad (16.37)$$

As yet no experimental verification of this relationship has been made.

Chapter 17

Comparison of AC and DC breakdown behaviour

17.1 Introduction to statistical differences

Polymeric materials are widely used as insulators for both AC and DC applications. For example power transmission systems may be either AC or DC powered and often use polyethylene formulations for the cable insulation[166]. Submarine optical telecommunication cables are similarly insulated and also DC powered. In these cases the average fields experienced by the insulation are relatively small[832] ($\leqq 3 \times 10^7$ V · m^{-1}). A different application that is attracting considerable interest is that of high capacitance polymer films[833,834] for compact energy storage systems in satellites[835] etc. and these will have to withstand much higher fields ($\geqq 5 \times 10^7$ V · m^{-1}) which may become even bigger as the thickness is decreased and capacitance increases. It is therefore rapidly becoming necessary that some understanding of the results of DC breakdown tests be achieved so that lifetime predictions can be made with some confidence over a wide range of operating fields[836].

Although the bulk of experience with lifetime prediction has been obtained on AC operating systems, the deterministic breakdown theories (see Part 3 of the book) have been developed in terms of applied DC fields. It is usually assumed that the DC breakdown field for a particular mechanism is equivalent either to the field amplitude[175] (E_b) or peak-to-peak value[120] ($2E_b$) of the characteristic breakdown stress in a distributed process under AC conditions ($E_b(t) = E_b \cdot e^{i\omega t}$). Taking this relationship into account the statistical analysis of accelerated and ramp DC breakdown tests have typically been made in the same way as for the AC case (see Chapter 14). This procedure is not always valid however, since alternating stress automatically introduces both physical and statistical factors that are not necessarily present in the DC case.

By definition an alternating field changes polarity twice each cycle, and will thus cause any field-dependent factor to range from zero to a maximum on each half-cycle. The system therefore experiences what is in effect a series of pulses in those factors that cause breakdown when a critical value is exceeded. Statistically each pulse represents an attempt to fail, and the overall survival probability is the joint survival probability for the number of pulses, i.e.:

$$P_S(N_f) = [P_S(1)]^{N_f} \tag{17.1}$$

where $P_S(1)$ is the probability of surviving one half-cycle and N_f is the number of half-cycles. Since $N_f = 2ft$, eqn. 17.1 introduces a time factor into the exponent of the survival probability which is more properly regarded as the number of half-cycles of stress experienced during the lifetime[702]. This argument applies to the case where the field amplitude is ramped (i.e. $E(t) = \dot{E} \cdot t \cdot e^{i\omega t}$) as well to that where it is kept constant and shows that in AC breakdown tests the time of stress enters as an independent variable

Under DC fields the system experiences no such repetitive stress pulses. Instead it might be expected that either an equilibrium is eventually established or that the system proceeds to breakdown in a given time, i.e. deterministic breakdown (see Part 3 of the book). However as is shown in Part 4 this is an oversimplification, stochastic events may be responsible for both initiation and propagation to breakdown leading to a statistical description. Here the equivalent factor to N_f in eqn. 17.1 will be the number of attempts to initiate a breakdown process, and thus, for constant DC stress conditions, time will appear as an independent variable in the same way as for 'constant stress' AC conditions. Now however the time-scale is that of a natural stochastic process which may or may not be field dependent, unlike the driven time-scale $(2f)^{-1}$ of an alternating field. When it comes to ramp conditions though a new time-scale is introduced; that of the ramp rate. The stochastic inception process is replaced by a deterministic mechanism in which breakdown is initiated once a field is reached for which the inception time is essentially zero. In other words the system is driven to breakdown by the field ramp. Ramp experiments with AC fields, in contrast, retain a stochastic inception component as long as the ramp rate is not fast enough to ensure that all breakdowns occur in less than one half-cycle.

The significance of the above considerations is that DC ramp tests do not necessarily contain sufficient information to define the statistics of breakdown under constant DC stress[836], whereas with ramped alternating fields this is always at least an *a priori* possibility. Development programmes for DC systems must therefore include some constant stress tests and reasonable simulations of service conditions[837,832,716].

17.2 The relationship to space charge and local currents

17.2.1 Homogeneous dielectrics in AC fields
In view of the difficulty in achieving a purely empirical analysis of DC breakdown statistics it is vital that a more fundamental understanding of the features specific to DC fields be obtained. Only by following this route can the statistical data be interpreted with any confidence.

A useful starting point for a comparison between the effect of AC and DC fields upon a dielectric are the appropriate Maxwell equations:

$$\text{curl } \mathbf{H} - \frac{\dot{\mathbf{E}}}{c} = \frac{4\pi}{c} \mathbf{j} + \frac{4\pi}{c} \dot{\mathbf{P}} \qquad (17.2a)$$

$$\text{div } \mathbf{E} = 4\pi(\rho - \text{div } \mathbf{P}) \qquad (17.2b)$$

where $\mathbf{H}$ is the magnetic field and $\mathbf{E}$ the electric field. The bulk current density $\mathbf{j}$ and polarisation $\mathbf{P}$ are related to $\mathbf{E}$ via the conductivity, σ, and susceptibility, χ, respectively, which may themselves be functions of E at high fields, i.e.

$$j = \sigma E \qquad (17.3a)$$

$$P = \chi E \qquad (17.3b)$$

Contact between a metal electrode and an insulator is sufficient to establish a space charge density, ρ, in the surface states[838] whose sign is dependent upon the relative magnitudes of the metal and insulator work functions[839,840]. Typically polymer work functions are close to those of most metals, and though it is usually electrons which are injected into the polymer, it is possible with some metals, gold for instance[841], that electrons are extracted. The application of a voltage to the contact results in the injection of homocharge into the polymer and in some cases both homocharge and heterocharge may be trapped[842] in the same region of the polymer in an unrecombined form[843].

When an alternating field is applied to a uniform dielectric eqns. 17.3 *a* and 17.3*b* show that both the current density and polarisation alternate in phase with the field. The time dependent polarisation also generates a displacement current ($\dot{P}$), $\pi/2$ out of phase with the field. The relationship between current density and space charge is contained in the continuity equation:

$$\text{div } j + \frac{\partial \rho}{\partial t} = 0 \qquad (17.4)$$

which can be derived by taking the divergence of eqn. 17.2a, using the operator rule div · curl ≡ 0, and substituting for E from eqn. 17.2b. Eqn. 17.4 applies to both interface injection currents and bulk current gradients and ignoring charge trapping shows that space charge can only build up under transient conditions when a current gradient exists. Eventually the resulting space-charge field will suppress further increase in ρ. In an alternating field j and ρ will oscillate at the same frequency as the applied field, with ρ lagging $\pi/2$ in phase behind j.

Most theories (Part 3 and Chapter 15) relate breakdown to charge displacement in one form or another, for example via the onset of a runaway current or by the build up of local space charge concentrations sufficient to cause failure through local electrostatic forces as in electrofracture. Alternating fields therefore place the dielectric at risk of failure once every half-cycle if the direction of charge flow is immaterial and every cycle otherwise. If the injection currents are polarity dependent (see Chapter 5) space charge may build up progressively, and in addition local damage and deformation may be accumulated thereby causing fatigue in the insulation. Therefore under alternating stresses a dielectric experiences repetitive attempts to fail as it ages.

17.2.2 Dielectrics in DC fields

In a DC field a uniform dielectric initially experiences a transient behaviour as the polarisation and current density relax to their static values[844,845]. Only if the current density locally exceeds a critical value will the system runaway to breakdown[162,588,589,428]. Otherwise an equilibrium current density will be reached[354]. During the transient period current gradients will exist at injecting and/or extracting contacts and space charge will build up[841,846–849] to a steady state value. The electric field gradient established as a result of the space charge (eqn. 17.2b) may be able to age the polymer via electromechanically induced creep (among other processes) and hence eventually cause failure at the applied DC stress. However in the absence of such aging mechanisms injected homocharge will reduce the dielectric thickness somewhat but not otherwise directly cause failure. Dielectric breakdown would not therefore occur at DC voltages below those required for deterministic mechanisms if it were not for the possibility of fluctuations.

Structural fluctuations may cause local variations in charge density leading to local current enhancement and damage generation. Random processes may cause the production of excess charges, which increase the current density transiently before being swept out of the system. Thermal fluctuations and thermal gradients[832] will also produce local current enhancements as de-trapped space charge is redistributed around the system. All of these factors combine to put the system at risk over a working lifetime, but with a hazard rate (see Chapter 14) which is independent of the mechanism whereby failure actually occurs as in the homogeneous breakdown model of Section 15.2.1. In addition to such fluctuations, which are an intrinsic feature of the system, other factors such as polarity reversal and voltage spikes (e.g. lightning strikes on cables) may place the insulation at risk when they occur during service[716,832,837]. In both cases large local currents may arise transiently which can lead to failure. Rapid polarity reversals may be particularly severe in this respect since here the space charge transiently increases the field near the electrodes to a considerable extent[166,434], thereby producing large injection currents which may even reverse before the system returns to equilibrium. A substantial risk of damage exists during this process[348,287].

17.2.3 Polymers as heterogeneous materials

Most polymeric materials cannot be regarded as uniform dielectrics since regions of different density, chemical composition, and morphology (see Chapter 1) introduce a heterogeneity on a scale from ~5 nm to 100 μm. Thus regions differing in their dielectric permittivity and conductivity will occur. As a result of the spatial variation in permittivity, gradients in the polarisation and hence electric field will exist when the material is electrically stressed (eqn. 17.2b). Furthermore differences in conductivity will lead to transient current gradients and the build up of space charge (eqn. 17.4) most particularly at the amorphous-crystalline boundaries of semicrystalline materials such as polyethylene and polypropylene, or around the boundaries of filler particles in EPR for example. Thus even when the applied field is 'uniform', and without allowing for space-charge injection, polymeric materials will contain local space charge concentrations[106] giving field gradients and a range of local field strengths. Such features will respond to structural fluctuations, as discussed in Section 15.2, giving local currents and fluctuating fields even for an applied DC voltage. Applications of an AC field serves to drive the local currents and polar fluctuations at its own frequency, rather than the natural time-scale of the material fluctuations.

Polymeric materials can thus be thought of as experiencing a range of slowly fluctuating local fields and space-charge currents on a sub-micron scale, which is typically the size on which damage is first initiated in cases such as free volume and cumulative damage breakdown. This places the material continually at risk of an unfavourable combination of fields and currents leading directly to failure. This risk though will be very low at service voltages, and it is more likely that electrical aging may cause an accumulation of local deep and shallow trapped space-charge concentrations[106] leading to a weakening of the dielectric and an increase in its failure probability at the applied field. All these processes are likely to be enhanced when space charge injection from the electrodes can occur leading to a reduced lifetime even in the absence of extraneous factors such as polarity reversals. Under DC fields the aging time-scale will be natural to the system though probably field dependent, whereas alternating fields will drive the system at their own frequency and hence introduce a fatiguing element into the aging process.

17.3 'Constant' stress conditions and space charge build-up

17.3.1 Space charge measurement

As pointed out in the previous section, AC fields ($E(t) = E \cdot \cos(\omega t)$) of constant amplitude (E) will drive local currents and space charge concentrations within the polymeric material. In the absence of an irreversible accumulation the average local values of j and ρ taken over a cycle will be zero. In contrast a constant DC stress will build-up local space charge concentrations which may possibly increase throughout the insulation lifetime[847,846].

Since the dissipation of accumulated space charge can be very long with times ranging from a few hours[348] to weeks[287,846] or even longer at room temperature, it is possible to investigate the space charge distribution in DC stressed samples by destructive measurements[850,851]. These methods involve microtome sectioning and the use of electrostatic attraction of powders for a charged surface or field probes[846,847,848] to identify regions of different charge density and sign. Using the dust figure method[851], bands of different sign were found in XLPE cable sections with the major contribution being that of heterocharges near the electrodes. Field probes however indicate homopolar injections[848,846].

The metal of a sectioning blade may however influence the charge distribution on the surface and new non-destructive methods have now been developed for investigating space-charge. These techniques probe the space charge distribution by means of either a heat pulse[852] or an acoustic pressure wave[853-856]. In the latter case the wavefront displaces the space charges as it passes through their location thereby inducing voltage (open circuit) or current (short circuit) pulses on the counter electrode[857]. Correlation of the time at which the induced pulses are detected with the position of the pressure wave in the material allows the space charge and field distribution[857,855] to be determined. In contrast the heat pulse technique and its sinusoidal variant[858] can only yield the total space charge and its centroid. Initially the pressure waves were generated either by shock tubes[853,854] or piezoelectrically[854] however these methods lacked reproducibility and now laser techniques[855,859] are mostly used, although the generation of pressure waves at space charge layers by means of an externally applied voltage pulse has also been proposed[860,861].

Of all these techniques that in which the pressure wave is generated by the ablation of a target electrode[855] (colloidal graphite) has been demonstrated to be the most versatile and effective. Difficult deconvolution procedures are not required, unlike the voltage pulse[860] method where they are essential, or the photo absorption pressure pulse method[859] where they are needed to obtain spatial resolution. Although initially confined to thin films[857] the laser ablation method was extended to thick (~ 8 mm) insulation by making allowance for the attenuation of the pressure wave along its propagation path[855]. Application to cylindrical geometry[862] has also proved possible by making use of the focussing effect to partially offset the attenuation. Although the spatial resolution is as yet insufficient to allow investigation of space charge around a needle electrode (1μm is required), this technique at present holds out the greatest promise for both research and commercial applications. The interested reader is referred to papers by Lewiner[857,842] and further details are expected to be published in a more-detailed review article[863].

Recent results show that either homocharge or heterocharge may be present in LDPE cables dependent upon a number of factors including both the material formulation and the nature of the semiconducting screens[842,864]. It is also clear that manufacturing history and the presence of 'impurities' can affect the space charge distribution significantly, which is thus a system rather than a polymer property. The earlier contradictory results[848,851] may therefore have originated with undisclosed system differences rather than problems of technique.

17.3.2 Homocharge

Homocharge is generated by charge injection at the electrodes and is therefore dependent upon the electrode-insulator contact, the temperature, and in DC fields, the polarity. Polarity will affect the situation because some metals may inject only carriers of one sign, while others will support bipolar injection. The field within the homocharge region will be reduced and eventually injection will be suppressed, however the reduction of the effective dielectric thickness will increase the field in the bulk, see Fig. 17.1a. Given sufficient time the carriers will be transported across the insulation under DC fields[353], where they may combine with injected charge of opposite polarity or be extracted. Replenishment of the homocharge density will then occur by means of charge injection from the electrode. However this may not be instantaneous, instead it is more likely to occur as a result of random processes with a time-scale which will probably be dependent upon the *local* field. Thus the space charge will be injected in bursts[387,452] with

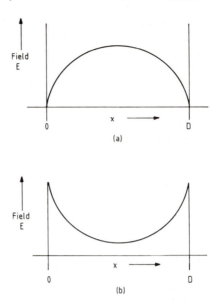

Fig. 17.1   A schematic representation of the field distribution in an insulator containing space charge regions extending from the electrodes at $x = 0$ and $x = D$. The field within a homocharge region (*a*) will be reduced and eventually injection will be suppressed whereas in a heterocharge region (*b*) the electrode field increases

some distribution of injection times. The resulting space-charge distribution will depend upon the relative magnitudes of the injection efficiency and time, trapping times, combination efficiency and times and transit times.

When the transit and combination times are long compared to the injection time, space charge may continually build up in the body of the material and the insulation will only approach the equilibrium condition very slowly. It may even be possible that no equilibrium may ultimately exist and the insulation will finally fail after a suitable (very long) time at service voltages. An analogy may be drawn here with the glassy state of matter which is unstable in the long-term. Since the pulses of injected charge will repeatedly place the insulation at risk and function as attempts at breakdown, the breakdown distribution in time will be governed by the distribution of injection times.

17.3.3 Heterocharge
A heterocharge may appear in polymeric insulation when one of the electrodes partially blocks the extraction of mobile charge[501] injected by the other[808], such as for example aluminium which appears to block electron extraction from polyethylene when acting as the positive terminal[857]. Other sources of heterocharge are ionically dissociable chemicals which are often added to polyethylenes to inhibit electrical treeing (see Chapter 8), and deep-trapped charges intrinsic to the insulation manufacturing processes.

In heterocharge conditions the electrode field increases and that within the bulk decreases, Fig. 17.1*b*. As space charge transits the insulation the field at the electrodes will build up. Detrapping from progressively deeper traps will occur as the time of voltage application increases[865] and will also contribute to an increase in the field. As a result of these processes the local electrode field may over a period of time become sufficiently large to initiate local damage through field-enhanced electrode-injection currents[82,501]. A similar mechanism has been proposed for breakdown in metal–oxide thin films, with electronic heterocharge exciting surface plasmons in the metal as it is extracted, causing hot hole injection[756]. With polymeric insulators an electrochemical reaction at the electrode is also possible[166].

In the heterocharge case the local electrode field will be proportional to the amount of charge that has crossed the insulation. A critical field for breakdown would thus be obtained by the transit of a given quantity of charge, as noted by Chniba and Tobazeon for the DC breakdown of thin polypropylene films[774,833]. Because of the short transit time (≤ 1 s) and the behaviour of the current transients these authors rule out field distortion as an explanation and attribute the effect to current-generated structural deformation, however electrode effects of the type discussed above[756] would be just as likely. Evidence for the development of a heterocharge in poly-propylene films[866] has also been obtained through measurements of the charging and discharging currents with an asymmetric electrode system (Al/PP/Au). The data of Kitagawa *et al.*[434] for polyethylene thin films are also consistent with this concept.

The distribution of breakdown times in the heterocharge case will depend on factors that govern the release and transit of the charge carriers (e.g. the distribution of detrapping times which is likely to be of the stretched exponential (Weibull) form[867,868]). Defects which are likely to scatter carriers or trap them for a substantial time will effectively increase the breakdown time (or field[869] in a ramp test). On the other hand ionically-dissociable chemical impurities in cable materials will *contribute* to the intrinsic heterocharge reservoir and reduce breakdown strengths. This will be par-ticularly the case for the crosslinking by-products of XLPE and may be a reason for its lower DC breakdown strength[870] compared to HDPE and LDPE. Another factor governing space-charge concentration in these materials is the degree of side-chain branching which is increased in the high-pressure manufacturing process for LDPE giving a type of polyethyl-ene with a lower breakdown strength[870].

17.3.4 Space charge aging
In general both heterocharge transport and homocharge injection may combine to cause damage in polymeric insulators during DC stressing[434]. However the greater thickness of cable insulation compared to thin films makes homocharge injection more important in the short term for such systems. Here field moderators which relieve the homocharge concentra-tions may be effective in increasing the field-reversal breakdown strength[871]. Increased conductivity[872] though will favour the build up of heterocharge, and may be damaging in the long term.

One possible origin for heterocharge in cable insulation is void discharge. In DC conditions such discharges will be extinguished rapidly by the build up of a counter field in the void, and the damage produced will be small. However transport of the induced charges from the void surface to the electrodes will both build up a heterocharge at the interface and allow further discharges to occur in the void, in contrast to charge recombination on the void surface which only allows discharging to be renewed. Thus, over a long period of time, void discharges will cause the electrode interface to be stressed at the discharge repetition rate, and the distribution of breakdown times will be determined by that of the occurrence of severe discharges.

Just as under constant amplitude AC fields, the DC stressed insulation will experience repeated attempts to achieve critical local conditions. However the time-scale will be that of the processes that build up space-charge concentration rather than that of the AC frequency. In both cases space charge will be accumulated during the time of stress[847,873] which will contribute to an aging of the dielectric[874]. As a result the probability of a successful breakdown attempt occurring within a given *time interval* (i.e. the hazard rate) increases with the period of stress. If the breakdown distribution is of the Weibull form this type of failure, termed 'wear out' when applied to capacitors, will possess a Weibull time exponent $a \geqq 1$ (see Chapter 14). The distribution density of breakdowns will be grouped around a most probable failure time $(a > 1)$. Since exceedingly high fields $(\geqq 10^9 \text{ V} \cdot \text{m}^{-1})$ are required to cause breakdown at the electrode-insulator interface, the characteristic failure time is likely to be enormously long at service voltages, particularly for DC cable insulation which is very thick compared to thin films $(\sim 1 \ \mu\text{m})$. Voids at the electrode interface[324,348], surface imperfections[434] and cumulative damage[131,132] are the most likely forms of risk for long-term failure. Such features may initiate breakdown when subjected to the transient high local fields produced by the voltage surges and polarity reversals which occur throughout service. These transient events may therefore be regarded as attempts at breakdown with the number experienced during a given period of service fulfilling the same role in the failure statistics as N_f in eqn. 17.1.

17.3.5   Weak spots

Along with aging a different situation may occur in which 'defects' or 'weak spots' cause the system to proceed inevitably to breakdown once it comes under stress. The Weibull exponent for this type of breakdown is typically $a < 1$ (i.e. an infant mortality function) and proof tests can be used to eliminate most of these failures prior to service (see Chapter 18). This in fact is the basis of the self-healing process used to prepare capacitors for service. Essentially the more severe defects achieve breakdown too rapidly for compensating factors to be brought into play whereas the time to breakdown of less severe defects is increased by such factors and the probability of failure within a given time interval decreases with the time of survival at a given stress. We have already discussed some models of this type in Chapter 15, particularly the conducting path breakdown of Section 15.2.3. In polyethylene thin films such weak spots have been directly observed to be regions in which the temperature was raised above ambient by local Joule heating[601]. Breakdown was assumed to occur via a combination of thermal and electromechanical factors (see Part 3 of the book). Here the process may be caused by the build up of space charge at a defect centre which establishes a current gradient (eqn. 17.4) sufficiently large to initiate self-sustained local heating. The less rapid the space charge build up the smaller the current gradient, and the more time available for compensating factors such as thermal conduction and local counter fields (eqn. 17.2b) to prevent the centre proceeding to runaway. Under these latter conditions the current can be expected to pass through a maximum and decrease as observed in samples which did not proceed to failure[601]. Other workers[875,876,877] have associated such current peaks with an increase of hopping carriers building up space charge at centres with low escape probabilities. Such centres may be carbonyl groups in oxidised polyethylene[875] although other defects produced by aging may be involved[878]. The presence of a peak was explained by Röhl[876,877] as due to a reduced mean mobility of the carriers as the space charge concentrations built up. These ideas are consistent with space charge initiating a hot spot but the data of Nagao *et al.*[601], which directly correlates the temperature and size of the hot spot with the current peak, clearly illustrates the role played by local heating in sustaining the process.

17.4 Progressively increasing stresses (ramps)

17.4.1 Statistical features
Constant stress tests aimed at determining the quality and lifetime of insulation systems often have large statistical errors and are time consuming as well. Consequently many test procedures rely on failure statistics obtained under ramp conditions. Here the applied field is time dependent with a form, $E(t) \cdot e^{i\omega t}$ $(=E_a(t))$ for AC fields and $E(t)$ for DC fields. For a linear ramp $E(t)$ is proportional to time through a constant ramp rate $\dot{E}$ (i.e. $E(t) = \dot{E} \cdot t$).

Statistically, constant stress and ramp conditions are interrelated via the hazard function (see Chapter 14) which defines the probability of failure per unit time after a given time of stress. Hence for a system following Weibull statistics we find that the *cumulative* survival probability for a given time t, i.e. $P_S(t)$, is given by:

$$P_S(t) = \exp\left\{-aC \int_0^t t^{a-1}[E_a(t)]^b \, dt\right\} \tag{17.5}$$

which in constant-stress conditions becomes:

$$P_S(t) = \exp\{-Ct^a E_a^b\} \tag{17.6}$$

and in ramp conditions:

$$P_S(t) = \exp\{-aCt^{a+b}\dot{E}^b/(a+b)\} \tag{17.7a}$$

$$= \exp\{-aCE^{a+b}/[(a+b)\dot{E}^a]\} \tag{17.7b}$$

A relationship therefore exists between the characteristic breakdown field and the ramp rate

$$E_b \propto \dot{E}^{a/(a+b)} \qquad\qquad (17.8)$$

which enables the exponents a and b to be derived from the failure distributions obtained for different ramp rates. Application of eqn. 17.6 then allows life predictions to be made. In many cases the system behaviour is expressed through a voltage life exponent $n(=b/a)$ but it should be noted that this is insufficient to characterise the distribution and without a value for a no lifetime predictions can be made, except for the characteristic lifetime which corresponds to $\sim$63% failure and is not what is usually desired in a service life guarantee.

As illustrated in Fig. 14.12a the Weibull distribution, eqn. 17.7, appears to be valid for ramped AC fields. In this case a log–log plot of the breakdown field as a function of the ramp rate, Fig. 14.12b shows that eqn. 17.8 is obeyed, and values of a and b can be determined. This behaviour appears to be typical of AC fields and hence eqn. 17.6 can be used with some confidence to predict AC constant amplitude lifetimes. However it should be noted that in practice the equivalence should be checked by at least one determination of a via a constant-stress test.

In DC ramp tests the ramp exponent $(a+b)$ is often very high (>10) and difficult to measure precisely even with a large set of failures. On the other hand the characteristic breakdown field tends to change only slightly with ramp rate, see for example the results of Kitagawa *et al.*[434], Fig. 15.6. Often the variation is so small that with the small data sets available for large-scale insulation ($< \sim 30$) it may not be possible to be sure that it is statistically significant. Only with self-healing breakdowns[435,434] can a dependence of E_b upon ramp rate $\dot{E}$ be ascertained with confidence and even here there may be some difficulty because of an increase in E_b at low ramp rates. For these reasons it is necessary to combine DC ramp tests with time-consuming DC constant-stress tests if lifetime predictions are required[836]. It should be remembered however that DC systems may see a number of powering up ramps during service, and the ramp test data can be used to predict the survival probability for a series of such operations.

17.4.2 Analysis of experimental data

Most detailed work on DC ramp failure has been carried out on metal oxide films such as used in device technology[879,704,378,880,705]. Here the data implies a weak but progressive increase in E_b with $\dot{E}$, but even with $\sim 10^3$ breakdowns per sample it is often difficult to demonstrate this conclusively[879]. Shatkes and Av-Ron[880,705] have suggested that in fact a is unity but two independent distributions of failures, one intrinsic and one of defects with different time-scales combine to give the observed behaviour. It is known that such combinations can be difficult to separate in ramp tests[879], and it is not clear whether the solution of Shatkes and Av-Ron[705] is unique, nor how sensitive it is to the assumption of a value of $a = 1$ for both distributions. Although this analysis reproduces limited life constant-stress test data reasonably well[705], it is known in other cases[879] that the 'defect' distribution has a value of $\sim 0\cdot 3$ for a whereas for the major distribution a may take values between $0\cdot 6$ and $1\cdot 5$. Such uncertainties will yield enormous differences in predicted lifetimes and are not really acceptable technologically. Furthermore given a defect to intrinsic number ratio of $\sim 0\cdot 01$ it is unlikely that such a defect distribution could be identified in the limited data sets usually available for polymeric systems. In view of the possible uncertainty in Shatkes and Av-Ron's[705] determination of a, it is interesting to note that they find an approximate field dependence of $t_b \propto E^{-50}$ for the breakdown time on the assumption of $a = 1$, whereas a Weibull plot of their ramp data gives $a + b \simeq 25$. These results may be reconciled if $a \simeq 0\cdot 5$ rather than unity.

Remarkably the life exponent $n(= b/a)$, which is typically 40—50 for thin oxide films, is much the same in thin polymer films[434] once samples with weak spots have been eliminated, although here the estimation is often based on limited data[836] and only a couple of ramp rates. For example the data of Kitagawa *et al.*[434] for polyethylene, shown in Fig. 15.6, gives $a \simeq 0\cdot 2$ and $n \simeq 52$. It might seem therefore that this field dependence is typical of a thin film breakdown mechanism which has usually been assumed to be avalanche dependent[146,434] although an electrode-interface mechanism cannot be ruled out[756], however the same type of effect is seen in thick cable insulation[881], see Fig. 17.2. Here the two sets of data at the higher ramp rates are almost statistically indistinguishable. This is in fact what would be expected if n were to have a value of ~ 50, since the values of $a + b = 2\cdot 8$—$3\cdot 7$ imply a to be less than $0\cdot 1$. Just as with the thin polyethylene films, Fig. 15.6, breakdown at low ramp rates shifts to higher stresses, and in this particular case breakdown of the cable was never encountered at the slowest ramp rate of $15\ \mathrm{V \cdot s^{-1}}$ before the maximum voltage of $500\ \mathrm{kV}$ attainable by the equipment was reached. The three points defining curve (c) in fact correspond to breakdowns which took place in the cable terminations. This curve should therefore not be taken as representing a true distribution of results but rather as its minimum boundary under these conditions.

It would generally be accepted on the basis of experience that a DC life exponent (n) of ~ 50 is far larger than that applying during long-time service, which for thin film capacitors[836] seems to be $n \simeq 15$. Such a value is not too different from values found for the ramp exponent $(a + b)$ in these experiments, see Fig. 15.6, and thus it appears that ramp tests give a reasonable estimate for the field exponent b. However their near independence of ramp rate magnitude implies that the equivalence of constant stress and ramp statistics given by eqns. 17.5 to 17.7 is not correct. Either the ramp probes a different mechanism, a different facet of the same mechanism, drives the mechanism in a different way, or measures a different distribution. This latter possibility may well be applicable since the low values of a ($< 0\cdot 3$) correspond to an infant mortality distribution whose members would be removed in a proof test, or early in service. Whatever the origins however the results imply that constant stress tests are required to supplement ramp tests, when attempting to determine DC service life-times, and this fact seems to have been recognised by the capacitor industry[836].

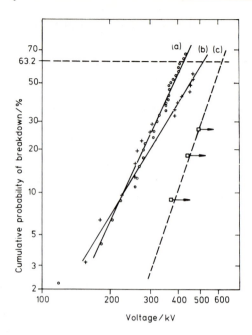

Fig. 17.2 A Weibull plot showing the results of DC progressive stress tests carried out on extruded polymeric high voltage power cables at three different rates of increase of voltage[881].

Test	a	b	c
Rate of voltage increase (kV · s^{-1})	5·0	1·1	0·015
Sample size	45	30	10
Number of breakdowns	31	18	3
Weibull slope	3·7	2·8	5·1
Characteristic breakdown voltage (kV)	423	505	>616

Note that the data of curve (b) almost exactly coincides with that of curve (a) up to around 300 kV (i.e. ~270 s) but shifts to higher stresses at longer times. This may just be a statistical artifact, or, imply either the presence of two distributions or a time delayed mechanism. The single straight line shown may not therefore be a completely valid representation of this data

17.4.3 Do ramps accelerate constant-stress breakdown?
In the previous section we pointed out that for DC ramp tests:

(i)   At typical ramp rates there is only a very weak dependence of the characteristic field (time) upon the ramp rate;

(ii) At low ramp rates (long times) there is a shift of the distribution towards higher breakdown fields. In some cases the shift is extremely large;

(iii) The shape of the breakdown distribution (exponent $a + b$) remains essentially unchanged regardless of ramp rate;

(iv)   No equivalence between constant stress and ramp statistics could be
obtained.

In contrast the breakdowns in AC ramp tests did shift with the ramp rate
as expected from eqn. 17.8 and an equivalence between 'life' and 'ramp'
statistics could be verified where the mechanism of breakdown remained
the same. In view of these AC results it seems pertinent to question whether
the difference between AC and DC behaviour might not be due to the way
in which the system is probed for breakdown in the two cases rather than
changes in mechanism, or defect contribution.

First let us take the AC ramp case. Here the current density j, space
charge density ρ, and polarisation $\boldsymbol{P}$ oscillate with the same frequency as
the applied field ($\dot{E} \cdot t \cdot \cos \omega t$), though their maximum amplitudes on each
half-cycle increase as the ramp progresses. There is also an extra oscillating
component to the displacement current ($\dot{\boldsymbol{P}}$) which does not increase in
amplitude, i.e.:

$$\chi'[E(t)]\dot{E} \cos (\omega t) \tag{17.9}$$

In addition to an increasing dissipative (absorption) current

$$\omega \chi''[E(t)]\dot{E}t \sin (\omega t) \tag{17.10}$$

The system is thus stressed repetitively just as under 'constant' stress conditions, with the maximum stress building up during the test in proportion
to $E(t)(= \dot{E} \cdot t)$. Furthermore the irreversible features such as local deformation, strains and residual space charge accumulated per cycle will also build
up in proportion to $E(t)$ in the same way as the dissipation, i.e. eqn. 17.10.
Thus an AC ramp test acts as a 'constant' stress (life) test that has been
accelerated by the ramping procedure.

Turning to the DC case we find that the current density and polarisation
both increase continuously with $E(t)$. The displacement current however is
constant and proportional to the ramp rate. Where divergences of current
exist, for example at injecting contacts, space charge will build up at a rate
proportional to the field $E(t)$. The accumulation of space charge is thus
driven by the ramp. This is in contrast to the constant DC stress condition
where space charge concentration is controlled by a variety of processes
with their own time-scales, such as charge detrapping etc. discussed in the
previous section. If the ramp rate is sufficiently high bulk processes may
not have time either to build up space charge at the electrodes or reduce
it by the transport of compensating bulk charges. Alternatively local currents
near the electrode may have time to runaway to failure[428,162,588,589] prior
to field moderation by accumulated homo space charge. Thus DC ramp
tests do not accelerate uniformly the processes that give rise to breakdown
in constant DC conditions. Instead they accelerate one component of the
synergistic process to the exclusion of the others, and in particular to the
exclusion of those processes which give the time-scale to constant DC stress
breakdown in polymeric systems. This can be seen most clearly if it is
assumed that in the experimental DC ramp results the actual value of the
Weibull time exponent, a, is zero. The *predicted* survival probability at

constant DC stress (eqn. 17.6) then becomes:

$$P_S(t) = \exp\{-CE_a^b\}$$

i.e. the system either fails within the ramp timescale at the applied field (E_a) or survives indefinitely with the given probability.

The above considerations indicate that in a DC ramp test breakdown originates through charge injection processes at or near the electrode-insulator interface, and show why similar results are observed for thin metal oxide films, thin polymer films and thick polymeric insulation. Oxide films however will usually have a low defect concentration and hence the electrode interface will continue to dominate breakdown under constant stress conditions, thereby allowing ramp tests to be used for life-time estimation. Polymeric materials though are much more internally irregular in morphology, composition and defect distribution and it is for this reason that a difference between ramp and constant-stress statistics will tend to occur.

17.5 Space charge as a critical parameter

The voltage applied to a system is a controlled variable and to some extent this has conditioned us to think of field as being the critical parameter for breakdown. In heterogeneous systems containing space charge however the local field will vary considerably from place to place in the material and be controlled substantially by space charge concentrations and their fluctuations. It may therefore be better to describe the state of the system through the distribution of local space charge concentrations that it contains. Breakdown in this picture would occur when a critical local space charge density is exceeded.

Some justification for this hypothesis exists when we consider the one-carrier impact ionisation (avalanche) model[146]. Although such a mechanism is unlikely to cause runaway breakdown in bulk polymeric insulation under constant DC conditions, it may well apply to thin films[434]. Using this model Klein[146] derives an expression for the time to current runaway, t_r, (eqn. 35 of Reference 146)

$$t_r = \frac{2\varepsilon_0\varepsilon_r E^2}{BD\alpha_t j_t} \qquad (17.11)$$

where E is the applied field, B the field parameter for Fowler–Nordheim emission, j_t the initial Fowler–Nordheim injection current density, α_t the impact ionisation coefficient, and D the sample thickness. This expression is valid for $D \cdot \alpha_t < 1$ under constant voltage conditions with the build up of hole concentrations opposed by hole drift. By noting that $t_r \cdot D \cdot \alpha_t \cdot j_t$ is just the total positive (hetero) space charge produced during the time t_r, we find that the breakdown (runaway) condition becomes:

$$\rho_h(t) = 2Q(E)E/B = Q_c(E) \qquad (17.12)$$

where $\rho_h(t)$ is the time-dependent heterocharge density and $Q(E)$ is the capacitive charge density appropriate to a field E applied uniformly to the material. $Q_c(E)$ therefore defines a critical charge density for breakdown under a DC applied field E.

The equivalent criterion for breakdown under DC ramp conditions (eqn. 46a of Reference 146) is:

$$\rho_h(t) = Q(E) \tag{17.13}$$

In this case the factor E/B is absent and the generation of $\rho_h(t)$ is driven by the time-dependent increase in E until it reaches $Q(E)$ and failure ensues.

Although it may be possible to relate the stress enhancement of injecting sites on a rough metal surface to a distribution of local charge densities, this has not yet been carried out for the avalanche model. However two theories for which such a distribution can be obtained are discussed in Chapter 15. Both of these are based on the contention of Bahder *et al.*[120] that an electric tree or filamentary conducting path initiated either by void discharges or charge injection (see Chapter 13) contains a quantity of charge proportional to its total branch length, S, eqn. 15.18. Using the fractal dimension d_f of the branched structure the charge can then be related to its overall length $L(t)$, by eqn. 15.18, i.e.:

$$S \propto [L(t, E_a)]^{d_f} \propto \rho(t, E_a) \tag{17.14}$$

Two criteria have been used for the critical condition:

(a) Breakdown occurs when a critical local field is exceeded for which eqn. 15.21 gives the cumulative probability of survival P_S (per tree) as:

$$P_S(t, E_a) = \exp\left\{ -\left[\frac{\rho(t, E_a)}{\rho_c} \right]^{\beta/d_f} \right\} \tag{17.15}$$

where ρ_c is the critical charge appropriate to the critical length L_c determined from eqn. 15.19, and β/d_f may range between 3 and 6 according to the model.

(b) Breakdown occurs when the path crosses the insulation. Here eqn. 15.13 can first be written in terms of the average tree length $\langle L(t, E_a) \rangle$ grown in time t in a field E_a, giving the survival probability (per initiating site) as

$$P_S(t, E_a) = \exp\left\{ -\left[\frac{\langle L(t, E_a) \rangle}{D} \right]^{1/ms} \right\} \tag{17.16}$$

If it is now assumed that since the conducting path is blocked by the insulator, it is in fact a discharge containing space charge according to the criterion of eqn. 17.14 until it crosses the insulation, $P_S(t, E_a)$ becomes:

$$P_S(t, E_a) = \exp\left\{ -\left[\frac{\rho(t, E_a)}{\rho_c} \right]^{1/msd_f} \right\} \tag{17.17}$$

where $(msd_f)^{-1}$ may typically take values in the range 0·3—1·5, see Table 15·2.

The two mechanisms described above which are more likely to apply to DC breakdown in polymeric insulation than the one-carrier impact ionisation model of Klein[146], show that the breakdown statistics have a Weibull form with space charge density as the critical variable. In fact the defect extension model of Section 15.4.2b can also be expressed in this form if the size of the defect ($l_\xi(t)$) in eqn. 15.38 is equated with local space charge via an expression such as eqn. 17.14. Thus it appears that we may be able to express $P_S(t, E_a)$ in the general form:

$$P_S(t, E_a) = \exp\left\{ -\left[\frac{\rho(t, E_a)}{\rho_c} \right]^\delta \right\} \qquad (17.18)$$

where the differences in detail between the different breakdown mechanisms lie in the value of δ, the form of ρ_c, and the dependence of $\rho(t, E_a)$ upon time, and field E_a.

17.6 A model for ramp and constant stress statistics

17.6.1 Constant stress (E_a) DC tests

Here it is reasonable to assume that space charge is accumulated by events repeated on a time-scale $\tau(E_a)$. Therefore we take the build up of space charge to be proportional to the number of contributing events, i.e.:

$$\rho(t, E_a) \propto f(E_a)t = t/\tau(E_a) \qquad (17.19)$$

where the type of event and the field dependence of its timescale, $\tau(E_a)$ determines the magnitude of $\rho(t, E_a)$.

Time thus appears as a stochastic factor in the space charge build up and will enter the failure statistics directly through the exponent δ appropriate to a specific mechanism of breakdown. It should be noted that for DC treeing $L(t)$ may be proportional to t (see Section 5.3) and hence d_f may not enter into the definitions of δ in eqns. 17.15 and 17.17. In this case the process may be close to deterministic ($\delta \sim 5$—10) for tree induced breakdowns or almost random ($\delta \sim 0.6$—2) for conducting path formation.

17.6.2 Progressive DC stress tests

Here we assume that injection continuously builds up homo-space-charge at the injecting point until sufficient is present to initiate a rapid runaway process causing near instantaneous failure. Contributing or compensating bulk processes are not allowed time to occur.

Under these circumstances the space-charge density is always in equilibrium with the field applied to the injecting point, though not necessarily with the bulk material, and

$$\rho[t, E_a(t)] \propto f[E_a(t)] \qquad (17.20)$$

Note that if avalanche or damage-producing events occur prior to breakdown then $\rho[t, E_a(t)]$ will also include an additional contribution (such as eqn. 17.19) with a time factor dependent upon the timescale of their initiation, τ, which may be independent of the field. Substituting for $\rho[t, E_a(t)]$ from eqn. 17.20 into eqn. 17.18 gives a characteristic breakdown strength which is independent of the ramp rate, and determined by solving the relationship:

$$f[E_a(t)] \propto \rho_c \qquad (17.21)$$

for the field. The observed behaviour of the breakdown statistics in DC ramp tests therefore indicates that in a ramp the build up of space charge is almost deterministic with very little contribution from the stochastic time processes that are likely to dominate constant stress DC failures.

17.6.3 AC 'constant' stress tests: half-cycle space-charge variation
In order to examine AC stress tests it is first necessary to establish the probability of survival for one cycle or half cycle and hence establish the probability of survival at time, t, in terms of the number of cycles in that time.

First we take $\rho(t, E_a(t))$ to vary sinusoidally with the field frequency so that it is zero at the beginning of the cycle, i.e.

$$\rho[t, E_a(t)] = \rho_p \sin(\omega t) \qquad (17.22)$$

Usually we can expect the peak space charge density, ρ_p, to be a function of E_p, the AC field amplitude, that is the 'peak' field. Using eqn. 17.18 the cumulative probability of survival for a half cycle, i.e. after $t = \pi/\omega$, is then given by:

$$P_S(\omega t = 0 \text{ to } \pi) = \exp\left\{-\left[\frac{\rho_p}{\rho_c}\right]^\delta \left(\frac{c}{\pi}\right) \int_0^\pi [\sin(\omega t)]^\delta \, d(\omega t)\right\} \qquad (17.23)$$

in which we have assumed a constant number of attempts at breakdown, c, to occur during each half cycle. For large values of δ which are usually found, the integral can be approximated to

$$\int_0^\pi [\sin(\omega t)]^\delta \, d(\omega t) \simeq \left[\frac{2\pi}{\delta}\right]^{1/2} \qquad \delta \gg 1 \qquad (17.24)$$

(see Appendix 3 for proof) so that

$$P_S(\omega t = 0 \text{ to } \pi) = \exp\left\{-\left[\frac{\rho_p}{\rho_c}\right]^\delta \left[\frac{2\pi}{\delta}\right]^{1/2} \left(\frac{c}{\pi}\right)\right\} \qquad (17.25)$$

By assuming that breakdown depends upon the space charge magnitude but not its sign the cumulative survival probability for the test time, t, can be obtained as the joint probability of the survival of N_ω ($= \omega t/\pi$) half cycles[120], i.e.:

$$P_S(t, E_p) = [P_S(\omega t = 0 \text{ to } \pi)]^{N_\omega}$$

$$= \exp\left\{-\left[\frac{\rho_p}{\rho_c}\right]^\delta \left(\frac{\omega t}{\pi}\right) \left(\frac{2\pi}{\delta}\right)^{1/2} \left(\frac{c}{\pi}\right)\right\} \qquad (17.26)$$

If we take the maximum value of ρ in each cycle (i.e. ρ_p) to be independent of the prior period of stress, the eqn. 17.26 for the cumulative survival probability has an exponential form in time, in which the time factor appears purely as a result of the repetitive stressing and no aging is allowed for. Aging can however be accommodated in this formalism as long as it occurs so slowly that ρ_p can be taken to be essentially constant over a single half-cycle, i.e. the aging time-scale is much slower than π/ω. Under these conditions ρ_p can still be taken outside the integral in eqn. 17.23 but now its magnitude will increase with the number of half cycles of stress previously experienced, i.e.

$$\rho_p = N_\omega f(E_p) = \frac{\omega t}{\pi} f(E_p) \qquad \omega t > 1 \tag{17.27}$$

In this case the Weibull time exponent becomes greater than unity (i.e. $a = 1 + \delta$, as in the aging models of Section 15.2).

17.6.4 AC progressive-stress tests: half-cycle space-charge variation
In a ramp or progressive-stress test the applied field amplitude (E_p) is increased at a given rate dE_p/dt $(=\dot{E}_p)$ until breakdown occurs. As long as the ramp rate is slow enough that no significant change occurs in E_p during a single half-cycle (i.e. $dE_p/dt < E_p\omega/\pi$) we can take eqn. 17.25 as the probability of survival during an 'infinitesimal' time interval of the ramp. The probability of survival for $N_\omega(=\omega t/\pi)$ half-cycles is thus:

$$P_S(N_\omega) = \exp\left\{ -\int_0^{N_\omega} \left[\frac{f(E_p(N_\omega))}{\rho_c} \right]^\delta KN_\omega^\delta \, dN_\omega \right\} \tag{17.28}$$

where we have used eqn. 17.27 for ρ_p and collected the other factors in the constant $K = (2\pi/\delta)^{1/2}(c/\pi)$. Since the applied field after N_ω half-cycles is given by

$$E_p(N_\omega) = \dot{E}_p t = \dot{E}_p(\pi/\omega)N_\omega \tag{17.29}$$

the integral in eqn. 17.28 can be converted to the variable E_p, and the characteristic breakdown field is given by the solution to

$$\int_0^{E_p} \left[\frac{f(E_p)}{\rho_c} \right]^\delta KE_p^\delta \, dE_p = [\dot{E}_p(\pi/\omega)]^{\delta+1} \tag{17.30}$$

The characteristic breakdown field in AC ramp tests will thus be a function of the ramp rate, $\dot{E}_p$. It should be noted that this will still be the case even when the time-dependent aging of ρ_p in eqn. 17.27 is neglected. In this latter case the dependence upon $\dot{E}_p$ results solely from the repetitive stressing of the material during the ramp.

An explicit form for the characteristic field can only be obtained if the form of $f(E_p)$ is known. In many cases however the integral in eqn. 17.30 can be approximated to a power of E_p at least over a limited range of fields, giving

$$A\frac{(E_p)^\Delta}{\Delta}\frac{K}{(\rho_c)^\delta} = [\dot{E}_p(\pi/\omega)]^{\delta+1} \tag{17.31}$$

where A is a proportionality constant that converts units of $(\text{field})^{\Delta-\delta-1}$ to $(\text{charge})^{\delta}$. This expression gives the characteristic value for E_p as

$$E_p = \left[\frac{\rho_c^{\delta}\Delta}{KA}\right]^{1/\Delta}[\dot{E}_p(\pi/\omega)]^{(\delta+1)/\Delta} \tag{17.32a}$$

$$= \left[\frac{\rho_c^{\delta}}{A}\right]^{1/(\Delta-\delta-1)}\left[\frac{\Delta}{K}\right]^{1/\Delta}\left[\left(\frac{A}{\rho_c^{\delta}}\right)^{1/(\Delta-\delta-1)}\dot{E}_p(\pi/\omega)\right]^{(\delta+1)/\Delta} \tag{17.32b}$$

where the last factor in square brackets in eqn. 17.32b has been written in a dimensionless form.

It is interesting to note that in the limit of $\Delta \to \infty$ an estimate of the breakdown field can be made which is independent of ramp rate. This result applies formally to a deterministic situation, i.e. breakdown only occurs when the characteristic field is applied and then instantaneously, and can be taken to apply to the impulse limit. The estimated field can therefore be used to determine the equivalent DC field strength. Using eqn. 17.32b we find:

$$E_{pbAC}\big|_{\Delta\to\infty} = \left[\frac{\rho_c^{\delta}}{A}\right]^{1/(\Delta-\delta-1)} = E_{bdc} \tag{17.33a}$$

The equivalence to the DC breakdown field shown above can be obtained by substituting the equivalent power law form of $f(E_a)$ (i.e. $f(E_a) = A^{1/\delta}E_a^{(\Delta-\delta-1)/\delta}$) in eqn. 17.21 which determines the DC characteristic breakdown strength. In terms of the RMS value the breakdown field is

$$E_{bAC}\big|_{\Delta\to\infty} = \frac{1}{\sqrt{2}}\left[\frac{\rho_c^{\delta}}{A}\right]^{1/(\Delta-\delta-1)} \tag{17.33b}$$

So in the case where there is no initial charge density, the ratio of equivalent DC field strength to RMS AC characteristic breakdown strength is $\sqrt{2}$. Note that in practice Δ is less than infinity, the AC breakdown strength is less than that of the impulse limit[166] eqn. 17.33(b) and hence the ratio may be bigger[716,718] than $\sqrt{2}$.

17.6.5 Full-cycle space-charge variation

So far we have considered the case where the space charge reverses sign with the field (i.e. eqn. 17.22). Breakdown was assumed to be a function of the magnitude that ρ attains on each half cycle independently of its sign. Such conditions could be expected to apply to bipolar injection or situations in which carriers of different sign and the same mobility are available for heterocharge formation. However in some cases space charge may build up over one half-cycle and then be neutralised in the succeeding half-cycle. It has been argued that the charge generated by a partial discharge in a void is of this type (note that the polarisation of the polymer surrounding the void is not regarded as contributing to the surface space charge of the void). In this case

$$\rho[t, E_a(t)] = \rho_p[1 - \cos \omega t] \tag{17.34}$$

and we must describe the AC constant stress tests through the probability of survival over a complete cycle, i.e.

$$P_S(\omega t = 0 \text{ to } 2\pi) = \exp\left\{-\left(\frac{\rho_p}{\rho_c}\right)^\delta \left(\frac{c}{\pi}\right) \int_0^{2\pi} [1 - \cos \omega t]^\delta \, d(\omega t)\right\} \quad (17.35)$$

The integral can be approximated for large δ just as in eqn. 17.24 as shown in Appendix 3, and gives:

$$\int_0^{2\pi} [1 - \cos \omega t]^\delta \, d(\omega t) \approx 2^{\delta+1}(\pi/\delta)^{1/2} \quad (17.36)$$

Now however the integral is dependent upon the power δ, so that:

$$P_S(\omega t = 0 \text{ to } 2\pi) = \exp\left\{-\left(\frac{2\rho_p}{\rho_c}\right)^\delta 2\left(\frac{\pi}{\delta}\right)^{1/2}\left(\frac{c}{\pi}\right)\right\} \quad (17.37)$$

Remembering that time must now be counted in units of complete cycles[120] (i.e. $N_\omega = \omega t/(2\pi)$) the derivations of the previous two sections can be carried out in the same way, but now the effective critical charge density is $\rho_c/2$.

If it is assumed that the $(1 - \cos \omega t)$ factor in eqn. 17.24 originates from the applied field it indicates that the space charge is a function of the peak-to-peak field. Thus the resulting RMS impulse breakdown field:

$$E_{bAC}\big|_{\Delta \to \infty} = \frac{1}{\sqrt{2}}\left[\frac{(\rho_c/2)^\delta}{A}\right]^{1/(\Delta-\delta-1)} \quad (17.38)$$

should be compared with eqn. 17.33a as the equivalent DC breakdown field. In this case the ratio will approach $2\sqrt{2}$ when the aging component of ρ (i.e. t in eqn. 17.27) is absent and the time dependence of $\rho[t, E_p(t)]$ is that of the peak-to-peak field, i.e.:

$$\rho[t, E_p(t)] \approx A[E_p(1 - \cos \omega t)]^{(\Delta-\delta-1)/\delta} \quad (17.39)$$

The factor of 2 that converts the equivalent DC field strength to a peak-to-peak value appears because of the power-dependent factor in the integral 17.36. It should be noted that a similar alteration to the definition of $\rho[t, E_a(t)]$ in subsections 17.6.3 and 17.6.4 i.e.:

$$\rho[t, E_p(t)] \approx A[E_p \sin \omega t]^{(\Delta-\delta-1)/\delta} \quad (17.40)$$

will have no effect upon the results obtained previously, because the integral of eqn. 17.24 is independent of the power of $\sin(\omega t)$.

17.7   The increase of DC breakdown strengths at low ramp rates

This effect seems to be a feature of ramp tests on polymeric systems as seen for example in the results of Kitagawa *et al.*[434,37,435], as shown in Fig. 15.6, and of Leach[881] shown in Fig. 17.2. Since the observations imply that the breakdown field increases if the time-scale of the experiment increases it is natural to assume that this is because more time has been allowed for competing processes to reach equilibrium. Under these circumstances processes which are not accelerated strongly by the ramp begin to compensate those components of breakdown that are driven by the ramp.

A variety of compensating processes may occur depending on the nature of the primary component of the breakdown mechanism. For example if the build-up of injected space charge is crucial bulk transport processes have a compensating effect. Alternatively if heterocharge development is the important feature then low-level injection processes may retard the approach to a critical concentration. Below we will briefly discuss two main categories of behaviour.

17.7.1 Charge sweeping by DC fields

In insulators space charges are formed from electrically-charged carriers (electrons and ions) which may be 'trapped' or localised within the dielectric[849]. These space charges may originate from the manufacturing process, for example the high extrusion rates may lead to tribo- and piezo-electrically formed charges. Chemical reactions such as those involved in the stabilisation and crosslinking processes may also lead to carrier generation. Additives and impurities may also be ionically dissociable thereby acting as a charge source, and the operating environment itself may supply charge carriers e.g. from radiation and from electron injection at the electrodes and partial discharges within the bulk.

In a DC field mobile charge carriers will be swept towards the electrodes, where they may concentrate as heterocharges if the interface in blocking[774], or be swept out of the system otherwise[354]. Both mechanisms may serve to modify local fields[166] and space charge concentrations in a DC field and thereby accelerate or retard the approach to breakdown depending on whether the local field is increased or decreased. In the AC case field enhancement is not so important because the charge displacement in a half-cycle is small, and carriers that may be swept out in one half-cycle can be returned in the succeeding one so that there will be very little depletion of the reservoir of mobile carriers, and next to no build up of space charge at the electrodes[354].

The effect of these many processes depends upon the nature of the breakdown process itself. Thus if a high local stress is required to produce a critical local current either in the form of a damage-producing injection current, or as an internal local current near the electrode, then field-moderating impurities will retard breakdown if given sufficient time. Here field-enhanced ionic dissociation increases the local conductivity at high stress sites thereby grading the local field. Homocharge injection below the damage-producing level also moderates the local field and retards further charge injection[287] if given sufficient time to occur. On the other hand the drift of heterocharge from the bulk material will serve to enhance the local field if the electrodes are blocking (as for example with ions). Thus slow ramp rates in this case may either accelerate or retard breakdown depending upon the relative time-scales of the various factors. When injection takes place from a metal asperity it is unlikely that bulk heterocharges will have much effect, and homocharge moderation can be expected to dominate[287,697,808] leading to an increase in characteristic breakdown voltage. However in uniform fields heterocharge build up at slow ramp rates will cause a reduction in breakdown voltage and since the electrodes may be blocking only to carriers of one sign a pronounced polarity effect may occur here.

In the absence of conducting asperities and inclusions charge concentration in the bulk is likely to be the crucial factor. Field-enhanced ionic dissociation may then increase the local space charge density as well as moderating high stress sites. Such charge carriers will be progressively released from traps[287] as the electric field is ramped up and the potential barriers are reduced by the Poole–Frenkel mechanism[149] (Section 9.3.3). Because they are likely to be re-trapped their average drift velocity will probably be low despite the presence of relatively high fields and even high mobilities. Thus their overall mobility is probably trap limited. At high ramp rates the carriers will not have time to be swept through the system before breakdown occurs. However at low enough ramp rates all but the space charges contained in the deepest traps escape[865] and can be swept from the system thereby relieving the space-charge concentration and retarding the approach to breakdown.

Precise calculations of the relative movement of charge under the different test conditions are difficult to carry out as the precise bulk and microscopic drift mobilities are not well known. In particular the spatial positions and depth of the traps, which are critical, are probably highly dependent upon impurities within the material and also its structure. There are also likely to be enormous variations in both these parameters dependent upon the manufacturing processes used for the material and the final product. It is not surprising then that there is a large range of trap-limited mobilities for polyethylene quoted in the literature. For example Zeller et al.[67] quote a figure as high as $\mu_n < 10^{-8} \, \text{m}^2 \, \text{V}^{-1} \, \text{s}^{-1}$ whilst Lewis[882] suggests a range of 10^{-11} to $10^{-14} \, \text{m}^2 \, \text{V}^{-1} \, \text{s}^{-1}$ for μ_n. Lewis notes a 'super-Ohmic' behaviour with increasing field which can be attributed to the lowering of potential barriers surrounding traps due to the Poole–Frenkel effect. It is therefore reasonable to assume that the effective mobility in the case of breakdown tests on polyethylene is likely to be at the high end of the range he quotes as the field reaches very high levels.

If a value of $10^{-11} \, \text{m}^2 \, \text{V}^{-1} \, \text{s}^{-1}$ is taken as the trap-limited (effective) value of mobility then the average distance moved by the charge in the examples of polyethylene cable failure considered in Fig. 17.2 would be 0.18 mm for the fastest ramp rate ($5 \times 10^3 \, \text{V} \cdot \text{s}^{-1}$), 1·16 mm for the intermediate ramp rate ($1·1 \times 10^3 \, \text{V} \cdot \text{s}^{-1}$), and 12·6 mm for the slowest ramp rate ($15 \, \text{V} \cdot \text{s}^{-1}$). As the cable insulation was approximately 6 mm thick, it is reasonable to assume that most of the charge would be swept out in the case of the slowest ramp rate whilst very little would be swept out in the case of the fastest ramp rate.

These figures however should be treated with some caution as the distribution of drift velocities is likely to be wide and the value taken for the mobility may be incorrect. For example if the same value of mobility ($10^{-11} \, \text{m}^2 \, \text{V}^{-1} \, \text{s}^{-1}$) is used for the thin films of Kitagawa et al.[434,435] (Fig. 15.6) then it can be shown that space charge would be completely swept out before breakdown occurred for both the 10^2 and 10^4 V · s^{-1} ramp rates. However it can be seen that in this case the retardation effect is only noticeable at 10^2 V · s^{-1} suggesting a much lower mobility of approximately 10^{-13} m^2 V^{-1} s^{-1}. This may be reasonable as one would expect thin film specimens to be dominated by surface effects where there are likely to be high concentrations of deep traps.

When breakdown is determined by the build up of a heterocharge which eventually causes a destructive injection current as for example in the avalanche breakdown model of Klein[146] charge transport from the bulk material at low ramp rates is likely to accelerate rather than retard break-down. It has been suggested by Kitagawa *et al.*[434] that slow ramp rates allow more homocharge injection which moderates the local field and thereby retards breakdown. In Section 15.5 it was shown that, in the absence of space charge fields, inception of electric trees by slow ramps occurs at lower fields (see Fig. 15.23) however slow ramps would allow compensating homocharge injection below the damage limit[287] and it is unclear which of these many factors give rise to the behaviour observed[434] in Fig. 15.6.

Overall the picture developed in this section is one in which several competing processes, namely: homocharge injection; injected charge trans-port; heterocharge transfer from the bulk material; sweeping out or blocking of charge at electrodes; all combine to govern the time-scale of the approach to breakdown. In order to determine the relative magnitudes of these processes detailed information will be required for each particular case, which is often not available. Most probably breakdown in cable sections is governed by space charge concentration with bulk transport being the major competing process. In the case of thin polyethylene films the situation is not clear cut[434] and the low dielectric thickness may allow several factors to play a role here.

17.7.2 Spatial distribution of space charge

If the distribution of the space charge is not spatially uniform then fast ramp rates may produce different space-charge structures than slow ramp rates. Such an alteration in the spatial distribution of the space charge will change the local electric fields with the more spiky structures produced by fast ramps having higher field enhancement than the more uniform structures produced when slow ramps allow sufficient time for full space-charge rearrangement to occur. Bahder *et al.*[120] have suggested that this type of behaviour applies to the distribution of space charge in electric trees and tree-like structures generated by discharging voids (Chapter 13).

In this case very close tree branching (as in bush-type trees with high fractal dimensions) will decrease the density of the space charge at the ends of the individual branches thus causing lower electrical stresses at the branch tips. In addition there is the mutual shielding effect of the branches that are in close proximity. The rate at which space charges propagate along the branches of the tree will depend upon the branch diameter and shape. Thus at high rates of voltage increase the depth of charge propagation within some branches will be much greater than within others; so that even in a tree with a high density of branches their mutual shielding may not be 'fully developed' and a higher field may occur at the branch ends compared to that existing in the slow ramp case. Furthermore the channels are 'capacitor like' so that a high rate of voltage increase implies that a higher current will flow in them. Therefore bush-type trees of high fractal dimension ($d_f \approx 2\cdot5$) may act as branch-like trees of low dimension ($d_f \approx 1\cdot6$) under a rapidly ramped voltage, and cause breakdown at lower fields (see Section 15.3 and Fig. 15.13) than those appropriate to slow ramps. When breakdown is the result of local fields generated by tree-like formations therefore, the breakdown voltage of polymeric insulation may be lower for DC voltages with short rise times than for voltages with long rise times.

Chapter 18

Cable assessment procedures

18.1 Cable validation tests

Validation tests are those carried out on each manufactured article before it leaves the manufacturer. Unlike the characterisation tests discussed in Chapter 16 there are no internationally agreed validation tests since each article is different. Generally customers and manufacturers will agree both a specification for the article and the tests to check that the specification is met. In this book we restrict our discussion at this point to cables as these form a major class of products using high-voltage polymer insulation and also because they fall within the direct experience of the authors. Various validation tests are widely used for cables although not all manufacturers worldwide use them.

The most common cable validation tests are voltage-withstand tests and partial discharge tests. These are briefly described in this chapter with some comments on their efficacy.

18.1.1 Voltage withstand tests

Voltage-withstand tests are the standard 'proof' tests carried out on completed cables. The AEIC describe voltage-withstand tests for polyethylene and crosslinked polyethylene insulated shielded cables in Reference 818. These documents give tables of voltages to be applied and other details depending on the cable ratings and insulation thickness. To a reasonable approximation they suggest a test voltage which corresponds to an average field of 3·4 kV/mm AC or 8·5 kV/mm DC across the minimum average insulation thickness for one hour.

Some British manufacturers advocate proof tests which are designed to give the same probability of failure of the cable as it would have in the first 25 years of service. The tests are often usually designed to last about one minute. Assuming a 2-parameter Weibull distribution of cumulative probability of breakdown, eqn. 14.19b:

$$P_F(V, t) = 1 - \exp\{-Ct^a V^b\}$$

then for this to be the case:

$$t_p^a V_p^b = t_s^a V_s^b \tag{18.1}$$

where the subscripts p and s represent proof test and service conditions respectively. A value of b/a is required to calculate an appropriate test voltage. The data given in Section 14.3 suggests values for polymers of 14·4 (Fig. 14.8b)[177], 12—13 (Fig. 14.11)[706], 6·8 (Fig. 14.12)[174], or 15·4 (Fig. 14.14)[702]. For a value of $b/a = 15$, which may be typical of a polymeric power cable, this suggests that a voltage of three times the rated (phase-to-ground) voltage would be suitable. Typical cables have working voltage stresses of 1·8—2·0 kV/mm so this proof test implies test stresses of up to about 6 kV/mm (AC). Note however that the ratio of V_p/V_s is highly dependent upon b/a. For example if b/a is only 7 (and it has been suggested it might be as low as 4) then a proof test voltage of about ten times the service stress would be appropriate.

The level of proof testing is therefore controversial. Its statistical basis is fraught with assumptions which are difficult to corroborate. As well as being unsure of the exponent b/a, no account is taken above of any threshold voltage which may exist and finally any degradation which does take place in the short time of the proof test is almost certainly entirely different from that which would take place over the much longer time period in which it is in service.

Aging the cable using a proof test may or may not be desirable. This can be understood by considering the breakdown hazard function of time. Fig. 14.1f shows this for the Weibull distribution for different values of the a parameter. The hazard function gives the most-likely failure rate and it can be seen that this decreases with time if $a < 1$, is constant if $a = 1$, and increases with time if $a > 1$. Whilst the proof test is primarily designed to verify the efficacy of the product, it should not significantly increase its failure rate. For a typical proof test the equivalent aging time is likely to be less than about 0·1 on the normalised time scale. Most cables are likely to have values of $a < 1$ and so an initial aging caused by a proof test has the desired effect of decreasing the subsequent failure rate. For values of a just above unity $(1 \leqq a \ll 2)$ the hazard function increases rapidly from zero in this initial time period and tends to stabilise at about one. Proof testing increases the subsequent failure rate in this case but because the hazard function is reasonably stable it is still a useful technique to find weak products and it will not be very deleterious in the long term. If the value of a is high, $a > 2$, then proof testing will not be useful in detecting weak samples as they are not likely to fail during the test and are much more likely to fail subsequently. In these cases, which are rare in polymeric insulation, a proof test is a positive menace.

18.1.2 Partial discharge tests

It is beyond the scope of this book to give details of the experimental arrangements used for detecting partial discharges. Excellent critical reviews on the various methods available for partial discharge detection and measurement in terms of their chronological development and applications for cable, capacitor, transformer, and machine insulation specimens have been given by Bartnikas[649]. The relevant standards and widely-used test procedures are given in References 818, 883–889. A useful summary is also given in Tanaka and Greenwood[166].

(a) Partial discharge detection

The principle of partial discharge detection in polymeric insulation generally involves applying a power voltage across a specimen and observing sudden changes in current through it, i.e. at frequencies much greater than the power frequency. Other methods have also been employed, such as 'listening' for partial discharges acoustically or with a radio receiver or observing light emission, but these are not widely used for polymeric insulation. The simplest method is to place an impedance in series with the specimen and to observe the voltage across the impedance using an oscilloscope with appropriate protection, buffering, filtering, and amplifying circuits.

Many refinements have been made to this basic technique mainly to improve the signal-to-noise ratio. The most common improvement is to use a bridge circuit in which the series specimen–impedance combination is balanced by an equivalent combination in which the equivalent specimen circuit is known to be discharge free. In the case of cable specimens this latter circuit is usually another piece of cable. A differential amplifier is then used to monitor the voltage across the bridge and partial discharges are readily observed. Noise introduced on the power line is cancelled out using this technique and, if the system is well balanced, it may not be necessary to have a high-pass filter to remove the effects of the charging current through the specimen.

Various types of impedances for detection have been used, the most common for a capacitive specimen being a resistor or series resistor–inductor combination. The former is the most straightforward since the voltage across the impedance is directly proportional to the current through the specimen and may therefore be the most useful for investigating the nature of the partial discharge. However pure resistors have a constant frequency response and therefore do not eliminate many types of noise. Furthermore, unless a bridge circuit is employed, it will be necessary to filter out the charging current component through the specimen.

Alternatively reactive impedances can be used such that the specimen forms part of a resonant circuit which is caused to oscillate by the harmonics of the current steps arising from partial discharges. For a capacitive specimen such as a cable, the impedance would have an inductive reactance. The technique inherently amplifies the effects of the partial discharges whilst blocking the effects of the power charging current. The circuit may be calibrated using calibration pulses from a pulse generator. Whilst the circuit is more sensitive than those which use a purely resistive impedance, it is more useful for routine industrial measurements than a research/development tool since the shape of the observed waveforms are more difficult to relate to the physical origins of the partial discharges. It is mainly used with low capacitance specimens such as small capacitors and transformers. Since the resonant circuit requires a high gain at its resonant frequency in order to be sufficiently sensitive, the experimental configuration may also be susceptible to radio frequency interference. This may be overcome to some extent using a balanced bridge arrangement. Typically the circuit is set to resonate at 30—70 kHz for polymeric insulation and should have a sensitivity of at least 5 pC[890] for industrial use. The time resolution of such partial discharge detection systems is generally limited by two factors: (i) RCL circuits have slowly exponentially decaying oscillations which may give overlapping discharge patterns if there are a lot of successive partial discharges; and (ii) cables are usually tested without matched terminations, and since they act as transmission lines this may limit the pulse resolution time to a few microseconds.

A wide variety of techniques are used to locate the source of partial discharges and are summarised in References 166 and 649.

For unscreened cables in production a continuous scanning system may be used in which the cable is drawn through a tubular electrode arrangement such that a high voltage is applied to a short length of the cable. The electrodes are also used to detect the partial discharges. This allows the fault position to be determined to within about 30 cm. For shielded cables Baghurst[891] has demonstrated a technique in which extra partial discharge activity is stimulated by illuminating successive sections of the cable with X-rays.

A more popular method for cables is to use travelling wave methods[892] in which pulses are applied to one end of the cable which acts as a transmission line, the time delay between the onset of the pulse and the observation of partial discharges may be used as an indication of their position. Conventionally this has been restricted to specimens with only one discharge site, however Beyer *et al.*[893] have used digital correlation techniques to extend this to specimens with many discharge sites. It may be that, with the advent of digital signal processing (DSP) IC technology, such techniques will become standard.

Computer-aided data processing is also becoming more common in partial-discharge instrumentation. As well as displays of repetition rate versus discharge magnitude, previously carried out using multichannel analysers, statistical analyses of the data is readily available[166,894] including statistical techniques to improve the signal-to-noise ratio. It is also likely that these techniques will improve current understanding of partial discharge phenomena.

(b) Routine partial discharge testing

In Chapter 13 it was shown that it is not possible to assess life expectancy from partial discharge measurements and so the standards give a 'go–no-go' approach. They specify permissible levels of partial discharge pulse magnitudes at various levels of overvoltage. For example AEIC 5–75[818] gives a partial discharge standard for extruded cables and the test apparatus and calibration procedure is given in Reference 890. AEIC 5–75 states that cables should be tested at overvoltages corresponding to 1·5, 2·0, 2·5, and 3·0 times their rated line-to-ground voltage provided this overvoltage is not greater than their voltage-withstand voltage (corresponding to an average field of 3·4 kV/mm AC, see above). Since these cables are usually used to transmit three-phase power, the line-to-ground voltage is equivalent to their rated (phase-to-phase) voltage/$\sqrt{3}$. At each level of overvoltage the maximum permissible partial discharge level is given in pico-coulombs by $5 + 30\,(R{-}1{·}5)$ where R is the ratio of the overvoltage to the rated line-to-ground voltage.

In order to estimate the likely deterioration of insulation due to partial discharging, frequency acceleration has been advocated in the past under the premise that the pulse discharge energy expended per cycle at the elevated test frequency is the same as that at the power frequency. However, more recently, pulse-height analysis has demonstrated[895] that the size of the partial discharge diminishes with frequency so that the aging rate is proportional to a power of the frequency with an exponent less than unity[649]. The interpretation of such frequency-acceleration tests is therefore not straightforward.

18.2 Lifetime prediction of cables subject to water tree degradation

The method of lifetime prediction or, more correctly, the estimation of the probability of failure as a function of time, described in Chapter 16 takes as one of its premises that the breakdown characteristics of the insulating system (e.g. the Weibull characteristics C, a, and b in eqn. 14.19) do not vary over the life of the cable. However in practice various low-level degradation mechanisms may cause changes in these parameters which lead to an increase in the probability of failure after any given time. For example it is known that the breakdown properties of nitrogen-crosslinked polyethylene cable insulation are substantially better when first made than they are after a few months. This is at least partly due to the outgassing of nitrogen and crosslinking reaction byproducts, which have good electrical breakdown characteristics, from the free volume of the insulation, thereby reducing the partial discharge inception voltage.

However the primary mechanism of long-term low-level degradation in polymeric power cable insulation is water treeing. Water treeing deteriorates the cable's breakdown characteristics[175] and failure may eventually occur via the generation of electrical trees[177] from the end of the water trees[119,134,137,276,408,409,411–414]. Whilst water tree degradation and indeed cable dielectric breakdown are claimed not to be major causes of modern cable systems failure (e.g. Bjørløw–Larsen *et al.*[184]) it is very important to be able to accurately predict their behaviour in order to avoid 'over-designing' cables with too great a safety margin and which would therefore be uneconomic to produce.

18.2.1 Considerations for predicting water treeing

In principle it should be possible to quantitatively predict the effect of water tree growth on the breakdown characteristics of cables from accelerated laboratory tests and thereby establish a 'lifetime'. In practice it may be difficult to do this accurately, unless cable samples are used in the tests, because of the significant differences in morphology and inhomogeneities (both in the bulk and at electrodes) between moulded laboratory specimens and cables. In order to assess the extent of deterioration it is usual to estimate the time taken for the longest tree in the system to exceed a critical length. Since it is known that water trees can span the insulation without the immediate onset of breakdown (Section 6.1, e.g. Bahder *et al.*[363], Karner *et al.*[408]) this critical length may be either less than or even equal to the thickness of the insulation. In order to carry out such an analysis it is necessary to estimate the length of the longest tree as a function of applied field, time and the many other parameters which may or may not vary during the course of the cable lifetime. The main influencing parameter which will vary is temperature due to changing load conditions and consequent conductor heating as well as environmental temperature changes, whilst frequency; mechanical strain; electrode type and shape; concentration and distribution of electrode surface inhomogeneities; bulk material inhomogeneities; and material additives, which may remain reasonably constant, could nevertheless be used as accelerating parameters in tests. For example, needle electrodes may be used in the laboratory to reduce the inception time, t_d, and scale parameter, t_1, but care needs to be taken as this may also reduce the maximum length of growth[206]. It is interesting to note that when using a point–plane geometry, tree growth continuously reduces the field enhancement at the tree tip as the increasing size of the tree reduces the field divergence. However for the case of a tree growing from the outer conductor of a coaxial cable, whilst this divergent field enhancement is initially reduced, after a certain tree length the weaker divergence of the cable will predominate and stress enhancement may *increase* as the tree approaches the inner conductor. This will maintain the tree growth and favour finger-like trees rather than bushy shapes (Section 4.2.3). This difference may be critical for cables and could be the reason for the differences in the shape and effect of trees growing in them.

An alternative to attempting to grow 'longest' trees and measuring their length as a function of time and other relevant parameters is to establish the distribution of tree lengths and their number density (per unit area in the case of vented trees or per unit volume in the case of bow-tie trees) in terms of these parameters and then to use a censoring technique to estimate the average length of the longest tree. Such an analytical technique has been proposed by Rowland *et al.*[229]. Whilst it is difficult to take into account analytically the daily, yearly, and random variations of temperature and other factors which affect the parameters controlling the rate of growth (eqn. 4.5), it is possible to accommodate these features by means of a complementary numerical technique which is briefly discussed later. In either case it is necessary to establish a function which describes the probability density distribution of tree lengths at any particular time. Experimentally it has been shown that bow-tie trees show an exponential length distribution[238] whereas for vented water trees the distribution is peaked[235]. In both cases a reasonable description of the distribution of lengths is obtained using the two-parameter Weibull function[700] where the probability of finding a tree of length less than or equal to length L is given by:

$$P(L) = 1 - \exp\left\{-\left(\frac{L}{L_c(t)}\right)^\alpha\right\} \qquad (18.2)$$

where L_c is the characteristic tree length and is a function of time, and α is the shape parameter for the distribution. (See also Section 4.2.2c.) The two-parameter Weibull distribution has been used on a purely empirical basis in that it has been found useful to describe the data. For example Fig. 4.8 shows vented water tree data and it can be seen that the distribution function gives a good fit. For the case of bow-tie trees the exponential distribution becomes the special case of eqn. 18.2 with the shape parameter $\alpha = 1$. In practice the shape parameter is found to be independent of time and of the order 2 to 5 for vented trees with some but only limited variation from sample to sample[234]. The value of L_c, the characteristic tree length may be determined using eqn. 4.5 once suitable experiments have been carried out in order to establish the parameters.

The longest tree is therefore proportional to the characteristic tree length in a ratio which is a function of the shape parameter, α, in eqn. 18.2 and the probability $P(L)$ which for the longest tree in a total of N trees is approximately $(N - 0.3)/(N + 0.4)$. Experiments therefore need to be carried out in order to estimate the total number of trees, N, present in the system which is itself a function which increases with time[229]. However, perhaps surprisingly, the ratio of the average length of the longest tree to the characteristic length (L/L_c) is not highly dependent on the total number of trees for typical shape parameter values of $2 < \alpha < 5$. This is shown in Fig. 18.1 from which it can be seen, for example, that for a shape parameter of $\alpha = 3$, the ratio L/L_c only changes from 2·1 to 3·0 with a change in the total number of trees, N, from 10^4 to 10^{12}. So that in a typical 1000 km length of cable, containing perhaps $N = 10^9$ trees, the actual value of N does not need to be evaluated very accurately. Tree concentration, i.e. the number of vented water trees per unit length of cable may however be important since this varies greatly along the cable. So that when estimating the water

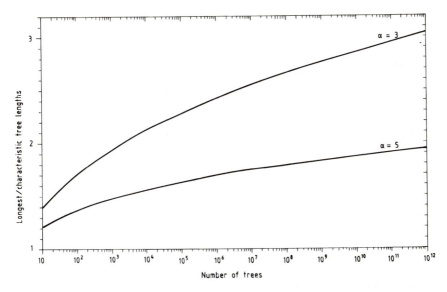

Fig. 18.1 Assuming a Weibull distribution of vented tree lengths it is shown that the ratio of the maximum tree length likely to be found as a function of the characteristic tree length is only weakly dependent upon the total number of trees for $N > 10^4$.

tree lifetime for 1 km we really need to know the maximum density rather than the average[172,239].

Rowland *et al.*[229] have shown that, using the deterministic growth law of eqn. 4.5 and assuming a form appropriate to the field-squared dependence of growth rate discussed in Section 4.2.2*b*, the value of the shape parameter, α, for $\alpha > 1$, did not affect the *time* taken for the longest vented tree to reach a critical length as long as the value of N was large ($N > \sim 1000$). In practice the number density of vented water trees is unlikely to be uniform along a section of cable insulation[239] and under these circumstances the value of α may assume more importance. On the other hand the number density of water trees becomes of increasing importance when $\alpha \lesssim 1$ such as may apply to bow-tie water trees. These results reflect the different behaviour of vented and bow-tie trees as regards breakdown. The results of these model calculations[229] are summarised in Fig. 18.2.

18.2.2 Accelerated testing techniques

Accelerated testing techniques are generally used to predict the long term behaviour and may be used to establish both the tree length distribution and the number density of trees. Methods for accelerating tree growth include the use of increased electrical stresses, temperatures and electrolyte concentration. In order to compare insulating materials, needle-shaped voids filled with salt water are sometimes used as electrodes to produce sites

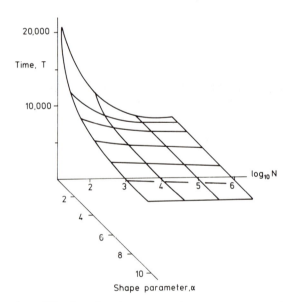

Fig. 18.2   The time (T) taken for the average longest tree to attain a given length (L). The shape of the curve is independent of L as long as $T \gg t_1, t_d$ (Fig. 4.3*a*). Here $t_1 = t_d$ for convenience. Note that N is more effective in reducing T in the case where $a \lesssim 1$ which is the situation for bow-tie trees[229]. After Rowland *et al.*[229] Copyright © 1986 The IEEE

of locally-high electric fields from which water trees start to grow immediately and rapidly. Whilst the length as a function of time from initiation has the same form for point-plane specimens and uncut cable specimens (Section 4.2.2) the technique gives virtually no information about inception times. Since it should be the aim of cable manufacturers to make the inception time longer than the cable life, uniform-field plaques may be better for placing materials in order of merit. The distribution of tree lengths as a function of time may be completely different in water needle tests and in service conditions. There is however some theoretical and limited experimental evidence in support of an acceleration technique using increased frequencies which may overcome both these problems. It is this technique which we discuss in more detail here.

The reason why the above acceleration techniques cannot be used for the purposes of accurate prediction is that the variable used for acceleration (temperature, electric field, mechanical strain, material additives, artificial experimental configurations such as needle electrodes and surface scratching) tend to produce different growth acceleration factors in the different stages of tree growth[206] (inception, rapid growth, long-time growth). For example an increase in temperature leads to an increase in the inception time but a decrease in the time taken to grow to a specified length thereafter. Increasing the electric field produces a very marked affect on the inception time ($t_d \propto E^{-n}$, $n \approx 4$) but the subsequent initial growth rate has a much weaker field dependence lying between a linear and square proportionality. Experiments which use needle-shaped water-filled electrodes or scratches on the electrode–insulator interfaces, aim to start tree growth immediately and are therefore incapable of giving any information about the inception time.

In the case of frequency being used as the acceleration parameter however, there is some experimental evidence that all three growth stages may be affected in a similar manner, at least up to a few kilohertz[206,234,246]. This would lead to the conclusion that increased frequencies may produce true linear acceleration so that $P(L) = f(\omega \cdot t)$. In many cases this experimental evidence has been inferred from experimental results designed to show other effects. It has been collated from a variety of workers using different measurement and presentation techniques. It is therefore rather weak and the results collated in Reference 206 show trends rather than definitive results. Most workers have not taken great care to ensure that their power supplies were free of distortion, (for example due to resonant effects), this is very important as even quite small amplitude harmonics need to be eliminated especially those of higher order. Assuming that water tree growth rate is proportional to frequency and a power of voltage and that superposition holds then the error caused by harmonics is:

$$\frac{\sum_{i=1}^{i=\infty} (f_i V_i^n) - f_1 V_1^n}{f_1 V_1^n}$$

where the subscripts refer to the harmonic number (1 is the fundamental),

f and V are the frequency and amplitudes of the harmonics, and n is the exponent of the power of the voltage which is proportional to tree growth rate. In the case of inception $n \sim 4$ and during growth $n \sim 2$. Since the amplitudes of the harmonics are likely to be less than that of the fundamental the smallest likely value of n should be taken to give the worst case situation. In practice it may be a good idea to assume $n = 1$.

A lack of appreciation of the distribution of tree lengths and other parameters may also have confused the findings. For example in practice, spatial inhomogeneities, such as local variations in polymer morphology, will lead to variations from tree to tree in the growth characteristics, such as inception time. This will lead to the existence of a distribution of tree lengths which changes with time[234].

Whilst the experimental evidence justifying the use of high frequency acceleration is not strong, there is a reasonably strong theoretical foundation for expecting such a technique to be useful. Summarising the theory of tree growth presented in Section 4.3.3 it can be seen that all phases of tree growth may be linearly accelerated by frequency. The inception of vented water trees is envisaged to follow the following stages:

(i)   Solvated ions from an external aqueous reservoir are forced into the interface by the action of an AC field. (Solvated ions of different signs have their attractive forces reduced and so are more easily separated whereas those with the same signs have their repulsive forces reduced and so can pack into a boundary layer.)

(ii)   Since the polymer is hydrophobic it tends to prevent the ingress of water droplets but it can trap desolvated ions particularly in the oxidised regions. This process is favoured if the ions are easily dehydrated (i.e. desolvated).

(iii)   Counter-ion ingress now takes place since, when dehydrated, the attraction between ions of opposite sign is increased. Thus ions tend to be trapped as pairs of opposite sign thereby forming a neutral group of ions rather like a weakly bound salt. (Note however that these are *not* salt crystals and the process cannot be simulated by salt particle seeding.)

(iv)   In the AC field the electrostatic forces between the ions cause repetitive local mechanical stresses.

(v)   In the AC field, electrochemistry may occur which leads to oxidation.

(vi)   In the AC field water molecules dissolved in the polymer may condense onto an ion group.

(vii)   Processes (iv), (v), and (vi), lead to the formation of cavities and strain in the region around the cavities.

(viii) The field enhances the water condensation into newly-formed cavities.

(ix)   The strained regions allow transport paths for solvated ions and so the process repeats.

During this transformation, structures greater than a critical size must be formed so that growth proceeds through nucleation on the phase boundary. This gives rise to a time lag, the inception time, before steady-state conditions are established during which break-up and aggregation of smaller structures is occurring. When the nucleated phase (water tree) is stabilised by a bulk free energy density proportional to E^2, the time lag has been calculated by Kaschiev[896] to be of the form:

$$t_d \propto f^{-1} E^{-4} \tag{18.3}$$

so it can be seen that the inception time is theoretically predicted to be inversely proportional to frequency. The application of Kaschiev's work requires some justification since he shows that a water-filled cavity is *less* stable than an air-filled one in an electric field whereas, in fact, the opposite is true[897]. However, since only the *sign* of the free energy change is reversed, but ΔG itself is still proportional to E^2 (assuming that the attempt rate is determined by the repetitive nature of the AC field) it should be possible to apply Kaschiev's theoretical expression for the time lag for nucleation of the favoured state (i.e. water in cavities) to get eqn. 18.3. This is similar to Zeller's idea[262] of the water tree as a favoured microphase (liquid water in cavities and associated strain and easy transport routes) compared to the water (molecular or droplet) and ions dissolved in the polymer phase which is homogeneous. From the picture of water tree inception given above the condensation of water (present as a solution of molecular vapour in the polymer) into cavities where it takes the form of a liquid is only part of the overall process. The details of Kaschiev's calculation may not apply unless each individual process is governed by an E^2 free energy density term. However all the processes are repetitive and one should always have $t_d \propto f^{-1}$ as a *leading* term.

At the end of the inception stage, the water tree is growing with a field-dependent steady-state nucleation rate. However as the tree penetrates the insulation interior, it must overcome the mechanical resistance of the polymer to penetration. It does this by exerting a strong local Maxwell stress, in phase with the applied field, causing higher density regions to deform and permitting entry of ions from the tree front prior to mechanical relaxation. Subsequent solvation reconnects the ions to the tree, thereby advancing the phase boundary. The growth rate, dL/dt, for this mechanism may be represented[137] as:

$$\frac{dL}{dt} \propto \omega \chi'(\omega) E^2 \tag{18.4}$$

where χ' is the real part of the complex dielectric susceptibility. So, during growth, the growth rate, dL/dt, is theoretically predicted to be proportional to $\omega \cdot \chi'(\omega)$ and since χ' is reasonably independent of frequency over the frequency range from a few hertz to a few kilohertz[137], the growth rate during this stage may be expected to be directly proportional to frequency. Since both nucleation and growth of the stable water tree phase are driven by the repetitive nature of an AC field, the frequency acceleration of inception and growth rate will be the same ($\propto f$) independent of the detailed mechanism given above.

Frequency acceleration has the further advantage that, within the test procedure, the simulation of periodically varying environmental conditions (such as temperature of the inner conductor of a power cable reflecting the daily variation in load) may be directly incorporated.

In conclusion there is considerable theoretical and some limited experimental evidence in support of the use of frequency as a linear accelerating parameter for water-tree degradation. Parameters other than frequency are not likely to produce representative aging accelerations since they accelerate tree growth in the different stages in different ways.

18.2.3 Modelling water tree growth

If the constants in the empirical growth-rate equation (eqn. 4.5) are known one can, in principle, numerically integrate this to establish tree length as a function of time. This can then take into account the effects of cycling of temperature and other parameters on the various constants and the effects of field enhancement at the tip.

In order to do this a form of diurnal temperature variation must be assumed and the resulting temperature gradient through the cable calculated. The field at the end of the tree needs to be calculated as it grows through the naturally diverging field of the coaxial cable. This could be carried out most simply by assuming the water tree to be conducting, or perhaps of a higher permittivity[137]. The inception time is particularly difficult to calculate for a temperature varying situation. Equations for the rate of growth are given in Reference 206 together with some estimates of the parameters for various experimental conditions.

It would also be necessary, if using this technique to estimate lifetime, to estimate the longest tree of an ensemble of such trees. So for example one might consider the longest tree to be approximately three times longer than the characteristic (e.g. Fig. 18.1).

Such a numerical technique may be useful for verifying that parameters estimated from experiments are indeed reasonable and that the tree growth characteristics measured are similar to those predicted. In this way it may be possible for parameters evaluated from experiments on particular materials to be extended to other systems of the same material with some expectancy of a reasonable result.

18.2.4 Discussion

When using any technique to quantitatively assess the resistance of an insulating system to water tree degradation it is important to use the actual insulating system itself where possible. Probably the simplest and most efficient technique to predict water tree growth in power cables would be to use frequency as the accelerating parameter. Up to about 10^3 Hz in polyethylene the parameters t_d, $(1/k)$, and t_1 all show an inverse proportionality to frequency whereas the parameters n, m, and L_∞ are all reasonably independent of frequency[206]. In the range 10^3—10^5 Hz in polyethylene, there appears to be considerable piezoelectric activity possibly associated with a mechanical loss dispersion[898] which may be responsible for the peculiar non-linear and often non-monotonic frequency dependence of the parameters in that region. Therefore it would not be useful to exceed about 1 kHz in such a test. The effects of temperature cycling could be incorporated into such a test again at a proportionally increased rate. Here however the situation is unclear and appears to be dependent upon the experimental context[119] (see Section 7.3). Thus although it seems that in the case of vented trees higher temperatures are likely to increase the inception time for the appearance of trees, the propagation rate constant, Q, does not necessarily behave monotonically. Straightforward temperature acceleration is therefore not feasible, although the technologically-important condition of temperature cycling would appear to generally reduce the time taken to grow bow-tie trees of a particular length. Electric field acceleration is a feasible approach but has the drawback that the various parameters have different field dependencies and also electrical trees may initiate from the surface and thus cause premature breakdown. In any such approach it would be necessary to establish the distribution of lengths and density of water trees in order to estimate the length of the longest tree. Bulinski and Densley[175] have given a measured distribution of vented tree lengths for cables as a function of voltage at constant temperature. Yoshimitsu *et al.*[231] have measured this distribution at different temperatures for wire-sandwich configurations. Various tree length distributions showing peaks for vented trees have also been observed by Doepken *et al.*[115]. Whilst the authors would advocate a systematic frequency-acceleration approach it should be pointed out that they know of no experimental results which conclusively demonstrate its applicability to real systems. That is to date that they know of no results showing the distribution of water tree lengths as a function of frequency in which the temperature has been either maintained constant or cycled at a rate proportional to the electric field.

Chapter 19
Detecting electrical degradation non-destructively

Laboratory and factory test programmes must, of necessity, be carried out under conditions which accelerate the pre-breakdown and breakdown processes, such that they can be completed in a reasonable time-scale. Even the most comprehensive programme therefore can at best yield only good estimates of the parameters defining the breakdown statistics under the chosen conditions. Prediction of the system life-time under operating conditions requires extrapolation of the test results, a procedure that relies for its validity on the assumption that the failure mechanism remains the same with an unchanged distribution and is merely slowed down. Even granting that these assumptions are valid the accuracy of the predicted lifetime depends crucially on the accuracy to which the distribution function can be determined in the test programme. For example percentage differences in the Weibull parameters *a* and *b* can lead to order of magnitude changes in lifetime on extrapolation to service voltages. In addition such prediction can only be made on a statistical basis, i.e. the number of units failed by a given time, or the probability of failure of a large composite system. This type of analysis cannot determine which particular unit or system component will degrade most rapidly and hence give the earliest failure.

It would clearly be of advantage to manufacturers and users to be able to detect the extent of degradation in individual units and hence determine those at imminent risk of failure so that they can be replaced. Such an ability is highly desirable in cases where the extrapolation assumptions are invalid (i.e. systems where low-field degradation processes dominate at service voltages) and would still be useful even when extrapolation is acceptable. For this reason some effort has been expended on trying to develop methods to detect electrical degradation *in situ*[126,899–901]. A review of those test procedures currently accepted in the industry is contained in References 166, 649 and 902–904 and the reader is referred to them for details. Here we will be concerned only with a selection of the methods proposed for the detection of the degradation processes covered by this book. The list is not intended to be exhaustive but rather to illustrate the types of approach adopted and to assess their effectiveness.

19.1 Water trees

As we have noted in Chapter 4 the water present in water trees produces an increase in the dielectric loss[244,137,134,905] at frequencies below about

10^3 Hz. This effect is clearly distinguishable from that due to water droplets which do not form part of a water tree, since this latter case gives a loss peak around MHz frequencies[906,907] due to Maxwell–Wagner processes. It would therefore seem possible to utilise dielectric measurements in the low frequency regime to identify cables containing water trees[235]. Furthermore since it appears that the magnitude of the loss ($\propto \varepsilon''$) increases with the time of aging[134,905] such measurements should reflect the extent of water tree degradation, and hence make it possible in principal to identify regions of insulation which are in a hazardous condition.

The dielectric loss of high voltage insulation is most conveniently measured via the loss tangent $(= \varepsilon''(f)/\varepsilon'(f))$ at a chosen frequency. However water trees, even long vented trees occupy only a very small fraction of the insulation volume, and it may not be possible to detect the resulting small differences[222] in loss tangent with sufficient accuracy to identify dangerous regions. For example experiments carried out by Swingler *et al.*[899] on cable samples in the laboratory show that while it is possible to detect water tree aging in highly contaminated silane crosslinked polyethylene cables, this was not possible with steam crosslinked samples even under these favourable conditions.

An alternative method proposed by Isshiki and Yamamoto[908] is to measure the discharge current ($i(t)$) following DC charging at a known voltage V. The amount of charge Q that relaxes during a given period is determined by integrating the current and this is compared to the static capacitive charge of the insulation to yield a degradation index (D.I.) with

$$D.I. = \frac{100Q}{CV}\% \tag{19.1}$$

where C is the static capacity. When the period of integration (2—30 s) is chosen to cover the frequency range ($f \equiv (2\pi \cdot \text{time})^{-1}$) in which water treeing produces an increase in the dielectric loss, the degradation index should be a measure of the extent of water tree degradation in the system[909]. Preliminary investigations[908] demonstrated that water tree aged XLPE cables steadily increased their D.I. from 1% (virgin samples) to ~20% (8 months aging at 70°C). Laboratory tests by Swingler *et al.*[899] showed that this method was more sensitive than the loss tangent measurement, although the silane crosslinked samples were again much more responsive (30% after 4 years) than LDPE (<5% after ~1 year). Field trials on aged LDPE cables in power stations, some of which were shown to contain a high density of water trees ($1 \cdot 4$ trees $\cdot$ mm^{-3}), gave no such indication of degradation via their D.I.s which were all ~0·1%. In all cases, including the samples cut from a failed cable with water trees a much higher level of noise was found than was observed in the laboratory test, and the signal had to be digitally filtered.

It seems reasonable to conclude from these studies that whereas identification of water tree aged samples may be made under favourable conditions in the laboratory, this is likely to be impossible in the field where the small size of water trees and their non-uniform densities in large lengths (~100 m) of cables combine to make the signal-to-noise ratio very small. As yet therefore there is no reliable detector for water trees in field aged insulation.

19.2 Partial discharges and electrical trees

These two subjects have been grouped together here because partial discharges become a hazard to the insulation when they start to generate degradation with the form of electrical trees[120,126]. At this point dielectric failure within a reasonably short time (<1 year at service voltages) becomes inevitable and the insulation will have to be replaced. Identification of the onset of electrical treeing would thus be of use in determining which parts of an insulation system need immediate replacement.

The simplest measurement that could be made to follow this type of degradation mechanism is that of the discharge magnitude, Q. Usually insulation standards require Q to lie below a specified level, Q_m, when the insulation system is first placed in service. For example:

$$Q_m = 5 + [(V_T/V_0) - 1]30 \tag{19.2}$$

is recommended for power cables[166], where V_T is the test voltage, V_0, the operating voltage, and Q_m is in pico-coulombs. A direct relationship between the discharge magnitude and the insulation life at operating voltages has not however been established analytically (see Chapter 13) and the value of Q_m in eqn. 19.2 is determined more by the limitations of the detection apparatus than consideration of what constitutes a dangerous level of discharge[666]. Therefore it cannot be guaranteed that the maximum permitted level of discharge will not lead to degradation over a long period of service. It would though appear reasonable to expect that if such degradation were to occur it would cause the discharge magnitude to increase[666] and hence be detectable by means of periodic tests. However laboratory experiments show that to the contrary the maximum discharge actually *drops* to almost undetectable levels[125–127,380] ($Q \leq 5$ pC) when the discharges start to form a tree and breakdown becomes imminent, Fig. 19.1a.

In the case of electrical trees laboratory studies show an increasing discharge (up to 100's pC) during the development of the primary tree, however, the discharge magnitude also drops to ≤ 5 pC when a secondary tree[317,384,126] is produced in the penultimate pre-breakdown stage, Fig. 19.1b (see Chapter 5). Furthermore in some materials such as epoxy resins the discharge activity observed during primary tree growth is intermittent with considerable periods of quiescence[373].

In view of these laboratory studies on partial discharge and electrical-tree initiated breakdowns it is clear that a field test showing a maximum partial discharge within the accepted level for initial entry into service (e.g. eqn. 19.2) is no guarantee that degradation has not taken place during the intervening period. Nor can it be said that a low discharge amplitude implies that any degradation present is inactive at that time and hence will not lead

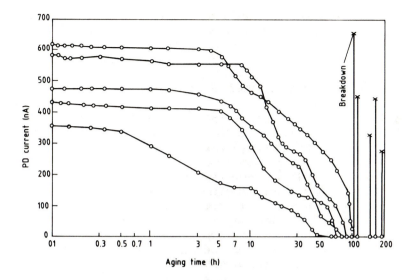

Fig. 19.1
(*a*) Variation with aging time of the maximum partial discharge current observed
for an artificial (crescent-shape) void (CIGRE method I) in a cured epoxy resin.
The breakdown times are also indicated. Taken from Reference 127
Copyright © 1985 The IEEE

to imminent failure. In fact the contrary might be the case as Fig. 19.1*a*
shows.

The above discussion shows that the partial discharge amplitude is not a
reliable means of detecting degraded insulation systems in danger of
failure[373]. Although increasing discharge magnitudes do tend to indicate
degradation and aging[910], the absence of detectable discharges may or may
not indicate the absence of deterioration. Therefore a different means for
assessing the state of the insulation must be sought.

A number of indicators have been used for the state of degradation of
high-voltage rotating machine insulation including the energy dissipated
per cycle per unit of capacitance, J_c; the RMS flow of charge produced in
a detection circuit during partial discharges, DQ; the mean ionisation
current detected, i_{ion}; and the change in tan δ and capacitance produced
by a defined increase in voltage (Δ tan δ and ΔC). None of these parameters
have however proved entirely reliable as a measure of the degradation state
of the system[901,911,912,913], although the voltage dependence of ΔC has had
some limited success with older types of insulation[914,901] (i.e. bitumen or
polyester resin). In view of these inadequacies Goffaux has proposed[901,915]
a more comprehensive set of measurements to act as a fingerprint of the
state of the insulation. This methodology which is available as a test package
has raised considerable interest recently because of its claim to be able to
predict the residual life of individual units.

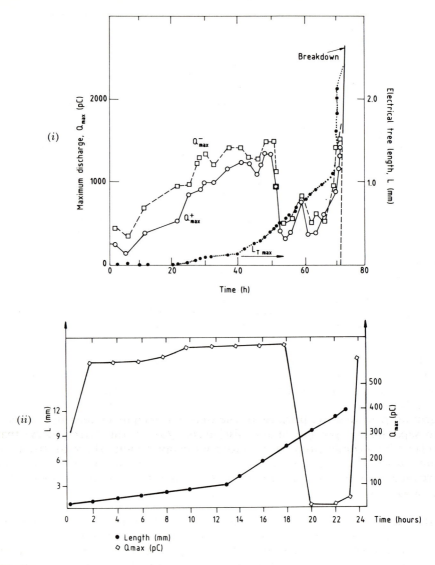

Fig. 19.1

(*b*)   (i)   Variation with aging time *t* of maximum discharge (Q_{max}^+, positive phase, Q_{max}^-, negative phase) and tree length (*L*) observed in the same type of artificial void in an epoxy resin. Taken from reference 126
Copyright © 1985 Japanese Journal of Applied Physics

(ii)   Time variation of maximum discharge and tree lengths for needle-electrode generated trees in polyethylene (bush-branch type tree). Taken from Reference 317
Copyright © 1980 The IEEE

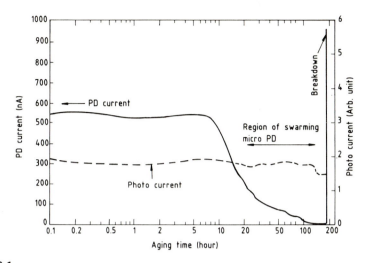

Fig. 19.1
(*c*)   Variation with aging time of maximum partial discharge current and photo-current up to breakdown[127], in the same type of void as in (*a*). Data taken from Reference 380

In Goffaux's test procedure Fourier components of the discharge decay current are measured by means of a Schering bridge at a high ($\sim 10^2$ kHz) and low ($< \sim 10$ Hz) frequency, together with the discharge repetition rate, as a function of the applied voltage U. According to the theoretical background provided the purpose of the different frequency measurements is to separate the portion of the discharge, $Q_M(U)$, (high frequency) that relaxes by recombination via surface conduction around the void, from the ion currents generated in the insulation neighbouring the void (low frequency). These measurements are then used to make predictions in the following way.

(i)   The relaxation time for the high-frequency component (θ) decreases as the applied voltage increases and hence a maximum in the $Q_M(U)$ curve will be found at the voltage U_{QMAX} for which the recombination time is equal to the inverse of the angular frequency ($2\pi f_l$) of the measurement (see Fig. 19.2). By means of laboratory measurements on stator bars a linear relationship was established between U_{QMAX} and the eventual breakdown voltage for the bar. This relationship has been used to predict breakdown voltages for machine units in factory ramp tests with some success[901]. The rationalisation of this result appears to be that U_{QMAX} defines a timescale for the duration of the maximum discharge, which is linearly related to the time available in the factory tests for such a discharge to initiate breakdown. Note that in ramp tests time is proportional to field.

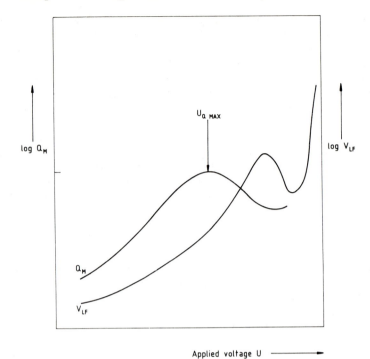

Applied voltage U ————▶

Fig. 19.2 A schematic representation of the variation of the discharge quantities Q_M and V_{LF} with applied voltage U, such as may be found in high-voltage machine insulation (see for example Reference 901). The reader is referred to the text for a definition of Q_M and V_{LF}

(ii)   The voltage generated by the ion current at low frequency, V_{LF}, may also possess a maximum as a function of U, though it will start to increase again at higher voltages (Fig. 19.2). Here it is considered that due to a number of processes including impact ionisation by the discharge and electrical dissociation of chemical impurities, the density of the ionic space charge increases with voltage. This effect gives a contribution to $V_{LF}(U)$ increasing with U. When combined with a relaxation time $(\tau_{SC}(U))$ which decreases with increasing U, a maximum may be produced in the $V_{LF}(U)$ curve providing $\tau_{SC}(U)$ is short enough. The presence of such a maximum is taken to indicate a state of the system just prior to the onset of electrical treeing at the applied voltage at which it occurs, i.e. large ion concentrations are generated rapidly on the void surface by the local field of the discharge leading to heterocharge field enhancement and local channel formation (see Chapter 5). The closer the voltage at which the maximum occurs is to the operating voltage (U_0) the nearer the system will be to failure. An estimation of the extent of the degradation can therefore be obtained by measuring V_{LF} at a chosen low voltage $(U = 1 \text{ kV})$ as a fraction of its value (V_{LFMAX}) at breakdown. Using measurements on old machines very close to failure values for V_{LFMAX} were obtained which appeared to depend solely on the material of insulation, e.g.: mica/epoxy, 100 mV; polyester, 150 mV; shellac, 350 mV. As a result a scaled aging curve could be constructed for V_{LF}/V_{LFMAX} as a function of the ratio of service time to lifetime. Thus measurement of V_{LF} (at 1 kV) for a given insulation of known period of operation, allows the residual life to be read off the plot[916].

(iii)   The $Q_M(U)$ plot may also show a maximum if discharges are extinguished due to the formation of a semiconducting or conducting void surface. In this case the discharge repetition rate will drop to a minimum for $U > U_{QMAX}$, and this measurement can be used to distinguish this relatively harmless situation from the highly dangerous one described in (i).

A major advantage of this test package is that the voltages used can be set to levels for which no further degradation is produced (i.e. $U < 1 \cdot 5 U_0$ for $Q_M(U)$), however in some cases this is insufficient to reach the maximum in Q_m and an empirical factor is used to convert the measured U_Q to U_{QMAX}. The use of other empirical factors necessary to allow for curve shape also detract from the attractiveness of this technique. Furthermore although the method has been accepted in some parts of continental Europe, field trials in North America have not been wholly favourable (EPRI project RP2577-1). Here some success was obtained in predicting the breakdown voltage of hydroelectric generator windings, but it was felt that the technique did not distinguish clearly enough between void discharges and discharge in slots (gaps in the outer insulation layers).

There must also be some doubts as to whether the theoretical model proposed by the authors is a unique interpretation of their measurements. For example damage to the void surface caused by ballistic impact[123,124] (pitting) seems to play no role in their model although it is known to presage the initiation of tree-like failure[380,383,166]. Surface erosion is introduced as one of the factors influencing[915] $\tau_{SC}(U)$, however this is the type of degradation resulting from low magnitude (Townsend-like) discharges rather than the harmful streamer-type discharges[381]. In fact the shape of a discharge pulse appropriate to the region $\tau_{SC} < \theta$ taken by Goffaux to indicate the immediate onset of treeing failure[901] is just what would be expected of a streamer-type discharge[917,381], and it is possible that what is being identified is the degree of insulation degradation that causes Townsend-type discharges to cross over to the streamer type. At present it must be concluded that although the test technique offers some possibilities for identification of high-voltage machine insulation degradation (in AC fields), it requires to be underpinned by a substantial amount of more detailed work.

The possibility of identifying dangerous discharge degradation in polymeric insulation systems seems brighter with the recent development of partial discharge fingerprinting[918] by means of their amplitude-phase[288,379,126] and number-phase[919,357,920] maps. This approach is based on the observation[379] that the amplitude-phase distribution of discharges

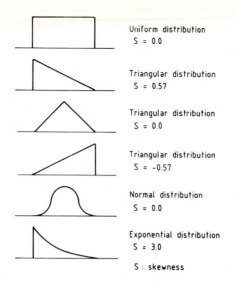

Uniform distribution
S = 0.0

Triangular distribution
S = 0.57

Triangular distribution
S = 0.0

Triangular distribution
S = -0.57

Normal distribution
S = 0.0

Exponential distribution
S = 3.0

S : skewness

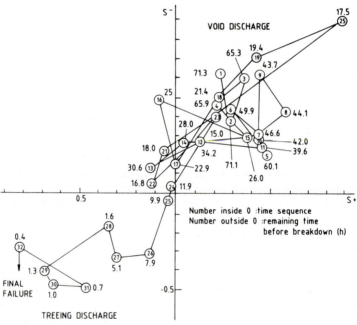

Fig. 19.3

(a) Typical distribution shapes and their skewnesses taken from Reference 126
Copyright © 1985 Japanese Journal of Applied Physics

(b) Plot of discharge distribution skewness in the negative phase, S^-, as a function of the skewness for discharges in the positive phase, S^+, measured in an epoxy resin[126]. Note the transfer of the trajectory into the 3rd quadrant when an electrical tree starts to propagate where it remains up to breakdown. The numbers in the circles show the sequence of the measurements. Taken from Reference 380
Copyright © 1986 The IEEE

produced by an AC field is different for the two cases of a long cylindrical needle-like void, and a flat disk-like void. Here it was found that discharges in the disk-like void exhibited a positive skew in the amplitude-phase distribution whereas those in the needle void had a negative skew, see Fig. 19.3a. Skewness is the third moment of the distribution about its mean value divided by the third power of its standard deviation[921]. Since channels in the tree-like structure generated by partial discharges preliminary to final breakdown are small and needle shaped while the initiating void is closer to a large disk-type void, the above distinction offers the possibility of identifying when partial discharges start to produce electrical trees and hence become dangerous. This is most clearly demonstrated in the skewness plot[921,126,909], Fig. 19.3b. Here the skewness of the discharge amplitude distribution on the negative phase is plotted as a function of that on the positive phase. When the locus of the moving point enters the third quadrant of the plane, discharges are occurring in tree-like channels and thereafter the locus should remain in that quadrant up to breakdown. These concepts have been checked with respect to electrical trees initiated by semiconducting needle electrodes inserted in polyethylene and as predicted the skewness locus remains in the third quadrant throughout the propagation of the tree[126]. Measurements made on standard voids (CIGRE method I) in epoxy resins[126,380,921] have also been able to identify the cross-over from void discharge to treeing discharge from the skewness plots. In this latter case the cross-over has also been correlated with the visual appearance of trees and the discharge amplitude performance which drops sharply from $\sim 10^3$ pC to ~ 50 pC as the first channel starts to propagate[126]. During this phase of behaviour the number of small discharges increases considerably[380] almost so as to compensate for their low maximum amplitude[127], Fig. 19.1c. Measurements of the number-phase distribution should thus show a change to positive skewness (larger numbers of smaller discharges earlier in the phase cycle) as electrical trees begin to propagate[919,357].

Laboratory exploration of the above technique vividly illustrates its potential however its application to field testing is by no means trivial. Several drawbacks can be immediately envisaged. For example the test rig must have phase discrimination and sensitivity to small discharges, <5 pC for polyethylene, and $< \sim 50$ pC for epoxy resins. Since the discharges that are involved in treeing damage are very small compared to those that have not yet initiated trees and which may even be relatively harmless, there is also the problem of detection. It is possible that this may be overcome if sufficiently small regions could be tested such that all discharges present were dangerous. Finally there is the problem of locating the regions of dangerous discharge in a cable or high-voltage machine in service. This is not simple since discharge pulses are reduced in amplitude and have their shape distorted as they travel through the insulation. In order to reconstitute the signal and locate the original discharging centres sophisticated analytical and computing techniques are required and some advances have recently been made in their development[649,922–928]. However it must be concluded that the discharge fingerprint technique has as yet to be shown to work in field detection of partial discharge degradation.

19.3 General types of degradation

Several of the factors discussed above have been advanced as indicators of the general state of insulation materials. Prominent among these are tan δ, Δ tan δ, C, ΔC, and the insulation resistance[909,929–931]. Another common measure, specifically designed for insulation in DC systems, is the polarisation index (PI), which is defined as the ratio of the charging currents 1 minute and 10 minutes after a step-up voltage is applied.

As in Section 19.2 it has been noted that whereas large changes in the above indicators usually, though not always, implied severe degradation[932], a lack of significant change did *not* conversely indicate an insulation system free of degradation[932,929]. In fact Braun[933] pointed out that in several ethylene propylene rubber systems studied no change was observed in tan δ, dielectric constant and resistance during aging to breakdown at 600 V, 60 Hz, and 90°C. Goffaux has noted the same behaviour with respect to the PI in stator windings[901]. These measures must therefore be regarded as unreliable indicators of the state of the insulation.

In keeping with his philosophy of AC testing Goffaux has suggested[915] that the charging current is composed of two significant contributions, the ion migration current, I_i, and the electronic relaxation current, I_e, which have different relaxation times, τ_i, and τ_e, respectively. It is considered that during aging morphological destructuring reduces τ_i and produces deep traps which increase τ_e. As a result the conductivity index defined as:

$$\frac{I_i \tau_i^2}{I_e \tau_e^2} \tag{19.3}$$

reduces considerably[915,901]. As yet however the data available is too limited to allow a judgment to be made as to the effectiveness of this measure, although some encouraging results have been reported.

A useful contribution to the development of a viable indicator of degradation has been made by Montanari and co-workers[900]. This group has developed the theoretical analysis of progressively censored breakdown tests[934] so that samples which are removed from test to investigate their change in indicator properties can be included in the test statistics. In this way any change in the indicator can be properly related to the progress of the sample towards failure. As yet only early results are available but these show a progressive increase in the DC conductivity of the samples as they approach failure. However even in this case it was noted that a significant fraction of the change (8—50%) was recovered when the sample was left to stand for 60 minutes, instead of being measured 1 minute after removal from aging in the breakdown test. It therefore cannot be taken that this approach has established an increase in the DC conductivity as an indication of an approach to breakdown. Rather instead these results should be taken as preliminary work which justifies further effort in this direction.

Part of the problem associated with all these indicators is that they are by and large global measures of bulk properties, and breakdown is often preceded by local rather than global degradation. In such a case any changes are effectively lost in the contribution of the major part of the material which remains unaltered.

Until indicators are identified which are especially sensitive to extreme changes in small volumes of the material, the search for a reliable index of pre-breakdown degradation in materials is unlikely to be successful.

Concluding remarks and future directions

It is said of Science in general that every conclusion should be regarded as no more than a platform from which to launch further investigations; dielectric breakdown is no exception to this rule. In this book we have developed a picture of breakdown in polymers which encompasses the totality of the process. Pre-breakdown phenomena such as water and electrical trees (Part 2) have been shown to arise as a result of composite mechanisms in which elements of physical and chemical aging (Part 1) are combined together with electrically-driven degradation, below the level for runaway (Part 3) to produce a local defect region. The overall picture is completed in Part 4 where it is shown how various local defect processes can extend to cause dielectric breakdown and give rise to the observed failure statistics discussed in Chapter 14. At present however the resulting picture is rather like 'The Adoration of the Magii' by Leonardo da Vinci: the overall design is starting to become apparent but the detail is missing in some important areas. In the following pages we wish to identify some points of detail that, in our opinion, need elucidation. We shall also consider the direction that might be taken by future developments of the overall breakdown models. In this discussion we shall follow closely the sequence of phenomena as they are dealt with in the book.

1 Polymer structure and electrical behaviour

As we have stated this area of the book is intended mainly to provide a suitable background against which pre-breakdown and breakdown processes can be discussed. However it has become clear that there are several gaps in our understanding that have a strong bearing upon the breakdown mechanisms that may occur and we have identified some of the main ones below:

- *Space charge accumulation.* Here we need to know the nature and location of trapping centres, the influence of polar structures and polymer inhomogeneity upon the generation of charge carriers and whether charge carriers are able to be transported to regions near the electrode. All these factors may be influenced by the polymer composition and its gross morphology in a way that is not completely clear. Indeed the transport process in polymeric insulation at high fields is very poorly understood. Much more work is needed in this area and the non-destructive laser pressure-pulse techniques now make it look extremely promising that many features of space charge build-up will soon be understood. It remains however to correlate the accumulation with premature failure, and it is hoped that experimental work will soon be performed to investigate this hypothesis.

- *Behaviour of electrons in extended free volume regions.* If the electrical transport properties of polymers are unclear the behaviour of charges in low density regions is almost unknown. Do these regions have more traps or is it easier to de-trap charge in them? Do the electrons have a longer free path for scattering and a higher ionisation probability per unit path length? Is the material in this region more susceptible to deformation? All these questions must be answered if the role of free volume in polymer aging and breakdown is to be elucidated.

- *Filamentary current formation.* The circumstances under which a uniform current breaks up into filaments needs to be examined both theoretically and experimentally. In particular it needs to be established whether morphological and chemical heterogeneity plays a role here, or do other factors such as local temperature and density fluctuations dominate? Can filamentary current formation occur when the charge source is uniform, as in photoexcitation, or is it solely dependent on localised charge injection associated with localised electrode or discharge driven processes. Also, if uniform space charge regions are formed, under what conditions could these become unstable and lead to filamentary currents.

- *Mechanical fracture.* Mechanical forces have been shown to influence the shape and direction of growth of both electrical trees and water trees, as well as altering the breakdown strength. A major factor appears to be the orientation of the amorphous phase in glassy polymers and the lamellae in semi-crystalline polymers. When these are aligned favourably channel formation is favoured. The relationship of mechanical fracture properties to electrical breakdown and pre-breakdown processes thus becomes of importance. In particular attention should be addressed to attempts to correlate specific pre-breakdown processes to the mechanical fracture properties of a material. For example is water tree void formation related to the tendency of polyethylene to ductile fracture? Will materials more prone to brittle fracture exhibit a different behaviour pattern in water and electrical treeing? Does free volume breakdown initiate ductile fracture and only involve brittle fracture in the final stages? Much more work is needed to resolve these questions.

2 Treeing degradation and failure

(i) Water trees

It is clear that the water treeing mechanism is not yet clearly understood, with researchers being roughly divided into two schools, i.e. those that favour a purely mechanical degradation process and those that prefer a substantial chemical component. The mechanism proposed in Part 2 is synergistic with both chemical and mechanical processes being essential.

Part of the problem for workers in this field has been an inadequate measure of the development and progress of the phenomena via factors such as the 'water tree growth rate' and 'treed area'. Here we have sought to remedy this difficulty through the development in Section 4.2 of an empirical propagation equation, which may be used to compare facets of water tree formation, such as number density, inception time, and growth-rate constant. In particular it would be useful to clarify the role of oxidation in inception, i.e. does it reduce inception times, or increase tree number density, or both, or neither. The reported data is at present ambiguous because it uses a measure which combines some aspects of both inception time and number density, and oxidation may affect these in different ways. Since electrochemistry in water trees may follow a different route (e.g. equation 4.20a—4.20m) to that of thermal oxidation some light may be thrown upon the role of oxidation by comparing the effects of different pre-oxidation treatments upon inception times and number density. A similar investigation of propagation could be made by allowing gaseous oxidising agents to access the damage boundary through the polymer rather than via the water reservoir where other effects may occur. Other questions that need answering with regard to water tree formation, concern the chemical identification of the fluorescing species in the water tree, and the role played by metallic ions, for example: can ionomer species be identified?

The ultimate aim of the theoretician as regards water trees is the establishment of a propagation equation containing only controlled parameters, and system and material variables. To achieve this it is necessary to know how the tree damage is produced and what its density within the tree is. Another question that is raised here is that of the relationship of the damage boundary to the visible tree boundary: does the damage boundary extend beyond the visible boundary as the ion concentration appears to, or does it in fact delineate the visible tree. Recently it has become clear that void coalescence can lead to the formation of disconnected channels in favourable morphologies, although in most cases only voids are formed. Such work may explain the observed formation of cracks when artificially produced water filled voids are electromechanically stressed to a high degree. The work should be extended to see if the channels contain high ion concentrations and whether their formation is enhanced if the water tree is produced in oxidising conditions. The damage density may also be the important factor leading either to low-field breakdown or conversion to an electrical tree at working stresses. Some estimate of its value should be made for very long or very old trees found in field-aged cables, which are generally thought to be extremely damaging, and if possible related to residual life. It is obviously not useful to adopt polymer morphologies which reduce propagation rates (Chapter 8) if they also increase the damage density and reduce the breakdown field. A further question to be answered here is whether the damage density increases with length and aging time.

(ii)   Electrical trees
The mechanism of electrical tree formation is substantially better understood than that of water trees (Chapter 5), however even here there are

details which require further work. For example it is not yet clear whether electrical trees can initiate at a voltage just above the threshold for luminescence (light emission) or whether their inception threshold should be taken as a higher voltage still, i.e. when avalanching starts to occur. The theory of the inception statistics developed here (Section 15.5) should prove very useful in investigating this kind of question since it allows the minimum field for damage generation and the cumulative work done in forming the first channel to be derived. Comparison of these factors with the voltage dependence of luminescence should settle the question. This also requires that luminescence associated with pre-discharge phenomena are clearly identified and not confused with microdischarges or partial discharges. However it is possible that the luminescence mechanism should be regarded as an aging phenomena, and comparison of tree inception in samples aged this way with 'unaged' samples will also yield useful information. Clearly both theoretical and experimental work is needed here in order to improve understanding of the processes involved in the generation of luminescence.

The theory of charge injection from needle points (Chapter 9) should also be investigated in greater detail, with due allowance for the field arising from the space charge. Particular attention should be paid experimentally and theoretically to the spatial distribution of space charge and the current density and the circumstances under which it becomes filamentary. Other factors which need to be investigated are the transient (time-dependent) behaviour of the injecting system, and the effect of very small localised injection spots. A specific point of interest is the development of regions of high energy density which may cause damage by local heating if they persist for long enough.

Although the propagation of electrical trees is understood in general terms some features still require to be cleared up. For example: how do the discharges extend the tree; is it simply a case of ballistic damage and erosion; are induced avalanches in the polymer dominant; or is some form of electrofracture involved; and at what stage of tree growth is it important? It would also be useful to know what the relative importance is of increasing gas pressure and wall charges in controlling the discharge activity and the continued propagation of electrical trees. Further work needs to be carried out on the relationship between discharge activity and the fractal dimension of the tree, in particular what is the origin for stochastic features in tree growth, and what causes tree passivation? Some discussion on these points is contained in Chapters 5 and 6 but further work is essential to clarify the situation. Also some estimation of the wall conductivity of the tree as compared to charge dissipation and the conductive response of the material would be useful in developing a picture of the discharge pattern and growth of the channels. Tree simulation models will be very helpful in developing ideas in this area, since they can be used as test beds to investigate the effect of varying a parameter in the sort of controlled manner that cannot be achieved experimentally. However analytical models should also be developed in order to determine the influence that the assumed stochastic features have on the result. An ultimate aim of this work would be the derivation of a propagation equation for electrical trees, at least at various stages of growth, as a function of external variables such as field and temperature, and material properties (e.g. yield stress and ductility threshold). Although such an expression would be very useful for predictive purposes, as yet none have been proposed other than the empirical expressions discussed here in Chapters 5 and 6. Some experimental investigation of the statistics of tree growth is also of use particularly if correlated with frozen mechanical stresses, polymer composition, morphology and temperature.

3   Deterministic mechanisms of breakdown

As Part 3 on deterministic breakdown models was not intended to be an in-depth study, the reader was referred throughout it to more detailed explanations and this may have given the impression that the understanding in this area was fairly complete. However, there are serious gaps and some of these are highlighted below. In most of the areas described, particularly charge injection, high-field transport and electronic breakdown, the theoretical work is probably as exhaustive as is reasonably possible and yet still it only yields order-of-magnitude results. The reasons for this must lie in the gross approximations and simplifications made regarding the electronic structure and properties of the polymer bulk and interfaces. It is difficult to see how to overcome this sort of problem without either embarking on complicated mathematical analyses with very restricted domains of applicability, or by introducing other simplifying assumptions. Perhaps the new mathematical modelling techniques afforded by the emerging fields of fractals and chaos may enable the diversity of such electronic structures to be modelled more realistically whilst maintaining a reasonably general applicability.

The injection mechanisms of Poole–Frenkel and, especially at higher fields, of Fowler–Nordheim, appear to be reasonable. The factors affecting injection however, such as the shape of the injection potential barrier, especially across a discharge-plasma polymer interface, are both difficult to evaluate and to characterise in a sufficiently general and meaningful way. Charge extraction is easy to understand in terms of spontaneous electron exo-emission (i.e. from a conduction band whose energy levels are higher than the vacuum level) but in most practical situations charge trapping is likely to dominate in non-predictable ways at the interface. Other problems associated with the understanding of space charge build-up at electrodes and charge transport have already been cited earlier in this concluding chapter.

The study of electromechanical failure in polymers seems worthy of further investigation. The combination of a low Young's modulus and toughness and even viscosity in polymers, especially as the melting temperature is approached, together with the natural applicability of these models to filamentary propagation appears to make the electromechanical mechanism a strong contender for failure in this class of materials. More experiments need to be carried out here, in particular an examination of filamentary crack growth as a function of Young's modulus and fracture toughness, and the formation of tree structures in terms of local morphologically-induced variations in mechanical properties and electric field. The influence of surface and bulk charge is not well understood or modelled and a micromechanical/molecular model of electromechanical initiation and tip extension is still required.

The different models of electronic breakdown seem to have advanced very little in recent years and it seems unlikely that it will be possible to distinguish them experimentally. O'Dwyer's considerations of local field enhancement by avalanche space-charge build-up leading to further electronic breakdown appears to be the best contender here. The avalanche model is reasonable and is known to occur in crystalline materials and the space-charge modification allows this mechanism to take place at observed breakdown fields rather than the higher ones which one might otherwise expect.

In spite of intensive investigation there is still a number of unsolved features relating to the degradation produced by partial discharges. The most important factor here is the onset of tree-like damage. Does this follow the occurrence of streamer discharges and if so what is the role played by the erosion produced by Townsend or glow-type discharges? What factors cause a streamer discharge to occur, is it a deterministic development or is it the result of stochastic factors? Is the exhaustion of oxygen from the cavity a necessary preliminary to pitting caused by nitrogen ions? There are many more questions of this type that need answering but we turn now to the tree structure produced. From large single voids these structures tend to be rather like branched trees, but is this always the case, are no bush-type structures produced? Breakdown from large single voids tends to show a single runaway branch from a small tree, to what extent are mechanical features such as a weak path involved in this behaviour, and what happens when many smaller voids are present, do they cooperate and link up or are they independent? These are probably just a few of the questions of a scientific nature that remain to be answered in an area that has received extensive study from an engineering application point of view.

4 Stochastic modelling and breakdown statistics

When it comes to modelling the overall breakdown processes as in Chapter 15, it is fair to say that the job has really only just begun. At present the framework of the models applicable to a number of situations has been outlined and their general features identified. However much of the detail essential to their predictive capability has yet to be determined. Many of these features such as the field dependence of the time-scale for defect generation or free volume extension can be obtained by a more quantified investigation of the relevant pre-breakdown processes. However there are also some features of the models themselves which require detailed investigation. For instance all the models require some criterion for the onset of a runaway breakdown process. In some cases this criterion is fulfilled by a pre-breakdown structure bridging the insulation, (Section 15.2.3) however in a number of cases runaway is initiated before this happens (Section 15.3.4). The local quantities governing the onset of the runaway are not yet known in any specific case, and, although in a number of the models considered charge density and hence local field is used as a critical quantity, this has still to be verified experimentally.

The breakdown statistics derived for the models considered have been determined as an analytical consequence of the self-similar (fractal) structures generated by the stochastic evolution of the pre-breakdown defects (i.e. electrical trees, extended free volume etc.). In most cases the statistical parameters, such as Weibull exponents, can be obtained from known analogies in other similar fractal systems such as sub-percolation clusters (Section 15.4). However the stochastic development of pre-breakdown structures in many of the models (e.g. the electrical tree and cumulative models of Chapter 15) are amenable to computer simulation. Their average geometric and spatial properties and failure statistics can thus be determined directly by repeated simulations once a breakdown criterion is chosen. Such studies can also reveal the effect upon the breakdown statistics of a systematic variation in any chosen factor (e.g. tree channel resistance, initial resistance distribution in the cumulative breakdown model of Section 15.4, or runaway criterion).

The major feature of these models is that they provide a means of connecting the physics and statistics of breakdown processes. Many variations of these basic constructions can be envisaged once more is known about the physical mechanism of formation of any particular pre-breakdown defect. The aim in the long term will be to relate specific breakdown processes to classes of statistical behaviour. It may even be possible in the case of electrical tree failure to deduce both fractal dimension and failure statistics from a knowledge of the polymer morphology and composition. In order to support the theoretical modelling more detailed experimental evidence on time and voltage thresholds is necessary. This will require large sample sets (to characterise accurately the statistical distributions) and long term experiments; in particular evidence for the 'flat-z' characteristics can only be acquired from service data as it is unrealistic to carry out laboratory experiments on the necessary time scale. Experiments designed to test the models should also be carried out, for example can the Weibull exponents predicted by the fluctuation model be related to the dynamic features of noise measurements? It is likely that substantial activity will occur in this area in the future.

5 Engineering aspects

The engineer's interest in breakdown is, of necessity, a practical one. During a development programme the merits of the insulation material and system must be assessed. Later manufacturers guarantees must be underpinned by predictions of lifetime at working stresses and factory tests designed to ensure that the production units remain true to specification. Finally monitoring of the system during service is desirable. Because of limitations upon the time available accelerated test procedures are routinely used with voltage, temperature and sometimes frequency as the accelerating factors. Economic restraints also often limit the size of the sample test set. As we have pointed out in Chapters 14 and 16 these restrictions limit the accuracy with which the breakdown distribution can be defined under the test conditions. Lifetime predictions for working stresses can therefore be substantially in error, sometimes being incorrect by an order of magnitude. Features such as threshold times or fields, which are difficult to establish unambiguously in 'simple' statistical tests, are likely to cause lifetime predictions to be pessimistic if no allowance is made for them. In this way perfectly viable systems may be rejected as unsuitable. It is here that a scientific understanding of the particular breakdown process concerned can guide the engineer's approach. If a mechanism is suspected that is likely to have a threshold, then a statistical procedure should be devised that will reveal the threshold more clearly. For example a destructive progressive-stress test can be interrupted by a constant-stress plateau at a voltage below the putative threshold (Chapter 16). If no failures are observed on the plateau and the failure distribution remains unchanged as the constant stress period is increased, there is reasonable grounds for assuming that the possible threshold is genuine. All such test procedures should be supported by a laboratory investigation of the breakdown mechanism.

When assessing competing materials it is common to look for changes in the characteristic (or average) breakdown strength. In this case statistical uncertainty due to the limited number of samples in the set will impose error limits on the determined value which have nothing to do with the accuracy of the measurements. It should be noted that this uncertainty is *not* the range of fields (times) at which breakdown occurred but the *confidence range* within which one is sure that the actual characteristic strength (time) lies. It is usual to use for this range the 90% confidence limits (see Chapters 14 and 16). Such bounds should be placed on all estimated values so that the validity of any trend can be assessed. In many cases it will be found that the quoted differences are statistically insignificant. This does not mean that they do not exist but instead that more samples should be tested in order to reduce the statistical error and other laboratory experiments should be carried out to verify the reported trends in performance.

An important point to keep in mind when analysing accelerated tests is that the characteristic breakdown strength applies to the breakdown mechanism pertinent to high-field conditions. In polymeric materials many mechanisms can lead to breakdown and the one appropriate to the characteristic may be dominant only under the accelerating conditions. Under service conditions a different mechanism may dominate, and in this case extrapolation of accelerated tests is not valid. It is for this reason that temperature is not recommended as an accelerating factor, since for polyethylene the dominant process is thought to change from avalanche-type breakdown at cryogenic temperatures to an unknown combination of electro-mechanical and thermal runaway (Part 3) at typical service temperatures closer to the softening point.

The possibility that a number of breakdown mechanisms may occur under the same accelerated conditions can lead to the presence of multiple distributions in the tests. This, however, may be difficult to discern, particularly where the low-field failures are concerned, since these will lie in the low-field tail of the distribution which is always poorly defined unless the sample set is very large. However it is just this region which is important at service stresses, and it is quite likely that it will not extrapolate according to the same rules as the mechanism which is dominant in the tests. Another difficulty associated with accelerated tests is that they may accelerate different features of the *same* breakdown mechanism to different extents, for example inception may dominate over propagation in ramp tests (Chapter 15) whereas the reverse may be the case in service. In the case of DC voltages it is even possible that ramp tests accelerate the breakdown mechanism in a totally different way to that occurring during service (Chapter 17). Slow aging processes may also alter the local conditions for breakdown and hence increase the probability of failure at service stresses (Section 6.1, 15.2.2, and 15.4). It is therefore clear that without a detailed understanding of the physical nature of the breakdown mechanisms applying in a particular case, and the way they are affected by service aging, an accelerated programme of destructive tests can only be used to compare new materials with older systems for which service experience obtains. Even here there will be the implicit assumption that both new and old materials will behave the same way during service aging, and this is by no means guaranteed even for very similar materials. Such a comparison also becomes invalid if a change in the system design accompanies the move to a different insulation material. Consequently, much more work is required to determine and then compare materials or model-system results with performance in-service where the materials used are the same.

When we turn to in-service monitoring (Chapter 19) the message remains the same, namely, successful techniques will only be developed once the detailed nature of the various breakdown and pre-breakdown mechanisms is understood. The obvious and simple indicators of degradation have not proved reliable, in many cases giving negative responses when pre-breakdown processes are already well advanced. Those techniques of recent provenance which show a promise of future applicability are all based on the better understanding of the relevant processes gained in laboratory experiments. Invariably they rely on subtle features of the process concerned which could not have been developed from engineering and service experience.

6 Summary

Our conclusion is therefore:

'There is no substitute for understanding.'

The ability of the design engineer to complete his appointed task with confidence rests on scientific knowledge of the physical processes of break-down obtained in the laboratory. In order that this information may be useful, the gap between the statistical approach of the engineer and the physical approach of the scientist must be bridged by the development of comprehensive analytical breakdown models. A start along this path has been made here but there is still a long way to go. In the process we believe we have answered some questions but raised many more which we expect to form the basis for many research programmes in the coming years.

Computer program for calculating Weibull parameters

A brief description of the operation of the program follows.

The first 43 lines deal with inputting the data, sorting it into order, and finding its logarithms which are used later on. Lines 44—67 use a least-squares technique to find a first estimate for the shape parameter. Lines 68—103 use the Newton–Raphson technique to find maximum-likelihood estimates of the distribution parameters. The program will run if only the first 103 lines are entered and produce correct maximum likelihood estimates. The remainder of the program estimates confidence limits in a rather unusual way. The graphs used for the graphical methods for confidence limit estimation described in References 723 and 820 are represented as several-order equations (polynomials, hyperbolas etc.). Values calculated from these are then used in the same way as in the graphical techniques. The equations are faithful representations of the graphs for values of n between 5 and 120 and the technique is reasonable provided the data is not highly censored. Also shown is a typical output from the program.

Listing of computer program to calculate Weibull and 1AEV parameters

```
1   REM   Calculates maximum likelihood estimates
2   REM      and 90% confidence limits for
3   REM      Weibull or 1AEV (Gumbel) parameters
4   REM   ----------------------------
5   REM   Select Weibull or 1AEV distributions
6   PRINT   "----------------------------
7   PRINT   "MAXIMUM LIKELIHOOD PARAMETER ESTIMATION"
8   PRINT   "FOR WEIBULL AND 1AEV DISTRIBUTIONS"
9   PRINT   "----------------------------
10  PRINT
11  PRINT   "Do you wish to fit data to either:"
12  PRINT   "  1.  Weibull distribution, or"
13  PRINT   "  2.  1AEV (Gumbel) distribution."
14  PRINT
15  INPUT   "Enter 1 or 2:  ";DISTRIBUTION
16  IF DISTRIBUTION<>1 AND DISTRIBUTION<>2 THEN 15
17  REM   ----------------------------
18  REM   Input Data
19  PRINT
20  INPUT   "How many specimens were tested:  ";NTEST
21  INPUT   "How many specimens broke down:  ";NFAIL
22  DIM TIME (NFAIL)
```

```
23   PRINT   "Input failure times or voltages:"
24   FOR SPECIMEN=1 TO NFAIL
25     INPUT TIME (SPECIMEN)
26     IF DISTRIBUTION=2 THEN TIME (SPECIMEN)=EXP (TIME (SPECIMEN))
27     NEXT SPECIMEN
28   REM    --------------------------
29   REM    Place data in order
30   PRINT
31   PRINT   "Placing data in order."
32   FOR POINTER1=1 TO NFAIL-1
33     FOR POINTER2=POINTER1+1 TO NFAIL
34       IF TIME (POINTER1)>TIME(POINTER2) THEN SWAP TIME(POINTER1),
TIME(POINTER2)
35       NEXT POINTER2
36     NEXT POINTER1
37   REM    ------------------------------
38   REM      Find natural logarithms of data
39   DIM   LNTIME(NFAIL)
40   FOR   SPECIMEN=1 TO NFAIL
41     LNTIME (SPECIMEN)=LOG(TIME(SPECIMEN))
42     NEXT SPECIMEN
43   REM    ------------------------------
44   REM    Find first estimate for beta using linear regression
45   PRINT   "Estimating initial value using linear regression."
46   REM
47   REM    Generate estimates for cumulative probability of failures
48   REM       using F(i, n)=(i-0·3)/(n+0·4)
49   DIM PROB(NFAIL), WEIBULL(NFAIL)
50   FOR SPECIMEN=1 TO NFAIL
51     PROB(SPECIMEN)=(SPECIMEN-.3)/(NTEST+.4)
52     WEIBULL(SPECIMEN)=LOG(LOG(1/(1-PROB(SPECIMEN))))
53     NEXT SPECIMEN
54   REM
55   REM    Do least squares
56   SUMTIME=0
57   SUMPROB=0
58   SUMTIMESQUARED=0
59   SUMPRODUCT=0
60   FOR SPECIMEN=1 TO NFAIL
61     SUMTIME=SUMTIME+LNTIME(SPECIMEN)
62     SUMPROB=SUMPROB+WEIBULL(SPECIMEN)
63     SUMTIMESQUARED=SUMTIMESQUARED+(LNTIME(SPECIMEN))^2
64     SUMPRODUCT=SUMPRODUCT+LNTIME(SPECIMEN)*WEIBULL(SPECIMEN)
65     NEXT SPECIMEN
66   BETA=(SUMPRODUCT-SUMTIME*SUMPROB/NFAIL)/(SUMTIMESQUARED
-SUMTIDE^2/NFAIL)
67   REM   ------------------------------
68   REM    Set up constants and variables for iteration
69   C=SUMTIME/NFAIL
70   DIM A(3)
71   TOLERANCE=.0001
72   PRINT "Iterating..."; 
73   REM    --------------------------4
74   REM    Iteration loop starts here
75   REM
76   REM      Calculate A1, A2, and A3 for this iteration
77   FOR K=1 TO 3
78     SUM=0
79     FOR SPECIMEN=1 TO NFAIL
```

```
80      SUM=SUM+(TIME(SPECIMEN)^BETA)*(LNTIME(SPECIMEN)^(K-1))
81      NEXT SPECIMEN
82    A(K)=SUM+(NTEST-NFAIL)*(TIME(NFAIL)^BETA)*(LNTIME(NFAIL)^(K-1))
83    NEXT K
84    REM
85    REM      Calculate values for f(beta) and its differential
86    F=A(2)/A(1)-1/BETA-C
87    DIF=A(3)/A(1)-(A(2)/A(1))^2+1/BETA^2
88    REM
89    REM      Find next estimate of beta
90    BETA=BETA-F/DIF
91    PRINT "...";
92    IF ABS(F)>TOLERANCE THEN 74
93    PRINT
94    REM    ----------------------------
95    REM    End of iterations
96    ALPHA=(A(1)/NFAIL)^(1/BETA)
97    U=LOG(ALPHA)
98    B=1/BETA
99    PRINT
100   PRINT "the maximum likelihood estimates are:"
101   IF DISTRIBUTION=1 THEN PRINT "Scale parameter, alpha=    ";ALPHA:PRINT
"Shape parameter, beta=    ";BETA
102   IF DISTRIBUTION=2 THEN PRINT "Location parameter, u=    ";U:PRINT "Scale
parameter, b=   ";B
103   REM    ----------------------------
104   REM    Calculate confidence limits of parameters
105   REM
106   IF NTEST>120 OR NFAIL>120 THEN PRINT "Too many samples to estimate
confidence limits with this program.":END
107   REM      ... for beta
108   WU=(1.63617-.0272229*NFAIL+5.78997E-04*NFAIL^2-5.39819E-06*NFAIL^3
+1.80269E-08*NFAIL^4)
109   WL=1/(1.08509+6.89715/NFAIL+13.53525/NFAIL^2+15.17619/NFAIL^3)
110   BETAUP=WU*BETA
111   BETALOW=WL*BETA
112   BL=B/WU
113   BU=B/WL
114   REM
115   REM      ... for alpha
116   REM          First find ZURN=Zu(r=n)
117   ZURN=.11254+7.03279/NFAIL-34.29964/NFAIL^2+135.38431/NFAIL^3
-90.73569/NFAIL^4
118   ZU=.17+(ZURN-.17)*(NTEST-1)/(NFAIL-1)
119   ZL=.10964+7.33402/NTEST-17.70106/NTEST^2+41.43168/NTEST^3
120   ALPHAUP=ALPHA*EXP(ZU/BETA)
121   ALPHALOW=ALPHA*EXP(-ZL/BETA)
122   UL=U-B*ZL
123   UU=U+B*ZU
124   PRINT
125   PRINT "The 90% confidence limits are approximately:"
126   IF DISTRIBUTION=1 THEN PRINT ALPHALOW;"<alpha<";ALPHAUP:PRINT
BETALOW;"<beta<";BETAUP
127   IF DISTRIBUTION=2 THEN PRINT UL;"<u<";UU:PRINT BL;"<b<";BU
128   IF NFAIL<NTEST/2 THEN PRINT "WARNING: Few samples failed: alpha limits
may be inaccurate."
129   REM    ----------------------------
130   REM    Calculate confidence limits for 1, 5, and 10 percentiles
```

```
131  IF NTEST>25 OR NFAIL>25 THEN PRINT "Too many samples for accurate
     percentile confidence limit estimation":GOTO 156
132  VU10=.9950281+.05331*NTEST-.0011307*NTEST^2
133  VL10=2.452405+20.3/NFAIL
134  VU5=1.461485+.0662451*NTEST-.0014737*NTEST^2
135  VL5=2.939888+28.81153/NFAIL
136  VU1=2.555121+.0724486*NTEST-.0013272*NTEST^2
137  VL1=4.42838+45.58383/NFAIL
138  PRINT
139  PRINT  "1, 5, and 10 percentiles and approximate confidence limits:"
140  IF DISTRIBUTION=2 THEN 149
141  REM Weibull distribution
142  PRINT  "1 percentile=  ";ALPHA*(-LOG(.99))^(1/BETA)
143  PRINT  "lower limit=  ";ALPHA*EXP(-VL1/BETA);"  upper limit=  "
     ;ALPHA*EXP(-VU1/BETA)
144  PRINT  "5 percentile=  ";ALPHA*(-LOG(.95))^(1/BETA)
145  PRINT   "lower limit=  ";ALPHA*EXP(-VL5/BETA);"  upper limit=  "
     ;ALPHA*EXP(-VU5/BETA)
146  PRINT  "10 percentile=  "; ALPHA*(-LOG(.9))^(1/BETA)
147  PRINT  "lower limit=  ";ALPHA*EXP(-VL10/BETA);"  upper limit=  "
     ;ALPHA*EXP(-VU10/BETA)
148  GOTO 156
149  REM  1AEV distribution
150  PRINT  "1 percentile=  ";U+B*LOG(LOG(1/.99))
151  PRINT  "lower limit=  ";U-B*VL1;" upper limit=  ";U-B*VU1
152  PRINT  "5 percentile=  ";U+B*LOG(LOG(1/.95))
153  PRINT   "lower limit=  ";U-B*VL5;"  upper limit=  ";U-B*VU5
154  PRINT   "10 percentile=  ";U+B*LOG(LOG(1/.9))
155  PRINT   "lower limit=  ";U-B*VL10;" upper limit=  ";U-B*VU10
156  END
```

Typical output from computer program

————————————————————————————

MAXIMUM LIKELIHOOD PARAMETER ESTIMATION
FOR WEIBULL AND 1AEV DISTRIBUTIONS

————————————————————————————

Do you wish to fit data to either:
1. Weibull distribution, or
2. 1AEV (Gumbel) distribution.

Enter 1 or 2: ? 1

How many specimens were tested: ? 12
How many specimens broke down: ? 10
Input failure times or voltages:
? 11.03
? 10.88
? 10.21
? 7.90
? 11.15
? 11.75
? 12.10
? 12.38

? 12.85
? 12.43

Placing data in order.
Estimating initial value using linear regression.
Iterating . . . .

The maximum likelihood estimates are:
Scale parameter, alpha= 12.23302
Shape parameter, beta= 10.01905

The 90% confidence limits are approximately:
  11.49682<alpha<13.11106
  5.2038<beta<14.19141

1, 5, and 10 percentiles and approximate confidence limits:
1 percentile= 7.729146
    lower limit= 4.988679   upper limit= 8.858829
5 percentile= 9.094626
    lower limit= 6.842418   upper limit= 9.975238
10 percentile= 9.772083
    lower limit= 7.820514   upper limit= 10.39136

Calculating the threshold value of a 3-parameter Weibull distribution using MathCAD[826]

FIND PARAMETERS OF 3-PARAMETER WEIBULL DISTRIBUTION USING LEAST SQUARES

Example data:

Number of samples tested: $n := 10$

Number of samples failed: $r := 9$

Labelling each breakdown "i" $i := 1, \ldots, r$

Find cumulative probability of breakdown, p, for each sample, i, using median rank approximation:

$$p_i := \frac{i - 0 \cdot 3}{n + 0 \cdot 4}$$

Define w for each "p" such that w values correspond to the linearised vertical axis scale on the Weibull plot:

Weibull $(p) := \log(-\ln(1-p))$ $w_i := $ Weibull $[p_i]$

The example data points, x, corresponding to nine times-to-breakdown are:

$x_1 := 240$ $x_2 := 300$ $x_3 := 340$ $x_4 := 390$ $x_5 := 490$

$x_6 := 530$ $x_7 := 590$ $x_8 := 750$ $x_9 := 900$

Now guess initial values for parameters:

Characteristic value: alpha $:= 400$
Shape parameter: beta $:= 1$
Threshold value: gamma $:= 200$

Putting these into a form of $y = mx + c$, we have $w = m \cdot \log(x - g) + c$. The values of m, c, and g are therefore:

$m := $ beta $c := -$ beta $\cdot \log$ (alpha) $g := $ gamma

Find the best values of the parameters m, c, and g by solving Normal equations

Given

$$\sum_i \left[m \cdot [\log [x_i - g]]^2 + \log [x_i - g] \cdot [c - w_i] \right] \approx 0$$

$$\sum_i [c - w_i + m \cdot \log [x_i - g]] \approx 0$$

$$\sum_i \left[\frac{\log (e)}{x_i - g} [w_i - [m \cdot \log [x_i - g]] - c] \right] \approx 0$$

$$g < x_1$$

parameters := find (m, c, g)   (This assigns the vector called "parameters" to contain the found values of m, c, and g)

m := parameters$_0$ m = 1·16   c := parameters$_1$ c = −3·03

g := parameters$_2$ g = 199

The best values of the parameters are:

alpha := $10^{-\left[\frac{c}{m}\right]}$ alpha = 413

beta := m   beta = 1·16

gamma := g   gamma = 199

The best fitting, least squares line is given by f:

$$f_i := \text{weibull} \left[1 - \exp \left[-\left[\frac{x_i - \text{gamma}}{\text{alpha}} \right]^{\text{beta}} \right] \right]$$

Plotting these on 2-parameter Weibull plot scales gives a good straight line if the threshold value is first subtracted from the data values. Otherwise a convex curve is obtained:

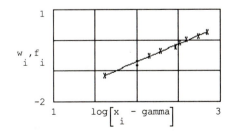 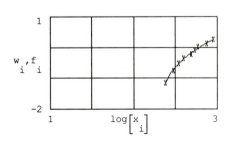

Mathematical proof

It is required to prove that

$$I_\delta = \int_0^\pi \sin^\delta x \, dx \simeq \sqrt{(2\pi/\delta)}$$

for large integral values of δ.

From Gradshteyn and Ryzik[935];

$$I_\delta = \frac{\delta-1}{\delta} \cdot \frac{\delta-3}{\delta-2} \cdot \cdots \cdot \frac{1}{2} \cdot \pi \quad \text{if } \delta \text{ is even}$$

and

$$I_\delta = \frac{\delta-1}{\delta} \cdot \frac{\delta-3}{\delta-2} \cdots \frac{2}{3} \cdot 2 \quad \text{if } \delta \text{ is odd}$$

The solution of I_δ using this formula is shown as the points in Fig. A3.1 as a function of δ. Therefore:

$$I_{\delta+1} \cdot I_\delta = \frac{\delta}{\delta+1} \cdot \frac{\delta-2}{\delta-1} \cdots \frac{1}{2} \cdot \pi$$

$$\cdot \frac{\delta-1}{\delta} \cdot \frac{\delta-3}{\delta-2} \cdots \frac{2}{3} \cdot 2$$

$$= \frac{2\pi}{\delta+1}$$

Similarly:

$$I_{\delta+2} \cdot I_{\delta+1} = \frac{2\pi}{\delta+2}$$

So that:

$$\frac{I_{\delta+2}}{I_\delta} = \frac{\delta+1}{\delta+2} \rightarrow 1 \quad \text{as } \delta \rightarrow \infty$$

and

$$I_{\delta+1} \simeq \sqrt{(2\pi/\delta)} \simeq \sqrt{(2\pi/(\delta+1))}$$

or

$$I_\delta \simeq \sqrt{(2\pi/\delta)} \cdot \quad \text{Q.E.D.}$$

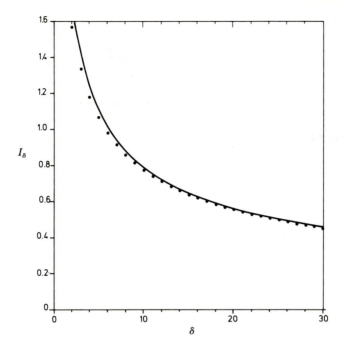

Fig. A3.1 A graph showing the function:

$$I_\delta = \int_0^\pi \sin^\delta x \, dx$$

for different integral values of δ. The points show the actual solutions of the function. The line shows that the approximation used, $I_\delta \simeq (2\pi/\delta)^{1/2}$ is reasonable for large values of δ

List of symbols

a	crystal lattice spacing (Chapter 2 and Chapter 9)
	or shape parameter for Weibull distribution as a function of time
A	energy gained by free electrons from field
b	shape parameter for Weibull distribution as a function of electric field or voltage
C	scaling constant Weibull distribution as function of time and field (e.g. eqns. 14.17, 14.18, and 14.19)
CED	cohesive energy density
C_s	effective capacitance of dielectric in series with partial discharge void
C_t	energy required to produce a channel forming rupture (see eqn. 4.28)
C_p	specific heat at constant pressure
	or effective capacitance of dielectric in parallel with partial discharge void
C_v	specific heat at constant volume
	or effective capacitance of partial discharge void
d	thickness of insulation or distance between electrodes
d_0	initial thickness of insulating film before electromechanical compression (Chapter 11)
d_f	fractal dimension
D	insulation thickness (Chapter 3)
	or density (Chapter 10)
	or electric flux density ($C \cdot m^{-2}$) (Chapter 11)
DI	degradation index (Chapter 19)
e	magnitude of charge on electron ($1 \cdot 602177 \times 10^{-19}$ C)
$\boldsymbol{E}$	electric field ($E = |\boldsymbol{E}|$)
$\mathbb{E}$	electron energy
$\mathbb{E}_c$	conduction band edge
$\mathbb{E}_f$	Fermi energy level
$\mathbb{E}_t$	trap energy level
$\mathbb{E}_v$	valence band edge
$\dot{E}$	rate of field increase in progressive-stress (ramp) tests
E_a	apparent field across void or discharge channel
	or applied field (Chapter 17)
E_{aval}	breakdown field predicted by avalanche breakdown mechanism
$E_c(p)$	characteristic breakdown strength for volume fraction of defects, p
E_{coll}	breakdown field predicted by collective breakdown mechanism
E_d	discharge inception field

E_{ef}	breakdown field predicted by electrofracture mechanism
E_{fem}	breakdown field predicted by filamentary electromechanical breakdown mechanism
E_{Fr}	breakdown field predicted by Fröhlich's high-energy criterion
E_h	longitudinal field within a discharging filament
E_{inj}	critical (minimum) field for injection of carriers into conduction band
E_0	field required for capacitive bond to fail in percolation model
E_l	local field along a bond in the random breakdown model
E_{mc}	critical field according to FLSC model above which charges may move through the insulation
E_{sc}	critical (minimum) field for space charge injection
E_{th}	threshold (minimum) field for discharge to occur
E_{vH}	breakdown field predicted by von Hippel's low-energy criterion
f	frequency
F	Coulombic force of attraction
$g(\ )$	probability density function
G	mechanical toughness
$G(\ )$	'initial' probability distribution
G_0	damage threshold energy in free volume breakdown
G_{th}	damage threshold energy in tree channel formation
h	Planck's constant ($6 \cdot 62607 \times 10^{-34}$ J $\cdot$ s)
$\hbar$	$= h/2\pi$
I	electric current
I_f	electric current through breakdown filament
J	current density ($A \cdot m^{-2}$) ($J = |J|$)
k	electron wave vector
k	rate constant in water tree growth equation indicating 1/(period of fast growth) (see Section 4.2)
k_B	Boltzmann constant ($1 \cdot 38066 \times 10^{-23}$ J $\cdot$ K^{-1} mol^{-1})
k_i	1/(time required to initiate 63% of trees) (see Section 4.2.1)
l	mean distance between electron collisions in direction of electric field
L	tree length
$L_c(E)$	critical void size such that when largest void is less than $L_c(E)$ no breakdown will occur
l_{max}	size of largest free-volume region
L_m	final length of the first branch of an electrical tree
L_{max}	maximum tree length
$\bar{L}_{max}$	characteristic length of longest tree distribuion
L_ξ	characteristic void size
L_∞	length parameter in water tree growth equation (see Section 4.2)
m	fractional exponent (eqn. 15.6) which is a measure of connection between motions in different clusters
M	equivalent mass of fractal figure if it were solid and of homogeneous density
m_e	rest mass of electron ($9 \cdot 10939 \times 10^{-31}$ kg)
m^*	electron effective mass (defined in Chapter 2)

n	degree of polymerisation
	or charge carrier number density (m^{-3})
	or field exponent for life equation (life $\propto E^{-n}$)
n_c	electron number density in conduction band
$N(\mathbb{E})$	number density of electron states between electron energies $\mathbb{E}$ and $\mathbb{E} + dE$
N_{eff}	effective density of electron states (see Chapter 2)
N_f	number of half cycles of alternating voltage
$N(r)$	number density of microvoids with radii between r and $r + dr$
$N(S)$	number density of electrons in speed range S to $S + dS$
n_t	number density of occupied traps
N_t	number density of traps (occupied and empty)
$N(v)$	number density of electrons in velocity range v to $v + dv$
p	field exponent defined in eqn. 4.7.
	or gas pressure (especially in Chapter 15)
	or electron momentum (Chapter 9)
$P(\ )$	probability
$P(\mathbb{E}, T)$	Fermi–Dirac distribution of an electron state, $\mathbb{E}$, being occupied at temperature, T
$P_F(\ )$	cumulative probability of failure
$P_S(\ )$	survivor (or reliability) function (see Chapter 14)
Q	electric charge
	or rate constant for water tree growth (eqn. 4.7b)
Q_a	Ashcraft's rate constant for water tree growth (eqn. 4.7a)
r	radius of electrode curvature
	or microvoid radius
	or relaxation rate exponent (Fig. 3.1b and 3.1c)
R	electrical resistance
	or reflection coefficient of electrons at polymer–metal interface
	or gas constant
r_f	radius of breakdown filament or tube
R_S	effective resistance of insulation in series with partial discharge void
s	field exponent for length of trees in eqn. 15.12
	or viscosity exponent (Fig. 3.1b and 3.1c)
S	sum of arc-lengths of the individual branches of a tree
	or electron speed
s_i	arc length of ith branch of tree
s_{th}	thermal speed (not velocity) of electrons
t	time
t_1	time parameter for tree growth (see Section 4.2)
T	temperature
T_0	ambient temperature
T_1	temperature of polymer–metal interface
t_b	time to breakdown
t_d	inception (or delay) time before start of water tree growth
t_f	formative time lag before onset of avalanching
T_g	glass-transition temperature (classical interpretation)

T'_g	precisely defined glass transition temperature (see Fig. 3.1b and 3.1c)
t_I	inception period for electrical trees
t_s	statistical time lag before onset of avalanching
t_v	time to form microvoid or unit of damage
V	voltage
$\dot{V}$	rate of voltage increase in progressive-stress (ramp) voltage tests
$\bar{V}$	molar volume
V_{app}	applied voltage
V_b	breakdown voltage
V_c	characteristic breakdown voltage
v_d	drift velocity of charge carriers
V_i	inception voltage for partial discharge breakdown
V_{ION}	ionisation potential of molecule
V_{TFL}	voltage required for trap-filled limit in space-charge limited current model (Chapter 9)
V_{th}	breakdown voltage predicted by thermal breakdown mechanism (Chapter 10 only) *or* threshold voltage below which breakdown cannot occur (Chapter 14 *et seq.*)
V_{tr}	transition voltage in space-charge limited current model (Chapter 9)
V_v	voltage appearing across a void
W_{es}	electrostatically stored energy
W_p	energy required to overcome plastic deformation
W_s	surface energy required to overcome toughness
y	cluster variable (e.g. density, polarisation, space-charge density, order parameter) scaled to its value in a characteristic cluster (see Chapter 15)
Y	Young's modulus of elasticity
α	Townsend's primary ionisation coefficient *or* characteristic value in general expressions for Weibull distribution (subscripts u and l represent upper and lower statistical confidence limits)
β	shape parameter in general expressions for Weibull distribution (subscripts u and l represent upper and lower statistical confidence limits) *or* in the expression for the distribution in size L of simulated fractal structures proportional to $[L/L_\xi]^{-(\beta+1)}$
γ	thermal conductance per unit electrode area *or* Townsend's secondary ionisation coefficient *or* threshold value in general expressions for Weibull distribution
$\Gamma(\ )$	gamma function ($\Gamma(j+1)=j!$ for integer values of j)
δ	$\tan(\delta)=$ dielectric loss $=\varepsilon''/\varepsilon'$ *or* exponent defined by eqn. 15.32b *or* shape parameter in Weibull distribution of charge densities
$\Delta \mathbb{E}$	width of band of shallow trap state energies
$\Delta \mathbb{E}_t$	electron energy difference between $\mathbb{E}_c$ and $\mathbb{E}_t$

ΔG	Gibbs free energy of activation
ΔH_{vap}	heat of vapourisation
ΔS	configurational entropy
ΔU	internal energy
ε	permittivity (ε' = real part of $\sim$, ε'' = imaginary part of $\sim$, ε_0 = $\sim$ of free space $= 8\cdot854187817 \times 10^{-12}\,\mathrm{F}\cdot\mathrm{m}^{-1}$, ε_r = relative $\sim$)
ε_m	mechanical strain
ζ_0	cosine of half angle subtended by hyperboloid of revolution
η	viscosity
	or exponent of field in dielectric breakdown model (DBM) such that probability of breakdown of a bond $\propto E_l^{\eta}$
θ	n_c/n_t
	or half angle of hyperboloid of revolution
κ	thermal conductivity
κ_0	temperature-independent thermal conductivity value
λ	electron wavelength
$\Lambda(\ )$	asymptotic, extreme value distribution function (1st. $\Lambda_1(\ )$; 2nd. $\Lambda_2(\ )$; and 3rd. $\Lambda_3(\ )$)
μ	mobility of charge carrier (drift velocity per unit electric field)
ν	photon, phonon, or lattice vibration frequency (Hz)
	or attempt to fail frequency (eqn. 15.9)
ν_0	attempt to escape frequency
ξ	characteristic cluster size (eqn. 15.11)
ρ	electrical resistivity
ρ_c	charge density
	(*or* characteristic value of charge density in distribution)
ρ_h	heterocharge density
σ	electrical conductance
σ_0	proportionality constant in Arrhenius equation (e.g. eqn. 9.11)
σ_K	constant of proportionality in Klein's equation for conductivity (eqn. 10.14)
σ_m	mechanical stress
τ	mean time between scattering events
	or growth period for water trees ($= t - t_d$) or electrical trees ($= t - t_I$)
	or length of impulse required to cause thermal breakdown
τ_c	mean time between collisions
	or characteristic time to breakdown in distribution
$\tau(E_a)$	timescale of DC pre-breakdown event (Chapter 17)
τ_s	statistical time lag before onset of partial discharging
ϕ	work function (subscript m implies metal)
	or exponent in Arrhenius equation (e.g. eqn. 9.11)
	or barrier height for injection
χ	electron affinity (subscript i implies insulation)
ψ	wave function ($\psi \cdot \psi^* \cdot dx$ = probability of electron existing between x and $x + dx$)
ω	angular frequency (radians per second) $= 2\pi f$
Ω	volume of system (eqn. 15.32a)

References

1 SOLYMAR, L., and WALSH, D.: 'Lectures on the electrical properties of materials' (Oxford University Press, 3rd edition, 1984)
2 FRANZ, W.: in 'Handbuch der physik', ed. FLÜGE, S. (Springer, Berlin, 1956) **17**, p. 155
3 SAECHTLING, H.: 'International plastics handbook/Saechtling' (Carl Hanser Verlag, Munich, 1983)
4 PERSSON, K.G. and THUNWALL, B.: *ITT Electrical Communication*, 1985, **59(4)**, pp. 442–445
5 MILLS, N.J.: 'Plastics: microstructure, properties and applications' (Edward Arnold, London, 1986) ISBN 0-7131-3565-4
6 HALL, C.: 'Polymer materials: an introduction for technologists and scientists' (Macmillan Press, London, 1981) ISBN 0-333-28907-2
7 McCRUM, N.G., BUCKLEY, C.P., and BUCKNALL, C.B.: 'Principles of polymer engineering' (OUP, Oxford, 1988) ISBN 0-19-856152-0
8 ASHBY, M.F., and JONES, D.R.H.: 'Engineering materials: an introduction to their properties and applications' (Pergamon Press, Oxford, 1980) p. 58
9 HOFFMAN, J.D., DAVIS, G.T., and LAURITZEN, J.I.: in 'Treatise on solid state chemistry', Vol. 3, ed. HANNEY, B. (Plenum Press, New York, 1976) Chapter 7
10 KELLER, A.: *Phil. Mag.*, 1957, **2**, pp. 1171–1175
11 KEITH, H.D., and PADDEN, F.J.: *J. Appl. Phys.*, 1963, **34(8)**, pp. 2409–2421
12 KEITH, H.D., and PADDEN, F.J.: *J. Appl. Phys.*, 1964, **35**, pp. 1270–1285
13 KEITH, H.D., and PADDEN, F.J.: *J. Appl. Phys.*, 1964, **35**, pp. 1286–1296
14 WUNDERLICH, B., and MEHTA, A.: *J. Polym. Sci.: Polym. Phys. Ed.*, 1974, **12(2)**, pp. 255–263
15 STEVENS, G.C., and SWINGLER, S.G.: 'Conf. Diel. Mats., Meas. & Applics.', 1984, IEE Conf. Pub. **239**, pp. 191–194
16 CAPPACIO, G., GÖLTZ, W., and ROSE, L.J.: *ETG Fachberichte*, 1985, **16**, pp. 123–126
17 PHILLIPS, P.J.: 'Conf. Diel. Mats., Meas. & Applics.', 1984, IEE Conf. Pub. **239**, pp. 187–190
18 MITSUI, T., KOBAYASHI, T., SHIMIZU, H., NAGASAKI, S., MATSUBARA, H., KUNO, M., KOMATSUBARA, H., and OKADA, T.: 'Conf. rec. Int. Symp. Elec. Insul. (Montreal)', 1984, pp. 22–26
19 BAMJI, S., BULINSKI, A., DENSLEY, J., and GARTON, A.: *IEEE Trans. Elec. Insul.*, 1983, **EI-18**, pp. 32–41
20 STEVENS, G.C.: in 'Structural adhesives: developments in resins and primers', ed. KINLOCK, A.J. (Elsevier Appl. Sci. Publ., London, 1986) Chapter 7

21 STEVENS, G.C., CHAMPION, J.V., and LIDDEL, P.J.: *J. Polym. Sci., Polym. Phys. Ed.*, 1982, **20**, pp. 327–344
22 STEVENS, G.C., PERKINS, E., and CHAMPION, J.V.: 'IEE Conf. Diel. Mats., Meas. & Applics.', 1988, IEE Conf. Pub. **289**, pp. 234–237
23 TANAKA, J., and LUTHER, R.: *Int. Symp. Elec. Insul.*, 1982, **14**, pp. 292–295
24 SHIBUYA, Y. SOLEDZIOSKI, S., and CALDERWOOD, J.H.: *IEEE Transactions on Power Apparatus and Systems*, 1977, **PAS-96**, pp. 198–207
25 HILL, R.M., DISSADO, L.A., and JACKSON, R.: *J. Phys. C.: Solid State Phys.*, 1981, **14**, pp. 3915–3926
26 DISSADO, L.A., and HILL, R.M.: *Nature*, 1979, **279**, pp. 685–689
27 DISSADO, L.A., and HILL, R.M.: *Phil. Mag.*, 1980, **B-41**, pp. 625–642
28 OLLEY, R.H., HODGE, A.M., and BASSETT, D.C.: *J. Poly. Sci.: Poly. Phys. Ed.*, 1979, **17**, pp. 627–643
29 BARNES, S.R.: *Polymer*, 1980, **21**, pp. 723–726
30 DISSADO, L.A., FOTHERGILL, J.C., and WOLFE, S.V.: *Ann. Rep. Conf. Elec. Insul. & Diel. Phenom.*, 1981, pp. 264–277
31 DISSADO, L.A., FOTHERGILL, J.C., and WOLFE, S.V.: *ITT Electrical Communication*, 1982, **57**, pp. 87–90
32 PHILLIPS, P.J.: *IEEE Transactions on Electrical Insulation*, 1978, **EI-13**, pp. 451–453
33 NAYLOR, K.L., and PHILLIPS, P.J.: *J. Poly. Sci.: Poly. Phys. Ed.*, 1983, **21**, pp. 2011–2026
34 WAGNER, H.: *Ann. Rep. Conf. Elec. Insul. & Diel. Phenom.*, 1976, pp. 363–371
35 SCHONHORN, H.: *J. Poly. Sci. B, Polym. Lett.*, 1964, **2**, pp. 465–467
36 JENCKEL, E., TEEGE, E., and HINRICHS, W.: *Kolloid Z*, 1952, **129**, pp. 19–24
37 KITAGAWA, K., SAWA, G., and IEDA, M.: *Jpn. J. Appl. Phys.*, 1980, **19(2)**, pp. 389–392
38 ROSE, L.J., ROSE, V., and DE BELLET, J.J.: 'Proc. 2nd Int. Conf. Cond. & Breakdown in Solid Dielectrics (Erlangen)', 1986, pp. 237–244
39 FILIPPINI, J.C., POGGI, Y., RAHARIMALALA, V., DE BELLET, J.J., and MATEY, G.: 'Proc. 2nd Int. Conf. Cond. & Breakdown in Solid Dielectrics (Erlangen)', 1986, pp. 250–256
40 WOLFE, S.V.: Personal communication
41 TANAKA, J.: *IEEE Trans. Elec. Insul.*, 1980, **EI-15**, pp. 201–205
42 IEDA, M., KOSAKI, M., and SUGIYAMA, K.: *Ann. Rep. Conf. Elec. Insul. & Diel. Phenom.*, 1970, pp. 11–17
43 TAYLOR, D.M., and LEWIS, T.J.: *J. Phys. D.*, 1971, **4**, pp. 1346–1357
44 SEANOR, D.A.: *J. Polym. Sci. A-2*, 1968, **6**, pp. 463–477
45 MIYAMOTO, T., and SHIBAYAMA, K.: *J. Appl. Phys.*, 1973, **44(12)**, pp. 5372–5376
46 DAS-GUPTA, D.K., and DOUGHTY, K.: *IEEE Trans. Elec. Insul.*, 1987, **EI-22(1)**, pp. 1–7
47 SEYMOUR, J.: 'Electronic devices and components' (Pitman, London, 1988) Chapters 1–3
48 PETHIG, R.: 'Dielectric and electronic properties of biological materials' (J. Wiley & Sons, Chichester, 1979) Chapter 8
49 KITTLE, C.: 'Introduction to solid state physics' (J. Wiley & Sons, New York, 6th edition, 1986)
50 PARK, D.: 'Introduction to quantum theory' (McGraw-Hill, New York, 1974)

51 SLATER, J.C.: 'Quantum theory of molecules and solids' (McGraw-Hill, New York, 1965)
52 MERZBACHER, E.: 'Quantum Mechanics' (J. Wiley & Sons, New York, 1970)
53 DEKKER, A.J.: 'Solid state physics' (Macmillan, London, 1971)
54 ANDRÉ, J.-M.: in 'Electronic structure of polymers and molecular crystals', eds. ANDRÉ, J.-M., and LADIK, J. (Plenum Press, New York, 1974) pp. 1–21
55 McCUBBIN, W.L., and GURNEY, I.D.C.: *J. Chem. Phys.*, 1965, **43**, pp. 983–987
56 McCUBBIN, W.L.: in 'Electronic structure of polymers and molecular crystals', eds. ANDRÉ J.-M., and LADIK, J. (Plenum Press, New York, 1974) pp. 171–198
57 MOTT, N.F.: *Adv. Phys.*, 1967, **16**, pp. 49–144
58 COHEN, M.H. and LEKNER, J.: *Phys. Rev.*, 1967, **158**, pp. 305–309
59 IEDA, M.: *IEEE Trans. Elec. Insul.*, 1984, **EI-19**, pp. 162–178
60 MOTT, N.F.: 'Conduction in non-crystalline materials' (Clarendon Press, Oxford, 1987)
61 RITSKO, J.J.: in 'Electronic properties of polymers', eds. MORT, J., and PFISTER, G. (J. Wiley & Sons, New York, 1982) Chapter 2, pp. 13–57
62 IEDA, M.: *IEEE Trans. Elec. Insul.*, 1986, **EI-21(5)**, pp. 793–802 (The 1986 Conf. Elec. Insul. & Diel. Phenom. Whitehead Memorial Lecture)
63 BLOOR, D.: *Chem. Phys. Lett.*, 1976, **40(2)**, pp. 323–325
64 FABISH, T.J.: *CRC Crit. Rev. Solid State Mater. Sci.*, December, 1979
65 CLARK, D.T.: in 'Electronic structure of polymers and molecular crystals', eds. ANDRÉ, J.-M., and LADIK, J. (Plenum Press, New York, 1974) pp. 259–387
66 ANDRÉ, J.-M., DELHALLE, J., DELHALLE, S., CAUDANO, R., PIREAUX, J.J., and VERBIST, J.J.: *Chem. Phys. Lett.*, 1973, **23**, pp. 206–210
67 ZELLER, H.R., PFLUGER, P., and BERNASCONI, J.: *IEEE Trans. Elec. Insul.*, 1984, **EI-19(3)**, pp. 200–204
68 CARTIER, E., and PFLUGER, P.: *Phys. Rev. B*, 1986, **34(12)**, pp. 8822–8827
69 ZELLER, H.R.: *IEEE Trans. Elec. Insul.*, 1987, **EI-22(2)**, pp. 115–122
70 CARTIER, E., and PFLUGER, P.: *IEEE Trans. Elec. Insul.*, 1987, **EI-22(2)**, pp. 123–128
71 SEANOR, D.A.: in 'Electrical properties of polymers', ed. SEANOR, D.A. (Academic Press, New York, 1982) Chapter 1, pp. 1–58
72 MORT, J., and PFISTER, G.: in 'Electronic Properties of Polymers', eds. MORT, J., and PFISTER, G. (J. Wiley & Sons, New York, 1982) Chapter 6, pp. 215–265
73 SKOTHEIM, T.A. (ed.): 'Handbook of Conducting Polymers' (Marcel Dekker Inc., New York, 1986)
74 SAWA, G.: *J. Appl. Phys.*, 1977, **48(6)**, pp. 2414–2418
75 BINKS, A.E., and SHARPLES, A.: *J. Polym. Sci. A-2*, 1968, **6**, pp. 407–420
76 WALLACE, R.A.: *J. Appl. Phys. (USA)*, 1971, **42(18)**, pp. 3121–3124
77 WALLACE, R.A.: *J. Appl. Polym. Sci.*, 1974, **18**, pp. 2855–2859
78 CROWLEY, J.R., WALLACE, R.A., and BUBE, R.H., *J. Polym. Sci. Polym. Phys. Ed.*, 1976, **14**, pp. 1769–1787
79 PRICE, J.R., and DANNHAUSER, W.: *J. Phys. Chem.*, 1967, **71**, pp. 3570–3575

80 HOOVER, M.F., and CARR, H.E.: *TAPPI*, 1968, **51**, 552
81 SAITO, S., SASABE, H., NAKAJIMA, T., and YADA, K.: *J. Polym. Sci. A-2*, 1968, **6**, pp. 1297–1315
82 O'DWYER, J.J.: 'The theory of electrical conduction & breakdown in solid dielectrics' (Clarendon Press, Oxford, 1973)
83 DUKE, C.B., and FABISH, T.J.: *Phys. Rev. Lett.*, 1976, **37**, pp. 1075–1078
84 MORT, J., PFISTER, G., and GRAMMATICA, S.: *Solid State Commun.*, 1976, **18**, pp. 693–696
85 SCHER, H., and MONTROLL, E.W.: *Phys. Rev. B.*, 1975, **12(6)**, pp. 2455–2477
86 MAXWELL, J.C.: 'A treatise on electricity and magnetism' (Clarendon Press, Oxford, 1881) Vol. 1, p. 435
87 LORD RAYLEIGH: *Phil. Mag.*, 1892, **34**, 481–502
88 WAGNER, K.W.: *Arch. Elecktrotech*, 1914, **2**, p. 378 and 1915, **3**, p. 100
89 WAGNER, K.W.: in 'Die isolierstoffe der elektrotechnik, ed. SCHERING, H. (Springer, Berlin, 1924)
90 PELISSOU, S., ST-ONGE, H. and WERTHEIMER, M.R.: *IEEE Trans. Elec. Insul.*, 1988, **EI-23**, 325–333
91 GARTON, C.G., and PARKMAN, N.: *Proc. IEE*, 1976, **123(3)**, pp. 271–276
92 KOSAKI, M., YODA, M., and IEDA, M.: *J. Phys. Soc. Japan*, 1971, **31**, p. 1598
93 CRINE, J.P., and VIJH, A.K.: *Appl. Phys. Commun.*, 1984, **4**, pp. 135–147
94 STRUIK, L.C.E.: 'Physical aging in amorphous polymers and other materials' (Elsevier Press, Amsterdam, 1978)
95 REICH, L., and STIVALA, S.A.: 'Elements of polymer degradation' (McGraw-Hill, New York, 1971)
96 HILL, N.E., VAUGHAN, W.E., PRICE, A.H., MANSEL DAVIES: 'Dielectric properties and molecular behaviour' (Van Nostrand Reinhold, London, 1969) pp. 422 *et seq.*
97 PATHMANATHAN, K., DISSADO, L.A., and HILL, R.M., *J. Mat. Sci.*, 1985, **20**, pp. 3716–3728
98 HILL, R.M., and DISSADO, L.A.: *J. Phys. C.*, 1982, **15**, pp. 5171–5193
99 HILL, R.M., and DISSADO, L.A.: *J. Polym. Sci., Polym. Phys. Ed.*, 1984, **22**, pp. 1991–2008
100 ISHIDA, Y.: *Kolloid Zeitschrift*, 1960, **168**, 29–36
101 BHATEJA, S.K.: *J. Appl. Polym. Sci.*, 1983, **28**, pp. 861–872
102 LUSTIGER, A., and EPSTEIN, M.M.: *Ann. Rep. Conf. Elec. Insul. & Diel. Phenom.*, 1986, pp. 351–357
103 BRANCATO, E.L.: *IEEE Trans. Elec. Insul.*, 1978, **EI-13**, pp. 308–317
104 BARLOW, A., HILL, L.A., and MARINGER, M.F.: *IEEE Trans. Power Appar. & Syst.*, 1983, **PAS-102(7)**, pp. 1921–1926
105 STEVENS, G.C., PERKINS, E., and CHAMPION, J.V.: 'IEE Conf. Diel. Mats., Meas. & Applics.', 1988, IEE Conf. Pub. **289**, pp. 234–237
106 NATH, R., and PERLMAN, M.M.: *IEEE Trans. Elec. Insul.*, 1989, **EI-24**, pp. 409–412
107 PERLMAN, M.M., and HARIDOSS, S.: 'Proc. 2nd Int. Conf. Cond. Breakdown in Solid Dielectrics (Erlangen)', 1986, pp. 494–499
108 SEDDING, H.G., and HOGG, W.K.: 'IEE Conf. Diel. Mats., Meas. & Applics.', 1988, IEE Conf. Pub. **289**, pp. 207–210

109 SMIT, J.J., and GUERTS, W.S.M.: 'IEE Conf. Diel. Mats., Meas. & Applics.', 1984, IEE Conf. Pub. **239**, pp. 68–71
110 UMEMURA, T., and COUDERC, D.: *Ann. Rep. Conf. Elec. Insul. & Diel. Phenom.*, 1982, pp. 586–591
111 COUDERC, D., CRINE, J.P., and UMEMURA, T.: 'IEE Conf. Diel. Mats., Meas. & Applics.', 1984, IEE Conf. Pub. **239**, pp. 128–131
112 CUDDIHY, E.F.: *IEEE Trans. Elec. Insul.*, 1987, **EI-22**, pp. 573–589
113 OKAMOTO, T., ISHIDA, M., and HOZUMI, N.: *IEEE Trans. Elec. Insul.*, 1989, **EI-24**, pp. 599–607
114 BURNLEY, K.G., and EXON, J.L.T.: 'IEE Conf. Diel. Mats., Meas. & Applics.', 1988, IEE Conf. Pub. **289**, pp. 383–386
115 SRINIVAS, N., DOEPKEN, H.C., McKEAN, A.L., and BISKEBORN, M.C.: EPRI report **EL-647**, 1978
116 ZHURKOV, S.N., ZAKREVSKYI, V.A., KORSUKOV, V.E., and KUKSENKO, V.S.: *J. Poly. Sci. A-2*, 1972, **10**, pp. 1509–1520
117 THUE, W.A., and LYLE, R.: *IEEE Trans. Power Appar. & Sys.*, 1983, **PAS-102**, pp. 2116–2123
118 BERNSTEIN, B.S.: *IEEE Int. Symp. Elec. Insul. (Montreal)*, 1984, pp. 11–21
119 SHAW, M.T., and SHAW, S.H.: *IEEE Trans. Elec. Insul.*, 1984, **EI-19**, pp. 419–452
120 BAHDER, G., GARRITY, T., SOSNOWSKI, M., EATON, R., and KATZ, C.: *IEEE Trans. Power Appar. & Syst.*, 1982, **PAS-101**, pp. 1379–1390
121 EICHHORN, R.M.: *IEEE Trans. Elec. Insul.*, 1977, **EI-12**, pp. 2–18
122 MASON, J.: *Proc. IEE*, 1981, **128A**, pp. 193–201
123 MAYOUX, C.J.: *IEEE Trans. Elec. Insul.*, 1976, **EI-11**, pp. 139–148
124 MAYOUX, C.J.: *IEEE Trans. Elec. Insul.*, 1977, **EI-12**, pp. 153–158
125 LAURENT, C., MAYOUX, C.J., and SERJENT, A.: *IEEE Trans. Elec. Insul.*, 1981, **EI-16**, pp. 52–58
126 OKAMOTO, T., and TANAKA, T.: *Jap. J. Appl. Phys.*, 1985, **24**, pp. 156–160
127 KITAMURA, Y., and HIRABAYASHI, S.: *Ann. Rep. Conf. Elec. Insul. & Diel. Phenom.*, 1985, pp. 485–490
128 EICHHORN, R.M.: *Union Carbide Kabelitems*, 1979, **155**, Treeing Update IV
129 NOTO, F., and YOSHIMURA, N.: *Ann. Rep. Conf. Elec. Insul. & Diel. Phenom.*, 1974, pp. 207–217
130 BAMJI, S., BULINSKI, A.T., and DENSLEY, R.J.: 'Conf. Rec. IEEE Int. Symp. Elec. Insul. (Montreal)', 1984, pp. 37–40
131 BUDENSTEIN, P.P.: *IEEE Trans. Elec. Insul.*, 1980, **EI-15**, pp. 225–240
132 LLOYD Jr., J.M., and BUDENSTEIN, P.P.: *Ann. Rep. Conf. Elec. Insul. & Diel. Phenom.*, 1977, pp. 339–346
133 JONSCHER, A.K., and LACOSTE, R.: *IEEE Trans. Elec. Insul.*, 1984, **EI-19**, pp. 567–577
134 NAYBOUR, R.D.: *Ann. Rep. Conf. Elec. Insul. & Diel. Phenom.*, 1982, pp. 620–627
135 NAYBOUR, R.D.: '1st. Int. Conf. Cond. & Break. in Solid Diels. (Toulouse)', 1983, pp. 380–383
136 CROSS, J.D., and KOO, J.Y.: *IEEE Trans. Elec. Insul.*, 1984, **EI-19**, pp. 303–306
137 FOTHERGILL, J.C., DISSADO, L.A., and WOLFE, S.V.: 'Conf. Diel. Mats., Meas. & Applics.', 1984, IEE Conf. Pub. **239**, pp. 179–182

138 GARTON, C.G., and STARK, K.H.: *Nature*, 1955, **176**, pp. 1225–1226
139 PARK., C.H., KANEKO, T., HARA, M., and AKAZAKI, M.: *IEEE Trans. Elec. Insul.*, 1982, **EI-17**, pp. 234–240
140 PARK, C.H., HARA, M., and AKAZAKI, M.: *IEEE Trans. Elec. Insul.*, 1982, **EI-17**, pp. 546–553
141 ZELLER, H.R., and SCHNEIDER, W.R.: *J. Appl. Phys.*, 1984, **56**, pp. 455–459
142 YASUI, T., and HAYAMI, T.: *Sumitomo Electric Technical Review*, 1968, **97**, pp. 44–59
143 MITA, S., and YAHAGI, K.: *Jap. J. Appl. Phys.*, 1975, **14**, pp. 197–201
144 SCARISBRICK, R.M.: 'IEE Conf. Diel. Mats., Meas. & Applics.', 1984, IEE Conf. Pub. **239**, pp. 115–117
145 McMAHON, E.J., and PERKINS, J.R.: 'Am. Inst. Elect. Engineers, Middle Eastern District Meeting, Wilmington', 1962, DP G2-858, Paper 625
146 KLEIN, N.: *J. Appl. Phys.*, 1982, **53**, pp. 5828–5839
147 LAMPERT, M.A.: 'Injection currents in solids' (Academic Press, New York, 1965)
148 SIMMONS, J.G.: *Phys. Rev.*, 1967, **155(3)**, pp. 657–659
149 FRENKEL, J.: *Phys. Rev.*, 1938, **54**, pp. 647–648
150 O'DWYER, J.J.: *Ann. Rep. Conf. Elec. Insul. & Diel. Phenom.*, 1982, pp. 319–327
151 ZENER, C.: *Proc. Roy. Soc. (London)*, 1934, **A145**, pp. 523–529
152 SEITZ, F.: *Phys. Rev.*, 1949, **76**, pp. 1376–1393
153 SINGH, A., and PRATAP, A.R.: *Thin Solid Films*, 1982, **87(2)**, pp. 147–150
154 KADARY, V., and KLEIN, N.: *J. Electrochem. Soc. (USA)*, 1980, **127(1)**, pp. 139–151
155 COOPER, R., and ELLIOT, T.: *Brit. J. Appl. Phys.*, 1966, **17**, pp. 481–488
156 BÄSSLER, H., VAUBEL, G., RASSKOPF, K., and REINKE, K.: *Z. für Naturforsch*, 1971, **26A**, pp. 814–818
157 DORLANNE, O., ANDRAINJOHANINARIVO, J., PAULIN-DANDURAND, S., WERTHEIMER, M.R., and YELON, A.: '1st. Int. Conf. Cond. Breakdown Sol. Diels.', 1983, pp. 236–243
158 FISCHER, P., NISSEN, K.W., and ROHL, P.: *Siemens Forsch.- & Entwicklungsber (Germany)*, 1981, **10(4)**, pp. 222–227
159 FISCHER, P.: in 'Electrical properties of polymers', ed. SEANOR, D.A. (Academic Press, 1982) Chapter 8, pp. 319–367
160 FISCHER, P., and NISSEN, K.W.: *IEEE Trans. Elec. Insul.*, 1976, **EI-11(2)**, pp. 37–40
161 KLEIN, N.: 'Adv. Electronics & Electron Phys., Vol. 26; ed. MARTON, L. (Academic Press, New York, 1969) pp. 309–424
162 HIKITA, M., NAGAO, M., SAWA, G., and IEDA, M.: *J. Phys. D.: Appl. Phys.*, 1980, **13**, pp. 661–666
163 IEDA, M.: *IEEE Trans. Elec. Insul.*, 1980, **EI-15**, pp. 206–224
164 IEDA, M.: 'Proc. 1st. Int. Conf. Cond. Breakdown Sol. Dielectrics', 1983, pp. 1–33. *IEEE Trans. Elec. Insul.*, 1984, **EI-19(3)**, pp. 162–177
165 ALSTON, L.L. ed.: 'High-voltage technology' (Oxford University Press, Oxford, 1968)
166 TANAKA, T., and GREENWOOD, A.: 'Advanced power cable technology. Vol. 1: Basic concepts and testing' (CCRC Press, Florida, USA, 1983)

167 NAMIKI, Y., SHIMANUKI, Y., AIDA, F., and MORITA, M., *Ann. Rep. Conf. Elec. Insul. & Diel. Phenom.*, 1979, pp. 490–499
168 MATSUOKA, S.: *J. Appl. Phys.*, 1961, **32(11)**, pp. 2334–2336
169 CHEO, P.K., LUTHER, R., and PORTER, J.W.: *IEEE Trans. Power Appar. & Syst.*, 1983, **PAS-102(3)**, pp. 521–526
170 MUCCIGROSSO, J., and PHILLIPS, P.J.: *IEEE Trans. Elec. Insul.*, 1978, **EI-13**, pp. 172–178
171 YODA, B., and MURAKI, K.: *IEEE Trans. Power Appar. & Syst.*, 1973, **PAS-92**, pp. 506–513
172 JACKSON, R.J., and SWINGLER, S.G.: 'Conf. Diel. Mats., Meas. & Applics.', 1984, IEE Conf. Pub. **239**, pp. 92–95
173 HILL, R.M., and DISSADO, L.A.: *J. Phys. C.: Solid State Phys.*, 1983, **16**, pp. 2145–2156
174 DISSADO, L.A., FOTHERGILL, J.C., WOLFE, S.V., and HILL, R.M.: *IEEE Trans. Elec. Insul.*, 1984, **EI-19(3)**, pp. 227–233
175 BULINSKI, A. & DENSLEY, R.J., *IEEE Trans. Elec. Insul.*, 1981, **EI-16**, pp. 319–326
176 RYNKOWSKI, A.W.: *IEEE Trans. Power Appar. & Syst.*, 1981, **PAS-100(4)**, pp. 1829–1837
177 STONE, G.C., KURTZ, M., and VAN HEESWIJK, R.G.: *IEEE Trans. Elec. Insul.*, 1979, **EI-14(6)**, pp. 315–326
178 McMAHON, E.J.: *IEEE Trans. Elec. Insul.*, 1978, **EI-13**, pp. 277–288
179 BOUQUET, F.L., and MAAG, C.R.: *Ann. Rep. Conf. Elec. Insul. & Diel. Phenom.*, 1982, pp. 216–221
180 PASSENHEIM, B.C., VAN LINT, V.A.J., RIDELL, J.D., and KITTERERT, R.: *Ann. Rep. Conf. Elec. Insul. & Diel. Phenom.*, 1982, pp. 200–206
181 BAJBOR, Z.Z.: *IEEE Trans. Elec. Insul.*, **EI-22**, 1987, pp. 485–487
182 Flame Retardants '90 British Plastics Federation (Elsevier Press, London, 1990)
183 PETTERSON, R.L.: *ITT Electrical Communication*, 1985, **59(4)**, pp. 434–436
184 BJØRLØW-LARSEN, K., HOLTE, T.A., and LARSEN, P.B.: *ITT Electrical Communication*, 1985, **59(4)**, pp. 446–450
185 LAWSON, J.H., and VAHLSTROM, W.: *IEEE Trans. Power Appar. & Syst.*, 1973, **PAS-92**, pp. 824–835
186 KING, J., and KIM, S.: *IEEE Trans. Elec. Insul.*, 1990, **EI-25**, pp. 1170–1173
187 MINZ, J.D.: *IEEE Trans. Power Appar. & Syst.*, 1984, **PAS-103**, pp. 3448–3453
188 SAPHIEHA, S., and CRINE, J.P.: *Ann. Rep. Conf. Elec. Insul. & Diel. Phenom.*, 1980, pp. 212–217
189 MATSUBA, H., and KAWAI, E.: *IEEE Trans. Power Appar. & Syst.*, 1976, **PAS-95**, pp. 660–670
190 KATZ, C., and BERNSTEIN, B.S.: *Ann. Rep. Conf. Elec. Insul. & Diel. Phenom.*, 1973, pp. 307–316
191 SLETBAK, J., and ILDSTAD, E.: *Conf. Rec. IEEE Int. Symp. Elec. Insul. (Montreal)*, 1984, pp. 29–32
192 KISS, K.D., DOEPKEN Jr., H.C., SRINIVAS, N., & BERNSTEIN, B.S.: 'Durability of macromolecular materials, Vol. 95; ed. EBY, R.K. (ACS Symp. Ser., 1979) pp. 433–466
193 BAHDER, G., KATZ, C., LAWSON, J., and VAHLSTROM, W.: *IEEE Trans. Power Appar. & Syst.*, 1974, **PAS-93**, pp. 977–990
194 TABATA, T., NAGAI, H., FUKUDA, T., and IWATA, Z.: *IEEE Trans. Power Appar. & Syst.*, 1972, **PAS-91**, pp. 1354–1360

195 KATO, H., and FUJITA, H.: *Ann. Rep. Conf. Elec. Insul. & Diel. Phenom.*, 1980, pp. 203–211
196 YOSHIMITSU, T., & NAKAKITA, T.: *IEEE Int. Symp. Elec. Insul. (Montreal)*, 1978, pp. 116–121
197 YOSHIMURA, N., & NOTO, F.: *IEEE Trans. Elec. Insul.*, 1982, **EI-17**, pp. 363–367
198 ASHCRAFT, A.C., EICHHORN, R.M., and SHAW, R.G.: *IEEE Int. Symp. Elec. Insul. (Montreal)*, 1976, pp. 213–218
199 ASHCRAFT, A.C.: 'Proc. World Electrochem. Congress (Moscow)', 1977, Section 3A, Paper 13, pp. 21–25 (reprinted as Union Carbide Kabelitem, 'Treing Update III', publication F-46652, **152**, July 1977)
200 NAYBOUR, R.D.: 'Conf. Diel. Mats., Meas. & Applics., 1979', IEE Conf. Pub. **177**, pp. 238–241
201 ROWLAND, S.M. & DISSADO, L.A.: 'Conf. Diel. Mats., Meas. & Applics.', 1988, IEE Conf. Pub. **289**, pp. 246–249
202 PATSCH, R.: Conf. Diel. Mats., Meas. & Applics., 1988, IEE Conf. Pub. **289**, pp. 242–245
203 ROGOWSKI, W.: *Archiv fur Electrotechnik*, 1923, **xii(1)**, pp. 1–15
204 SLETBAK, J., & BOTNE, A.: *IEEE Trans. Elec. Insul.*, 1977, **EI-12**, pp. 383–388
205 ROWLAND, S.M., and WOLFE, S.V.: '2nd. Int. Conf. Power Cables & Acc. (London)', 1986, IEE Conf. Pub. **270**, pp. 129–133
206 DISSADO, L.A., FOTHERGILL, J.C., & WOLFE, S.V.: *IEEE Trans. Elec. Insul.*, 1983, **EI-18**, pp. 565–585
207 FILIPPINI, J.C., MEYER, C.T., and EL. KAHEL, M.: *Ann. Rep. Conf. Elec. Insul. & Diel. Phenom.*, 1982, pp. 629–637
208 FOURNIE, R., PERRET, J., RECOUPE, P., & LE-GALL, Y.: *IEEE Int. Symp. Elec. Insul. (Montreal)*, 1978, pp. 110–115
209 LAURENT, C., and MAYOUX, C.: 'Conf. Diel. Mats., Meas. & Applics.', 1979, IEE Conf. Pub. **177**, pp. 223–224
210 FORSTER, E.O.: *IEEE Trans. Elec. Insul.*, 1985, **EI-20**, pp. 891–896
211 MEYER, C.T.: *IEEE Trans. Elec. Insul.*, 1983, **EI-18**, pp. 28–31
212 BAMJI, S., BULINSKI, A., DENSLEY, J., GARTON, A., and SHIMIZU, N.: *Ann. Rep. Conf. Elec. Insul. & Diel. Phenom.*, 1984, pp. 141–147
213 ABDOLALL, K., ORTON, H.E., and REYNOLDS, M.W.: *Ann. Rep. Conf. Elec. Insul. & Diel. Phenom.*, 1985, pp. 302–311
214 KOO, J.Y., CROSS, J.D., EL. KAHEL, M., MEYER, C.T., and FILIPPINI, J.C.: *Ann. Rep. Conf. Elec. Insul. & Diel. Phenom.*, 1983, pp. 301–305
215 DORLANNE, O., WERTHEIMER, M.R., YELON, A., and DENSLEY, J.R.: *Ann. Rep. Conf. Elec. Insul. & Diel. Phenom.*, 1980, pp. 136–143
216 CRICHTON, B.H., GIVEN, M.J., FARISII, O., & BAMFORD, H.M.: *J. Phys. D.*, 1983, **16**, pp. L223–L226
217 GIVEN, M.J., FOURACRE, R.A., and CRICHTON, B.H.: *IEEE Trans. Elec. Insul.*, 1987, **EI-22**, pp. 151–155
218 BAMJI, S., BULINSKI, A., DENSLEY, J., GARTON, A., and SHIMIZU, N.: *CIGRE Int. Conf. Large High Voltage Elec. Syst.*, 1984, Paper 15-07
219 NUNES, S.L., and SHAW, M.T.: *IEEE Trans. Elec. Insul.*, 1980, **EI-15**, pp. 437–450
220 HENKEL, H.J., and MÜLLER, N.: *Ann. Rep. Conf. Elec. Insul. & Diel. Phenom.*, 1985, pp. 281–289

221 ABDOLALL, K., ORTON, H.E., REYNOLDS, M.W., ROBERT, B.D., KENNEDY, R., and CLAYMAN, B.P.: *Ann. Rep. Conf. Elec. Insul. & Diel. Phenom.*, 1982, pp. 604–614

222 TANAKA, T., FUKUDA, T., and SUZUKI, S.: 'IEEE/PES Winter Meeting & Tesla Symp.', 1976, Paper no. F76178–4. *IEEE Trans. Power Appar. & Syst.*, 1976, **PAS-95**, pp. 1892–1900

223 NUNES, S.L., and SHAW, M.T.: *Ann. Rep. Conf. Elec. Insul. & Diel. Phenom.*, 1983, pp. 312–317

224 GARTON, A., BAMJI, S., BULINSKI, A., and DENSLEY, J.: *IEEE Trans. Elec. Insul.*, 1987, **EI-22**, pp. 405–412

225 ROSS, R., GEURTS, W.S.M., and SMIT, J.J.: 'Conf. Diel. Mats., Meas. & Applics.', 1988, IEE Conf. Pub. **289**, 313–317

226 DISSADO, L.A.: *IEEE Trans. Elec. Insul.*, 1982, **EI-17**, pp. 392–398

227 KARAKELLE, M., and PHILLIPS, P.J.: *IEEE Trans. Elec. Insul.*, 1989, **EI-24**, pp. 1083–1093 and 1101–1108; PHILLIPS, P.J. and KARAKELLE, M.: *IEEE Trans. Elec. Insul.*, 1989, **EI-24**, pp. 1093–1100

228 DISSADO, L.A., FILIPPINI, J.C., FOTHERGILL, J.C., MEYER, C.T., and WOLFE, S.V.: *IEEE Trans. Elec. Insul.*, 1988, **EI-23(3)**, pp. 345–356

229 ROWLAND, S.M., DISSADO, L.A., and WOLFE, S.V.: 'Proc. 2nd. Int. Conf. Cond. & Breakdown in Solid Dielectrics (Erlangen)', 1986, pp. 280–284

230 KATZ, C., SRINIVAS, N., and BERNSTEIN, B.S.: *Ann. Rep. Conf. Elec. Insul. & Diel. Phenom.*, 1974, pp. 279–287

231 YOSHIMITSU, T., MITSUI, H., KENJO, S., and NAKAKITA, T.: *IEEE Trans. Elec. Insul.*, 1983, **EI-18**, pp. 23–27

232 NOTO, F.: *IEEE Trans. Elec. Insul.*, 1980, **EI-15**, pp. 251–258

233 HOSSAM-ELDIN, A.A., and EL. SHAZLI, A.: *Ann. Rep. Conf. Elec. Insul. & Diel. Phenom.*, 1982, pp. 638–643

234 DISSADO, L.A., WOLFE, S.V., FOTHERGILL, J.C., JONES, T., and ROWLAND, S.M.: *IEEE Trans. Elec. Insul.*, 1986, **EI-21**, pp. 651–655

235 BAHDER, G., EAGER, G.S., and LUKAC, R.: *Ann. Rep. Conf. Elec. Insul. & Diel. Phenom.*, 1974, pp. 289–310

236 DISSADO, L.A., ROWLAND, S.M., FILIPPINI, J.C., FOTHERGILL, J.C., WOLFE, S.V., and MEYER, C.T.: *Ann. Rep. Conf. Elec. Insul. & Diel. Phenom.*, 1986, pp. 417–425

237 SOMA, K., and KUMA, S.. *IEEE Int. Symp. Elec. Insul.*, 1980, pp. 212–215

238 SLETBAK, J., and ILDSTAD, E.: *IEEE Trans. Power Appar. & Syst.*, 1983, **PAS-102**, pp. 2069–2076

239 GROEGER, J.H.: 'Proc. 1st. Int. Conf. Cond. Breakdown Sol. Dielectrics', 1983, pp. 360–363

240 MORITA, M., HANAI, M., SHIMANUKI, H., and AIDA, F.: *Ann. Rep. Conf. Elec. Insul. & Diel. Phenom.*, 1975, pp. 335–343

241 PATSCH, R., and PAXIMADAKIS, A.: 'IEEE/PES Transmission & Distribution Conf. & Exposition, (New Orleans)', 1989, Paper 89, TD350-0 PWRD

242 DISSADO, L.A.: *IEEE Trans. Elec. Insul.*, 1986, **EI-21**, pp. 657–658

243 KLINGER, Y.: *Ann. Rep. Cond. Elec. Insul. & Diel. Phenom.*, 1981, pp. 293–298

244 DENSLEY, R.J., BULINSKI, A., ROBERT, J.E.P., & SUDARSHAN, T.S.: *Can. Elec. Eng. J.*, 1980, **5**, pp. 9-14

245 YOSHIMURA, N., NOTO, F., and KIKUCHI, K.: *IEEE Trans. Elec. Insul.*, 1977, **EI-12**, pp. 411–416

246 FILIPPINI, J.C., and MEYER, C.T.: *IEEE Trans. Elec. Insul.*, 1988, **EI-23**, pp. 275–278

247 FILIPPINI, J.C., POGGI, Y., and RAHARIMALALA, V.: *Conf. Rec. Int. Conf. Prop. & Applic. Diel. Mats. (Xi'an)*, 1985, **1**, pp. 388–390

248 GUMBEL, E.J.: 'Statistics of extremes' (Columbia University Press, New York, 1958)

249 ILDSTAD, E.: *Proc. Nord-IS 88* (Nordic Insulation Symposium) The Norwegian Institute of Technology, Trondheim, June 1988, Paper 22

250 FILIPPINI, J.C., and MARTEAU, C.: *IEEE Trans. Elec. Insul.*, 1986, **EI-21**, pp. 161–164

251 NITTMAN, J., DACCORD, G., and STANLEY, H.E.: *Nature*, 1985, **314**, pp. 141–144

252 GÖLZ, W.: *Colloid & Polymer Sci.*, 1985, **263**, pp. 286–292

253 STEENIS, E.F., and BOONE, W.: 'Proc. IEE 2nd. Int. Conf. Power Cables and Accessories 10 kV to 180 kV', 1986, IEE Conf. Pub. **270**, pp. 162–166

254 FRANKE, E.A., and CZEKAJ, E.: *Ann. Rep. Conf. Elec. Insul. & Diel. Phenom.*, 1975, pp. 287–295

255 FRANKE, E.A., STAUFFER, J.R., and CZEKAJ, E.: *IEEE Trans. Elec. Insul.*, 1977, **EI-12**, pp. 218–223

256 COX, B.M., and WILLIAMS, W.T.: *J. Phys. D.*, 1977, **10**, pp. L5–L9

257 FILIPPINI, J.C., and JANTZEN, A.: 'Proc. 2nd. Int. Conf. Cond. & Breakdown in Solid Dielectrics (Erlangen)', 1986, pp. 245–249

258 YAMANOUCHI, S., SHIGA, T., and MATSUBARA, H.: *Ann. Rep. Conf. Elec. Insul. & Diel. Phenom.*, 1976, pp. 386–393

259 TU, D.M., and KAO, K.C.: *Ann. Rep. Conf. Elec. Insul. & Diel. Phenom.*, 1983, pp. 307–311

260 MEYER, C.T., and FILIPPINI, J.C.: *C.R. Acad. Sci. (Paris)*, 1980, 24 Nov., p. 291

261 BULINSKI, A.T., BAMJI, S.S., DENSLEY, R.J., CRINE, J.P., NOIRHOMME, B., and BERNSTEIN, B.: 'Proc. 3rd. Int. Conf. Cond. Break. in Solid Dielectrics (Trondheim)', 1989, pp. 422–426

262 ZELLER, H.R.: *IEEE Trans. Elec. Insul.*, 1987, **EI-22**, pp. 677–684

263 PATSCH, R.: *Colloid & Polymer Sci.*, 1981, **259**, pp. 885–893

264 TANAKA, T., FUKUDA, T., SUZUKI, S., NITTA, Y., GOTO, H., and KUBOTA, K.: *IEEE Trans. Power Appar. & Syst.*, 1974, **PAS-93**, pp. 693–702

265 ISSHIKI, S., YAMAMOTO, M., CHABATA, S., MIZOGUCHI, T., and ONO, M.: *IEEE Trans. Power Appar. & Syst.*, 1974, **PAS-93**, pp. 1419–1429

266 MIZUKAMI, T., KUMA, S., and SOMA, K.: *Ann. Rep. Conf. Elec. Insul. & Diel. Phenom.*, 1977, pp. 316–323

267 WILKENS, W.D.: *Ann. Rep. Conf. Elec. Insul. & Diel. Phenom.*, 1973, pp. 317–321

268 DISSADO, L.A., and WOLFE, S.V.: *Ann. Rep. Conf. Elec. Insul. & Diel. Phenom.*, 1984, pp. 200–207

269 MORITA, M., HANAI, M., SHIMANUKI, H., AIDA, F., and SHIONO, T.: *Ann. Rep. Conf. Elec. Insul. & Diel. Phenom.*, 1980, pp. 195–202

270 FEDORS, R.F.: *Polymer*, 1980, **21**, pp. 863–865

271 NITTA, Y.: *IEEE Trans. Elec. Insul.*, 1974, **EI-9**, pp. 109–112

272 SLETBAK, J.: *IEEE Trans. Power Appar. & Syst.*, 1979, **PAS-98**, pp. 1358–1365

273 MEYER, C.T., FILIPPINI, J.C., and FELICI, N.: *Ann. Rep. Conf. Elec. Insul. & Diel. Phenom.*, 1978, pp. 374–381
274 MINNEMA, L., BARNVELD, H.A., and RINKEL, P.D.: *IEEE Trans. Elec. Insul.*, 1980, **EI-15**, pp. 461–472
275 BAMJI, S., BULINSKI, A., DENSLEY, R.J., and SHIMIZU, N.: *Ann. Rep. Conf. Elec. Insul. & Diel. Phenom.*, 1982, pp. 592–597
276 CHEN, J.L., FILIPPINI, J.C., and POGGI, Y.: *Conf. Rec. Int. Conf. Prop. & Applic. Dielec. Mat. (Xi'an)*, 1985, **1**, pp. 366–369
277 GARTNER, E., and TOBAZEON, R.: *IEEE Symp. Elec. Insul. (Philadelphia)*, 1978, pp. 209–211
278 GOSSE, B., GOSSE, J.P., and TOBAZEON, R.: *Proc. IEE*, 1981, **128**, pp. 165–173
279 YAMADA, Y., YAMANOUCHI, S., and MIYAMOTO, S.: *Ann. Rep. Conf. Elec. Insul. & Diel. Phenom.*, 1979, pp. 500–510
280 TANAKA, T., NITTA, Y., & FUKUDA, T.: *Ann. Rep. Conf. Elec. Insul. & Diel. Phenom.*, 1972, pp. 216–221
281 YOSHIMITSU, T., MITSUI, H., HISHIDA, K., & YOSHIDA, H.: *IEEE Trans. Elec. Insul.*, 1983, **EI-18**, pp. 396–401
282 BOGGS, S.A., and RIZZETO, S.: *Ann. Rep. Conf. Elec. Insul. & Diel. Phenom.*, 1985, pp. 441–447
283 GENT, A.N., and LINDLEY, P.B.: *Proc. Roy. Soc. A*, 1958, **249**, pp. 195–205
284 GLUCHOWSKI, S.: *IEEE Trans. Elec. Insul.*, 1984, **EI-19**, pp. 362–363
285 MEYER, C.T., and CHAMEL, A.: *IEEE Trans. Elec. Insul.*, 1980, **EI-15(5)**, pp. 389–393
286 HENKEL, H.J., MÜLLER, N., NORDMANN, J., ROGLER, W., and ROSE, W.: *IEEE Trans. Elec. Insul.*, 1987, **EI-22**, pp. 157–161
287 IEDA, M., and NAWATA, M.: *IEEE Trans. Elec. Insul.*, 1977, **EI-12**, pp. 19–25
288 OKAMOTO, T., and TANAKA, T.: *IEEE Trans. Elec. Insul.*, 1985, **EI-20**, pp. 643–645
289 NIEMEYER, L., PIETRONERO, L., and WIESMANN, H.J.: *Phys. Rev. Lett.*, 1984, **52**, 1033–1036
290 KOO, J.Y., and FILIPPINI, J.C.: *IEEE Trans. Elec. Insul.*, 1984, **EI-19**, pp. 217–291
291 GOLZ, W.: 'EPRI Treeing Workshop (Monterey, California)', 1985, Report No: R8000-61
292 KATO, H., and UEDA, A.: *Dainichi-Nippon Cables Review*, 1979, **64**, pp. 23–31
293 GARTON, A., GROEGER, J.H., and HENRY, J.L.: *IEEE Trans. Elec. Insul.*, 1990, **EI-25**, pp. 427–434
294 STEENIS, E.F., and KREUGER, F.H.: *IEEE Trans. Elec. Insul.*, 1990, **EI-25**, pp. 989–1028
295 ROSS, R., GUERTS, W.S.M., SMIT, J.J., v.D. MASS, J.H., and LUTZ, E.T.G.: 'Proc. 3rd. Int. Conf. Cond. Break. in Solid Dielectrics (Trondheim)', 1989, pp. 512–516
296 SUH, K.S., DAMON, D., and TANAKA, J.: 'Conf. Diel. Mats., Meas. & Applics.', 1988, IEE Conf. Pub. **289**, pp. 17–20
297 MAURITZ, K.A., and ROGERS, C.E.: *Macromolecules*, 1985, **18**, pp. 483–491
298 DATYE, V.K., TAYLOR, P.L., and HOPFINGER, A.J.: *Macromolecules*, 1984, **17**, pp. 1704–1708

299 WEISS, R.A., and SHAW, S.H.: *SPE ANTEC*, 1983, **29**, pp. 434–435
300 BARTNIKAS, R.: EPRI 'Treeing Workshop, (Monterey, California)', 1985, Report No: R8000-61
301 HANAI, T.: *Kolloid-Z*, 1960, **171**, pp. 23–31
302 CLAUSSE, M.: in 'Encyclopedia of Emulsion Technology', Vol. 1, ed. BECHER, P.: (M. Dekker, New York, 1983) pp. 481–715
303 NIKLASSON, G.A.: *J. Appl. Phys.*, 1987, **62**, pp. R1–R14
304 HEUMANN, H., PÄTSCH, R., SAURE, M., and WAGNER, H.: *CIGRE Int. Conf. Large High Volt. Syst.*, 1980, Paper 15–06
305 FUKUDA, T., NAKANO, K., TANABE, T., SUDO, A., and KUHARA, T.: *Ann. Rep. Conf. Elec. Insul. & Diel. Phenom.*, 1974, pp. 218–228
306 NAYBOUR, R.D.: 'Proc. 2nd Int. Conf. Cond. Break. in Solid. Dielectrics (Erlangen)', 1986, pp. 232–236
307 LIU, S.H.: *Solid St. Physics*, 1986, **39**, pp. 207–273
308 MEAKIN, P.: *Phys. Rev.* 1983, **A27**, pp. 2616–2623
309 SHAW, M.T.: 'EPRI Water Treeing Workshop (Monterey, California)', 1985, Report No: R8000-61
310 PATERSON, L.: *Phys. Rev. Lett.* 1984, **52**, pp. 1621–1624
311 PATERSON, L.: *J. Fluid Mechs.* 1981, **113**, pp. 513–529
312 NITTMAN, J., and STANLEY, H.E.: *Nature*, 1986, **321**, pp. 663–668
313 MALOY, K.J., FEDER, J., and JOSSANG, T.: *Phys. Rev. Lett.*, 1985, **55**, pp. 2688–2691
314 OXAAL, U., MURAT, M., BOGER, F., AHARONY, A., FEDER, J., & JOSSANG, T.: *Nature*, 1987, **329**, pp. 32–37
315 COOPER, D.E., FARBER, M., and HARRIS, S.P.: *Ann. Rep. Conf. Elec. Insul. & Diel. Phenom.*, 1984, pp. 32–37
316 KOSAKI, M., SHIMIZU, N., and HORII, K.: *IEEE Trans. Elec. Insul.*, 1977, **EI-12**, pp. 40–45
317 LAURENT, C., and MAYOUX, C.: *IEEE Trans. Elec. Insul.*, 1980, **EI-15**, pp. 33–42
318 SHIMIZU, N., and HORII, K.: *IEEE Trans. Elec. Insul.*, 1985, **EI-20**, pp. 561–566
319 SHIMIZU, N., KATSUKAWA, H., MIYAUCHI, M., KOSAKI, M., and HORII, K.: *IEEE Trans. Elec. Insul.*, 1979, **EI-14**, pp. 256–263
320 MITSUI, H., YOSHIMITSU, T., MITZUTANI, Y., and UMEMOTO, K.: *IEEE Trans. Elec. Insul.*, 1981, **EI-16**, pp. 533–542
321 KITANI, I., and ARII, K.: *IEEE Trans. Elec. Insul.*, 1984, **EI-19**, pp. 281–287
322 NAKANISHI, K., HIRABAYASHI, S., and INUISHI, Y.: *IEEE Trans. Elec. Insul.*, 1979, **EI-14**, pp. 306–314
323 ADRIANJOHANINARIVO, J., WERTHEIMER, M.R., and YELON, A.: *IEEE Trans. Elec. Insul.*, 1987, **EI-22**, pp. 709–714
324 YOSHIMURA, N., HAMMAN, M.S.A.A., NISHIDA, M., and NOTO, F.: *Ann. Rep. Conf. Elec. Insul. & Diel. Phenom.*, 1978, pp. 342–351
325 WASILENKO, E.: 'Conf. Diel. Mats., Meas. & Applics.', 1988, IEE Conf. Pub. **289**, pp. 387–390
326 HOZUMI, N., OKAMOTO, T., and FUKUGAWA, H.: *Ann. Rep. Conf. Elec. Insul. & Diel. Phenom.*, 1987, pp. 531–535
327 STANLEY, H.E., and MEAKIN, P.: *Nature*, 1988, **335**, pp. 405–409
328 JOCTEUR, R., FAVRIE, E., and AUCLAIR, H.: *IEEE Trans. Power Appar. & Syst.*, 1977, **PAS-96**, pp. 513–523
329 KAO, K.C., and TU, D.M.: *Ann. Rep. Conf. Elec. Insul. & Diel. Phenom.*, 1982, pp. 598–603

330 FORSTER, E.O.: 'Proc. 1st Int. Conf. Cond. Breakdown in Solid Dielectrics', 1983, pp. 306–311
331 FORSTER, E.O.: 'Conf. Diel. Mats., Meas. & Applics., 1984', IEE Conf. Pub. **239**, pp. 1–5
332 MORITA, M., HANAI, M., & SHIMANUKI, H.: *Ann. Rep. Conf. Elec. Insul. & Diel. Phenom.*, 1973, pp. 299–306
333 WAGNER, H., and GOLZ, W.: '3rd. Int. Symp. High Volt. Eng. (Milan)', 1979, Paper 21-08
334 KAGAWA, T., and YAMAZAKI, S.: *IEEE Trans. Elec. Insul.*, 1982, **EI-17**, pp. 314–318
335 FUJITA, H., NAKANISHI, T., and YAMAGUCHI, K.: *IEEE Trans. Elec. Insul.*, 1983, **EI-18**, pp. 520–527
336 BAMJI, S.S., BULINSKI, A.T., and DENSLEY, R.J.: *IEEE Trans. Elec. Insul.*, 1986, **EI-21**, pp. 639–644
337 HIBMA, T., and ZELLER, H.R.: *J. Appl. Phys.*, 1986, **59**, pp. 1614–1620
338 ZELLER, H.R., BAUMANN, TH., CARTIER, E., DERSCH, H., PFLUGER, P., and STUCKI, E.: *Adv. Solid St. Phys.*, 1987, **27**, pp. 223–240
339 BAUMANN, TH., HIBMA, T., PETHICA, J.B., PFLUGER, P., and ZELLER, H.R.: 'Proc. 2nd Int. Conf. Cond. Breakdown in Solid Dielectrics (Erlangen)', 1986, pp. 131–137
340 KRAUSE, G., GOTTLICH, S., MOLLER, K., and MEURER, D.: 'Proc. 3rd Int. Conf. Cond. Breakdown in Solid Dielectrics (Trondheim)', 1989, pp. 560–564
341 SAKAMOTO, H., and YAHAGI, K.: *Jap. J. Appl. Phys.*, 1980, **19**, pp. 253–262
342 TANAKA, T., and GREENWOOD, A.: *IEEE Trans. Power Appar. & Syst.*, 1978, **PAS-97**, pp. 1749–1759
343 LAURENT, C., MAYOUX, C., and NOEL, S.: 'IEEE Int. Symp. Elec. Insul. (Montreal)', 1984, pp. 76–79
344 DORLANNE, O., SAPHIEHA, S., WERTHEIMER, M.R., and YELON, A.: *IEEE Trans. Elec. Insul.*, 1982, **EI-17**, pp. 199–202
345 PFLUGER, P., BAUMANN, TH., CARTIER, E., DERSCH, M., STUCKI, F., and ZELLER, H.: *Ann. Rep. Conf. Elec. Insul. & Diel. Phenom.*, 1987, pp. 180–192
346 ZELLER, H.R.: 'Proc. 2nd Int. Conf. Cond. Breakdown in Solid Dielectrics (Erlangen)', 1986, pp. 98–109
347 HIBMA, T., ZELLER, H.R., PFLUGER, P., and BAUMANN, TH.: *Ann. Rep. Conf. Elec. Insul. & Diel. Phenom.*, 1985, pp. 259–265
348 YOSHIMURA, N., and NOTO, F.: *IEEE Trans. Elec. Insul.*, 1984, **EI-19**, pp. 135–140
349 YOSHIMURA, N., and NOTO, F.: *IEEE Trans. Elec. Insul.*, 1983, **EI-18**, pp. 120–124
350 LAURENT, C., MAYOUX, C., and NOEL, S.: *J. Appl. Phys.*, 1983, **54**, pp. 1532–1539
351 SHIBUYA, Y., ZOLEDIOWSKI, A., and CALDERWOOD, J.H.: *Proc. IEE*, 1978, **125**, pp. 352–354
352 GLUCHOWSKI, S., and JUCHNIEWICZ, J.: 'Proc. 1st Int. Conf. Cond. Breakdown in Solid Dielectrics', 1983, pp. 230–235
353 BAMJI, S.S., BULINSKI, A.T., and DENSLEY, R.J.: *Ann. Rep. Conf. Elec. Insul. & Diel. Phenom.*, 1988, pp. 173–179
354 VON OLSHAUSEN, R., and SACHS, G.: *IEE Proc.*, 1981, **A128**, pp. 183–192
355 SANO, N.: *J. Phys.*, 1988, **D21**, pp. 1025–1027

356 CRINE, J.P., and VIJH, A.K.: *Ann. Rep. Conf. Elec. Insul. & Diel. Phenom.*, 1988, pp. 424–429
357 FRUTH, B., BAUMANN, TH., and STUCKI, F.: 'Proc. 3rd Int. Conf. Cond. Breakdown in Solid Dielectrics (Trondheim)', 1989, pp. 35–39
358 NOEL, S., DAS-GUPTA, D.K., LAURENT, C., and MAYOUX, C.: 'Conf. Diel. Mats. Meas. & Applics.', 1984, IEE Conf. Pub. **239**, pp. 61–63
359 MORI, J., SHIMIZU, N., and HORII, K.: *IEE Japan Proc. 14th. Symp. Elec. Insul. Materials*, 1981, pp. 59–61
360 SHIMIZU, N., UCHIDA, K., and HORII, K.: *Ann. Rep. Conf. Elec. Insul. & Diel. Phenom.*, 1987, pp. 419–424
361 BUERGER, K.G.: *Ann. Rep. Conf. Elec. Insul. & Diel. Phenom.*, 1984, pp. 175–180
362 WU, C.Y., WERTHEIMER, M.R., YELON, A., BOGGS, S.A., and DENSLEY, R.J.: *Ann. Rep. Conf. Elec. Insul. & Diel. Phenom.*, 1976, pp. 354–362
363 BAHDER, G., DAKIN, T.W., and LAWSON, J.H.: 'CIGRE Int. Conf. Large High Voltage Elec. Syst.', 1974, Paper 15-05
364 COCKS, A.C.F., and ASHBY, M.F.: *Proc. Mat. Sci.*, 1982, **27**, pp. 189–244
365 XIE, H-K., TU, D-M., and KAO, K.C.: 'Proc. 1st Int. Conf. Cond. Breakdown in Solid Dielectrics', 1983, pp. 225–229
366 ZELLER, H.R., HIBMA, T., and PFLUGER, P.: *Ann. Rep. Conf. Elec. Insul. & Diel. Phenom.*, 1984, pp. 85–88
367 ARBAB, M.N., AUCKLAND, D.W., and VARLOW, B.R.: 'Conf. Diel. Mats., Meas. & Applics.', 1988, IEE Conf. Pub. **289**, pp. 231–233
368 BAMJI, S., BULINSKI, A.T., and DENSLEY, R.J.: *Ann. Rep. Conf. Elec. Insul. & Diel. Phenom.*, 1987, pp. 425–431
369 FORSTER, E.O.: *IEEE Trans. Elec. Insul.*, 1984, **EI-19**, pp. 524–528
370 HOZUMI, N., and OKAMOTO, T.: 'Proc. 3rd Int. Conf. Cond. Breakdown in Solid Dielectrics (Trondheim)', 1989, pp. 543–547
371 HOZUMI, N., OKAMOTO, T., and FUKUGAWA, H.: *Jap. J. Appl. Phys.*, 1988, **27**, pp. 572–576
372 HOZUMI, N., OKAMOTO, T., and FUKUGAWA, H.: *Jap. J. Appl. Phys.*, 1988, **27**, pp. 1230–1233
373 BAMMERT, U., and BEYER, M.: *IEEE Trans. Elec. Insul.*, 1988, **EI-23**, pp. 215–225
374 TANAKA, T., REED, C.W., DEVINS, J.C., and GREENWOOD, A.: *Ann. Rep. Conf. Elec. Insul. & Diel. Phenom.*, 1978, pp. 333–341
375 ESCAICH, J.J., LAURENT, C., and MAYOUX, C.: *Ann. Rep. Conf. Elec. Insul. & Diel. Phenom.*, 1983, pp. 279–285
376 AUCKLAND, D.W., BORISHADE, A.B., and COOPER, R.: 'Conf. Diel. Mats., Meas. & Applics.', 1975, IEE Conf. Pub. **129**, pp. 15–18
377 BORISHADE, A.B., COOPER, R., and AUCKLAND, D.W.: *IEEE Trans. Elec. Insul.*, 1977, **EI-12**, pp. 348–354
378 KLEIN, N.: *Adv. Phys.*, 1972, **21**, pp. 605–645
379 OKAMOTO, T., and TANAKA, T.: *Ann. Rep. Conf. Elec. Insul. & Diel. Phenom.*, 1985, pp. 498–513
380 TANAKA, T.: *IEEE Trans. Elec. Insul.*, 1986, **EI-21**, pp. 899–905
381 DEVINS, J.C.: *IEEE Trans. Elec. Insul.*, 1984, **EI-19**, pp. 475–495
382 FUJITA, H.: *IEEE Trans. Elec. Insul.*, 1987, **EI-22**, pp. 277–285
383 MASON, J.H.: *IEEE Trans. Elec. Insul.*, 1976, **EI-11**, pp. 211–238

384 LAURENT, C., MAYOUX, C., NOEL, S., and SINISUKA, N.I.: *IEEE Trans. Elec. Insul.*, 1983, **EI-18**, pp. 125–130
385 RÖHL, P., and NISSEN, K.W.: *Ann. Rep. Conf. Elec. Insul. & Diel. Phenom.*, 1979, pp. 520–528
386 DENSLEY, R.J.: *IEEE Trans. Elec. Insul.*, 1979, **EI-14**, pp. 148–158
387 SUEDA, H., and KAO, K.C.: *Ann. Rep. Conf. Elec. Insul. & Diel. Phenom.*, 1979, pp. 61–68
388 KNAUR, J.A., and BUDENSTEIN, P.P.: *IEEE Trans. Elec. Insul.*, 1980, **EI-15**, pp. 313–321
389 FUJIMORI, S.: *Ann. Rep. Conf. Elec. Insul. & Diel. Phenom.*, 1984, pp. 374–380
390 WIESMANN, H.J., and ZELLER, H.R.: *J. Appl. Phys.*, 1986, **60**, pp. 1770–1773
391 FUKAI, J., and SANO, N.: *Ann. Rep. Conf. Elec. Insul. & Diel. Phenom.*, 1987, pp. 432–439
392 MANDELBROT, B.B.: 'The fractal geometry of nature' (W. Freeman, New York, 1982)
393 MEAKIN, P.: *Phys. Rev.*, 1983, **A27**, pp. 604–607
394 SATPATHY, S.: in 'Fractals in physics', eds. PIETRONERO, L., and TOSATTI, E. (North Holland, Amsterdam, 1986) pp. 173–176
395 TANG, C.: *Phys. Rev.*, 1985, **A31**, pp. 1977–1979
396 KERTESZ, J., and VICSEK, T.: *J. Phys. A.*, 1986, **19**, pp. L257–L262
397 TAKAYASU, H.: *Phys. Rev. Lett.*, 1985, **54**, pp. 1099–1101
398 DUXBURY, P.M., LEATH, P.L., and BEALE, P.D.: *Phys. Rev. B.*, 1987, **36**, pp. 367–380
399 STEVENS, G.C., and DISSADO, L.A.: 'CEGB Partial Discharge Project', unpublished work
400 BERNSTEIN, B.S., SRINIVAS, N., and LEE, P.N.: *Ann. Rep. Conf. Elec. Insul. & Diel. Phenom.*, 1975, pp. 296–302
401 WILKENS, W.D.: *Ann. Rep. Conf. Elec. Insul. & Diel. Phenom.*, 1974, pp. 260–269
402 MATSUBARA, M., and YAMANOUCHI, S.: *Ann. Rep. Conf. Elec. Insul. & Diel. Phenom.*, 1974, pp. 270–278
403 FAVRIE, E., and AUCLAIR, H.: *IEEE Trans. Power Appar. & Syst.*, 1980, **PAS-99**, pp. 1225–1234
404 HAGEN, S.T., ILDSTAD, E., SLETBAK, J., and FAREMO, H.: 'Proc. 3rd Int. Conf. Cond. Breakdown in Solid Dielectrics (Trondheim)', 1989, pp. 569–573
405 McKEAN, A.L., DOEPKEN, H.C., TSUJI, K., and ZIDON, A.: *IEEE Trans. Power Appar. & Syst.*, 1978, **PAS-97**, pp. 1167–1176
406 HARIDOSS, S., CRINE, J.P., BULINSKI, A., DENSLEY, J., BAMJI, S., and BERNSTEIN, B.S.: *Ann. Rep. Conf. Elec. Insul. & Diel. Phenom.*, 1988, pp. 86–93
407 GAMEZ-GARCIA, M., BARTNIKAS, R., and WERTHEIMER, M.R.: *IEEE Trans. Elec. Insul.*, 1987, **EI-22**, pp. 199–205
408 KÄRNER, H., STIETZEL, U., SAURE, M., and GÖLZ, W.: 'CIGRE Int. Conf. Large High Volt. Elec. Syst.', 1984, Paper 15-02
409 LANCTOE, T.P., LAWSON, J.H., and McVEY, W.L.: *IEEE Trans. Power Appar. & Syst.*, 1979, **PAS-98**, pp. 912–925
410 DOEPKEN, H.C., McKEAN, A.L., and SINGER, M.L.: 'IEEE/PES Conf. Transmission & Distribution, (Atlanta)', 1979, IEEE Conf. Rec. 79CH1399-5-PWR, pp. 299–304
411 MIYASHITA, T.: *IEEE Trans. Elec. Insul.*, 1971, **EI-6**, pp. 129–135

412 KALKNER, W., MÜLLER, U., PESCHKE, E., HENKEL, H.J., & von OLSHAUSEN, R.: 'CIGRE Int. Conf. Large High Volt. Syst.', 1982, Paper 21–07

413 TABATA, T., FUKUDA, T., and IWATA, Z.: *IEEE Trans. Power Appar. & Syst.*, 1972, **PAS-91**, pp. 1361–1370

414 TANAKA, T., FUKUDA, T., SUZUKI, S., NITTA, Y., GOTO, H., & KUBOTA, K.: *IEEE Trans. Power Appar. & Syst.*, 1974, **PAS-93**, pp. 216–221

415 BULINSKI, A.T., BAMJI, S.S., and DENSLEY, R.J.: *Ann. Rep. Conf. Elec. Insul. & Diel. Phenom.*, 1987, pp. 440–447

416 McMAHON, E.J.: *IEEE Trans. Elec. Insul.*, 1981, **EI-16**, pp. 304–317

417 SALVAGE, B.: *Proc. IEE*, 1964, **111**, pp. 1173–1176

418 PELISSOU, S.: *Ann. Rep. Conf. Elec. Insul. & Diel. Phenom.*, 1988, pp. 101–108

419 WONG, P.P., and FORSTER, E.O.: *IEEE Trans. Elec. Insul.*, 1982, **EI-17**, pp. 203–220

420 DENSLEY, R.J.: 'IEEE Canadian Communications & Power Conf.', 1974, pp. 269–270

421 HEBNER, R.E., KELLEY, E.F., FORSTER, E.O., and FITZPATRICK, G.J.: *IEEE Trans. Elec. Insul.*, 1985, **EI-20**, pp. 281–292

422 YAMASHITSU, H., and AMANO, H.: *IEEE Trans. Elec. Insul.*, 1988, **EI-23**, pp. 739–750

423 VAHLSTROM, W.: *IEEE Trans. Power Appar. & Syst.*, 1972, **PAS-91**, pp. 1023–1035

424 MARTIN, M.A., and HARTLEIN, R.A.: *IEEE Trans. Power Appar. & Syst.*, 1980, **PAS-99**, pp. 1597–1605

425 McKENNY, P.J., and McGRATH, P.B.: *Ann. Rep. Conf. Elec. Insul. & Diel. Phenom.*, 1982, pp. 504–509

426 FORSTER, E.O.: *IEEE Trans. Elec. Insul.*, 1985, **EI-20(6)**, pp. 905–912

427 FORSTER, E.O.: *Ann. Rep. Conf. Elec. Insul. & Diel. Phenom.*, 1982, pp. 395–411

428 HIKITA, M., TAJIMA, S., KANNO, I., ISHINO, I., SAWA, G., and IEDA, M.: *Jpn. J. Appl. Phys.*, 1985, **24**, pp. 988–996

429 FOTHERGILL, J.C.: *IEEE Trans. Elec. Insul.*, 1, **EI-26(6)**, pp. 1124–1129

430 BOLTON, B., COOPER, R., and GUPTA, K.: *Proc. IEE*, 1965, **112**, pp. 1215–1220

431 KELEN, A.: *Acta Polytechnica Scand. Elec. Eng. Ser.*, 1967, pp. 3–86

432 PARR, D., and SCARISBRICK, R.M.: *Proc. IEE*, 1965, **112**, pp. 1625–1632

433 CRINE, J.P., HARIDOSS, S., HINTRICHSEN, P., HOUDAYER, A., and KAJRYS, G.: *Ann. Rep. Conf. Elec. Insul. & Diel. Phenom.*, 1988, pp. 94–100

434 KITAGAWA, K., SAWA, G., and IEDA, M.: *Jap. J. Appl. Phys.*, 1981, pp. 87–94

435 KITAGAWA, K., SAWA, G., and IEDA, M.: *Jap. J. Appl. Phys.*, 1982, **8(21)**, pp. 1117–1120

436 BRAUN, J.M.: *IEEE Trans. Elec. Insul.*, 1980, **EI-15**, pp. 120–123

437 EICHHORN, R.M.: *IEEE Trans. Elec. Insul.*, 1981, **EI-16**, pp. 469–482

438 KUDO, K., TANAKA, J., and DAMON, D.: *Ann. Rep. Conf. Elec. Insul. & Diel. Phenom.*, 1987, pp. 209–214

439 YAMADA, Y., YAMANOUCHI, S., and MIYAMOTO, S.: *Sumitomo Technical Review*, 1977, **17**, pp. 72–75

440 CHAN, J.C.: *IEEE Trans. Elec. Insul.*, 1978, **EI-13(6)**, pp. 444–447
441 PHILLIPS, P.J., KARAKELLE, M., and VATANSEVER, A.: *Polym. Comms.*, 1984, **25**, pp. 204–206
442 NAMIKI, Y., SHIMANUKI, Y., AIDA, F., and MORITA, M.: *IEEE Trans. Elec. Insul.*, 1980, **EI-15**, pp. 473–480
443 SHAW, S.H., DURAIRAJ, B., and SHAW, M.T.: *Ann. Rep. Conf. Elec. Insul. & Diel. Phenom.*, 1986, pp. 411–416
444 MORITA, M., HANAI, M., and SHIMANUKI, H.: *Ann. Rep. Conf. Elec. Insul. & Diel. Phenom.*, 1974, pp. 303–312
445 GROEGER, J., MIO, K., DAMON, D., and TANAKA, J.: 'Proc. 1st Int. Conf. Cond. Breakdown in Solid Dielectrics', 1983, pp. 244–249
446 FAVA, R.A.: *Proc. IEE*, 1965, **112**, pp. 819–823
447 DE BELLET, J.J., MATEY, G., ROSE, L., ROSE, V., FILIPPINI, J.C., POGGI, Y., and RAHARIMALALA, V.: *IEEE Trans. Elec. Insul.*, 1987, **EI-22**, pp. 211–217
448 DISSADO, L.A., and HILL, R.M.: *Physica Scripta*, **T1**, pp. 110–114
449 NAGASAKI, S., MATSUBARA, H., YAMANOUCHI, S., YAMADA, M., MATSUIKE, T., and FUKUNAGA, S.: *IEEE Trans. Power Appar. & Syst.*, 1984, **PAS-103**, pp. 536–544
450 ARBAB, M.N., and AUCKLAND, D.W.: *IEE Proc. A.*, 1989, **136**, pp. 73–78
451 NURY, J.: 'Proc. 1st Int. Conf. Cond. Breakdown in Solid Dielectrics (Toulouse)', 1983, pp. 373–379
452 XIE, H.K., and KAO, K.C.: *IEEE Trans. Elec. Insul.*, 1985, **EI-20**, pp. 293–297
453 WAGNER, H.: *Ann. Rep. Conf. Elec. Insul. & Diel. Phenom.*, 1974, pp. 62–70
454 IKEDA, M., TANAKA, Y., MATSUO, K., OTSUBO, T., OHKI, Y., HOZUMI, N., and HARASHIGE, M.: *Ann. Rep. Conf. Elec. Insul. & Diel. Phenom.*, 1988, pp. 305–311
455 EILHARDT, B., KINDIJ, E., RUMMEL, TH., and STENZEL, H-D.: *ETZ–A*, 1971, **92**, pp. 138–140
456 SHIRR, J.: *Int. Symp., High Volt. Tech.*', 1972, pp. 457–464
457 HEDVIG, P.: 'Dielectric spectroscopy of polymers' (A. Hilger, Bristol, 1977)
458 LABANOV, A.M., SPAKOVSKAYA, G.B., ROMANOVSKAYA, O.S., and SAZHIN, B.I.: *Vysokomol, Soed.*, 1969, **11**, p. 75
459 OKAMOTO, T., KANAZASHI, M., and TANAKA, T.: *IEEE Trans. Power Appar. & Syst.*, 1977, **PAS-96**, pp. 166–177
460 McMAHON, E.J., MALONEY, D.E., and PERKINS, J.R.: 1958, Dupont Paper no, 593, AIEE Summer Meeting, AIEE 58-981, 26 pages
461 POGGI, Y., FILIPPINI, J.C., and RAHARIMALALA, V.: 'Proc. 3rd Int. Conf. Cond. Breakdown in Solid Dielectrics (Trondheim)', 1989, 517–521
462 ANDRESS, B., FISCHER, P., REPP, H., and RÖHL, P.: 'Conf. Rec. IEEE Int. Symp. Elec. Insul. (Montreal)', 1984, pp. 65–67
463 WALTER, R.B., and JOHNSON, J.F.: *J. Polym. Sci. Macromol. Rev.*, 1980, **15**, pp. 29–53
464 WAGNER, H., and WARTUSCH, J.: *IEEE Trans. Elec. Insul.*, 1977, **EI-12**, pp. 395–401
465 TU, D.M., WU, L.H., WU, X.Z., CHENG, C.K., and KAO, K.C.: *IEEE Trans. Elec. Insul.*, 1982, **EI-17**, pp. 539–545
466 PATSCH, R.: *IEEE Trans. Elec. Insul.*, 1979, **EI-14**, pp. 200–206

467 ITO, T., HAMMAN, M.S.A.A., SCHIOMI, M., and SAKAI, T.: *Ann. Rep. Conf. Elec. Insul. & Diel. Phenom.*, 1980, pp. 227–233
468 LAURENT, C., and MAYOUX, C.: *J. Phys. D.*, 1981, **14**, pp. 1903–1910
469 LEBEY, T., LAURENT, C., and MAYOUX, C.: *Ann. Rep. Conf. Elec. Insul. & Diel. Phenom.*, 1988, pp. 109–115
470 MANDELKERN, L.: 'Crystallisation of polymers' (McGraw-Hill, New York, 1964) Chapter 8
471 LYLE, R., and KIRKLAND, J.W.: *IEEE Trans. Power Appar. & Syst.*, 1981, **PAS-100**, pp. 3764–3771
472 WILKENS, W.D.: *IEEE Trans. Elec. Insul.*, 1981, **EI-16(6)**, pp. 521–527
473 KAWAHARA, K., YOSHIMITSU, T., ISHIKAWA, Y., AKAHORI, H., and YOSHIDA, H.: 'IEEE Int. Symp. Int. Elec. Insul. (Montreal)', 1984, pp. 33–36
474 MATEY, G., NICOULAZ, F., FILIPPINI, J.C., POGGI, Y., and BOUZERARA, R.: 'Proc. 3rd Int. Conf. Cond. Breakdown in Solid Dielectrics (Trondheim)', 1989, pp. 500–506
475 IEDA, M., and NAWATA, M.: *Ann. Rep. Conf. Elec. Insul. & Diel. Phenom.*, 1972, pp. 143–150
476 PELISSOU, S., ST-ONGE, H., and WERTHEIMER, M.R.: 'Proc. 2nd Int. Conf. Cond. Breakdown in Solid Dielectrics (Erlangen)', 1986, pp. 116–125
477 NELSON, J.K., and SABUNI, H.: *Ann. Rep. Conf. Elec. Insul. & Diel. Phenom.*, 1980, pp. 499–507
478 LESSARD, G., PELISSOU, S., ST-ONGE, H., and WERTHEIMER, M.R.: *Ann. Rep. Conf. Elec. Insul. & Diel. Phenom.*, 1986, pp. 546–551
479 DERRINGER, G.C., EPSTEIN, M.M., GAINES, G.B., McGINISS, V.D., and THOMAS, R.E.: *Ann. Rep. Conf. Elec. Insul. & Diel. Phenom.*, 1978, pp. 182–186
480 NUNES, R.W., JOHNSON, J.F., and SHAW, M.T.: *Ann. Rep. Conf. Elec. Insul. & Diel. Phenom.*, 1980, pp. 188–194
481 SIMONI, L.: *IEEE Trans. Elec. Insul.*, 1981, **EI-16**, pp. 277–289
482 McKEAN, A.L., OLIVER, F.S., and TRILL, S.W.: *IEEE Trans. Power Appar. & Syst.*, 1967, **PAS-86**, pp. 1–10
483 COPPARD, R.W., BOWMAN, J., RAKOWSKI, R.T., DURHAM, R.T., and ROWLAND, S.M.: 'Proc. 3rd Int. Conf. Cond. Breakdown in Solid Dielectrics (Trondheim)', 1989, pp. 55–60
484 MANGARAJ, D., HASSELL, J., NIXON, J., and EPSTEIN, M.: *Ann. Rep. Conf. Elec. Insul. & Diel. Phenom.*, 1984, pp. 181–188
485 RZAD, S.J., and DEVINS, J.C.: *Ann. Rep. Conf. Elec. Insul. & Diel. Phenom.*, 1978, pp. 352–360
486 TU, D.M., WU, L.H., WU, X.Z., CHENG, C.K., and KAO, K.C.: *IEEE Trans. Elec. Insul.*, 1982, **EI-17**, pp. 539–545; and Discussion with RZAD, S.J., and DEVINS, J.C.: *IEEE Trans. Elec. Insul.*, 1983, **EI-18**, pp. 465–466
487 KIM, B-H., and LIM, K-J.: *Ann. Rep. Conf. Elec. Insul. & Diel. Phenom.*, 1986, pp. 398–403
488 ITO, T., OSONO, M., EHARA, Y., SAKAI, T., SATO, K., & HAMMAN, M.S.A.A.: *Ann. Rep. Conf. Elec. Insul. & Diel. Phenom.*, 1985, pp. 491–497
489 EICHHORN, R.M., and TURBETT, R.J.: *IEEE Trans. Power Appar. & Syst.*, 1979, **PAS-98**, pp. 2215–2221

490 MATSUBA, H., KAWAI, E., and SATO, K.: *IEEE Conf. Rec. Int. Symp. Elec. Insul.*, 1976, pp. 224–227
491 CHAN, J.C.: *Ann. Rep. Conf. Elec. Insul. & Diel. Phenom.*, 1985, pp. 296–301
492 BAMJI, S., BULINSKI, A., and DENSLEY, R.J.: *Ann. Rep. Conf. Elec. Insul. & Diel. Phenom.*, 1985, pp. 312–317
493 NITTA, Y., and FUNAYAMA, M.: *IEEE Trans. Elec. Insul.*, 1978, **EI-13**, pp. 130–133
494 MANGARAJ, D., KISS, K.D., MALAWER, E., and DOEPKEN Jr., M.C.: *Ann. Rep. Conf. Elec. Insul. & Diel. Phenom.*, 1978, pp. 195–205
495 NUNES, S.L., and SHAW, M.T.: *Ann. Rep. Conf. Elec. Insul. & Diel. Phenom.*, 1981, pp. 238–243
496 ALLARA, D.L., and WHITE, C.W.: in 'Stabilization and degradation of polymers', eds. ALLARA, D.L., and HAWKINS, W.L. (ACS Adv. Chem. Ser., 1978) **169**, p. 273
497 BLATT, E., GRIESSER, H.J., LODER, J.W., and MAU, A.W-H.: *Polym. Deg. & Stab.*, 1988, **21**, pp. 335–343
498 BLATT, E., HODGKIN, J., LODER, J., MAU, A.W-H., SASSE, W.H.F., and GHIGGINO, K.P.: *Polym. Deg. & Stab.*, 1988, **20**, pp. 75–88
499 BLATT, E., GRIESSER, H.J., HODGKIN, J.H., and MAU, A.W-H.: *Polym. Deg. & Stab.*, 1989, **25**, pp. 19–29
500 VON GENTZKOW, W., and WIEDENMANN, R.: *J. Appl. Polym. Sci.: Appl. Polym. Symp.*, 1979, **35**, pp. 173–182
501 O'DWYER, J.J.: *IEEE Trans. Elec. Insul.*, 1984, **EI-19**, pp. 1–9
502 GARTON, C.G., *Proc. IEE*, 1976, **123**, pp. 271–276
503 INUISHI, Y., *IEEE Trans. Elec. Insul.*, 1980, **EI-15**, pp. 139–151
504 WHITEHEAD, S.: 'Dielectric phenomena, Vol. 3: Breakdown of solid dielectrics'. Ed. WEDMORE, E.B., (Ernest Bern Ltd., 1932)
505 WHITEHEAD, S.: 'Dielectric breakdown of solids' (Clarendon Press, Oxford, 1951)
506 KLEIN, N., *Thin Solid Films*, 1983, **100(4)**, pp. 335–340
507 COELHO, R.: 'Fundamental studies in engineering, Vol. 1: Physics of dielectrics for the engineer' (Elsevier Scientific Pub. Co., 1979)
508 PARK, C.H., YOSHINO, K., OKUYAMA, K., and ICHIHARA, S., *Electr. Eng. Jpn, (USA)*, 1989, **109(4)**, pp. 1–6
509 STRÄB, H.: *Ver. Deut. Elek.* (Fachberichte, 1954), **18**, pp. I/28–I/33
510 CHAMPION, J.: Private communication
511 BRADWELL, A. ed.; 'Electrical insulation' (Peter Peregrinus, 1983)
512 COOPER, R.: *Brit. J. Appl. Phys.*, 1966, **17**, pp. 149–166
513 BALL, I.D.L.: *Proc. IEE*, 1951, **98(1)**, pp. 84–86
514 OAKES, W.G.: *Proc. IEEE*, 1949, **96**, 37–43
515 MORT, J., and PFISTER, G. eds.: 'Electronic properties of polymers' (Wiley Interscience, 1982)
516 BÖTTGER, H., and BRYKSIN, V.V.: 'Hopping conduction in solids' (VCH Publishers, Berlin, 1985)
517 VAN DER ZEIL, A.: 'Solid state physical electronics' (Prentice-Hall Inc., 1968)
518 FRÖHLICH, H., and SEWELL, G.L.: *Proc. Phys. Soc.*, 1959, **74**, pp. 643–647

519 BELMONT, M.R.: *J. Non-Cryst. Sol.*, 1975, **17**, pp. 284–288
520 LEWIS, T.J.: *Faraday Discussion Chem. Soc.*, 1989, **88**, pp. 189–201
521 LEWIS, T.J.: 'Proc. 1st Int. Conf. Cond. Breakdown in Solid Dielectrics (Toulouse)', 1983, pp. 207–224
522 LEWIS, T.J.: *IEEE Trans. Elec. Insul.*, 1986, **EI-21(3)**, pp. 289–295
523 SHOCKLEY, W.: 'Electrons and holes in semiconductors, with applications to transistor electronics' (Van Nostrand, New York, 1951)
524 GLARUM, S.H.: *J. Phys. & Chem. Solids*, 1963, **24**, pp. 1577–1583
525 STREITWIESER, A.: 'Molecular orbital theory for organic chemists' (J. Wiley and Sons, New York, 1961)
526 FRIEDMAN, L.: *Phys. Rev.*, 1964, **133A**, pp. 1668–1679
527 IOFFE, A.F., and REGEL, A.R.: *Progr. Semiconductors*, 1960, **4**, 237–291
528 BAYLISS, N.S.: *J. Chem. Phys.*, 1948, **16(4)**, pp. 287–292
529 HANKIN, A.G., and NORTH, A.M.: *Trans. Farad. Soc.*, 1967, **63**, pp. 1525–1536
530 McCUBBIN, W.L., and MANNE, R.: *Chem. Phys. Lett.*, 1968, **2**, pp. 230–232
531 DAS-GUPTA, D.K.: *IEE Conf. Diel. Mats., Meas. & Applics.*, 1988, **289**, pp. 29–32
532 MARTIN, E.H., and HIRSCH, J.: *J. Appl. Phys.*, 1972, **43**, pp. 1001–1007
533 DAVIES, D.K.: *Static Electrification Inst. Phys. Conf. Ser.*, 1975, **27**, pp. 74–83
534 TOOMER, R., and LEWIS, T.J.: *J. Phys. D.*, 1980, **13**, pp. 1343–1356
535 WINTLE, H.J.: *Jap. J. Appl. Phys.*, 1971, **10**, pp. 659–660
536 SONNONSTEIN, J.J., and PERLMAN, M.M.; *Inst. Phys. Conf. Ser.*, 1975, **27**, pp. 74–83
537 TANAKA, T., and CALDERWOOD, J.H.: *J. Phys. D: Appl. Phys.*, 1974, **7**, pp. 1295–1302
538 DAS-GUPTA, D.K., and MOON, T.: *J. Phys. D: Appl. Phys.*, 1975, **8**, pp. 1336–1340
539 DAVIES, D.K.: *J. Phys. D: Appl. Phys*, 1972, **5**, pp. 162–168
540 TANAKA, T., and CALDERWOOD, J.H.: *Trans. IEE Japan*, 1973, **93-A**, pp. 473–480
541 REDDISH, W.: *Pure & Appl. Chem.*, 1962, **5**, pp. 723–742
542 DAVIES, J.M., MILLER, R.F., and BUSSE, W.F.: *J. Amer. Chem. Soc.*, 1941, **63**, pp. 361–369
543 SEANOR, D.A.: in 'J. Polym. Sci. C17'; eds. REMBAUM, A., and LANDEL, R.F., 1967, pp. 195–212
544 BAIRD, M.E.: *J. Polym. Sci.*, Part A-2, 1970, **8**, pp. 739–745
545 HATANO, M., NOMORI, H., and KAMBARA, S.: *Amer. Chem. Soc., Div. Polym. Chem., Preprints*, 1964, **5(2)**, pp. 849–854
546 FOWLER, J.F., and FARMER, F.T.: *Nature*, 1953, **171**, pp. 1020–1021
547 AMBORSKI, L.E.: *J. Polym. Sci.*, 1962, **62(174)**, pp. 331–346
548 FOTHERGILL, J.C.: 'Electronic properties of biopolymers', PhD Thesis, University College of North Wales, Bangor, UK
549 POHL, H.A., REMBAUM, A., and HENRY, A.: *J. Amer. Chem. Soc.*, 1962, **84**, pp. 2699–2704
550 POHL, H.A., and OPP, D.A.: *J. Phys. Chem.*, 1962, **66**, pp. 2121–2126
551 BUI AI, H-T-G., SESTRUAL, P., BENDAOUD, M., and SAIDI, M.: *Ann. Rep. Conf. Elec. Insul. & Diel. Phenom.*, 1984, pp. 448–454

552 SAWA, G., INAYOSHI, Y., NISHIO, Y., NAKAMURA, S., and IEDA, M.: *J. Appl. Phys.*, 1977, **48**, pp. 2414–2418
553 FUOSS, R.M.: *J. Amer. Chem. Soc.*, 1939, **61**, pp. 2329 and 2335
554 O'SULLIVAN, J.B.: *J. Textile Inst.*, 1947, **38**, pp. T271, T285, T291 and T298
555 MURPHY, E.J.: *Canad. J. Phys.*, 1963, **41**, pp. 1022–1035
556 WARFIELD, R.W., and PETREE, M.E.: *Trans. Soc. Plastics Eng.*, 1961, **1**, pp. 3–8
557 KALLWEIT, J.H.: *Kunststoffe*, 1957, **47**, pp. 651–655
558 BINKS, A.E., and SHARPLES, A.: *J. Polym. Sci.*, 1968, **A2–6**, pp. 407–420
559 CAO, X., FOURACRE, R.A., GEDEON, S., TEDFORD, D.J., and BANFORD, H.M.: *Ann. Rep. Conf. Elec. Insul. & Diel. Phenom.*, 1988, pp. 481–486
560 FOWLER, J.F., and FARMER, F.T.: *Nature*, 1954, **173**, pp. 317–318
561 FOWLER, J.F., and FARMER, F.T.: *Nature*, 1955, **175**, pp. 590–591
562 FOWLER, J.F.: *Proc. Roy. Soc. A*, 1956, **236**, pp. 464–480
563 SMITH, F.S., and SCOTT, C.: *Brit. J. Appl. Phys.*, 1966, **17**, pp. 1149–1154
564 BARKER, R.E.: *Pure & Appl. Chem.*, 1976, **146(2–4)**, pp. 157–170
565 BADEN FULLER, A.J.: 'Engineering field theory' (Pergamon Press, 1973)
566 ALLISON, J.: 'Electronic engineering semiconductors and devices' (McGraw-Hill, New York, 1990)
567 LEWIS, T.J.: *Proc. Phys. Soc. (London)*, 1954, **B67**, pp. 187–200
568 SOMMERFIELD, A.: 'Optics' (Academic Press, 1954)
569 FOWLER, R.H., and NORDHEIM, L.: *Proc. Roy., Soc. London*, 1928, **A119**, pp. 173–181
570 WENTZEL, G.: *Zeitschrift fur Physik*, 1926, **38(6–7)**, pp. 518–529
571 KRAMERS, H.A.: *Zeits. f. Physik*, 1926, **39(10–11)**, pp. 828–840
572 BRILLOUIN, L.; *Comptes Rendus hebdomadaires des séances de l' Académie des Sciences*, 1926, **183**, pp. 24–26
573 GOOD, R.H., and MÜLLER, W.: 'Field emission' Handbuch der Physik, 1956, Vol. 21 (Springer Verlag)
574 BREHEMER, L., PLATEN, E., FANTYER, D., and LIEMANT, A.: *IEEE Trans. Elec. Insul.*, 1987, **EI-22(3)**, pp. 245–248
575 LAMB, D.R.: 'Electrical conduction mechanisms in thin insulating films' (Methuen, London, 1967)
576 MOTT, N.F., and GURNEY, R.W.: 'Electronic processes in ionic crystals' (OUP, New York, 1940)
577 CHUTIA, J., and BARUA, K.: *J. Phys. D: Appl. Phys.*, 1980, **13**, pp. L9–L13
578 EMIN, D.: *Adv. Phys.*, 1973, **22**, pp. 57–116
579 SUMMERFIELD, S., and BUTCHER, P.N.: *J. Non-Cryst. Sols.*, 1985, **77/78**, pp. 135–138
580 HARE, R.: Private communication
581 WAGNER, K.W.: *Trans. AIEE*, 1922, **41**, pp. 1034–1044
582 VON KÁRMÁN, T.: *Archiv. f. El.*, 1924, **13**, pp. 174–180
583 ROGOWSKI, W.: *Archiv. f. El.*, 1924, **13**, pp. 153–174
584 DREYFUS, L.: *Schweiz. Elek. Verein. Bull.*, 1924, **15**, pp. 321–344 and 577–597
585 FOCK, V.: *Archiv. f. El.*, 1927, **19**, pp. 71–81
586 MOON, P.H.: *Trans. IEEE*, 1931, **50(3)**, pp. 1008–1021

587 COPPLE, C., HARTREE, D.R., PORTER, A., and TYSON, H.: *Proc. IEE*, 1939, **85**, pp. 56–66
588 HIKITA, M., KANNO, I., SAWA, G., and IEDA, M.: *Ann. Rep. Conf. Elec. Insul. & Diel. Phenom.*, 1985, pp. 511–517
589 HIKITA, M., KANNO, I., SAWA, G., and IEDA, M.: *Jpn. J. Appl. Phys.*, 1985, **24**, pp. 1619–1622
590 HIKITA, M., KANNO, I., SAWA, G., and IEDA, M.: *Jap. J. Appl. Phys.*, 1985, **24(8)**, pp. 984–987
591 HIKITA, M., ISHINO, I., SAWA, G., and IEDA, M.: *Jap. J. Appl. Phys.*, 1986, **25(3)**, pp. 500–501
592 HIKITA, M., HIROSE, T., MIZUTANI, T., and IEDA, M.: 'Proc. 21 IEEE Symp. Elec. Insul. Mats.', 1988, pp. 41–44
593 HIKITA, M., KANNO, I., IEDA, M., ISHINO, I., DOI, S., and SAWA, G.: *IEEE Trans. Elec. Insul.*, 1987, **EI-22(2)**, pp. 175–179
594 MIZUTANI, T., KANNO, I., HIKITA, M., IEDA, M., and SAWA, G.: *IEEE Trans. Elec. Insul.*, 1987, **EI-22(4)**, pp. 473–477
595 SLETBAK, J., GJELSTEN, N.G., HENRIKSEN, E.E., and LANOUE, T.J.: *Nord-IS 88*, 1988, 19/1–10
596 KLEIN, N., and BURSTEIN, E.: *J. Appl. Phys.*, 1969, **40(7)**, pp. 2728–2740
597 NAGAO, M., FUKUMA, M., KOSAKI, M., and IEDA, M.: 'Proc. 5th. Int. Symp. Electrets (Heidelberg)', 1985, pp. 416–421
598 NAGAO, M., YAMAYUCHI, F., TOKUMARU, K., SUGIYAMA, I., KOSAKI, M., and IEDA, M.: *Electrical Engin. in Japan*, 1985, **105(5)**, pp. 11–17 (Transl of *Denki Gakkai Rowbunshi*, 1985, **105A(4)**, pp. 177–182)
599 WINKEINKEMPER, H., and KALKNER, W.: *Elektrotech Z.*, 1974, **A-95**, pp. 261–265
600 KALKNER, W., and WINKEINKEMPER, H.: *Int. High Volt. Symp. Zurich*, 1975, pp. 529–597
601 NAGAO, M., KITAMURA, T., MIZUNO, Y., KOSAKI, M., and IEDA, M.: 'Proc. 3rd Int. Conf. Cond. Breakdown in Solid Dielectrics (Trondheim)', 1989, pp. 77–81
602 GARTON, C.G., and STARK, K.H.: *Nature*, 1955, **176**, pp. 1225–1226
603 CHARLESBY, A., and HANCOCK, N.H.: *Proc. Roy. Soc.* 1953, **A218**, pp. 245–255
604 MOLL, H.W., and LEFEVRE, W.J.: *Industr. & Engin. Chem.*, 1948, **40**, pp. 2172–2179
605 McKEOWN, J.J.: *Proc. IEE*, 1965, **112**, pp. 824–828
606 LAWSON, W.G.: *Nature*, 1965, **206**, pp. 1248–1249
607 PARKMAN, N., GOLDSPINK, G.F., and LAWSON, W.G.: *Electron. Lett.*, 1965, **1**, pp. 98–100
608 BLOK, J., and LEGRAND, D.G.: *J. Appl. Phys.*, 1969, **40(1)**, pp. 288–293
609 GRIFFITH, A.A.: *Phil. Trans. Roy. Soc. London*, 1920, **A221**, pp. 163–198
610 WINTLE, H.J.: *J. Electrostat.*, 1987, **19**, pp. 257–274
611 IEDA, M., KITAGAWA, K., and SAWA, G.: 'Proc. 1st Int. Conf. Cond. Breakdown in Solid Dielectrics', 1983, IEEE number 83CH1836-6-EI, pp. 333–337
612 WAGNER, W.: *Z. Elektrotech, A. Ausg. A*, 1973, **94**, pp. 436–437
613 KOLESOV, S.N.: *IEEE Trans. Elec. Insul.*, 1980, **EI-15**, pp. 382–388
614 YAMADA, Y., KIMURA, S., and SATO, T.: 'Proc. 3rd. Int. Conf. Cond. Breakdown in Solid Dielectrics (Trondheim)', 1989, pp. 87–91

615 STRATTON, R.: *Progr. Dielectrics*, 1957, **3**, pp. 235–292
616 VON HIPPEL: *Ergeb. Exakten. naturwiss.* 1935, **14**, pp. 79–129
617 GOODMAN, B., LAWSON, A.W., and SCHIFF, L.I.: *Phys. Rev.*, 1947, **71(3)**, pp. 191–194
618 FRÖLICH, H.: *Proc. Roy. Soc., London*, 1937, **A160**, pp. 230–241
619 MAKANO, T., FUKUYAMA, M., HAYASHI, H., ISHII, K., and OHKI, Y.: Proc. 3rd Int. Conf. Cond. Breakdown in Solid Dielectrics (Trondheim)', 1989, pp. 82–86
620 IEDA, M.: *IEEE Trans. Elec. Insul.*, 1986, **EI-21**, pp. 793–802
621 MIZUTANI, T., SUZUOKI, Y., and IEDA, M.: *J. Appl. Phys.*, 1977, **48**, pp. 2408–2413
622 FRÖHLICH, H.: *Proc. Roy. Soc., London*, 1947, **A188**, pp. 521–532
623 FRÖHLICH, H.: *Phys. Rev.*, 1942, **61**, pp. 200–201
624 AUSTEN, A.E.W., and PELZER, H.: 'Electric strength of hydrocarbons' (Electrical Research Association (now ERA Technology Ltd., 1943)) Report LT/138
625 AUSTEN, A.E.W., and PELZER, H.: 'Electric strength of vinylite and its temperature dependence' (Electrical Research Association (now ERA Technology Ltd., 1944)) Report LT/149
626 BARAFF, G.A.: *Phys. Rev.*, 1962, **128**, pp. 2507–2517
627 MIZUTANI, T., HIKITA, M., UMEMURA, A., and IEDA, M.: *Ann. Rep. Conf. Elec. Insul. & Diel. Phenom.*, 1989, pp. 315–320
628 WIJSMAN, R.A.: *Phys. Rev.*, 1949, **75(5)**, pp. 833–838
629 KITANI, I., and ARII, K.: *IEEE Trans. Elec. Insul.*, 1981, **EI-16**, pp. 134–139
630 KITANI, I., and ARII, K.: *IEEE Trans. Elec. Insul.*, 1982, **EI-17**, pp. 571–576
631 O'DWYER, J.J.: *J. Phys. & Chem. Sols.*, 1967, **28**, pp. 1137–1144
632 PARACCHINI, C.: *Phys. Rev.*, 1971, **B4**, pp. 2342–2347
633 DI-STEFANO, T.H., and SHATZKES, M.: *J. Vac. Sci. Technol.*, 1975, **12**, pp. 37–47
634 BRADWELL, A., COOPER, R., and VARLOW, B.: *Proc. IEE*, 1971, **118**, pp. 247–254
635 AMAKAWA, K., and INUISHI, Y.: *Jap. J. Appl. Phys.*, 1973, **12**, pp. 755–756
636 SAWA, G.: *IEEE Trans. Elec. Insul.*, 1986, **EI-21(6)**, pp. 841–846
637 YOSHINO, K., HARADA, S., KYOKANE, J., and INUISHI, Y.: *Jap. J. Appl. Phys.*, 1979, **18**, pp. 679–680
638 ARII, K., KITANI, I., and INUISHI, Y.: *Tech. Rep. Osaka Univ. Jpn*, 1973, **24**, pp. 95–103
639 COOPER, R., VARLOW, B.R., and WHITE, J.P.: *J. Phys. D: Appl. Phys.*, 1977, **10**, pp. 1521–1529
640 GRIEMSMANN, J.W.E.: in 'Plastics for electrical insulation', ed. BRUINS, P.F. (Wiley Interscience, 1968) pp. 1–35
641 CHANG, D.D., SUDARSHAN, T.S., and THOMPSON, J.E.: *IEEE Trans. Elec. Insul.*, 1986, **EI-21**, pp. 213–219
642 MALIK, N.H., AL-ARAINY, A.A., KAILANI, A.M., and KHAN, M.J.: *IEEE Trans. Elec. Insul.*, 1987, **EI-22(6)**, pp. 787–793
643 PETERSEN, W.: *Archiv für Electrotechnik*, 1912, **1**, pp. 233–254
644 SCHERING, H., Ed.: 'Die isoliedtoffe der electrotechnik' (Verlag Julius Springer, Berlin, 1924)
645 GEMANT, A.: 'Die verlustkurve luftiger isolierstoffe' (Zietschrift für Technische Physik, 1932) **13**, pp. 184–189
646 DAWES, C.L., and HOOVER, P.L.: 1926, *AIEE Trans*, **45**, pp. 337–347

647 DAVIES, C.L., BEICHARD, H.H., and HUMPHRIES, P.H.: *AIEE Trans.*, 1929, **48**, pp. 382–395

648 DAVIES, C.L., and HUMPHRIES, P.H.: *AIEE Trans.*, 1930, **49**, pp. 766–776

649 BARTNIKAS, R.: The Whitehead Memorial Lecture given at the Conf. Elec. Insul. and Diel. Phenom. 1987 (Conf. Report pages 13–14) reprinted in *IEEE Trans. Elec. Insul.*, 1987, **EI-22(5)**, pp. 629–653

650 SEDDING, H.G., STONE, G.C., BRAUN, J.M., and HOGG, W.K.: *IEE Conf. Diel. Mats., Meas. & Applics.*, 1988, **289**, pp. 211–214

651 PEDERSON, A.: 'Proc. 3rd Int. Conf. Cond. Breakdown in Solid Dielectrics (Trondheim)', 1989, pp. 107–116

652 STONE, G.C., and Van HEESWIJK, R.G.: *Ann. Rep. Conf. Elec. Insul. & Diel. Phenom.*, 1988, pp. 376–381

653 TOWNSEND, J.S.: 'The theory of ionization of gases by collision' (Constable & Co. Ltd., London, 1910)

654 PASCHEN, F.: *Weid. Ann.*, 1889, **37**, pp. 69–96

655 REES, J.A. ed.: 'Electrical breakdown in gases' (MacMillan, 1973)

656 HARRISON, M.A., and GEBALLE, R.: *Phys. Rev.*, 1953, **91**, pp. 1–7

657 PENNING, F.M., and NED, T.: *Natuurkde*, 1938, **5**, pp. 33–56

658 MASON, J.H.: *Proc. IEE(IIA)*, 1953, **100**, pp. 149–158

659 HALL, H.C., and RUSSEK, R.M.: *Proc. IEE*, 1954, **101(II)**, pp. 47–55

660 FISCHER, P., and NISSEN, K.: *Elektrotech A*, 1978, **A-99**, pp. 475–480

661 WENIGER, M., and KUBLER, B.: *Elektrotech A*, 1976, **A-97**, pp. 477–480

662 YASUI, T., and YAMADA, Y.: *Sumitomo Electr. Tech. Rev.*, 1967, **10**, pp. 60–72

663 DAVIES, D.K.: in 'Electrical properties of polymers', ed. SEANOR, D.A. (Academic Press, 1982)

664 BEZBORODKO, P., LESAINT, O., and TOBAZÉON, R.: 'Proc. 3rd Int. Conf. Cond. Breakdown in Solid Dielectrics (Trondheim)', 1989, pp. 392–396

665 MORSHUIS, P.H.F., and KREUGER, F.H.: 'Proc. 3rd Int. Conf. Cond. Breakdown in Solid Dielectrics (Trondheim)', 1989, pp. 117–121

666 ROBINSON, G.: 'Conf. Diel. Mats., Meas. & Applics.', 1988, IEE Conf. Pub. **289**, pp. 215–218

667 MASON, J.H.: in 'Progress in dielectrics' eds. BIRKS, J.B., and SCHULMAN, J.H. (Heywood, London, 1959) **1**, pp. 1–58

668 HOWARD, P.R.: *Proc. IEE*, 1951, **98(2)**, pp. 365–370

669 PARKMAN, N.: ERA Report L/T231 (1955)

670 PARKMAN, N.: *Soc. Chem. Ind. Monograph*, 1959, **5**, p. 95

671 KREUGER F.H.: *Rev. Gen. Elect. (France)*, 1968, **77(5)**, pp. 540–542

672 REPP, H., NISSEN, K.W., and RÖHL, P.: *Siemens Forsch-u-Entwickl-Ber. Bd*, 1983, **12**, pp. 101–106

673 WOLTER, K.D., JOHNSON, J.F., and TANAKA, J.: *IEEE Trans. Elec. Insul.*, 1978, **EI-13**, pp. 327–330

674 WOJTAS, S.: *IEE Conf. Diel. Mats., Meas. & Applics.*, 1988, **289**, pp. 191–193

675 DENSLEY, R.J., and SALVAGE, B.: *IEEE Trans. Elec. Insul.*, 1971, **EI-6**, pp. 54–62

676 ROGERS, E.C.: *Proc. IEE*, 1958, **105(A)**, pp. 621–630

677 NISSEN, K.W., and RÖHL, P.: *Siemens Forsch. Entwicklungsber*, 1981, **10**, pp. 215–221

678 KRANZ, H.G.: *IEEE Trans. Elec. Insul.*, 1982, **EI-17**, pp. 151–155

679 SHAHIN, M.M.: *J. Chem. Phys.*, 1967, **45**, pp. 2600–2605

680 LU ZIBIN, GOLDMAN, M., GOLDMAN, A., and GATELLET, J.: 'Proc. 3rd Int. Conf. Cond. Breakdown in Solid Dielectrics (Trondheim)', 1989, pp. 122–125

681 HOLLAHAN, J.R., and BELL, A.T.: 'Techniques and applications of plasma chemistry' (Wiley, New York, 1974)

682 VENUGOPALAN, M.: 'Reactions under plasma conditions' (Wiley Interscience, New York, 1971)

683 YASUDA, H.: 'Plasma polymerization' (Academic Press, New York, 1985)

684 HILEY, J., NICHOLL, G.R., PEARMAIN, A.J., and SALVAGE, B.: *Ann. Rep. Conf. Elec. Insul. & Diel. Phenom.*, 1973, pp. 116–124

685 MONTANARI, G.C., PATTINI, P., and SIMONI, L.: 'Conf. Rec. IEEE Int. Symp. Elec. Insul. (Montreal)', 1984, pp. 54–57

686 SIMONI, L.: *IEEE Trans. Elec. Insul.*, 1982, **EI-17(4)**, pp. 373–375

687 SIMONI, L.: *IEEE Trans. Elec. Insul.*, 1984, **EI-19(1)**, pp. 45–52

688 DAKIN, T.W., and STUDIARZ, S.A.: 'Conf. Rec. Int. Symp. Elec. Insul.', 1978, IEEE Pub., **78CH1287-2-EI**, pp. 216–221

689 AMAKAWA, K., MORIUCHI, T., YOSHIDA, T., and INUISHI, Y.: *JIEE of Japan* (in Japanese), 1964, **84**, pp. 129–135

690 ARTBAUER, J.: *Kolloid Z. uZ für Polymere*, **202**, pp. 15–25

691 ARTBAUER, J.: *J. Polym. Sci. C*, 1967, **16**, pp. 477–484

692 WILLIAMS, M.L., LANDEL, R.F., and FERRY, J.D.: *J. Amer. Chem. Soc.*, 1955, **77**, pp. 3701–3707

693 PARK, C.H., OKAJIMA, K., HARA, M., and AKAZAKI, M.: *IEEE Trans. Elec. Insul.*, 1983, **EI-18**, pp. 380–389

694 CRINE, J.P., and VIJH, A.K.: *Appl. Phys. Comm.*, 1985, **5**, 139

695 TAKAHASHI, T., OHTSUKA, H., TAKLEHANA, H., and NUILA, T.: 'IEEE PES Winter Meeting', 1985, Paper 85-WM-004–7

696 SONE, M., ISONO, H., and MITSUI, H.: *Ann. Rep. Conf. Elec. Insul. & Diel. Phenom.*, 1986, pp. 528–533

697 KAO, K.C.: *J. Appl. Phys.*, 1984, **55**, pp. 752–755

698 SANO, N.: *J. Phys. D: Appl. Phys.*, 1989, **22**, pp. 309–315

699 HILL, R.M.: Personal communication

700 WEIBULL, W.: *J. Appl. Mech.*, 1951, **18**, pp. 293–297

701 FOTHERGILL, J.C.: *IEEE Trans. Elec. Insul.*, 1990, **EI-25(3)**, pp. 489–492

702 BAHDER, G., SOSNOWSKI, M., KATZ, C., EATON, R., and KLEIN, N.: *IEEE Trans. Power Appar. & Syst.*, 1983, **PAS-102(7)**, pp. 2173–2185

703 HILL, R.M., and DISSADO, L.A.: *J. Phys. C.: Solid State Phys.*, 1983, **16**, pp. 4447–4468

704 SOLOMON, P., KLEIN, N., and ALBERT, M.: *Thin Solid Films*, 1976, **35**, pp. 321–326

705 SHATZKES, M., and Av-RON, M.: *Thin Solid Films*, 1982, **91(3)**, pp. 217–230

706 BAHDER, G., KATZ, C., EAGER, G.S., LEBER, E., CHALMERS, S.M., JONES, W.H., & MANGRUM, W.H.: *IEEE Trans. Power Appar. & Syst.*, 1981, **PAS-100(4)**, pp. 1581–1590

707 BEYER, M., von OLSHAUSEN, R., and SACHS, G.: *Kunstoffe*, 1976, **66(1)**, pp. 35–38

708 DAVIES, A.E., and WEEDY, B.M.: *IEE Proc. A (GB)*, 1982, **129(8)**, pp. 607–610

709 FISCHER, P.: *Ann. Rep. Conf. Elec. Insul. & Diel. Phenom.*, 1980, pp. 661–670

710 GOULDSON, E.J.: *IEEE Trans. Power Appar. & Syst.*, 1971, **PAS-90**, pp. 2679–2682

711 HAKAMADA, T.: *IEEE Trans. Elec. Insul.*, 1984, **EI-19(2)**, pp. 114–118
712 KANEKO, T., and SUGIYAMA, K.: *IEEE Trans. Power Appar. & Syst.*, 1975, **PAS-94(2)**, pp. 367–377
713 LACOSTE, R., MUHAMMED, A., SEGUI, Y., and VOUMBO-MATOUMONA, L.: *IEEE Trans. Elec. Insul.*, 1984, **EI-19(3)**, pp. 234–240
714 MITSUI, H., and INQUE, Y.: *IEEE Trans. Elec. Insul.*, 1977, **EI-12(3)**, pp. 237–247
715 NELSON, W.B.: *IEEE Trans. Reliab.*, 1972, **R-21(1)**, pp. 2–11
716 OHATA, K., KOJIMA, H., SHIMOMURA, T., and ASAHI, K.: *IEEE Trans. Power Appar. & Syst.*, 1983, **PAS-102(7)**, pp. 1935–1941
717 TRINH, N.G., and VINCENT, C.: *IEE Trans. Power Appar. & Syst.*, 1980, **PAS-99(2)**, pp. 711–719
718 BULINSKI, A., BAMJI, S., DENSLEY, J., and SHIMIZU, N.: *Ann. Rep. Conf. Elec. Insul. & Diel. Phenom.*, 1983, pp. 294–300
719 ROWLAND, S.M.: 'Conf. Diel. Mats. Meas. & Applics.', 1984, IEE Conf. Pub. **239**, pp. 88–91
720 MASSETTI, C., PIGINI, A., BASSI, A., FARNETI, F., and METRA, P.: *Proc. 4th. BEAMA Int. Elec. Insul. Conf.*, 1982, pp. 29–33
721 DAKIN, T.W., STUDNIARZ, S.A., and HUMMERT, G.T.: *Ann. Rep. Conf. Elec. Insul. & Diel. Phenom.*, 1972, pp. 411–417
722 ZOLEDZIOWSKI, A.: 'Conf. Diel. Mats. Meas. & Applics.', 1984, IEE Conf. Pub. **239**, pp. 84–87
723 'Guide for the statistical analysis of voltage endurance data for electrical insulation', (1987) ANSI/IEEE Std 930-1987
724 FREI, C.J.: *IEEE Trans. Elec. Insul.*, 1985, **EI-20(2)**, pp. 303–307
725 WOLTERS, D.R., and ZEGERS-VAN DUYNHOVEN, A.T.A.: in 'Reliability technology: theory and applications', eds. MOLTCROFT, J., and JENSEN, F. (North Holland, Amsterdam, 1986) pp. 315–323
726 LAWLESS, J.F.: 'Statistical models and methods for lifetime data' (J. Wiley and Sons, 1982)
727 CROOK, D.L.: IEEE/IRPS Conf. on Reliability Phys. (1979), 1; CROOK, D.L., and MAYER, W.K., 'IEEE/IRPS Conf. on Reliability Phys.', 1981, 1
728 NELSON, R.A., and HAHN, G.J.: *Technometrics*, 1972, **14**, pp. 247–269
729 METZLER, R.A.: *IEEE/IRPS Proc. on Reliability Phys.*, 1979, p. 233
730 DISSADO, L.A., and HILL, R.M.: *Proc. Roy. Soc. (London)*, 1983, **A390**, pp. 131–190
731 PEEK, F.W.: *Trans AIEE*, 1919, **38(ii)**, pp. 1137–1177
732 ANOLICK, E.S., CHEN, L., and MALIK, S.K.: *IBM Tech. Discl. Bull. (USA)*, 1982, **24(11A)**, pp. 5730–5732
733 DAKIN, T.W.: 'Proc. 4th. Symp. Elec. Insul. Mats., (Tokyo, Japan IEE)', 1971
734 STONE, G.C.: *Ann. Rep. Conf. Elec. Insul. & Diel. Phenom.*, 1985, pp. 234–249
735 MORCOS, M.M., and CHERUKUPALLI, S.E.: 'Proc. 3rd Int. Conf. Cond. Breakdown in Solid Dielectrics (Trondheim)', 1989, pp. 574–581
736 DISSADO, L.A.: *J. Phys. D.*, 1990, **23**, pp. 1582–1591
737 SCHIFANI, R.: 'Proc. 3rd Int. Conf. Cond. Breakdown in Solid Dielectrics (Trondheim)', 1989, pp. 456–460
738 HIROSE, H.: *IEEE Trans. Elec. Insul.*, 1987, **EI-22(6)**, pp. 745–753
739 ENDICOTT, H.S., and WEBER, K.H.: *AIEE Trans. III*, 1956, **75**, pp. 371–381

740 DISSADO, L.A.: 'Proc. 3rd Int. Conf. Cond. & Breakdown in Solid Dielectrics (Trondheim)', 1989, pp. 528–532
741 LAWSON, J.H., and THUE, W.A.: 'Conf. Rec. 1980 IEEE Int. Symp. Elec. Insul., Boston, MA, USA', 9–11 June 1980, pp. 100–104
742 DAVID, H.A.: 'Order statistics' (J. Wiley & Sons, New York, 1970)
743 HERRMANN, H.J., and STANLEY, H.E.: *Phys. Rev. Lett.*, 1984, **53**, pp. 1121–1124
744 STANLEY, H.E.: *J. Stat. Phys.*, 1984, **36**, pp. 843–859
745 HALSEY, T.C., JENSON, M.H., KADANOFF, L.P., PROCACCIA, I., and SHRAIMAN, B.I.: *Phys. Rev. A*, 1986, **33**, pp. 1141–1151
746 BLUMEN, A., KLAFTER, J., and ZUMOFEN, G.: in 'Fractals in physics', eds. PIETRONERO, L., and TOSATTI, E. (Elsevier, New York, 1986) pp. 399–408
747 TAMOR, M.A.: *Phys. Rev.*, 1987, **B36**, pp. 2879–2882
748 BARCLAY, A.L., SWEENEY, P.J., DISSADO, L.A., and STEVENS, G.C.: *J. Phys. D.*, 1990, **23**, pp. 1536–1545
749 DISSADO, L.A., NIGMATULLIN, R.R., and HILL, R.M.: in Adv. in Chem. Phys., Vol. 63, 'Dynamic Processes in Condensed Matter', ed. EVANS, M.W. (J. Wiley & Sons, New York, 1985), Chapter 3
750 DISSADO, L.A., and HILL, R.M.: *J. Mat. Sci.*, 1981, **16**, pp. 633–648
751 GUPTA, M.S.: 'Electrical noise, fundamentals and sources' (IEEE Reprint Services, Wiley, New York, 1977)
752 DUTTA, P., and HORN, P.M.: *Rev. Mod. Phys.*, 1981, **53**, pp. 497–516
753 DISSADO, L.A., and HILL, R.M.: *J. Appl. Phys.*, 1989, **66**, pp. 2511–2524
754 DISSADO, L.A., and HILL, R.M.: *Chem. Phys.*, 1987, **111**, pp. 193–207
755 FISCHETTI, M.V., GASTALDI, R., MAGGIONI, F., and MODELLI, A.: *J. Appl. Phys.*, 1982, **53**, pp. 3129–3135
756 FISCHETTI, M.V.: *Phys. Rev. B.*, 1985, **31**, pp. 2099–2113
757 WINTLE, H.J.: *J. Appl. Phys. (USA)*, 1981, **52(6)**, pp. 4181–4185
758 DE GENNES, P.: 'Scaling concepts in polymer physics' (Cornell Univ. Press, Ithaca, 1979)
759 PELISSOU, S., St-ONGE, H., and WERTHEIMER, M.R.: 'Proc. 1st. Int. Conf. Cond. Breakdown Solid Dielectrics', 1983, pp. 338–344
760 HILL, R.M.: *Nature*, 1978, **275**, pp. 96–99
761 WARD, I.M.: 'Mechanical Properties of Solid Polymers' (J. Wiley, London, 1971)
762 PENDER, L.F., and WINTLE, H.J.: *J. Appl. Phys.*, 1979, **50**, pp. 361–368
763 HILL, R.M.: *J. Mat. Sci.*, 1981, **16**, pp. 118–124
764 HILL, R.M.: *J. Mat. Sci.*, 1982, **17**, pp. 3630–3636
765 FOTHERGILL, J.C., DISSADO, L.A., and WOLFE, S.V.: data from unpublished paper presented to the *Ann. Rep. Conf. Elec. Insul. & Diel. Phenom.*, 1983
766 STONE, G.C., and VAN HEESWIJK, R.G.: *IEEE Trans. Elec. Insul.*, 1977, **EI-12(4)**, pp. 253–261
767 PIETRONERO, L., and WIESMANN, H.J.: *J. Stat. Phys.*, 1984, **36**, pp. 909–916
768 PIETRONERO, L., and TOSATTI, E., eds.: 'Fractals in Physics' (North Holland, Amsterdam, 1986)
769 KAPLAN, T., and GRAY, L.J.: *Phys. Rev.*, 1985, **B32**, pp. 7360–7366
770 REYNOLDS, P.T.: 'Microscopy and analysis', 1990, pp. 43–46
771 CHADBAND, W.G.: *IEEE Trans. Elec. Insul.*, 1988, **EI-23**, pp. 697–706

772 NELSON, J.K.: *IEEE Trans. Elec. Insul.*, 1989, **EI-24**, pp. 835–847
773 MURAT, M.: in 'Fractals in Physics', eds. PIETRONERO, L., and TOSATTI, E. (North Holland, Amsterdam, 1986) pp. 169–171
774 CHNIBA, S., and TOBAZEON, R.: 'Conf. Rec. 1984 IEEE Int. Symp. Elec. Insul. (Montreal),' 1984, IEEE pub. no. **84CH1964-6-EI**, pp. 191–193
775 LEWIS, A.G., HALL, E.L., and CALDWELL, F.R.: *J. Res. Bureau Standards*, 1931, **7**, Res. Paper 347, p. 403
776 AUSTEN, A.E.W., and WHITEHEAD, S.: *Proc. Roy. Soc. A. (London)*, 1940, **176**, pp. 33–50
777 CLARK, P.M.: 'Engineering materials for design & engineering practice' (J. Wiley & Sons, 1962)
778 VERMEER, J.: *Physica*, 1954, **20**, pp. 313–326
779 LAMONT, Jr., L.T.: *J. Vac. Sci. Tech.*, 1977, **14(1)**, p. 122
780 HARARI, E.: *J. Appl. Phys.*, 1978, **49**, pp. 2478–2489
781 FERRY, D.K.: *J. Appl. Phys.*, 1979, **50**, pp. 1422–1427
782 MORSE, C.T., and HILL, G.J.: *Proc. Brit. Ceram. Soc.*, 1970, **18**, pp. 23–35
783 VON HIPPEL, A., and MAURER, R.J.: *Phys. Rev.*, 1941, **59**, pp. 820–823
784 VON HIPPEL, A., and ALGER, R.S.: *Phys. Rev.*, 1949, **76(1)**, pp. 127–133
785 WATSON, D.B., HEYES, W., KAO, K.C., and CALDERWOOD, J.H.: *IEEE Trans. Elec. Insul.*, 1965, **EI-1**, pp. 30–37
786 WATSON, D.B., HEYES, W., and KWAN, C.K.: *IEEE Trans. Elec. Insul.*, 1970, **EI-5(3)**, pp. 58–63
787 HANSCOMB, J.R.: *J. Appl. Phys.*, 1970, **41**, pp. 3597–3603
788 ARTBAUER, J., & GRIAC, J.K: *Proc. IEE*, 1965, **112**, p. 818
789 NELSON, J.K.: in 'Engineering dielectrics, Vol. IIA: Electrical properties of solid insulating materials (molecular structure & electrical behavior)' eds. BARTNIKAS, R., and EICHHORN, R.M. (ASTM Press, STP783, Philadelphia, 1983)
790 KAKO, Y., TSUKUI, T., HIRABAYASHI, S., KIMURA, K., MITSUI, H., and NATSUNE, F.: 'CIGRE (Int. Conf. Large High Volt. Elect. Syst.)', 1978, Paper 15-04, 13 pages.
791 JONSCHER, A.K.: *J. Phys. D.*, 1980, **13**, pp. L143–L148
792 HALPERIN, B.I., FENG, S., and SEN, P.: *Phys. Rev. Lett.*, 1985, **54**, pp. 2391–2394
793 DUXBURY, P.M., BEALE, P.D., and LEATH, P.L.: *Phys. Rev. Lett.*, 1986, **57**, pp. 1052–1055
794 DUXBURY, P.M., and LEATH, P.L.: *J. Phys. A.*, 1987, **20**, pp. L411–L415
795 BEALE, P.D., and DUXBURY, P.M.: *Phys. Rev.*, 1988, **B37**, pp. 2785–2791
796 DE ARCANGELIS, L., REDNER, S., and CONIGLIO, A.: *Phys. Rev. B*, 1985, **31**, pp. 4725–4728
797 STAUFFER, D.: *Phys. Rept.*, 1979, **54**, pp. 1–74
798 CHAKRABARTI, B.K., ROY, A.K., and MANNA, S.S.: *J. Phys. C.*, 1988, **21**, pp. L65–L68
799 MANNA, S., and CHAKRABARTI, B.K.: *Phys. Rev.*, 1987, **B36**, pp. 4078–4081
800 BENGUIGUI, L.: *Phys. Rev.*, 1988, **B38**, pp. 7211–7214
801 COPPARD, R.W., DISSADO, L.A., ROWLAND, S.M., and RAKOWSKI, R.: *J. Phys. Cond. Matter.*, 1989, **1**, pp. 3041–3045
802 KENT, R., and RAT, R.: *J. Electrostat.*, 1985, **17**, pp. 299–312

803 COPPARD, R.W., BOWMAN, J., DISSADO, L.A., ROWLAND, S.M., and RAKOWSKI, R.T.: *J. Phys. D.*, 1990, **23**, pp. 1554–1561
804 BAHDER, G., EAGER, G.S., SILVER, D.A., and LUKAC, R.G.: *IEEE Trans. Power Appar. & Sys.*, 1976, **PAS-95**, pp. 1552–1566
805 FISCHER, P., NISSEN, K.W., and RÖHL, P.: *Ann. Rep. Conf. Elec. Insul. & Diel. Phenom.*, 1979, pp. 539–545
806 DISSADO, L.A., and HILL, R.M.: *IEEE Trans. Elec. Insul.*, 1990, **EI-25**, pp. 660–666
807 EYRING, C.F., MACKEOWN, S.S., and MILLIKAN, R.A.: *Phys. Rev.*, 1928, **31**, pp. 900–909
808 TU, D.M., LIU, W.B., ZHUANG, G.P., LU, Z.Y., and KAO, K.C.: *IEEE Trans. Elec. Insul.*, 1989, **EI-24**, pp. 581–589
809 SEKII, Y., and YODA, B.: *Hitachi Review*, **21**, pp. 376–382
810 GOTOH, H., OKAMOTO, T., SUZUKI, S., and TANAKA, T.: *IEEE Trans. Power Appar. & Sys.*, 1984, **PAS-103**, pp. 2428–2434
811 BULINSKI, A.T., BAMJI, S.S., and DENSLEY, R.J.: *IEEE Trans. Elec. Insul.*, 1986, **EI-21**, pp. 645–650
812 British Standard: BS 2782 (1970) 'Methods of Testing Plastics'; Part 2 'Electrical Properties'; Methods 201a–201G 'Electric Strength'
813 Deutscher Normenausschuss (DNA): DIN 53481: VDE 0303 Part 2 (1974) 'Specifications for electrical tests of insulating materials; breakdown voltage, electric strength.' A subsequent draft (Entwurf) to replace this was published in 1987 'Testing of insulating materials; breakdown voltage and electric strength at power frequencies' which is identical to IEC 15A (central office) 52 (1986) which is a revision of IEC 243
814 International Electrotechnical Commission (affiliated to the International Organisation for Standardisation, ISO): IEC publication 243 (1967, 1st edition): 'Recommmended methods of test for electric strength of solid insulating materials at power frequencies'
815 PEEK, Jr., F.W.: 'Dielectric Phenomena in High-Voltage Engineering' (McGraw Hill, Inc., New York, 1920) p. 179
816 American Society for Testing and Materials: ASTM D149-81 (1981) 'Dielectric breakdown voltage and dielectric strength of solid electrical insulating materials at commercial power frequencies' (Replaces USA Federal Test Method Standard 406, Method 4031)
817 British Standards: BS 923: Part 1: 1989 and International Electrotechnical Commission: IEC 60-1:1989 'Guide on High-voltage testing techniques—General Definitions and Requirements' (Draft revisions of parts 2 and 3 are also available.)
818 Association of Edison Illumination Companies (AEIC): 'Specifications for Polyethylene and Crosslinked Polyethylene Insulated Shielded Power Cables Rated 5 through 69 kV' (AEIC, 5th edition, 1975), No. 1.
819 TEAM, TEAM easy analysis methods, 1976, 3(1), available from TEAM, Box 25, Tamworth, New Hampshire, NH 03886, USA
820 STONE, G.C., and ROSEN, H.: *IEEE Trans. Reliab.*, 1984, **R-33(50)**, pp. 362–368
821 TRUSTRUM, K., and JAYATILAKA, A. DE S.: *J. Mat. Sci.*, 1979, **14**, pp. 1080–1084
822 LAWLESS, J.F.: 'Statistical Models and Methods for Lifetime Data' (J. Wiley & Sons, 1982)
823 COHEN, A.C.: *Technometrics*, 1965, **7(4)**, pp. 579–589
824 LAWLESS, J.F.: *Technometrics*, 1978, **20(4)**, pp. 355–364

825 MANN, N.R., FERTIG, K.W., and SCHEUER, E.W.: 'Confidence and tolerance bounds and a new goodness-of-fit test for the two parameter Weibull or extreme-value distributions with tables for censored samples of size 3–25' (Aerospace Research Laboratories Report ARL 71–0077, Wright-Patterson AFB, Ohio, 1971)

826 'MathCAD', (Software Package) (Addison-Wesley Publishing Company Inc. and Benjamin/Cummings Publishing Company Inc., 1988)

827 SIMONI, L., and PATTINI, G.: *IEEE Trans. Elec. Insul.*, 1975, **EI-10(1)**, pp. 17–27

828 MONTANARI, G.C., and CACCIARI, M.: *IEEE Trans. Elec. Insul.*, 1985, **EI-20(3)**, pp. 519–522

829 CACCIARI, M., and MONTANARI, G.C.: *J. Phys. D.*, 1990, **23**, pp. 1592–1598

830 BREKHUNOV, V.I., MOTOSHKIN, V.V., and MUKHACHEV, V.A.: *Trans. Izvestiya Vysshikh Uchebnykh Zavedenii, Fizika*, 1977, **4**, pp. 120–124

831 PELISSOU, S., St-ONGE, H., and WERTHEIMER, M.R.: *IEEE Trans. Elec. Insul.*, 1984, **EI-19**, pp. 241–244

832 LUONI, G., OCCHINI, E., and PARMIGIANI, B.: *IEEE Trans. Power Appar. & Syst.*, 1981, **PAS-100(1)**, pp. 174–183

833 CHNIBA, S., and TOBAZEON, R.: 'Proc. 1st Int. Conf. Cond. Breakdown in Solid Dielectrics (Toulouse)', 1983, pp. 433–438

834 SHAW, D.C., CHICHANOWSKI, S.W., and YIALIZIS, A.: *IEEE Trans. Elec. Insul.*, 1981, **EI-16**, pp. 399–414

835 CYGAN, P., KRISHNAKUMAR, B., and LAGHARI, J.R.: *IEEE Trans. Elec. Insul.*, 1989, **EI-24**, pp. 619–625

836 COLLA, F.: *IEE Conf. Diel. Mats., Meas. & Applics.*, 1984, **239**, pp. 124–127

837 SAKAMOTO, Y., FUKAGAWA, H., NINOMIYA, K., YAMADA, Y., YOSHIDA, S., and ANDO, N.: *IEEE Trans. Power Appar. & Syst.*, 1982, **PAS-101**, pp. 1352–1362

838 LEWIS, T.J.: *IEEE Trans. Elec. Insul.*, 1984, **EI-19**, pp. 210–216

839 DAVIES, D.K.: *Proc. IEE*, 1981, **A128**, pp. 153–158

840 DAVIES, D.K.: *J. Phys. D.*, 1969, **2**, pp. 1533–1537

841 HIRSCH, J., KO, A.Y-Y., and IRFAN, A.Y.: *IEEE Trans. Elec. Insul.*, 1984, **EI-19**, pp. 190–192

842 LEWINER, J.: 'Proc. 3rd Int. Conf. Cond. Breakdown in Solid Dielectrics (Trondheim)', 1989, pp. 548–554

843 KITANI, I., TSUJI, Y., and ARII, K.: *Jpn. J. App. Phys. I*, 1984, **23**, pp. 855–860

844 HANSCOMB, J.R., and GEORGE, E.P.: *J. Phys. D.*, 1981, **14**, pp. 2285–2294

845 GEORGE, E.P., HANSCOMB, J.R., and HO., J.: *J. Phys. D.*, 1984, **17**, pp. 1423–1432

846 KHALIL, M.S., and HANSEN, B.S.: *Int. Symp. Elec. Insul. (Montreal)*, 1984, pp. 73–75

847 KHALIL, M.S., and HANSEN, B.S.: *IEEE Trans. Elec. Insul.*, 1988, **EI-23**, pp. 441–445

848 KHALIL, M.S., and ZAKY, A.A.: *IEEE Trans. Elec. Insul.*, 1988, **EI-23**, pp. 1043–1046

849 FUHRMANN, J., and HOFFMAN, R.: *IEEE Trans. Elec. Insul.*, 1984, **EI-19**, pp. 187–189

850 THIESSEN, V., WINKEL, A., and HERRMAN, K.: *Phys. Z.*, 1936, **37**, pp. 511–520

851 ANDO, N. & NAMAJIRI, F.: *IEEE Trans. Elec. Insul.*, 1979, **EI-14,** pp. 870–879

852 SUZUOKI, Y., MUTO, H., MITZUTANI, T., and IEDA: M.: *Jap. J. Appl. Phys.*, 1985, **24**, pp. 604–609

853 LAURENCEAU, P., DREYFUS, G., and LEWINER, J.: *Phys. Rev. Lett.* 1977, **38**, pp. 46–49

854 EISENMENGER, E., and HAARDT, M.: *Solid. St. Commun.*, 1982, **41**, pp. 917–920

855 ALQUIE, C., DREYFUS, G., and LEWINER, J.: *Phys. Rev. Lett.*, 1981, **47**, pp. 1483–1487

856 SESSLER, G.M., WEST, J.E., GERHARD-MULTHAUPT, R., and VON-SEGGERN., H.: *Ann. Rep. Conf. Elec. Insul. & Diel. Phenom.*, 1982, pp. 58–64

857 LEWINER, J.: *IEEE Trans. Elec. Insul.*, 1986, **EI-21**, pp. 351–360

858 LANG, S.B.: *J. Appl. Phys.*, 1983, **54**, pp. 5598–5602

859 ANDERSON, R.A., and KURTZ, S.R.: *J. Appl. Phys.*, 1984, **56**, pp. 2856–2863

860 MAENO, T., FUTAMI, F., KUSHIBE, H., TAKADA, T., and COOKE, C.M.: *IEEE Trans. Elec. Insul.*, 1988, **EI-23**, pp. 433–439

861 COOKE, C.M., WRIGHT, K.A., MAENO, T., KUSHIBE, H., and TAKADA, T.: *Ann. Rep. Conf. Elec. Insul. & Diel. Phenom.*, 1986, pp. 444–447

862 MAHDAVI, H., ALQUIE, C., and LEWINER, J.: *Ann. Rep. Conf. Elec. Insul. & Diel. Phenom.*, 1989, pp. 296–302

863 LEWINER, J.: to be published in *J. Phys. D*

864 DITCHI, T., ALQUIE, C., LEWINER, J., FAVRIE, E., and JOCTEUR, R.: *IEEE Trans. Elec. Insul.*, 1989, **EI-24**, pp. 403–408

865 WATSON, P.K.: *IEEE Trans. Elec. Insul.*, 1987, **EI-22**, pp. 129–132

866 NAGAO, M., KOSAKI, M., and HASE, Y.: *Ann. Rep. Conf. Elec. Insul. & Diel. Phenom.*, 1988, pp. 448–453

867 WATSON, P.K.: 'Proc. 3rd Int. Conf. Cond. Breakdown in Solid Dielectrics (Trondheim)', 1989, pp. 282–286

868 BENDLER, J.T., and SCHLESINGER, M.F.: *J. Mol. Liq.*, 1987, **36**, pp. 37–46

869 MIZUTANI, T., MORI, T., and IEDA, M.: 'Proc. 3rd Int. Conf. Cond. Breakdown in Solid Dielectrics (Trondheim)', 1989, pp. 377–381

870 SUZUKI, T., KATAKI, S., KANAOKA, M., and SEKII, Y.: 'Proc. 3rd Int. Conf. Cond. Breakdown in Solid Dielectrics (Trondheim)', 1989, pp. 66–71

871 KHALIL, M.S., HENK, P.O., and HENRIKSEN, M., 'Proc. 3rd Int. Conf. Cond. Breakdown in Solid Dielectrics (Trondheim)', 1989, pp. 192–196

872 KHALIL, M.S., HENK, P.O., HENRIKSEN, M., and JOHANNESSON, H.: *Ann. Rep. Conf. Elec. Insul. & Diel. Phenom.*, 1988, pp. 460–465

873 LEBEY, T., LAURENT, T., and MAYOUX, C.: 'Proc. 3rd Int. Conf. Cond. Breakdown in Solid Dielectrics (Trondheim)', 1989, pp. 40–44

874 MAHMOODIAN, A., KEMP., I.J., EASTWOOD, A.R., and FOURACRE, R.A.: 'Proc. 3rd Int. Conf. Cond. Breakdown in Solid Dielectrics (Trondheim)', 1989, pp. 77–81

875 PELISSOU, S., St-ONGE, H., and WERTHEIMER, M.R.: *IEEE Trans. Elec. Insul.*, 1988, **EI-23**, pp. 325–333

876 RÖHL, P.: *Siemens Forsch-u Entwickl-Ber.*, 1985, **14**, pp. 104–113

877 RÖHL, P.: *Ann. Rep. Conf. Elec. Insul. & Diel. Phenom.*, 1984, pp. 186–189

878 VLASTOS, A.E.: 'Proc. 3rd Int. Conf. Cond. Breakdown in Solid Dielectrics (Trondheim)', 1989, pp. 287–293

879 ROWLAND, S.M., HILL, R.M., and DISSADO, L.A.: *J. Phys. C.: Solid State Phys.*, 1986, **19**, pp. 6263–6385

880 SHATZKES, M., and AV-RON, M.: *IBM J. Res. Develop.*, 1981, **25(3)**, pp. 167–175

881 LEACH, A.: Private communication

882 LEWIS, T.J.: *Ann. Rep. Conf. Elec. Insul. & Diel. Phenom.*, 1976, pp. 533–561

883 BS 727 'Characteristics and performance of apparatus for measurement of radio interference' (1954)

884 American Society for Testing and Materials: ASTM Standard D1868, 'Detection and Measurement of Partial (Corona) Discharge Pulses in Evaluation of Insulation System' ASTM Book of Standards, Vol. 10,02, (Philadelphia, 1987)

885 American Society for Testing and Materials: ASTM Standard D3382, 'Methods for measurement of energy and integrated charge due to partial (Corona) discharges', ASTM Book of Standards, Vol. 10,02, (Philadelphia, 1987)

886 Association of Edison Illumination Companies (AEIC): 'Specifications for Ethylene Propylene Rubber Insulated Shielded Power Cables Rated 5 through 69 kV' (AEIC, 6th edition, 1975) No. 1

887 International Electrotechnical Commission: IEC Publication 270, 'Partial Discharge Measurements', 2nd edition (1981)

888 NEMA Publication 107–1964, 'Methods of measurements for radio influence voltage of high voltage apparatus'

889 CIGRE Working Group 12-01, Electra, 1971, No. 19, pp. 13–65

890 IPCEA Standard T–24–380, 'Guide for partial discharge test procedure' (Insulated Power Cables Engineers Association)

891 BAGHURST, A.H.; *Ann. Rep. Conf. Elec. Insul. & Diel. Phenom.*, 1985, pp. 471–476

892 GALAND, L.: *Revue général d'électricité*, 1971, **80**, pp. 399–405

893 van HOVE, C., LIPPERT, A., and WIZNEROWICZ, F.: *Electrizität-swirstschaft*, 1974, **73(26)**, pp. 776–780

894 AUSTIN, J., and JAMES, R.E.: *IEEE Trans. Elec. Insul.*, 1976, **EI-11(4)**, pp. 129–139

895 KURTZ, M.: Ontario Hydro Research Quarterly, 1973, **25(1)**, pp. 1–4

896 KASHCHIEV, D.: *Phil. Mag.*, 1972, **25**, pp. 459–470

897 ISARD, J.O.: *Phil. Mag.*, 1977, **35**, pp. 817–819

898 DAS-GUPTA, D.K., and DOUGHTY, K.: *Thin Solid Films*, 1982, **90(3)**, pp. 247–252

899 SWINGLER, S.G., JACKSON, R.J., and DRYSDALE, J.: 'Conf. Diel. Mats. Meas. & Applics.', 1984, IEE Conf. Pub. **239**, pp. 183–186

900 MOTORI, A., SANDROLINI, F., and MONTANARI, G.C.: 'Proc. 3rd Int. Conf. Cond. Breakdown in Solid Dielectrics (Trondheim)', 1989, pp. 352–358

901 GOFFAUX, R.: *Bulletin Scientifique de l'AIM*, 1987, **100**, (2 & 3), 104 pages

902 KELEN, A.: 'Proc. Int. Conf. Large High Tension Electric Systems (Paris)', 1976, Paper 15-03
903 BARTNIKAS, R.: in 'Engineering dielectrics, corona measurement and interpretation', eds. BARTNIKAS, R., and McMAHON, E.J. (ASTM Press, Philadelphia, 1979)
904 KURTZ, M., STONE, G.S., FREEMAN, D., MULHALL, V.R., and LONSETH, P.: *Proc. CIGRE*, 1980, Paper 11-09, 8 pages
905 NAKAMURA, Y., TAKAHASHI, T., NAKAYAMA, S., and NIWA, T.: 'Proc. 13th Symp. Elect. Insul. Mats. (Tokyo)', 1980, pp. 149–152
906 AYERS, S.: *Proc. IEE.*, 1979, **125**, pp. 711–715
907 JOHNSON, G.E., BAIR, H.E., ANDERSON, E.W., and DAANE, J.H.: *Ann. Rep. Conf. Elec. Insul. & Diel. Phenom.*, 1976, pp. 510–516
908 ISSHIKI, S., and YAMAMOTO, M.: *Fujikura Tech. Rev.*, 1979, pp. 19–24
909 SOMA, K., AIHARA, M., and KATAOKA, Y.: *IEEE Trans. Elec. Insul.*, 1986, **EI-21**, pp. 1027–1032
910 KURTZ, M., and STONE, G.C.: *IEEE Trans. Elec. Insul.*, 1979, **EI-14**, pp. 94–100
911 GOFFAUX, R.: *IEEE Trans. Elec. Insul.*, 1978, **EI-13**, pp. 1–8
912 KELEN, A.: *IEEE Trans. Elec. Insul.*, 1978, **EI-13**, pp. 14–21
913 JOHNSTON, D.R., and GAJA, N.V.: *IEEE Trans. Elec. Insul.*, 1978, **EI-13**, pp. 9–13
914 TERASE, H. & MOTOMURA, A.: *Jap IEE*, 1973, **93**, pp. 14–22
915 DACIER, J., and GOFFAUX, R.: 'Proc. 3rd Int. Conf. Cond. Breakdown in Solid Dielectrics (Trondheim)', 1989, pp. 602–607
916 KRECKE, M., and GOFFAUX, R.: *CIGRE*, 1988, Paper 11-12
917 STEINER, J.P., WEEKS, W.L., and FURGASON, E.S.: *Ann. Rep. Conf. Elec. Insul. & Diel. Phenom.*, 1985, pp. 477–484
918 JAMES, R.E., and JONES, S.L.: *IEEE Trans. Elec. Insul.*, 1988, **EI-23**, pp. 297–306
919 OKAMOTO, T.: *Ann. Rep. Conf. Elec. Insul. & Diel. Phenom.*, 1987, pp. 126–131
920 FRUTH, B., LIPTAK, G., ULLRICH, L., DUNZ, T., and NIEMEYER, L.: 'Proc. 3rd Int. Conf. Cond. Breakdown in Solid Dielectrics (Trondheim)', 1989, pp. 597–601
921 OKAMOTO, T., and TANAKA, T.: *IEEE Trans. Elec. Insul.*, 1986, **EI-21**, pp. 1015–1019
922 DUNBAR, W.G.: *Ann. Rep. Conf. Elec. Insul. & Diel. Phenom.*, 1987, pp. 45–50
923 STEINER, J.P., and WEEKS, W.L.: *Ann. Rep. Conf. Elec. Insul. & Diel. Phenom.*, 1987, pp. 73–78
924 BEYER, M.: *IEEE Trans. Power Appar. & Sys.*, 1982, **PAS-101**, pp. 3431–3438
925 WEEKS, W.L., and STEINER, J.P.: *IEEE Trans. Power Appar. & Sys.*, 1985, **PAS-104**, pp. 754–760
926 WIERINGA, L.: *IEEE Trans. Power Appar. & Sys.*, 1985, **PAS-104**, pp. 2–8
927 HUANG, S.J., LOWDER, S.M., SARKINEN, S.H., and CHARTIER, V.L.: *IEEE Trans. Power Appar. & Sys.*, 1985, **PAS-104**, pp. 407–415
928 JAMES, R.E., TRICK, F.E., PHUNG, B.T., and WHITE, P.A.: *IEEE Trans. Elec. Insul.*, 1986, **EI-21**, pp. 629–638
929 YOSHIDA, H., and UMEMOTO, K.: *IEEE Trans. Elec. Insul.*, 1986, **EI-21**, pp. 1021–1025

930 INOUE, Y., and YOSHIDA, Y.: *IEEE Trans. Elec. Insul.*, 1986, **EI-21**, pp. 1033–1036
931 KAMATA, Y.: *IEEE Trans. Elec. Insul.*, 1986, **EI-21**, pp. 1045–1048
932 BLODGETT, R.B., WADE, R.M., and WILKENS, W.D.: *IEEE Trans. Elec. Insul.*, 1981, **EI-16**, pp. 564–566
933 BRAUN, M.: *IEEE Trans. Elec. Insul.*, 1981, **EI-16**, p. 567
934 MONTANARI, G.C., and CACCIARI, M.: *IEEE Trans. Elec. Insul.*, 1988, **EI-23**, pp. 365–372
935 GRADSHTEYN, I.S., and RYZHIK, I.E.: 'Table of integrals, series & products' (Academic Press, Orlando, Florida, USA, 1980)
936 WYND, G.: *Electra*, 1974, **32**, pp. 70–74
937 HOZUMI, N., ISHIDA, M., OKAMOTO, T., and FUKAGAWA, H.: *IEEE Trans. Elec. Insul.*, 1990, **EI-25**, pp. 707–714
938 HIKITA, M., IEDA, M., and SAWA, G. *J. Appl. Phys.*, 1983, **54**, 2025–2029
939 KIRKLAND, J. W., THIEDE, R. S., and REITZ, R. A.: *IEEE Trans. Power. Appar. & Syst.*, **PAS-101**, 2128–2136

Alphabetical list of authors

(**Bold** references indicate the author is the first-named or the sole author.)

A

ABDOLALL, K.	**213, 221**
AHARONY, A.	314
AIDA, F.	167, 240, 269, 442
AIHARA, M.	909
AKAHORI, H.	473
AKAZAKI, M.	139, 140, 693
AL-ARAINY, A.A.	642
ALBERT, M.	704
ALGER, R.S.	784
ALLARA, D.L.	**496**
ALLISON, J.	**566**
ALQUIE, C.	**855**, 862, 864
ALSTON, L.L.	**165**
AMAKAWA, K.	**635, 689**
AMANO, H.	422
AMBORSKI, L.E.	**547**
ANDERSON, E.W.	907
ANDERSON, R.A.	**859**
ANDO, N.	837, **851**
ANDRAINJOHANINARIVO, J.	157, **323**
ANDRÉ, J.-M.	**54, 66**
ANDRESS, B.	**462**
ANOLICK, E.S.	**732**
ARTBAUER, J.	**788**
ARBAB, M.N.	**367, 450**
ARII, K.	321, 629, 630, **638**, 843
ARTBAUER, J.	**690, 691**
ASAHI, K.	716
ASHBY, M.F.	**8**, 364
ASHCRAFT, A.C.	**198, 199**
AUCKLAND, D.W.	367, **376**, 377, 450
AUCLAIR, H.	328, 403
AUSTEN, A.E.W.	**624, 625, 776**
AUSTIN, J.	**894**
AV-RON, M.	705, 880
AYERS, S.	**906**

FUKUGAWA, H.	326, 371, 372, 837, 937
FUKUMA, M.	597
FUKUNAGA, S.	449
FUNAYAMA, M.	493, 619
FUOSS, R.M.	**553**
FURGASON, E.S.	917
FUTAMI, F.	860

G

GAINES, G.B.	479
GAJA, N.V.	913
GALAND, L.	**892**
GAMEZ-GARCIA, M.	**407**
GARRITY, T.	120
GARTNER, E.	**277**
GARTON, A.	19, 212, 218, **224**, **293**
GARTON, C.G.	**91**, **138**, **502**, **602**
GASTALDI, R.	755
GATELLET, J.	680
GEBALLE, R.	656
GEDEON, S.	559
GEMANT, A.	**645**
GENT, A.N.	**283**
GEORGE, E.P.	844, **845**
GERHARD-MULTHAUPT, R.	856
GEURTS, W.S.M.	225
GHIGGINO, K.P.	498
GIVEN, M.J.	216, **217**
GJELSTEN, N.G.	595
GLARUM, S.H.	**524**
GLUCHOWSKI, S.	**284**, **352**
GOFFAUX, R.	**901**, **911**, 915, 916
GOLDMA, A.	680
GOLDMAN, M.	680
GOLDSPINK, G.F.	607
GOLTZ, W.	16
GÖLZ, W.	**252**, **291**, 333, 408
GOOD, R.H.	**573**
GOODMAN, B.	**617**
GOSSE, B.	**278**
GOSSE, J.P.	278
GOTO, H.	264, 414
GOTOH, H.	**810**
GÖTTLICH, S.	340
GOULDSON, E.J.	**710**
GRADSHTEYN, I.S.	**935**
GRAMMATICA, S.	84

N

NAGAI, H.	194
NAGAO, M.	162, **597**, **598**, **601**, **866**
NAGASAKI, S.	18, **449**
NAKAJIMA, T.	81
NAKAKITA, T.	196, 231
NAKAMURA, S.	552
NAKAMURA, Y.	**905**
NAKANISHI, K.	**322**
NAKANO, K.	305
NAKANISHI, T.	335
NAKAYAMA, S.	905
NAMAJIRA, F.	851
NAMIKI, Y.	**167**, **442**
NATH, R.	**106**
NATSUNE, F.	790
NAWATA, M.	287, 475
NAYBOUR, R.D.	**134**, **135**, **200**, **306**
NAYLOR, K.L.	**33**
NELSON, J.K.	**477**, **772**, **789**
NELSON, R.A.	**728**
NELSON, W.B.	**715**
NICHOLL, G.R.	684
NICOULAZ, F.	474
NIEMEYER, L.	**289**, 920
NIGMATULLIN, R.R.	749
NIKLASSON, G.A.	**303**
NINOMIYA, K.	837
NISHIDA, M.	324
NISHIO, Y.	552
NISSEN, K.W.	158, 160, 385, 660, 672, **677**, 805
NITTA, Y.	264, **271**, 280, 414, **493**
NITTMAN, J.	**251**, **312**
NIWA, T.	905
NIXON, J.	484
NOEL, S.	343, 350, **358**, 384
NOIRHOMME, B.	261
NOMORI, H.	545
NORDHEIM, L.	569
NORDMANN, J.	286
NORTH, A.M.	529
NOTO, F.	**129**, 197, **232**, 245, 324, 348, 349
NUILA, T.	695
NUNES, R.W.	**480**
NUNES, S.L.	**219**, **223**, **495**
NURY, J.	**451**

Index